# Environmental Assessment and Habitat Evaluation of the Upper Great Lakes Connecting Channels

# Developments in Hydrobiology 65

*Series editor*

H. J. Dumont

# Environmental Assessment and Habitat Evaluation of the Upper Great Lakes Connecting Channels

*Edited by*
M. Munawar and T. Edsall

*Reprinted from Hydrobiologia, vol. 219 (1991)*

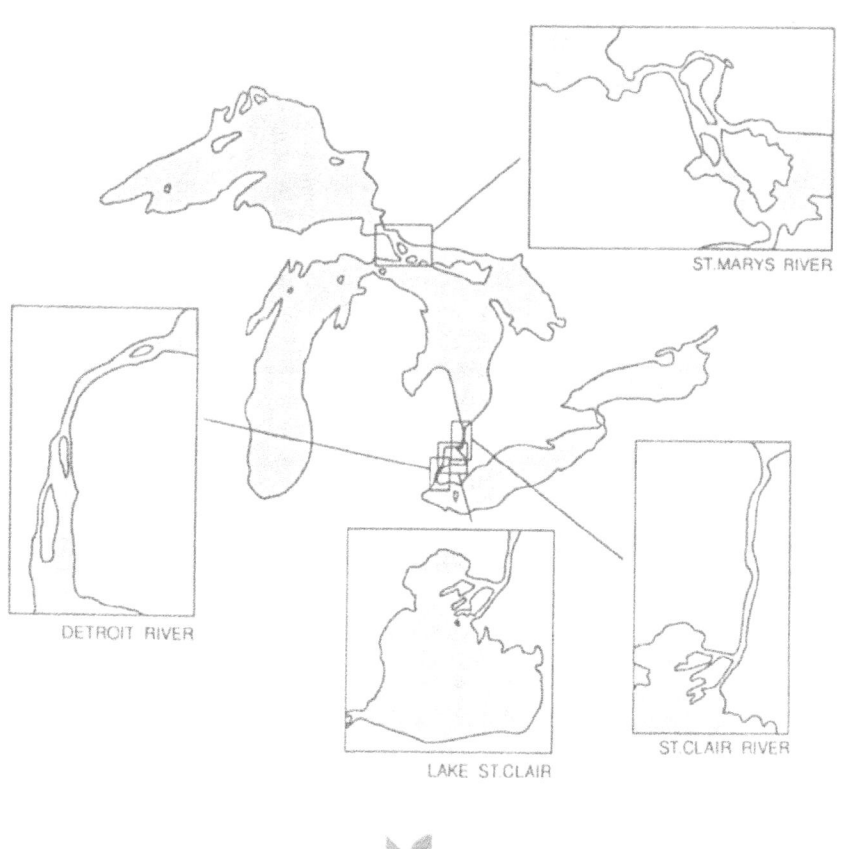

ST.MARYS RIVER

DETROIT RIVER

LAKE ST.CLAIR

ST.CLAIR RIVER

Springer Science+Business Media, B.V.

## Library of Congress Cataloging-in-Publication Data

Environmental assessment and habitat evaluation of the Upper Great
  Lakes connecting channels / edited by M. Munawar and T. Edsall.
        p.   cm. -- (Developments in hydrobiology ; 65)
      "Reprinted from Hydrobiologia, vol. 219 (1991)"
     ISBN 978-0-7923-1206-2      ISBN 978-94-011-3144-5 (eBook)
     DOI 10.1007/978-94-011-3144-5
     1. Stream ecology--Great Lakes--Congresses.  2. Stream fauna-
-Great Lakes--Congresses.   I. Munawar, M.  II. Edsall, T.
III. Hydrobiologia.  IV. Series.
   QH104.5.G7E58  1991
   574.5'26323'0977--dc20                                91-2725

ISBN 978-0-7923-1206-2

*Printed on acid-free paper*

# Contents

*Hydrobiologia* **219**: vii–viii, 1991.
M. Munawar & T. Edsall (eds),
*Environmental Assessment and Habitat Evaluation of the Upper Great Lakes Connecting Channels.*

# Foreword

My first experience with an Upper Great Lakes Connecting Channel was on a family tour of the Soo* in the mid-1930's. I clearly remember the closeness of the towering ships in the locks, the awesome rapids, and my impression that shooting the rapids in our rowboat wold be impossible.

My next experience with Great Lakes Connecting Channels was in the early 1950s crossing the Niagara River between my Connecticut home and Michigan State University. I shall always remember the majesty, beauty, and power of the falls, and also the chemical aroma of the mist.

In the mid-1950's I recall the massive waterfowl losses on the lower Detroit River due to oil pollution. Conservationists carried oil-soaked ducks to the steps of the Michigan state capital in a protest against weak laws and insignificant penalties.

My next experience was in the late 1950's when I returned to Michigan as an aquatic biologist. My assignment was to determine if the water quality and biota of the Detroit River was affected by the discharges of the municipal-industrial complex that had contributed so significantly to victory in World War II. Concern for the environment had not been important during that period and the consequences were obvious. Foamy green discharges from detergent manufacturers, red rivers from steel mills, fetid grey sewage dotted with condoms, square miles of oily surface sheen, and acres of putrid sludge overshadowed the analytical data. Pockets of clean water with thriving communities of diverse, pollution-intolerant plants and animals persisted, but were largely restricted to refugia on island shorelines in mid-river that were remote from polluted discharges. These surviving clean-water habitats and biota... reminiscent of conditions in bygone days when commercial bank seiners captured whitefish, and a whitefish hatchery operated successfully on the river... offered hope that the productive potential of the river could be restored if the pollution inputs could be reduced or halted.

My water quality and biota work on the Upper Great Lakes Connecting Channels (supplemented by hunting and fishing experiences) continued intermittently through the 1960's. Coworkers such as John Neal (Ontario Water Resources) with interests in the channels were scarce, and published information on the channels' biota and habitats was virtually non-existent. During this period, George Hunt from the University of Michigan and Tom Beak of T. W. Beak, Ltd., produced the only connecting channel biota papers of which I am aware. What a treasure of information the present volume would have been then! Securing judgements against polluters would have been a lot easier with the help of the information provided in this volume and with the assistance of the many agencies represented by the authors.

Immense changes have occurred in the connecting channels in recent years as we adapt to or correct our past mistakes, municipal and industrial discharges have been reasonably controlled to allow recovery of most parts of the connecting channels aquatic ecosystem. Low flows in the St. Marys Rapids caused by hydropower diversions that by-pass the rapids have been partially remediated by constructing berms to constrict part of the remaining flow to ensure there is spawning habitat for rainbow trout and other indigenous fishes that contribute to the river's valuable fishery. International salmon fishing derbies have become festive occasions in the Soo. The whitefish and herring fisheries have recovered from their lows. Both cities are enjoying tourism booms which focus on water and recreation.

On the downside, perhaps because of habitat quality improvements in the area, sea lamprey populations are out of control and contributing to serious mortalities of chubs, salmon, and lake trout in northern

---

* Soo is a colloquial name for the Sault Ste. Marie area that lies in both Michigan and Ontario where the outlet of Lake Superior forms the St. Marys Rapids and St. Marys River.

Lake Huron. Because techniques for sea lamprey control in rivers this large are not effective, the Great Lakes Fishery Commission must develop new methods to protect the fisheries.

In the Detroit River the billions of dollars spent on cleanup of industrial and municipal effluents have made the waterfront a focus for parks, marinas, and a host of recreational activities: e.g. 1,000,000-plus spectators for the fireworks display celebrating Dominion Day and Independence Day; 600,000 for hydroplane races. The walleye fishing, supported entirely by natural reproduction, is world class. Who could have imagined twenty years ago that city dwellers would be catching Pacific salmon from the shoreline parks of Windsor and Detroit?

Part of the charge from Canada and the United States to the Great Lakes Fishery Commission in the 1955 Convention on Great Lakes Fisheries is to determine the measures which will make possible the optimum sustained productivity of Great Lakes fish. The Commission has long recognized that habitat quality and quantity relate directly to this charge. The Commission has repeatedly confronted a wide range of habitat issues and concluded that its role in these matters is as an advocate for fishery resources. The Commission also acts as a catalyst for the development of improved habitat assessment and management capabilities among the agencies with programs involving the welfare of the Great Lakes ecosystem.

Progress has been made in recent years in improving water quality and the related benthic conditions in the connecting channels. Biota have responded favourably, but fishery agencies are handicapped in managing the resource for optimum sustained productivity. These agencies need the capability to identify critical habitat components that are still missing and to be able to create, rehabilitate, or restore them. This volume is a major contribution toward such assessment and evaluation. Perhaps the next volume will focus on identifying the habitat needed to sustain the intensively managed aquatic communities in the Great Lakes ecosystem and on providing insight into how to better develop and create that habitat.

Threats to the aquatic resources of the Great Lakes and their connecting channels are greater than any one agency, jurisdiction, or country can defend against. The opportunities to optimize management of those aquatic resources are also greater than any one entity can meet. Future management will continue in the direction being taken now with the strengthening of partnerships among governments, academia, and the public that will better integrate air, land, water, and fisheries and wildlife management.

Recall that in 1956, not long after the oil-soaked ducks were carried to the capitol steps, Woodie Guthrie wrote and sang, 'This land is your land, this land is my land, this land was made for you and me'. His message caught on and the people realized that their lands and waters were being badly used. The people started to care about their resources and gradually the improvements came through legislation, negotiation, regulation, and voluntary action. The public began to participate in the rediscovered, rejuvenated resources of the Great Lakes and the connecting channels. And they began to care. Caring is something one acquires through exposure and understanding. The public gained a sense of belonging. That sense is something that cannot be legislated or commanded, it seeps into your being as one becomes a part of things. The public developed a strong emotional tie with the aquatic resources and the feeling evolved that this is 'my resource', 'my river', 'my bay', 'my lake'. The resulting message is also strong and clear: the public will not allow the ecosystem insults of the past.

I am honored by the request of Editors Munawar and Edsall to contribute to this volume. Perhaps I have provided a perspective on the past and a feeling for our future direction if we are to achieve our goal of optimum sustainable aquatic communities. To become knowledgeable on the current situation in the connecting channels, and to prepare ourselves to take the next steps in ecosystem recovery, simply turn to the following text.

CARLOS M. FETTEROLF, Jr.
*Executive Secretary*
*Great Lakes Fishery Commission*
*and*
*First Vice-President*
*American Fisheries Society*

*Hydrobiologia* **219**: ix–xiii, 1991.
*M. Munawar & T. Edsall (eds),*
*Environmental Assessment and Habitat Evaluation of the Upper Great Lakes Connecting Channels.*
© 1991 *Kluwer Academic Publishers.*

# Preface

M. Munawar[1] & T. Edsall[2]
[1]*Department of Fisheries & Oceans, Great Lakes Laboratory for Fisheries, & Aquatic Sciences, Canada Centre for Inland Waters, Burlington, Ontario, L7R 4A6, Canada*; [2]*National Fisheries Research Centre-Great lakes, U.S. Fish and Wildlife Service, 1451 Green Road, Ann Arbor MI, 48105, USA*

*Key words:* Upper Great Lakes Connecting Channels, environmental assessment

The Connecting Channels of the Upper Great Lakes – the St. Marys River, the St. Clair River, Lake St. Clair, and the Detroit River – are large riverine waterways that carry the outflow of Lake Superior into Lake Huron and the combined outflows of Lakes Michigan and Huron into Lake Erie (Fig. 1; Table 1). These channels are important fish and wildlife habitats, support a wide variety of recreational uses, are major navigation routes for interlake and ocean-going vessels, and are the municipal and industrial water supplies for major population centres in the region. The channel waters are also used to dilute and carry away the effluents from municipal sewage treatment plants and from industry, including those produced by wood pulp processing and paper manufacturing, electrical power production, steelmaking and casting, mineral extraction, chemical manufacturing, petrochemical production and refining, and automobile manufacturing. The channels also receive and redistribute pollutants from urban and agricultural run-off and from the atmosphere (Tables 1 & 2).

Procedures for environmental assessment and habitat evaluation in the Upper Great Lakes Connecting Channels have not yet been rigorously developed and tested (Lawrence, 1986; Munawar, 1988; Dodge, 1989) even though concerns about these connecting channels extend back to at least the turn of the century, when the International Joint Commission (IJC) included the St. Marys, St. Clair, and Detroit rivers in a list of polluted Great Lakes waters. Since 1973, the Great Lakes Water Quality Board of the International Joint Commission has annually identified 'Areas of Concern' where the Great Lakes Water Quality Agreement guidelines (signed in 1972 by United States and Canada), have been violated and beneficial uses of the ecosystem have been impaired (IJC, 1987; Hartig & Thomas, 1988).

In 1983, the U.S. Environmental Protection Agency announced plans to initiate a study on the channels, and invited the Canadian Government to participate in the study. A joint U.S. and Canadian planning workshop was convened in 1984, and in 1985 the Upper Great Lakes Connecting Channels Study (UGLCCS) Management Committee drew on the results of the planning workshop to outline the following study objectives (Limno-Tech, 1985):

1. ascertain the existing environmental condition of the study area to determine information gaps;
2. identify and quantify pollutant impacts to human and aquatic life, and their beneficial uses in the study area;
3. determine the adequacy of existing or proposed control programs to ensure or restore beneficial uses; and
4. recommend appropriate control and surveillance programs to protect and monitor the waterways and the downstream reaches.

*Table 1.* Physical characteristics and water usage of the Upper Great Lakes Connecting Channels (modified from EC & EPA, 1988.)

| | St. Mary's River | St. Clair River | Lake St. Clair | Detroit River |
|---|---|---|---|---|
| **PHYSICAL CHARACTERISTICS** | | | | |
| Land Drainage Area (km × 1000) km² | 49.3 | 146.6 | 159 | 160.9 |
| Area km² | 101–121 | 64 | 1115 | 51 |
| Width km | 0.30–6.4 | 0.25–1.2 | 39 | 0.66–3.0 |
| Depth m | Shallow–30 | 9–21 | 3.4 average | 6–15 |
| Flow m³ s⁻¹ (× 10³) | 2.2 | 5.2 | – | 7.1 |
| **SURFACE WATER SUPPLIES TO:** | | | | |
| Drinking Water Intake | | | | |
| – Municipal | + | + | + | + |
| – Communal/Private | + | + | + | + |
| Industrial Intakes | | | | |
| – Iron & Steel | + | | | + |
| – Pulp & Paper | + | | | |
| – Petrochemical | | + | | + |
| – Refining | | + | | + |
| – Thermal Generating | | + | | |
| – Hydroelectric | + | | | |
| – Navigation (Locks) | + | | | |
| – Mineral (Salt & Lime) | + | | | |
| **RECEIVING WATER FOR:** | | | | |
| Municipal STP | + | + | + | + |
| Industrial | | | | |
| – Iron & Steel | + | | | + |
| – Pulp & Paper | + | | | |
| – Petrochemical | | + | | + |
| – Refining | | + | | |
| – Thermal Generating | | + | | + |
| – Mineral (Salt & Lime) | | | | + |
| – Fabrication (Auto) | | | | + |
| Ship Ballast | + | | + | + |

$$Flow\ m^3\ s^{-1}\ (\times 10^3)$$

WATER USAGE

Shipping
Commercial Fishing
Sport Fishing
Boating/Sailing
Swimming

Legend: Shipping, Commercial Fishing, Sport Fishing, Boating/Sailing, Swimming

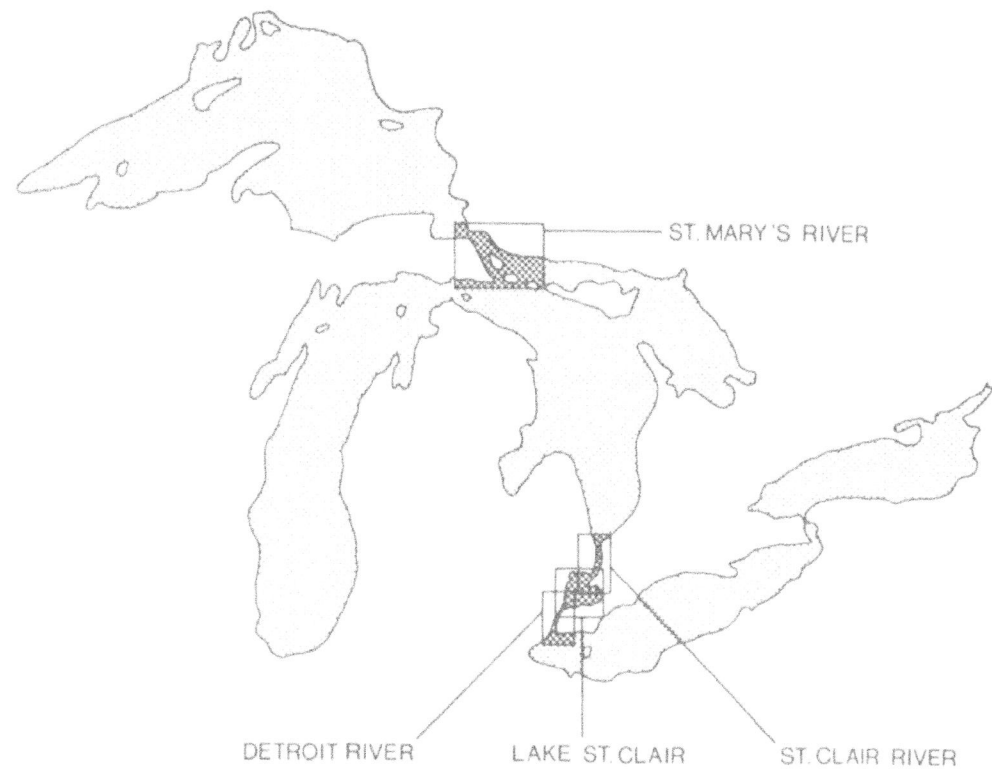

*Fig. 1.* Upper Great Lakes Connecting Channels.

*Table 2.* Summary of contaminant concerns in the Upper Great Lakes Connecting Channels (from EC & EPA, 1988.)

| Contaminants | St. Mary's River | St. Clair River | Lake St. Clair | Detroit River |
|---|---|---|---|---|
| Phosphorus | W | | W | WS |
| Ammonia | WB | | | WSB |
| Bacteria | W | WS | | W |
| Chlorides | | WB | | WB |
| Oil and Grease | WSB | SB | SB | WSB |
| Phenols | WB | | | WSB |
| Pesticides | | | WSB | WSB |
| PCBs | S | SB | SB | WSB |
| PAHs | WSB | SB | SB | WSB |
| Other Organics | WSB | S | WSB | |
| Heavy Metals | WSB | SB | WSB | WSB |
| Mercury | SB | SB | SB | WSB |
| Cyanide | W | | | S |

W: Water
S: Sediment
B: Biota

The UGLCCS Management Committee then appointed various work groups to initiate, implement, and coordinate research on the channels. One of these, the Biota Work Group, was charged with developing information on aquatic biota indigenous to the channels, to permit assessment of the effects of contaminants and other anthropogenic stresses on the biota and their habitats. Members of the Biota Work Group were T. Edsall (Chair), D. Kenaga and T. Nalepa (U.S.) and P. Kauss, J. Leach, M. Munawar and S. Thornley (Canada). The Biota Work Group provided nominal guidance for 38 research projects conducted in 1985–1987, summarized the results of this research, and prepared status reports on each of the four channels. These reports are included as attachments to volume 3 of the final report of the study (EC & EPA, 1989).

To facilitate dissemination of the new information generated by the binational effort, we convened a one-day symposium at the 31st Conference on Great Lakes Research (1988), McMaster University, Hamilton, Ontario, Canada. This symposium focused on environmental assessment and habitat evaluation in the Upper Great Lakes Connecting Channels. More than a dozen papers were presented at the symposium and the panel discussion which followed dealt with these papers and their use in environmental assessment and habitat evaluation. Encouraged by the interest expressed at the symposium, we decided to publish the symposium papers in a single, refereed volume. This volume contains most of the papers presented at the symposium. Several other papers representing research that was completed in the channels in 1988–1990 were added to this volume to provide a current and more complete overview.

To facilitate presentation of 24 papers we divided this volume into five sections. The first four sections each begin with a review paper that describes the status of each channel ecosystem. Each review is then followed by several papers describing the results of original research performed in that channel as part of the UGLCCS program. The fifth section includes papers that deal with original research conducted in two or more of the channels.

In addition to the review papers, this volume contains papers concerning various aspects of the channel ecosystem including channel bottom sediments, suspended sediments, fringing wetlands, bacteria, phytoplankton, zooplankton, plankton size spectra, macrophytes, benthic invertebrates, fish, the exotic zebra mussel and the Asiatic clam, navigationally induced stress on primary productivity, toxicity testing by algal fluorescence, phosphorus cycling by mussels, polycyclic aromatic hydrocarbons in caged mussels, deformities in larval *Procladius* and *Chironomini*, and tumours in fish.

The review process for the manuscripts included in this volume was strict and extensive. Each manuscript was reviewed by two referees, the co-editors, and a technical editor and the galley proofs were checked by the authors, the co-editors, and the technical editor.

We thank the referees (> 50 in number) for their meticulous and constructive reviews of the manuscripts, the authors for their careful revisions and cheerful cooperation and H.F. Nicholson, our technical editor, for his efforts, which enhanced the quality of this publication. We are grateful to C. Mayfield, I.F. Munawar, and G. Leppard for their advice and suggestions in the publication of this volume. Thanks are also due to L. McCarthy, W. Page, A. Aujla, S. Nielsen, M. Donnelly, and J. Wotherspoon for their assistance in various aspects of the publication of this volume. The International Association for Great Lakes Research convened the Symposium and the Aquatic Ecosystem Health and Management Society sponsored the publication of this volume. Finally, we wish to express our gratitude to W. Peters and H. Dumont of Kluwer Academic Publishers for their encouragement, and assistance throughout the publication process.

We hope this assemblage of research papers dealing with the connecting channels of the Upper Great Lakes will be useful not only to researchers and resource managers in the Great Lakes basin but also to readers interested in the functioning and management of other large, stressed, riverine ecosystems.

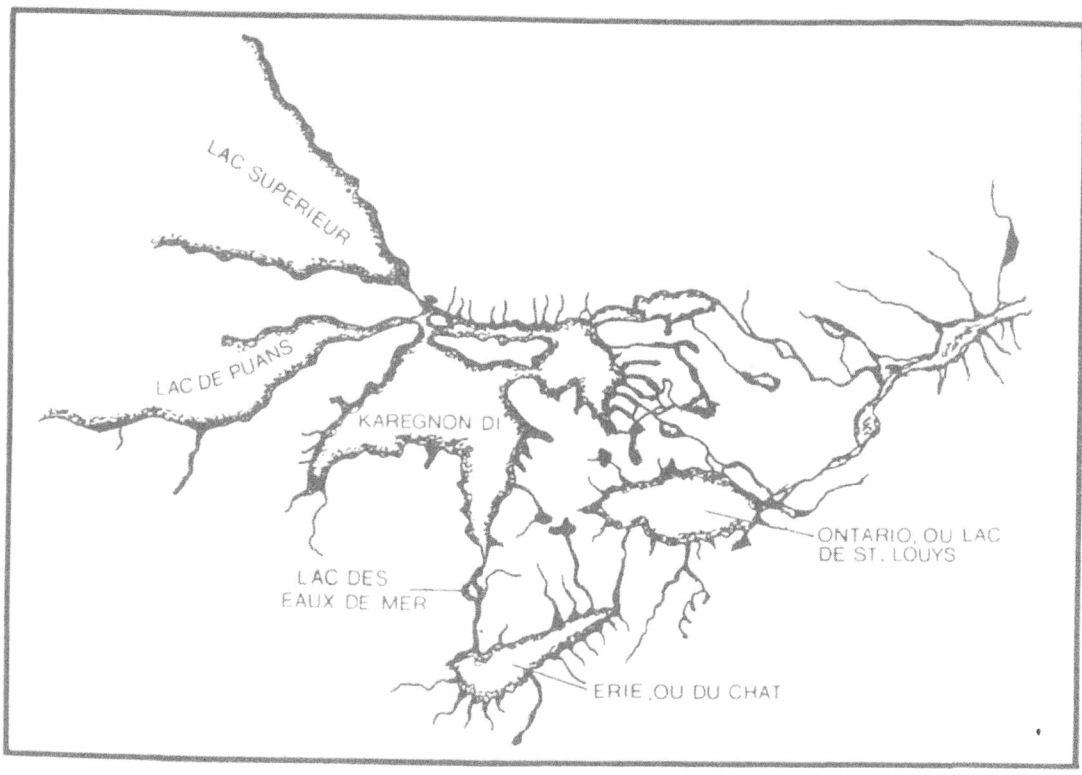

*Fig. 2.* Modified from Historical Map of Canada: *Le Canada ou Nouvelle France, 1656.* Courtesy: Public Archives of Canada.

## References

Dodge, D. P. (Ed), 1989. Proc. Internat. Large River Symp., Can. Spec. Publ. Fish. aquat. Sci. 106: 629 pp.

EC & EPA (Environment Canada & U.S. Environmental Protection Agency), 1988. Upper Great Lakes Connecting Channels Study. Final Report, Vol. 2. Chicago, Illinois & Toronto, Ontario, 626 pp.

EC & EPA, 1989. Upper Great Lakes Connecting Channels Study. Vol. 3. Appendices. Daryl W. Cowell & Associates Inc. Burlington, Ontario, 58 pp. + 40 appendices.

Hartig, J. H. & R. L. Thomas, 1988. Development of plans to restore degraded areas in the Great Lakes. Environmental Management 12 (3): 327–347.

IJC (International Joint Commission), 1987. Guidance on Characterization of Toxic Substances Problems in Areas of Concern in the Great Lakes Basin. Report to the Great Lakes Water Quality Board. Windsor, Ontario. 179 pp.

Lawrence, J. (Ed), 1986. The St. Clair River Pollution. Wat. Pollut. Res. Can. 21 (3). 459 pp.

Limno-Tech, Inc., 1985. Summary of the existing status of the Upper Great Lakes Connecting Channels data. Prepared for Upper Great Lakes Connecting Channels Study. Ann Arbor, Mich. 156 pp. + app.

Munawar, M. (ed.), 1988. Limnology and Fisheries of Georgian Bay and the North Channel Ecosystems. Developments in Hydrobiology 46. Kluwer Academic Publishers, Dordrecht. xii + 222 pp. Reprinted from Hydrobiologia 163.

*Hydrobiologia* **219**: 1–35, 1991.
*M. Munawar & T. Edsall (eds),*
*Environmental Assessment and Habitat Evaluation of the Upper Great Lakes Connecting Channels.*
© *1991 Kluwer Academic Publishers.*

1

# Biota of the St. Marys River: habitat evaluation and environmental assessment

P.B. Kauss
*Ontario Ministry of the Environment, Water Resources Branch, 1 St. Clair Avenue W. Toronto, Ontario M4V 1K6, Canada*

*Key words:* aquatic environment, plants, invertebrates, fish, birds, man's impact, pollution

## Abstract

The St. Marys River provides vital habitat for many species of plants, invertebrates, fish, and birds. It is also subject to many at times conflicting uses, including recreational boating, sport and subsistence fishing, municipal and industrial withdrawals and inputs, as well as commercial ship traffic and hydro-electric power generation.

In 1984, the United States and Canada jointly initiated the Upper Great Lakes Connecting Channels Study to identify and quantify the impacts of contaminants on these channels and their biota and to develop recommendations for more effective pollution control and surveillance programs. Results of the study in the St. Marys River showed that water entering the river from Lake Superior is of excellent quality. Industrial and municipal discharges in the Sault Ste. Marie, Ontario area have resulted in heavily contaminated sediments and a severely impaired benthic invertebrate community in this area and downstream; however, no major impacts on fish have been demonstrated. Nevertheless, major impacts of man on fish spawning and rearing habitats and on benthic and fish productivity have resulted from the alteration of channels, construction of navigation locks, and regulation of flow in the St. Marys Rapids. Increases in the productivity of such raptors as osprey during the 1980s suggest a reduction in organochlorine contaminants levels in their diet; however, the increasing numbers of gulls with low concentrations of PCBs, p,p′-DDE and other organochlorines in their eggs may adversely affect gull young or their predators.

## 1. Introduction

French missionaries to the upper Great Lakes area gave the St. Marys River its name in 1641. At that time, the permanent native residents of the rapids area were often referred to as 'Saulteur' or 'people of the rapids'. The word 'sault' is derived from the 17th century French word 'saut' which meant 'waterfall' or 'rapids' (Duffy *et al.*, 1987). From this was derived the name Sault Ste. Marie,

given to the twin cities bordering the St. Marys Rapids, or Falls. Long before the influx of Europeans, the native peoples called this area of the river the 'Bawatig', the Algonkan word for rapids (Osborne & Swainson, 1986).

Today, the St. Marys River is an international waterway or connecting channel between Lakes Superior and Huron that is shared by Canada and the United States (Fig. 1). It extends from latitudes 46° 00′ N to 46° 35′ N and from longitudes

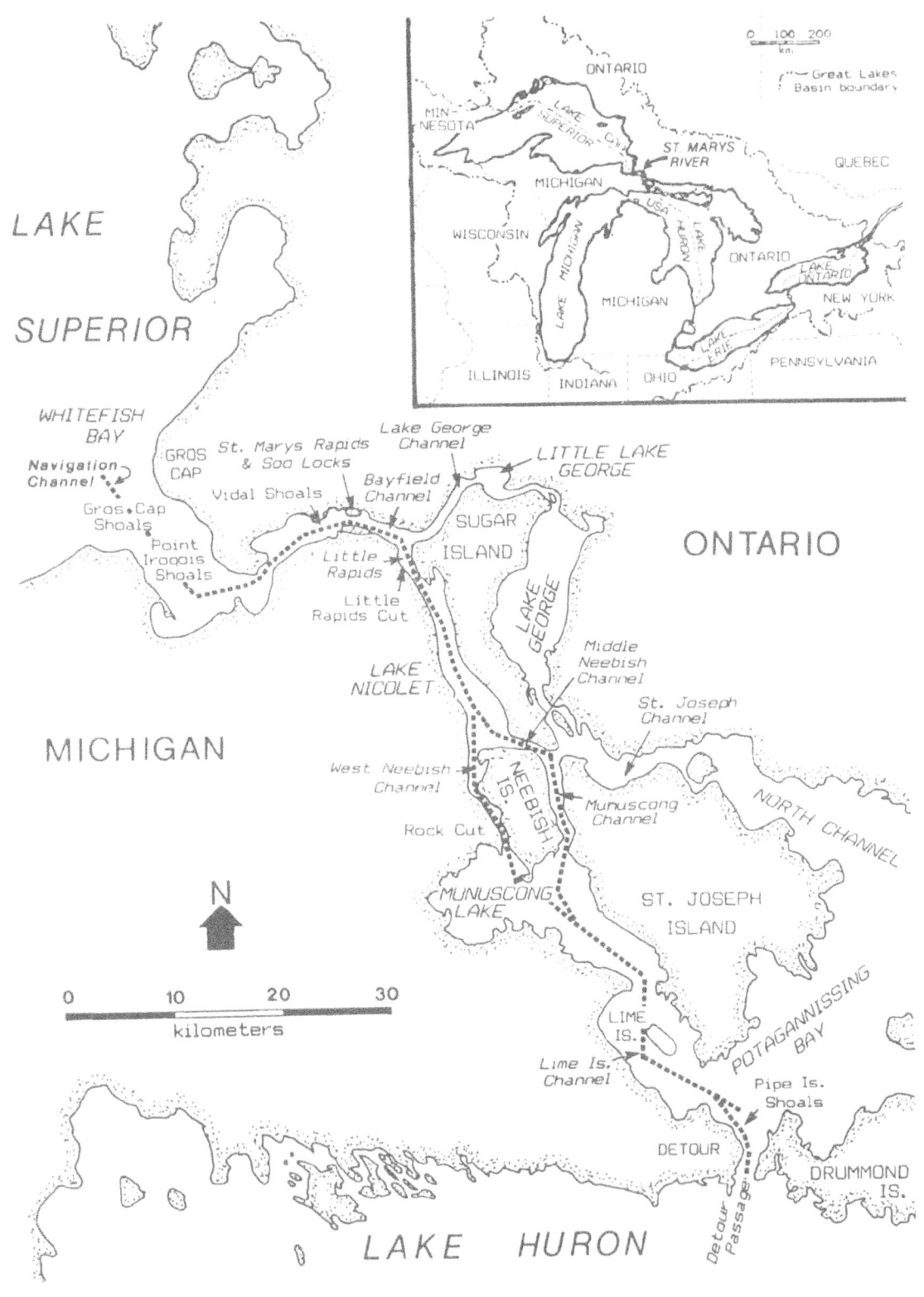

*Fig. 1.* The St. Marys River and its location in the Great Lakes Basin (inset).

*Table 1.* Summary of physical characteristics of the Great Lakes Connecting Channels.

| Connecting channel | Length (km) | Net elevation change (m) | Average flow ($m^3 s^{-1}$) | Watershed area ($\times 10^3$ km$^2$) | | |
|---|---|---|---|---|---|---|
| | | | | Land | Water | Total |
| St. Marys River | 112 | 6.7 | 2100 | 127.7 | 82.1 | 209.8 |
| St. Clair River | 43 | 1.5 | 5300 | 379.8 | 199.5 | 579.3 |
| Detroit River | 51 | 1.0 | 5400 | 397.5 | 200.7 | 598.2 |
| Niagara River | 59 | 99.3 | 5700 | 457.8 | 225.2 | 683.0 |
| St. Lawrence River | 808 | 74.0 | 6700 | 521.8 | 244.2 | 766.0 |

Sources: Upchurch (1976); Botts & Krushelnicki (1987)

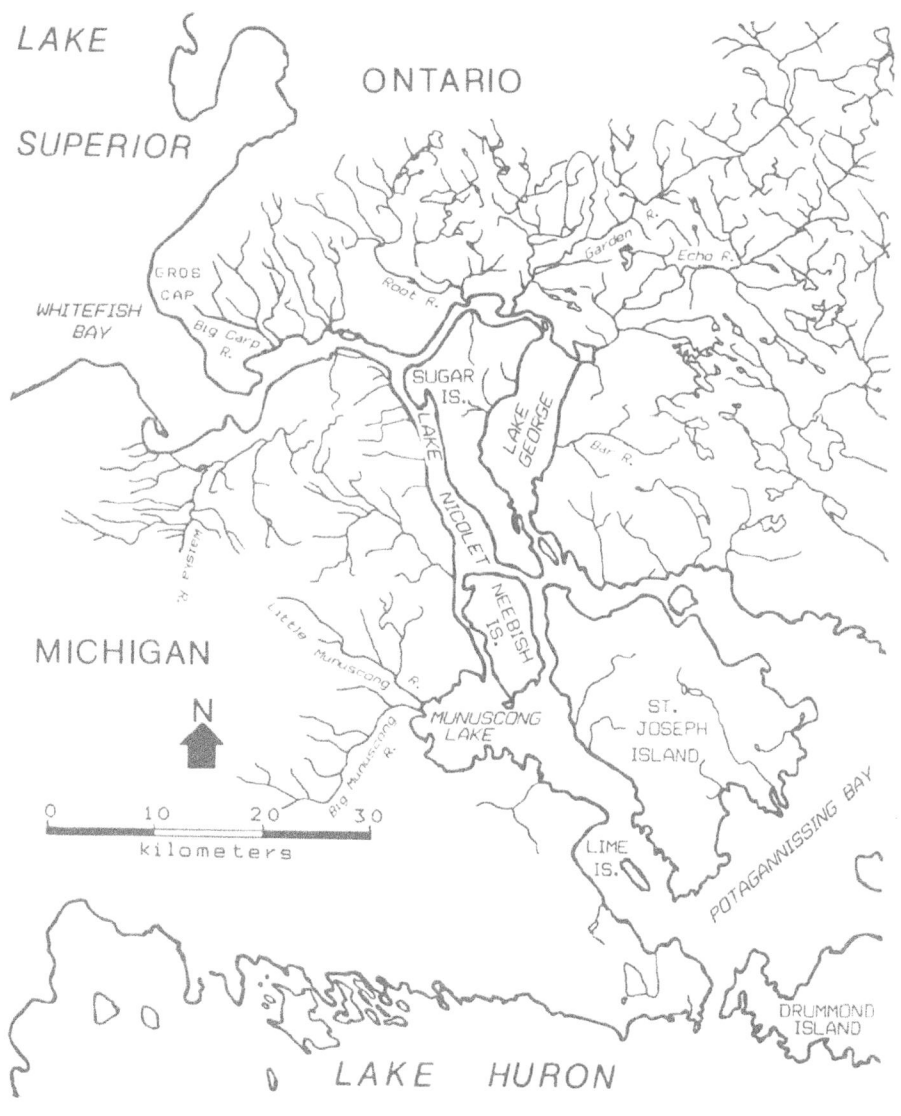

*Fig. 2.* Tributary watershed of the St. Marys River.

4

83° 40′ W to 84° 30′ W. Throughout its 120 km length, it provides a vital, diverse habitat for numerous species of plants and animals.

The river is extensively used for recreational boating, fishing, and swimming. It also provides drinking and process water for a sizeable urban and industrial area, the main centres of which are the twin cities of Sault Ste. Marie, Ontario, and Sault Ste. Marie, Michigan. These have a combined population of over 95 000 people. Ship locks, power plants, and compensating works span the river between these two population centres. These structures are used to regulate the outflow of Lake Superior, control the discharge of the river, and also to maintain water levels sufficient for commercial navigation, hydroelectric power generation, the fishery at the rapids, and other municipal and industrial uses. Discharge of the St. Marys River is regulated by the International Lake Superior Board of Control, with minimum outflow set at $1558 \text{ m}^3 \text{ s}^{-1}$ (USACE, 1988).

Numerous investigations have recently been conducted in the river by Canadian and U.S. Agencies as part of the binational Upper Great Lakes Connecting Channels Study, with most field work being conducted during 1985–86. This research was designed to update the existing data base on water quality, fill data gaps, follow up on new and emerging issues, and to provide information necessary for additional remedial actions. The results of this research are incorporated in a report that summarizes the status of biota, describes the impacts of contaminants, details sources and inputs of contaminants, outlines contaminant modelling efforts, and recommends remedial measures. Edwards *et al.* (1989) also provide a useful comparison of all of the connecting rivers between the Great Lakes.

The present paper focuses mainly on the aquatic biota of the St. Marys River and the effects of physical and chemical factors on that biota and their habitats. This paper draws freely on the report of the binational study (UGLCCS, 1989), and on the work of Duffy *et al.* (1987).

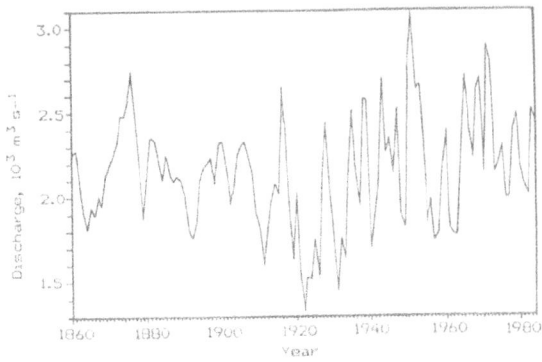

*Fig. 3.* Yearly average discharge of the St. Marys River at Sault Ste. Marie during the period 1860–1984. Sources: Quinn & Kelley (1983); U.S. Army Corps of Engineers Detroit District, unpubl. data; figure from Duffy *et al.* (1987).

## 2. Status of the St. Marys River

Compared with the other rivers that make up the connecting channels between the Great Lakes, the St. Marys River is intermediate in length, but has the smallest upstream drainage basin (i.e., Lake Superior and its watershed) and hence, the lowest average flow rate (Table 1). Nevertheless, the most important drainage basin is that of Lake Superior (Duffy *et al.*, 1987). Although the watershed adjacent to the river is comprised of 2 600 km² of land and 230 km² of water (CCGL, 1977) and contains numerous tributaries (Fig. 2), the latter are relatively small and contribute only a fraction of a percent to the river's flow (USACE,

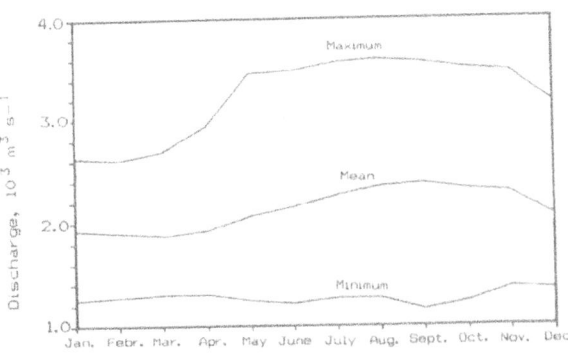

*Fig. 4.* Monthly average discharge of the St. Marys River at Sault Ste. Marie during the period 1900–78. Source: U.S. Army Corps of Engineers Detroit District, unpubl. data; figure from Duffy *et al.* (1987).

1988). The relatively small annual and seasonal variability in the flow of the river is mainly due to changes in Lake Superior outflow and to water level changes downstream in Lake Huron (Fig. 3). Minimum river flow occurs in March (mean: $1869 \, m^3 \, s^{-1}$) when Lake Superior water level is at its lowest, and maximum flow is in September (mean: $2379 \, m^3 \, s^{-1}$), coincident with highest lake level (Fig. 4). The river is divided by a number of large islands (Sugar, Neebish, St. Joseph, and Drummond), and 74 percent of the total flow passes through Lake Nicolet (Liston et al., 1986).

Current velocities vary considerably in the river, being swiftest in constricted areas and navigation channels (up to $1.0 \, m \, s^{-1}$) and essentially nil along some nearshore wetland areas. Consequently, flushing time or throughput varies considerably, from about 2 days to more than 13 days. Furthermore, Lake Superior and the St. Marys River are subject to wind-driven waves and seiches (Duffy et al., 1987), which can result in rapid and large changes in water level (more than 1.5 m over a 3 h period) that are greater than the average yearly variation (0.3 m). Passage of large vessels can also cause similar (0.01 to 0.7 m) short-term changes in water level as well as current speed increases (up to $1.0 \, m \, s^{-1}$) (McNabb et al., 1986).

The natural depth of the river varies from about 6 m at the inlet to over 40 m at its outlet into Lake Huron; however, certain areas of the river have

*Table 2.* Chronology of engineering changes associated with the development of the St. Marys River, 1797–1986.

| Year | Event |
| --- | --- |
| 1797 | Navigation lock 11.5 m long constructed on Canadian side. |
| 1822 | Raceway and sawmill built on American side by U.S. Army. |
| 1839 | Navigation canal started on American side, construction later aborted. |
| 1855 | Navigation lock completed on American side; construction begun in 1853. |
| 1859 | Dredging of lower Lake George Channel completed. |
| 1881 | Weitzel Lock on American side completed. |
| 1888 | International railway bridge completed. |
| 1894 | Dredging of Lake Nicolet Channel completed. |
| 1895 | Canadian government canal and lock completed. |
| 1896 | Old state locks on American side replaced by Poe Lock. |
| 1901 | Construction of compensating works begun. |
| 1902 | Edison Sault Hydroelectric Canal and power plant completed; canal diverted enough water to operate 41 turbines, each using approximately $10.6 \, m^3 \, sec^{-1}$ (total capacity: $435 \, m^3 \, sec^{-1}$). |
| 1908 | Ship canal through West Neebish Rapids (rock cut) completed. |
| 1914 | Davis Lock on American side completed. |
| 1915 | Additional 37 turbines added to Edison Sault hydroelectric plant. |
| 1916 | Hydroelectric canal and plant completed on Canadian side. |
| 1919 | Sabin Lock on American side completed. |
| 1921 | Construction of 16-gate compensating works completed and monthly river discharges set by IJC. |
| 1927 | Widening of Middle Neebish Channel completed. |
| 1933 | Widening of canal through West Neebish Rapids completed. |
| 1943 | MacArthur Lock on American side completed, replacing Weitzel Lock. |
| 1969 | Abitibi Paper Company water use reduced from approximately 198 to $1 \, m^3 \, s^{-1}$ permanently. |
| 1982 | Great Lakes Power hydroelectric power plant on Canadian side redeveloped and capacity increased from 510 to $1,076 \, m^3 \, s^{-1}$. |
| 1985 | Berm constructed to maintain water level over rapids along Canadian shore. (St. Marys Rapids-Whitefish Island Remedial Works for Fishery.) |

Sources: Adapted from Koshinsky & Edwards (1983), Edsall et al. (1988), Edwards et al. (1989) and Osborne & Swainson (1986).

Note: See Figs. 1 and 6 for locations.

6

been considerably altered by man since the late 18th century (Table 2). These engineering works included dredging, blasting and rock-cutting, and the construction of locks, canals and compensating works such as dams and berms to improve navigation and develop hydroelectric power. In all, some 101 km of the St. Marys River have been dredged to a depth of 8.3 m below low water datum, with widths ranging from 91 m to 457 m (Larson, 1981) to allow passage of commercial deep-draft vessels (Figs. 1 and 5). In recent years, both shipping traffic and tonnage have decreased. For example, in 1970 a total of 12 712 vessels carrying $39 \times 10^6$ tonnes passed through the locks, whereas in 1986 only 8 345 larger vessels carried about $34 \times 10^6$ tonnes (Burt *et al.*, 1988).

Almost all of the 6.7 m drop of the St. Marys River occurs in the 1.2 km-long St. Marys Rapids area. Consequently, both hydroelectric power plants and navigation locks are concentrated there and their construction has resulted in large-scale changes in flow distribution. For example,

an average of 93 percent of the outflow from Lake Superior is presently used for hydroelectric power generation (Fig. 6) and this diversion resulted in the dewatering of up to 25 ha of the rapids. In 1985, Canadian and U.S. hydroelectric interests funded the construction of a berm which prevents dewatering of the aquatic habitat in the main rapids area. Significant shoreline alteration has also occurred above the rapids along the Canadian shoreline as a result of infilling by Algoma Steel Corp., largely with slag from the manufacture of iron and steel (Fig. 7).

## 3. Chemical environment

Water entering the St. Marys River from Lake Superior is of exceptionally high quality and river water and sediment quality is good except where impacted by discharges (Edsall *et al.*, 1988). The major point source dischargers are congregated in the Sault Ste. Marie area (Fig. 7). With the new Sault Ste. Marie, Ontario west end municipal wastewater treatment plant brought on line in early 1986, the average flows of municipal and industrial facilities total about $9 \ \text{m}^3 \ \text{s}^{-1}$, with Algoma Steel Corp. contributing the majority (about 80%). While the above total is only 0.4 percent of the average river flow, the localized nature of these discharges, as well as those from storm and combined sewer overflows, shipping traffic and power generation have contributed to the degradation of water quality and benthic habitat in some areas. Studies have generally shown that water and sediment quality and biota were impacted in a zone along the Canadian shoreline below point sources, whereas the U.S. side was relatively unimpacted. Consequently, the Sault Ste. Marie portion of the river was classified as one of 42 Areas of Concern in the Great Lakes (IJC, 1985), where desirable water uses are restricted due to concentrations of certain contaminants in water, sediments and fish that exceed provincial water quality objectives (OME, 1984), Great Lakes Water Quality Agreement objectives (IJC, 1978), or both. The following paragraphs provide examples of the spatial extent

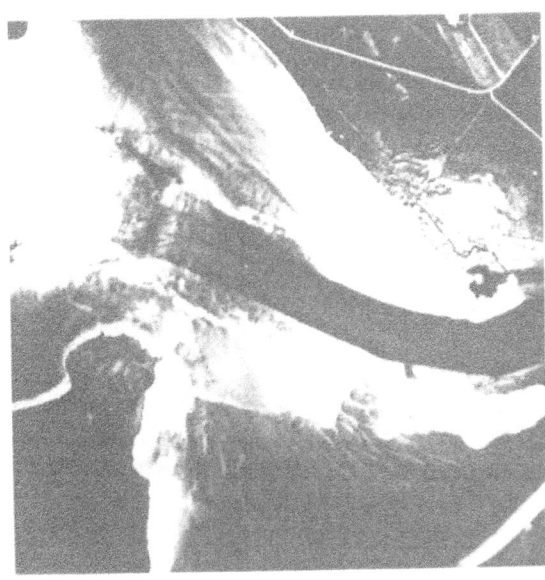

*Fig. 5.* Aerial photograph of the upper St. Marys River showing dredged shipping channel (dark horizontal band). Izaak Walton Bay is at bottom. Source: Ontario Ministry of Natural Resources; photograph taken June 15, 1984 from an altitude of 3048 m.

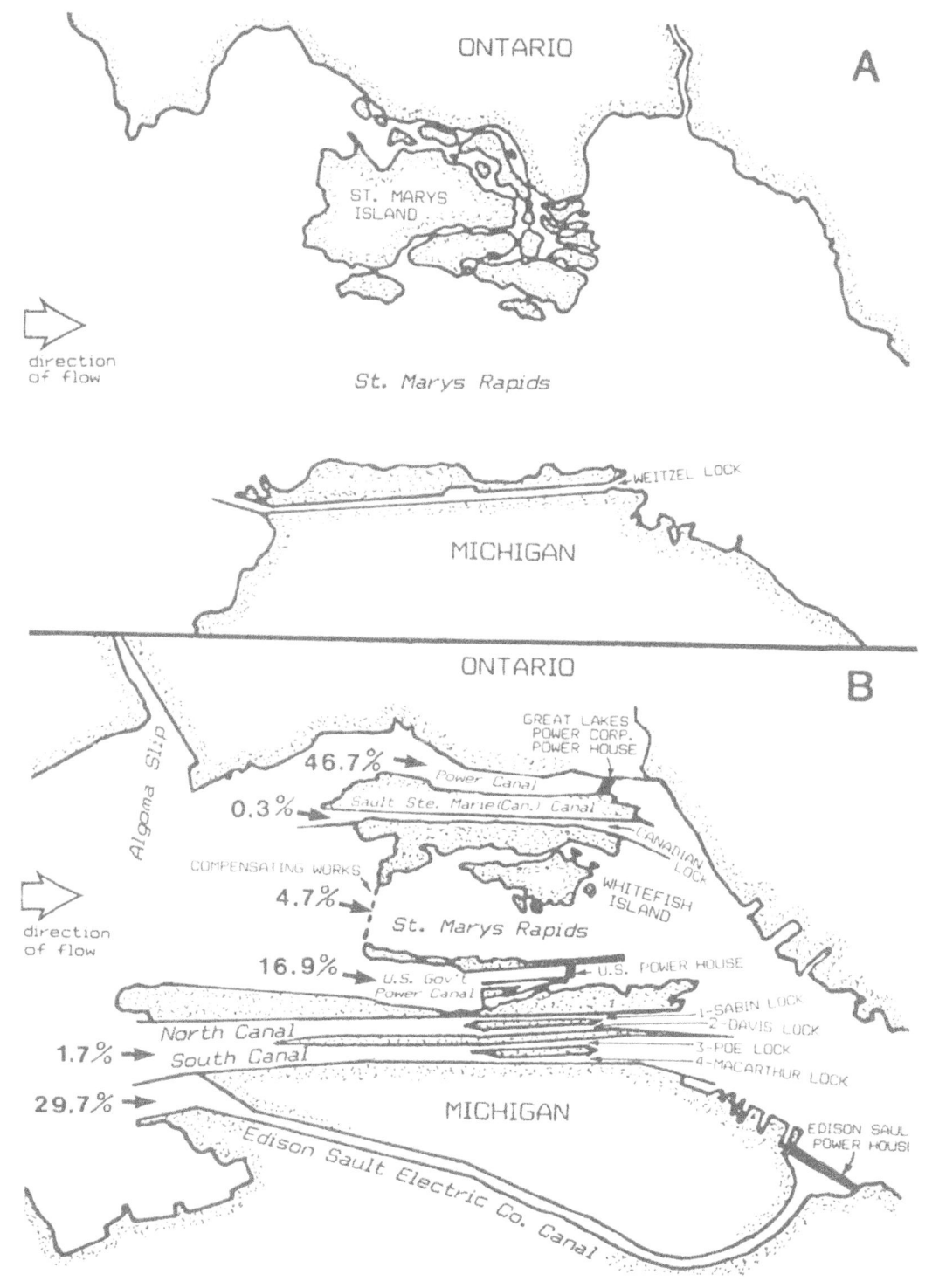

*Fig. 6.* The Rapids area of the St. Marys River during the period 1860–88 (A) and in 1983 (B). Percentages next to arrows indicate average flow distribution. Adapted from Koshinsky & Edwards (1983); figures from Duffy *et al.* (1987).

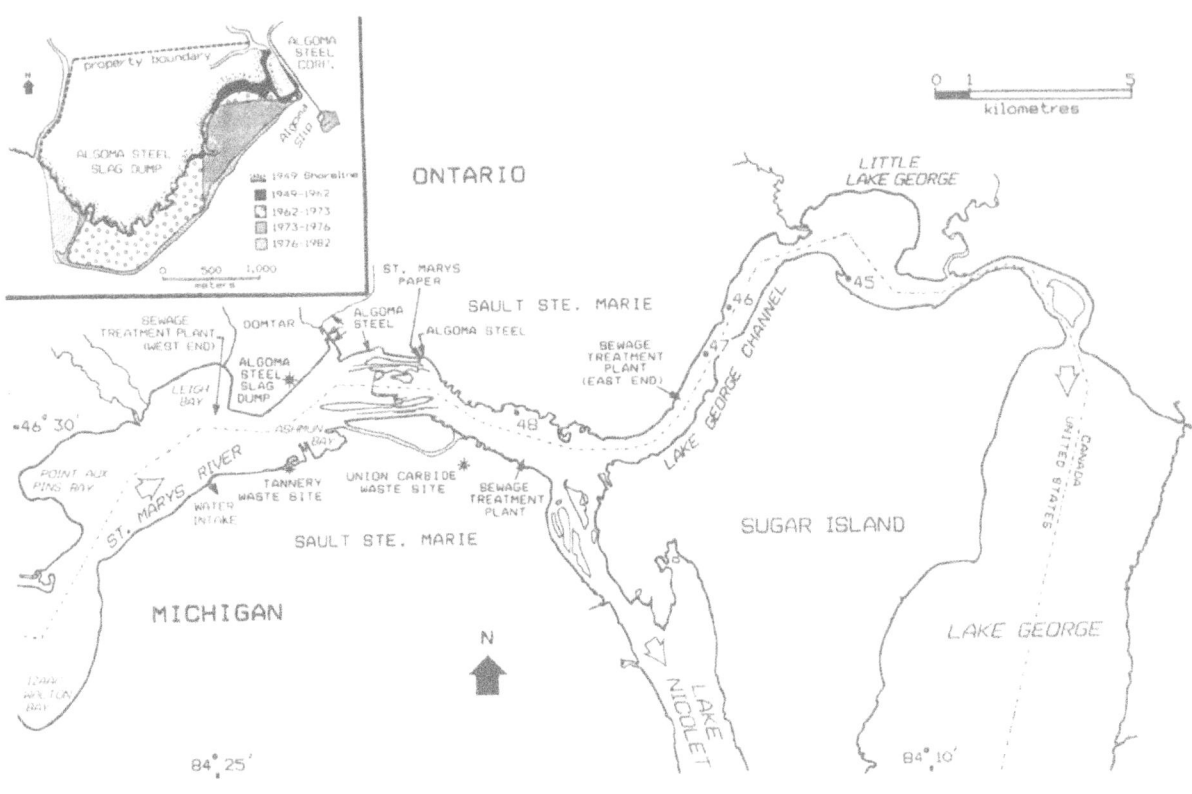

*Fig. 7.* Location of municipal water intakes and major industrial and municipal point source discharges and waste disposal sites in the St. Marys River. Numbers indicate station locations for Table 6. Inset shows historical development of Algoma Steel Corporation's slag dump. Source: Ontario Ministry of Natural Resources archives.

*Table 3.* Summary of physical and chemical characteristics of the St. Marys River.

| Parameter | Mean (range) | Source |
|---|---|---|
| Temperature (°C) | (0–21) | c |
| Oxygen saturation (%) | >90 (0–100) | c, b |
| Turbidity (NTU) | (<1–50) | c |
| pH | (6.7–8.4) | c, b |
| Alkalinity (mg l$^{-1}$ CaCo$_3$) | 40 | c |
| Conductivity ($\mu$S cm$^{-1}$) | 96 (85–188) | a, b |
| Chloride (mg l$^{-1}$) | 1.13 | a |
| Total phosphorus (mg l$^{-1}$) | 0.013 (0.001–0.031) | c |
| Total nitrogen (mg l$^{-1}$) | 0.413 (0.262–0.668) | c |
| Silica (mg l$^{-1}$) | 2.18 (0.98–3.66) | c |
| Chlorophyll *a* ($\mu$g l$^{-1}$) | 0.88 | c |
| Algal volume (mm$^3$ l$^{-1}$) | 0.332 (0.233–0.613) | a |

Sources: a–Hopkins (1986); b–Kauss & Hamdy (1991); c–Liston *et al.* (1986).

of these contaminants problems in water and sediments and also of temporal trends, where such data is available.

Tributaries to the St. Marys River can have significant local impacts on nearshore turbidity during major runoff events, particularly in Munuscong Lake in the lower river, due to tributary input of fine clays (USACE, 1988). Overall, however, it is Lake Superior water quality that has the major influence on the aquatic habitats and biota of the river, as will be seen below.

The mean and range of concentrations for nutrients, chlorophyll *a*, algal volume, other basic chemical characteristics, and annual temperature range in the St. Marys River (Table 3 & Fig. 8) generally indicate oligotrophic waters that are capable of supporting not only warmwater fish but also important coldwater fish species includ-

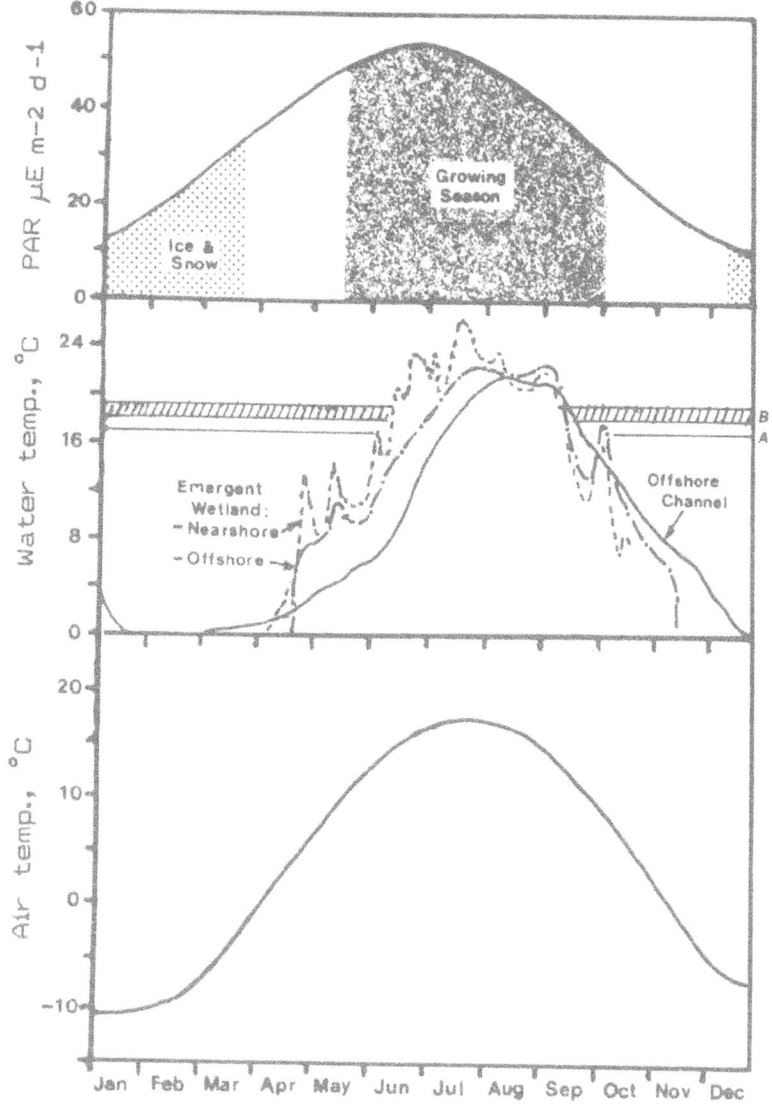

*Fig. 8.* Seasonal trends of air temperature at Sault Ste. Marie, Michigan, and of water temperature and photosynthetically active radiation (PAR) in Lake Nicolet. Figures adapted from Duffy *et al.* (1987); A = upper temperature limit for juvenile lake whitefish, B = upper temperature limit range for adult rainbow trout.

ing rainbow trout, whitefish, and lake herring in some areas for most of the year (Edsall *et al.*, 1988).

Water quality surveys of the St. Marys River date as far back as 1909 (IJC, 1914), when high bacterial densities were identified. Bacteria continued to be of concern after the surveys of 1947–48. The latter also showed that phenols were a problem (IJC, 1951). Data gathered since

1967 (Veal, 1968; Hamdy *et al.*, 1978; OME, unpubl. data) indicated elevated concentrations of phenols, iron, cyanide, ammonia, zinc, and sulphide in surface waters downstream of Canadian industrial and municipal sources. Bacterial densities were also above the Great Lakes Water Quality Agreement (IJC, 1978) and Ontario provincial objectives (OME, 1984) downstream of the east end wastewater treatment

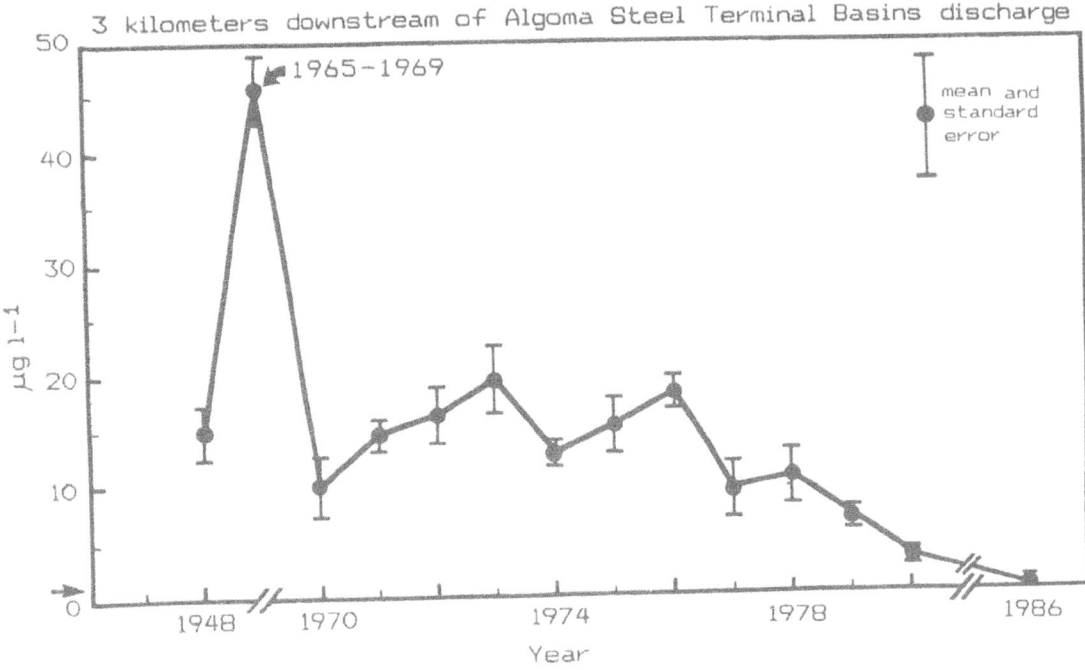

*Fig. 9.* Temporal trends in phenolics concentrations in St. Marys River surficial waters (1.5 m depth) along the Sault Ste. Marie, Ontario shoreline during the period 1948–86. Source: Ontario Ministry of the Environment unpubl. data. Arrow on X-axis indicates provincial (OME, 1984) and Great Lakes Water Quality Agreement (IJC, 1978) objective of 1 $\mu g \, l^{-1}$ to protect against tainting of fish flesh and taste and odour problems in domestic water supplies, respectively.

plant in the Lake George Channel and along the Sault Ste. Marie, Ontario, waterfront. Continued monitoring has shown that concentrations of contaminants associated with industrial discharges, such as phenols, ammonia, and cyanide, declined steadily along the Ontario shoreline between 1969 and 1980 (Kauss, 1986). By 1986, their concentrations approached objectives for the protection of aquatic life (Fig. 9). This improvement is mainly attributable to: (i) reduced point source loadings; and (ii) the doubling of river flow (and increased dilution of wastes) along the Ontario shoreline since 1982 resulting from increased diversion to Great Lakes Power (Table 2). However, bacterial contamination was still evident along both the Sault Ste. Marie, Ontario and Michigan waterfronts and in the Lake George Channel (UGLCCS, 1989).

Physical characteristics of the river bottom are an important habitat factor for benthic flora and fauna. In the upper river above the St. Marys Rapids, sediments are mainly composed of sand, along with rocks, cobble, and gravel in Whitefish Bay. The Rapids area separating the upper from the lower river is characterized by gravel, rocks, boulders, and exposed bedrock. Sediments in slower moving and less exposed areas of the lower river are composed mainly of sand and silt, or clay and silt (Fig. 10). As one gets closer to shore, the sediments tend to be mainly clay with organic detritus, whereas the proportion of sand increases with distance offshore. Also, more protected shores tend to have finer sediments than the course sand, gravel, and rock of exposed (usually easterly) shores. Finally, large portions of the dredged channel sediments consist mainly of clay (USACE, 1988). Disposal areas for dredged material are located along the shipping channel in both the upper and lower portions of the river, but constitute a relatively minor part of the total river substrate.

In 1947, river sediments near both cities of

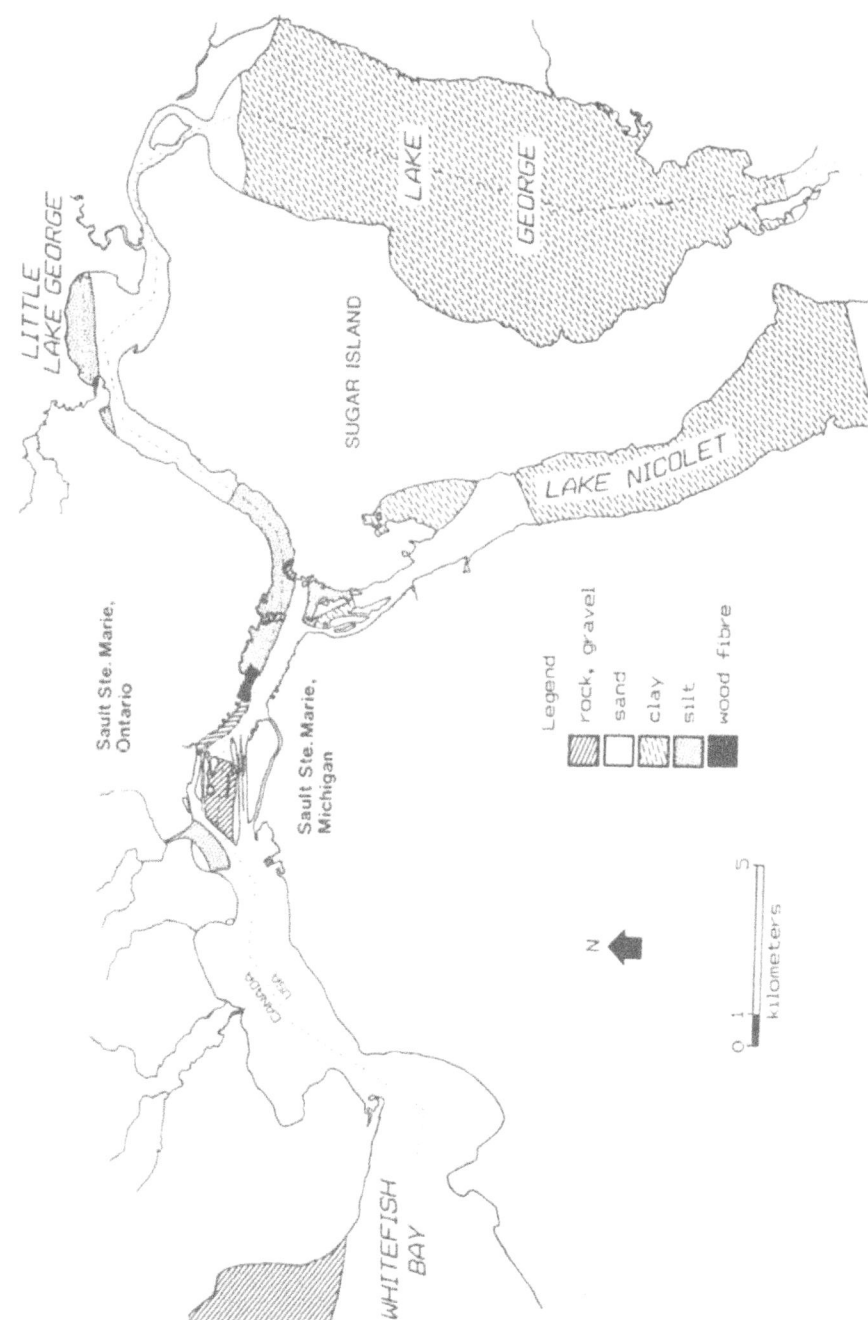

*Fig. 10.* Characteristics of St. Marys River surficial sediments (upper 3 cm). Source: Hamdy *et al.* (1978).

12

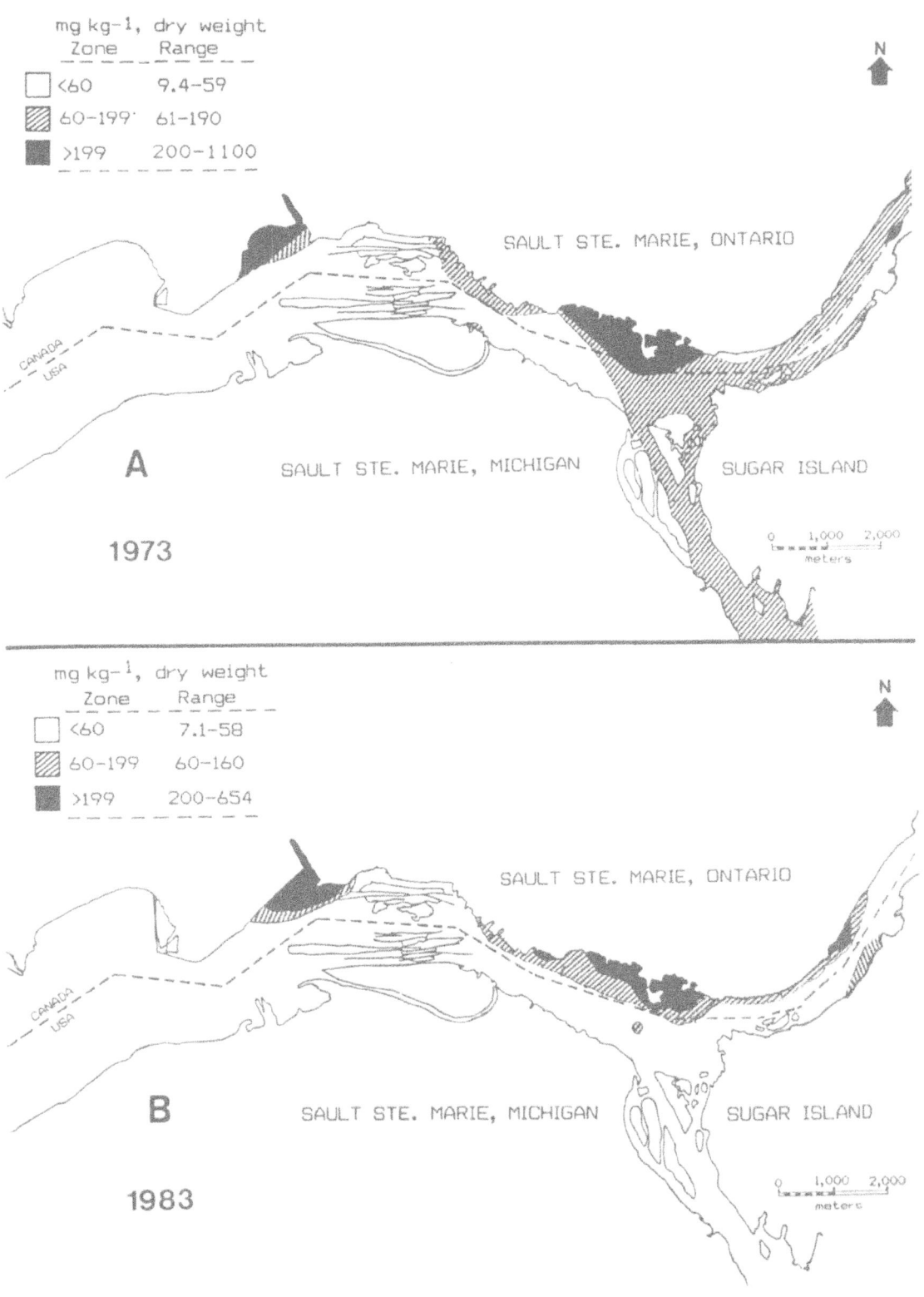

*Fig. 11.* Zinc in St. Marys River surficial sediments (upper 3 cm) in 1973 (A) and 1983 (B). Source: Kauss (1986).

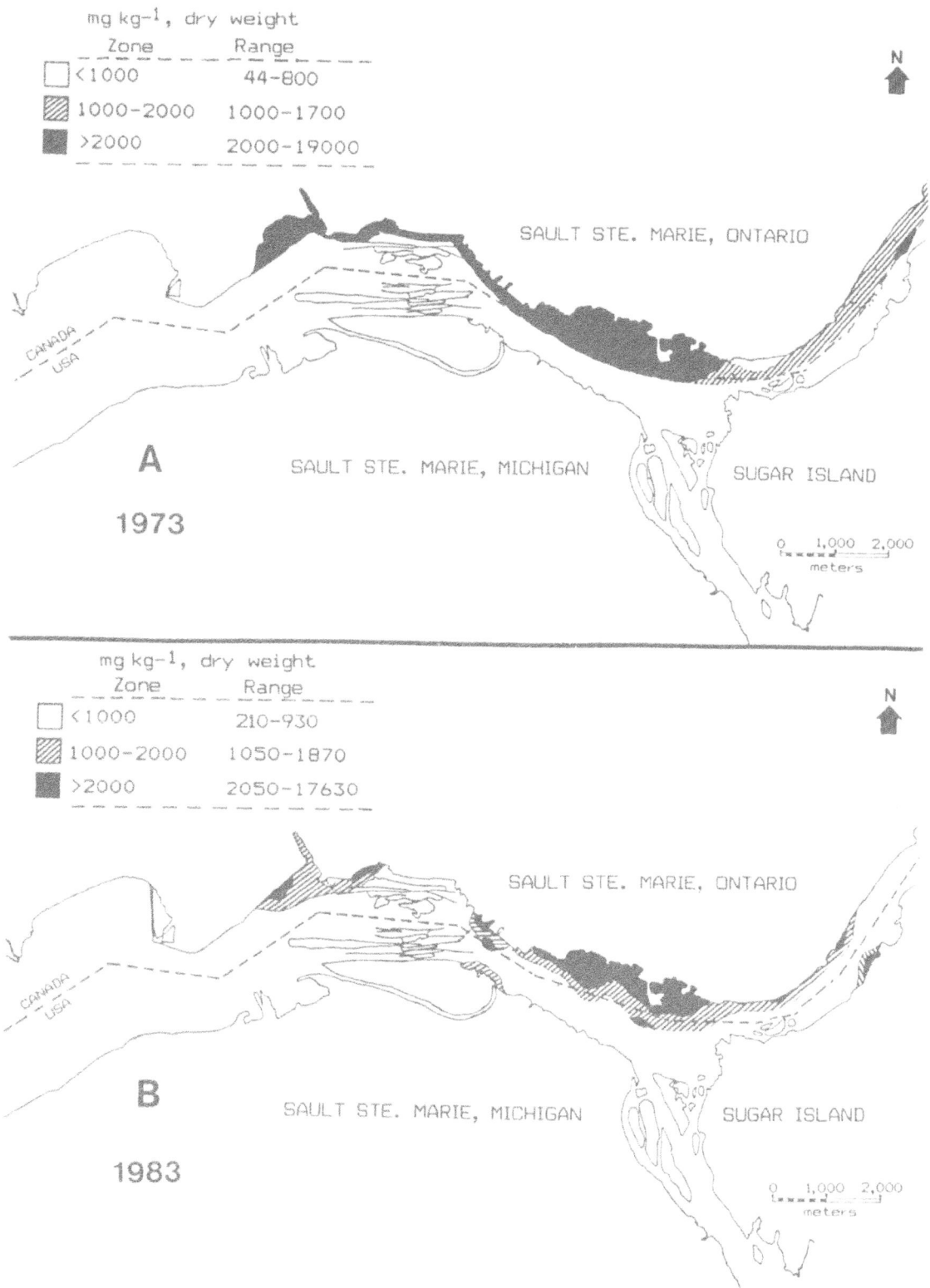

*Fig. 12.* Oil and grease in St. Marys River surficial sediments (upper 3 cm) in 1973 (A) and 1983 (B). Source: Kauss (1986).

14

Sault Ste. Marie were seriously polluted (IJC, 1951). On the U.S. side, chrome wastes contaminated the riverbed near the outfalls of a tannery on Ashmun Bay (Fig. 7). On the Canadian side, deposits of fine iron ore and coal dust were found near the Algoma Steel Corp. dock (Fig. 7) and bark from log booms also littered the bottom. In the lower river, sediments below Algoma Steel and Abitibi Pulp and Paper Co. (now St. Marys Paper) contained numerous paper fibres and iron ore solids, and were covered with a heavy gelatinous and filamentous growth. This pollution extended downstream into the Lake George Channel (IJC, 1951).

Subsequent studies revealed sediment contamination with phenols, wood particles, cyanide, metals (iron, zinc, chromium, copper, lead, mercury), oil and grease, and polychlorinated biphenyls (PCBs), particularly along the Sault Ste. Marie, Ontario, shoreline where levels exceeded OME guidelines for the disposal of dredged materials in open water (Persaud & Wilkins, 1976). River sediments adjacent to the tannery waste site are also contaminated with cyanide, copper, and lead in addition to chromium (Edwards et al., 1989; Kenaga, 1980).

The spatial distribution of contaminants in river surficial sediments sampled in 1983 was similar to that observed in 1973 (Hamdy et al., 1978; Kauss, 1986). Nevertheless, maximum concentrations of pollutants such as iron, zinc, and cyanide had decreased over the 10-year period, as had the areal extent of high concentrations (Fig. 11). The latter effect was also noted for oil and grease, although maximum levels were similar in both years (Fig. 12). These decreases in the degree of contamination of surficial sediments are generally substantiated by changes in the vertical concentration profiles of zinc and oil and grease in a core obtained from downstream Lake George in 1986 (Fig. 13), suggesting that some progress has been made in reducing loadings of these as well as other contaminants over the years.

The patterns of contamination by PCBs and polycyclic aromatic hydrocarbons (PAHs) were similar to that of metals and oil and grease, although elevated levels were also found along the Sault Ste. Marie, Michigan, shore in 1983 and 1985 (Kauss, 1986; Kauss & Hamdy, 1991). Although information on the distribution of these contaminants in earlier surficial samples is limited, data on total PCBs and PAHs obtained

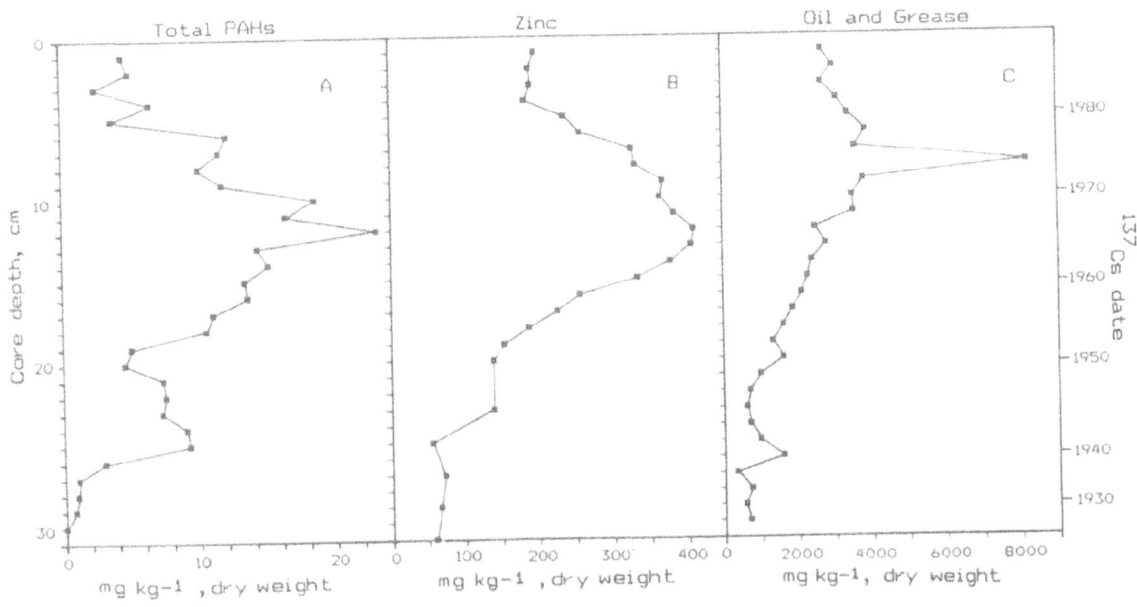

*Fig. 13.* Temporal trends (1930–86) of oil and grease, zinc and total polycyclic aromatic hydrocarbons (PAHs) in a Lake George sediment core. Sources: A and B – UGLCCS (1989); C – Ontario Ministry of the Environment unpubl. data.

from the Lake George core indicates a similar historical trend of declining concentrations in recent years from maxima observed in the late 1960s to early 1970s (Fig. 13 & UGLCCS, 1989).

## 4. Biota

### 4.1. Algae

Lake Superior is the source of most of the phytoplankton in the St. Marys River, as shown by the similarity in relative taxonomic composition (Fig. 14). Diatoms, chlorophytes, and chrysophytes are dominant in both water bodies. Biomass and productivity of phytoplankton are also similar in the river and in the open waters of Lake Superior. Little work has been done on the nanno- and micro-phytoplankton of the river, but

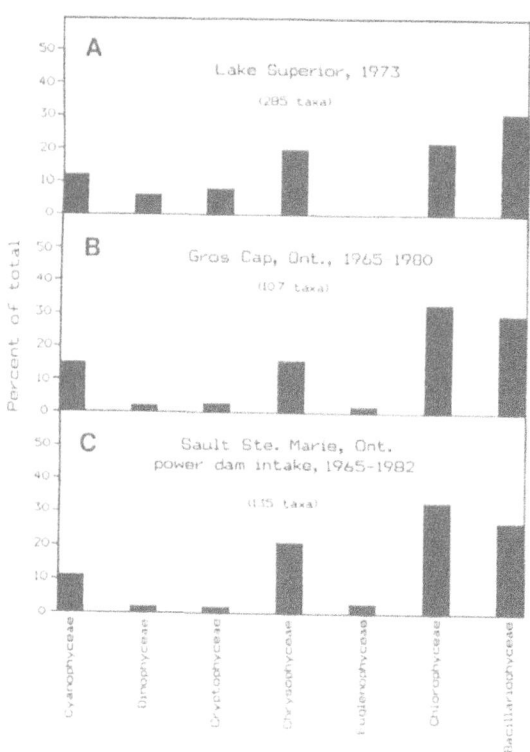

*Fig. 14.* Total number of phytoplankton taxa identified in Sault Ste. Marie, Ontario municipal water intakes and in Lake Superior. Sources: A – Munawar & Munawar (1978), identification to species level; B and C – Hopkins (1986), identification to genus level.

of the macrophytoplankton, diatoms typical of oligotrophic waters are dominant. In all, 72 diatom species were identified in Lake Nicolet (Liston *et al.*, 1986). Many of these were planktonic, while the remainder were benthic. The river community (particularly that of open waters) is transient, since it is continually being transported downstream (Duffy *et al.*, 1987). Benthic diatoms constituted up to 40 percent of the total algal cell volume in the St. Marys River plume entering Lake Huron (Kreis *et al.*, 1983).

Bioassays conducted with indigenous phytoplankton in Munuscong Lake revealed a significant enhancement of ultra-, micro-, and macrophytoplankton productivity during sediment resuspension resulting from the passage of a large vessel (Munawar, 1987). This suggests the absence of pronounced sediment-bound toxicity in the lower river, but not necessarily in the upper river, where sediments are more contaminated (Figs. 11 and 12).

Diatoms constitute the majority of periphyton growing on shoots of emergent macrophytes (Edwards *et al.*, 1989). The periphyton community develops rapidly after ice-out but fluctuates considerably as a result of dessication, ice development, and shoot decay (Duffy *et al.*, 1987).

### 4.2. Macroalgae and macrophytes (wetlands)

Most of the information concerning these communities is derived from studies in Lake Nicolet and downstream reaches of the river. These studies showed that submersed wetlands spread as meadows of low-growing plants at depths of 2 to 16 m in broad areas of the river that have silty-clay or sandy-clay substrate and good water clarity (Liston *et al.*, 1986; Duffy *et al.*, 1987). Of the 22 documented species (Liston *et al.*, 1986; McNabb *et al.*, 1986), biomass is dominated by the macroalgae *Chara globularis* and *Nitella flexilis* and the macrophyte *Isoetes riparia*. These species sometimes occur in monotypic stands. Other species common to submersed wetlands include *Eleocharis acicularis*, *Myriophyllum tenellum* and various species of *Potamogeton*.

Fig. 15. Emergent wetlands (shaded areas) in the St. Marys River.

The boundaries, species composition, and biomass of these wetlands have remained stable due to the perennial nature of the dominant species. Biomass of the three dominant species is relatively constant throughout the growing season (May to September), owing largely to the *in situ* mineralization of the over-wintering biomass which occurs as the new shoots develop (Duffy *et al.*, 1987).

Suitable substrates along unmodified shorelines of the river, particularly those protected from wind and waves, usually have well-developed stands of emergent vegetation (Fig. 15).

Some 42 species of submersed and emergent

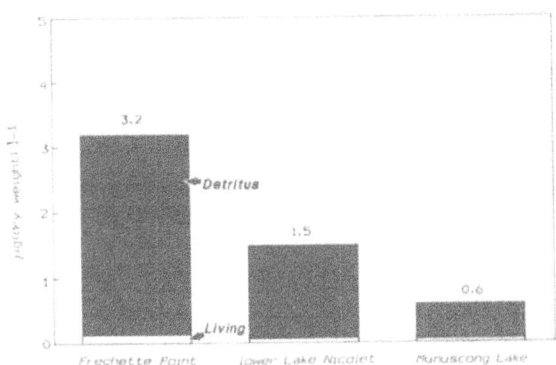

Fig. 16. Biomass of drifting aquatic plant material in littoral waters of the St. Marys River, February–March, 1985. Source: Jude *et al.* (1986).

macrophytes have been identified in the river (Liston *et al.*, 1986). Again, biomass is dominated by three main emergent species (*Eleocharis smallii*, *Scirpus acutus* and *Sparganium eurycarpum*), with submersed species forming a diffuse understory of low biomass (Duffy *et al.*, 1987).

The production and maintenance of emergent wetlands in the river results from vegetative and clonal growth. This type of growth tends to result in monotypic stands. The persistent nature of such stands has maintained a relatively permanent structure in emergent wetlands for at least 30 years (Duffy *et al.*, 1987).

As with the submersed wetland dominants, the rootstocks of dominant emergent macrophytes are present in the substratum year-round. However, the onset of ice-out, warmer temperatures, and increased light during spring (Fig. 8) results in the generation and rapid growth of new shoots during June and July followed by rootstocks in August–September. This annual cycle of plant growth and the persistent nature of rootstocks helps to protect the nearshore sediments from erosion (Duffy *et al.*, 1987).

Table 4 compares the net primary production of phytoplankton, submersed macrophytes, and emergent wetlands in Lake Nicolet. This shows that, despite the larger water area inhabited by phytoplankton, submersed and emergent wetlands produce 96 percent of the annual dry weight biomass because of their high productivity per unit area. Furthermore, the shoots and rootstocks

of emergent wetlands compose the majority of the macrophyte biomass. These relative productivities are probably representative of other broad, lake-like areas of the river with low turbidity (Duffy *et al.*, 1987).

There is a tight cycling of nutrients in wetlands, because dead shoot material is rapidly mineralized in spring and partially utilized by new growth (Duffy *et al.*, 1987). Nevertheless, there is some loss of detritus and, to a minor degree, living material to both offshore and downstream areas of the river during ice-out and as a result of wave action. Drifting plant detritus in littoral waters was greater at Frechette Point just downstream of Sault Ste. Marie, Michigan at the inlet to Lake Nicolet than at the head of the river near Whitefish Bay or in the lower reaches of the river (Poe & Edsall, 1982; Jude *et al.*, 1986) and was lower in late winter than in spring. This biomass drift (Fig. 16) may constitute an important mechanism for the redistribution of nutrients as well as any incorporated or adsorbed contaminants, both within and out of the system (Manny *et al.*, 1989). For example, the net weight of plant drift leaving the river via the St. Joseph Channel during the period April–October 1986 was 74.1 tonnes, wet weight (Manny *et al.*, 1987).

Although some of the nutrients released during the decomposition of plant litter are recycled into new plant biomass, much of the remainder supports secondary production of zooplankton and benthic macroinvertebrates. For example, benthic

*Table 4.* Annual net primary production in the Lake Nicolet reach of the St. Marys River.

| Community type | Hectares occupied | Productivity | |
|---|---|---|---|
| | | g AFDW m$^{-2}$ y$^{-1}$ | tonnes AFDW y$^{-1}$ |
| Phytoplankton | 3598 | 5 | 198 |
| Submersed macrophytes* | 2100 | 35 | 735 |
| Emergent wetlands: | 298 | | |
| shoots | | 650 | 1937 |
| periphyton | | 12 | 36 |
| rootstocks | | 930 | 2771 |

Source: McNabb, unpubl. data (in Duffy *et al.*, 1987).
Notes: AFDW = ash-free dry weight; * = periphyton of submersed macrophytes not included.

macroinvertebrate increases observed in emergent wetlands occur during the pulses of macrophyte decay in early spring and of periphyton production in late summer. In contrast, the maximum biomass of zooplankton coincides with peaks in water temperature and phytoplankton availability (Duffy *et al.*, 1987).

### 4.3. Zooplankton

Zooplankton may form an important link between phytoplankton and plant detritus as well as with higher trophic levels, but details of their role in the St. Marys River are limited.

Some 30 species of zooplankton enter the river from upstream Whitefish Bay. Of these, calanoid copepods, cyclopoid copepods, and cladocerans accounted for 48, 39, and 13 percent, respectively (Selgeby, 1975). The zooplankton community of open waters in the lower river was very similar in species composition to the summer community of the upper river (Thomas & Liston, 1986) but not as abundant. In contrast to pelagic waters, 87 percent of the zooplankton of an emergent wetland were composed of cladocerans (Fig. 17) and densities were at least an order of magnitude greater (Duffy, 1985).

### 4.4. Benthic invertebrates

Little quantitative data exists on the micro- and meio-benthos of the river, but nematodes, ostracods, and *Hydra* species are probably the most abundant components of these two groups (Schirripa, 1983; Duffy, 1985; Burt *et al.*, 1988). In contrast, considerable information is available on benthic macroinvertebrates. The St. Marys River community is diverse, with 303 individual taxa identified (Duffy *et al.*, 1987). Community composition is mainly influenced by substrate, depth, water temperature, currents and wave action, the presence and density of aquatic vegetation, and in certain areas, point source pollution (Duffy *et al.*, 1987; Burt *et al.*, 1988).

The high energy environment evidenced by the

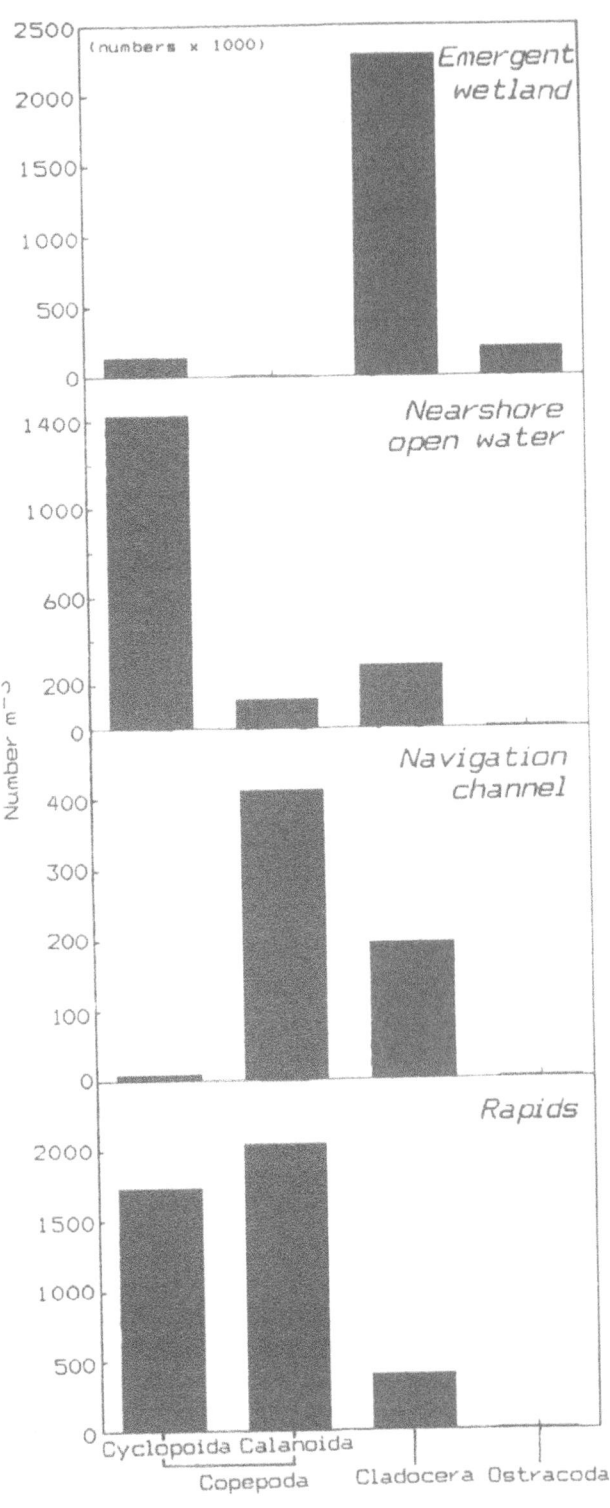

*Fig. 17.* Abundance of zooplankton by taxonomic group in four different habitats of the St. Marys River. Sources: Duffy (1985); Selgeby (1975); Thomas & Liston (1986).

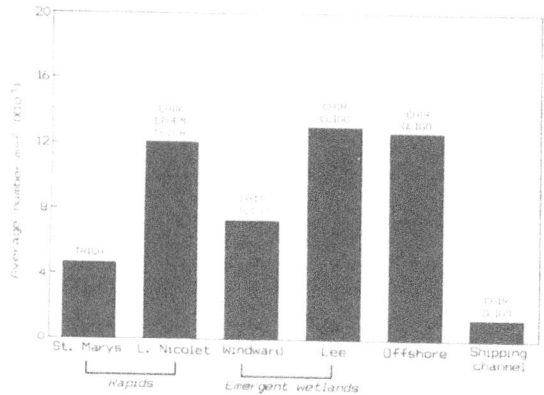

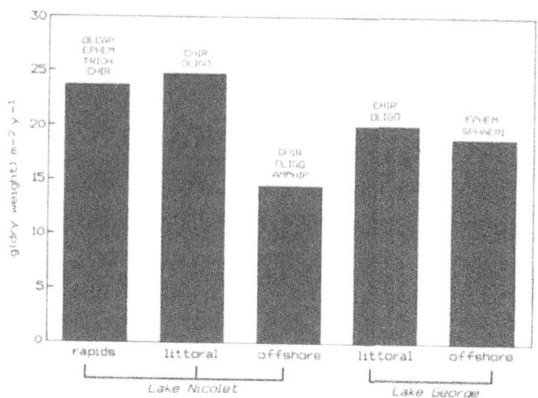

*Fig. 18.* Abundance and dominant groups of benthic macroinvertebrates in different habitats of the St. Marys River. CHIR = Chironomidae; EPHEM = Ephemeroptera; OLIGO = Oligochaeta; TRICH = Trichoptera. Sources: Duffy *et al.* (1987); Koshinsky & Edwards (1983); Schirripa (1983).

*Fig. 19.* Estimated annual benthic macroinvertebrate production in different habitats of Lakes Nicolet and George in the St. Marys River. AMPH = Amphipoda; CHIR = Chironomidae; DECAP = Decapoda; EPHEM = Ephemeroptera; OLIGO = Oligochaeta; SPHAERI = Sphaeriidae; TRICH = Trichoptera. Source: Duffy *et al.* (1987).

loose, coarse substrate in shallow waters of the upper river generally hinders colonization. Nevertheless, Chironomidae dominate the macroinvertebrate community in sandy substrates near the head of the river; however, chironomids are most numerous in the middle reaches of the river where substrates are composed of finer silts and clays. As one moves farther downstream, Oligochaeta tend to compose an increasing proportion of the benthic communities (Duffy *et al.*, 1987).

Both chironomids and oligochaetes are numerically abundant in soft substrate, emergent wetland, rapids, and shipping channel habitats (Duffy *et al.*, 1987), composing 60 to 90 percent of the benthos at a given site (Edwards *et al.*, 1989). Ephemeroptera and Trichoptera, because of their larger size and biomass, are also important in

some habitats (Fig. 18). Navigation channels have a low diversity and density of macroinvertebrates, probably as a result of vessel-induced turbulence and the loss of soft substrates through dredging (Duffy *et al.*, 1987). Ship traffic has also caused an increase in the number of benthic invertebrates and zooplankton, as well as the amount of plant material and detritus, in the water column of wetlands (Poe & Edsall, 1982; Jude *et al.*, 1986), and these can be subsequently carried out into the river and downstream (Duffy, 1985).

In Lake Nicolet, soft-bottom benthos production is dominated by Oligochaeta, Chironomidae, and Amphipoda, whereas Ephemeroptera and Sphaeriidae are dominant in Lake George (Fig. 19).

*Table 5.* Annual secondary production in the Lake Nicolet reach of the St. Marys River.

| Habitat type | Hectares occupied | Productivity | |
|---|---|---|---|
| | | g dry wt m$^{-2}$ y$^{-1}$ | tonnes dry wt y$^{-1}$ |
| Soft bottom benthos | 2647 | 14 | 382 |
| Emergent wetland benthos | 298 | 25 | 74 |
| Emergent wetland zooplankton | 298 | 0.56 | 1.67 |
| Rapids benthos | 1 | 24 | 0.24 |

Source: Duffy, unpubl. data (in Duffy *et al.*, 1987).

The productivity of benthic macroinvertebrate communities in soft bottom, deeper offshore waters is generally less than in littoral areas or the rapids (Fig. 19). However, the annual production of benthos alone in the former outweighs by a factor of five that of zooplankton and benthos combined in the latter (Table 5), because the soft bottom habitats occupy a much larger portion of the river bed than littoral or rapids habitats.

Contaminants have also affected soft-bottom benthic communities in the river (Veal, 1968; Hamdy *et al.*, 1978; Kenaga, 1979; Burt *et al.*, 1988). Although the benthos on the U.S. side of the river is indicative of clean water communities, sediments downstream of Sault Ste. Marie, Ontario municipal and industrial discharges are contaminated and dominated by oligochaetes such as *Tubifex tubifex* and *Limnodrilus hoffmeisteri* which are tolerant of organic pollution. Also, the burrowing mayfly *Hexagenia limbata* has largely been eliminated from these areas as far downstream as Lake George (Fig. 20). Other studies (Schloesser *et al.*, 1990; Burt *et al.*, 1988) have shown that the abundance of *Hexagenia* nymphs is reduced in areas where visible oil is present in sediments (Hiltunen & Schloesser, 1983). Furthermore, dry weight production of nymphs was reduced in sediments that contained concentrations of metals, cyanide, and oil and grease in excess of guidelines for the classification of polluted sediments (Edsall *et al.*, 1991). This adverse impact is potentially of great significance, due to the central role of Ephemeroptera such as *Hexagenia* in trophic interactions in the river (Duffy *et al.*, 1987).

Current data indicates decreases in contaminants levels in water and sediments and some improvement in the benthic communities along the Ontario shoreline. Nevertheless, in 1985, a zone of severe to moderate benthic impairment, similar to that shown previously for contaminants in surficial sediments (Figs. 11 and 12), was still evident along the Sault Ste. Marie waterfront, as well as a moderate impairment zone in Little Lake George and upper Lake George (Burt *et al.*, 1988; McKee *et al.*, 1991).

Recent investigations into contaminant levels

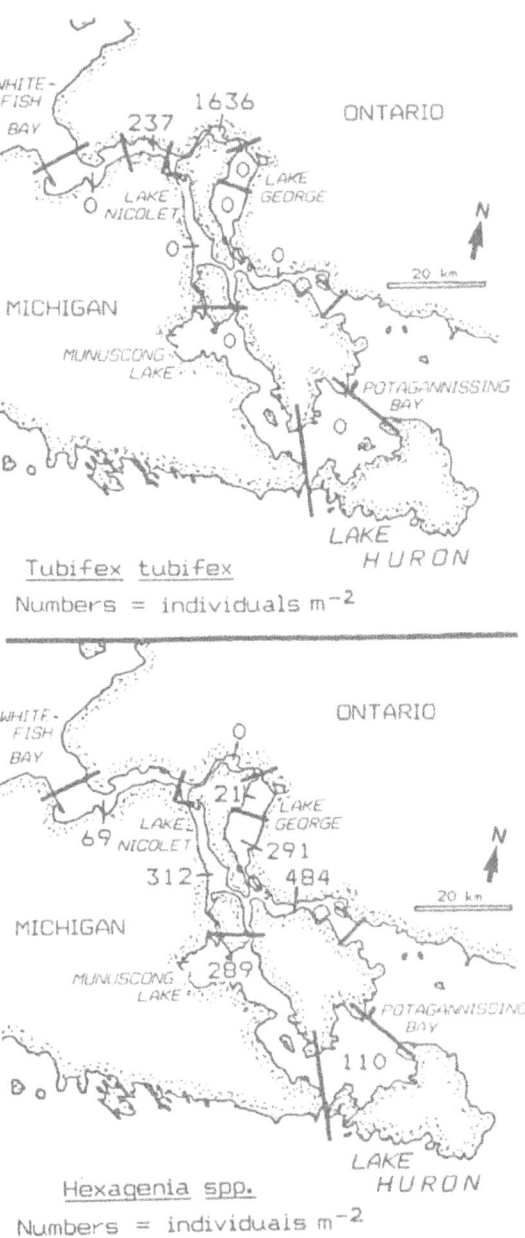

*Fig. 20.* Abundance of *Tubifex tubifex* (Oligochaeta) and *Hexagenia* (Ephemeroptera) in the St. Marys River in 1968. Sources: Veal (1968); figures from Duffy *et al.* (1987).

in benthic macroinvertebrates have focused on Oligochaeta (*T. tubifex* and *L. hoffmeisteri*) because of their numerical abundance in the river. A study in 1983 showed that, although more than 75 percent of the cadmium, copper, lead, and zinc in sediments were in potentially available forms,

*Table 6.* Concentrations of arsenic and metals in sediment (mg kg$^{-1}$, dry wt) and benthic macroinvertebrate (Oligochaeta) tissue (mg kg$^{-1}$, dry wt, gut-corrected) from different locations in the St. Marys River, 1983.

| Station | Copper | | | Zinc | | | Lead | | | Cadmium | | |
|---|---|---|---|---|---|---|---|---|---|---|---|---|
| | Sediment | Benthic tissue | BCF | Sediment | Benthic tissue | BCF | Sediment | Benthic tissue | BCF | Sediment | Benthic tissue | BCF |
| 45 | 23.1 | 22.5 | 1.0 | 118.7* | 131.3 | 1.1 | 54.4* | 2.4 | 0.0 | 1.0 | 0.3 | 0.3 |
| 46 | 166.9* | 24.0 | 0.1 | 730.3* | 115.8 | 0.2 | 217.4* | 0.2 | 0.0 | 3.5* | 0.3 | 0.1 |
| 47 | 89.4* | 11.4 | 0.1 | 493.8* | 107.5 | 0.2 | 157.1* | 1.2 | 0.0 | 2.0* | 0.3 | 0.2 |
| 48 | 168.1* | 17.0 | 0.1 | 947.3* | 110.7 | 0.1 | 619.3* | 6.8 | 0.0 | 4.5* | 0.3 | 0.1 |

| Station | Iron | | | Manganese | | | Mercury | | | Arsenic | | |
|---|---|---|---|---|---|---|---|---|---|---|---|---|
| | Sediment | Benthic tissue | BCF | Sediment | Benthic tissue | BCF | Sediment | Benthic tissue | BCF | Sediment | Benthic tissue | BCF |
| 45 | 6272.3 | 1030.2 | 0.2 | 64.9 | 39.6 | 0.6 | 0.1 | 0.5 | 5.0 | NA | NA | NA |
| 46 | 15043.8 | 258.2 | 0.0 | 197.1 | 12.0 | 0.1 | 0.6 | 0.3 | 0.5 | 34.3 | 18.0 | 0.5 |
| 47 | 8949.2 | 715.1 | 0.1 | 128.0 | 8.5 | 0.1 | 0.4 | 0.1 | 0.3 | NA | NA | NA |
| 48 | 13280.3 | 104.2 | 0.0 | 362.4 | − 1.2** | 0.0 | 0.6 | 0.0 | 0.0 | NA | NA | NA |

Source: Persaud *et al.* (1987).

Notes: Station locations shown on Fig. 7.

* Exceeds Ontario Ministry of the Environment guideline for the open water disposal of dredged materials.

** Negative value is the result of extremely low levels of Mn in benthic tissues compared with the sediment concentration.

BCF = biological concentration factor.

NA = not available.

oligochaetes from most locations had low contaminant concentrations. In fact, the only location with a biological concentration factor of 1 or more for copper, zinc, and mercury had the lowest bulk sediment concentrations of these contaminants (Station 45, Table 6). This lower accumulation of metals in polluted areas may be related to the higher organic carbon and oil and grease content of these sediments (Persaud *et al.*, 1987).

As noted previously (e.g., Fig. 13), PAHs are also of concern in the river. Data from 1984 and 1985 studies (Kauss, 1986; Kauss & Hamdy, 1991) showed that numerous PAHs were biologically available to filter-feeding organisms such as unionid mussels, particularly in the vicinity of industrial point sources such as Algoma Steel (Fig. 21).

## 4.5. Fish

A total of 75 fish species representing 22 families have been identified in the river (Duffy *et al.*, 1987), although many of these were considered transients or rare. These species include both coldwater and warmwater fish. This diverse community results from the river's connections with the fish communities of oligotrophic Lake Superior and Lake Huron, as well as from the large variety of habitats available in the river. Generally, the fish community of the St. Marys River may be described as percid because it contains walleye, northern pike, yellow perch, and white sucker, species that contribute to the persistence of the community (Duffy *et al.*, 1987). A number of species including rainbow smelt, coho salmon, chinook salmon, pink salmon, alewife, and sea lamprey have been introduced to the Great Lakes accidentally or intentionally, or through gradual invasion from the Atlantic Ocean. Rainbow smelt is an important forage fish in the St. Marys River (USACE, 1988) and the sea lamprey is abundant in the St. Marys Rapids. The latter species is parasitic and predatory on larger fish, and has had a major negative impact on Great Lakes fisheries. It has been recorded in

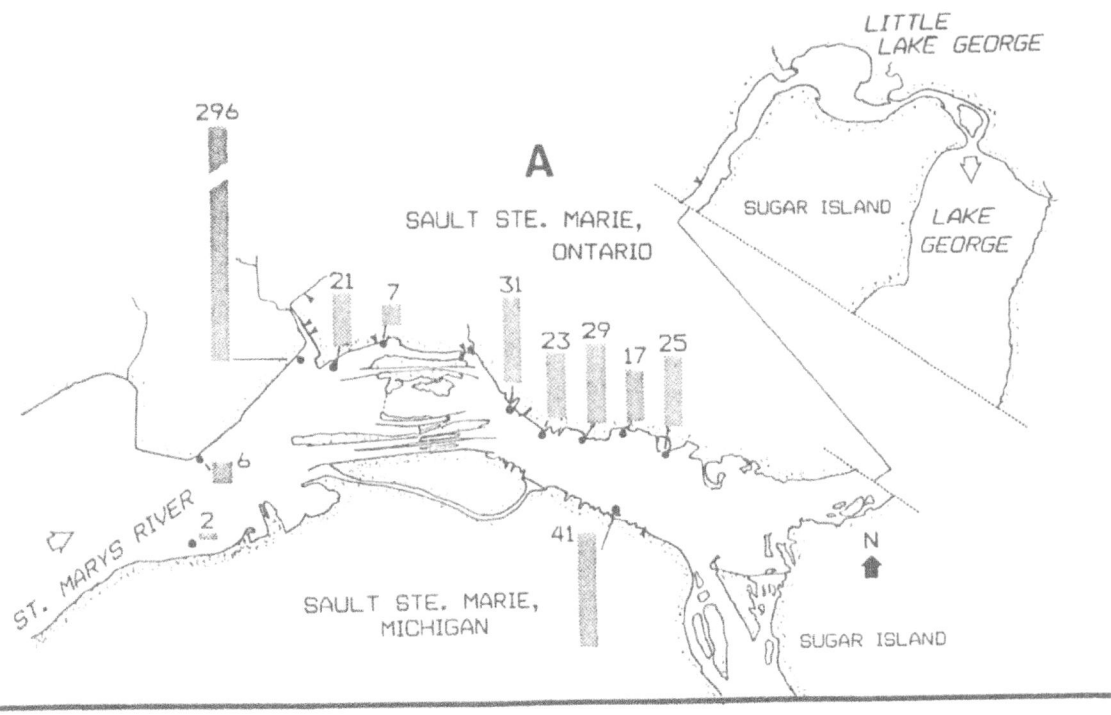

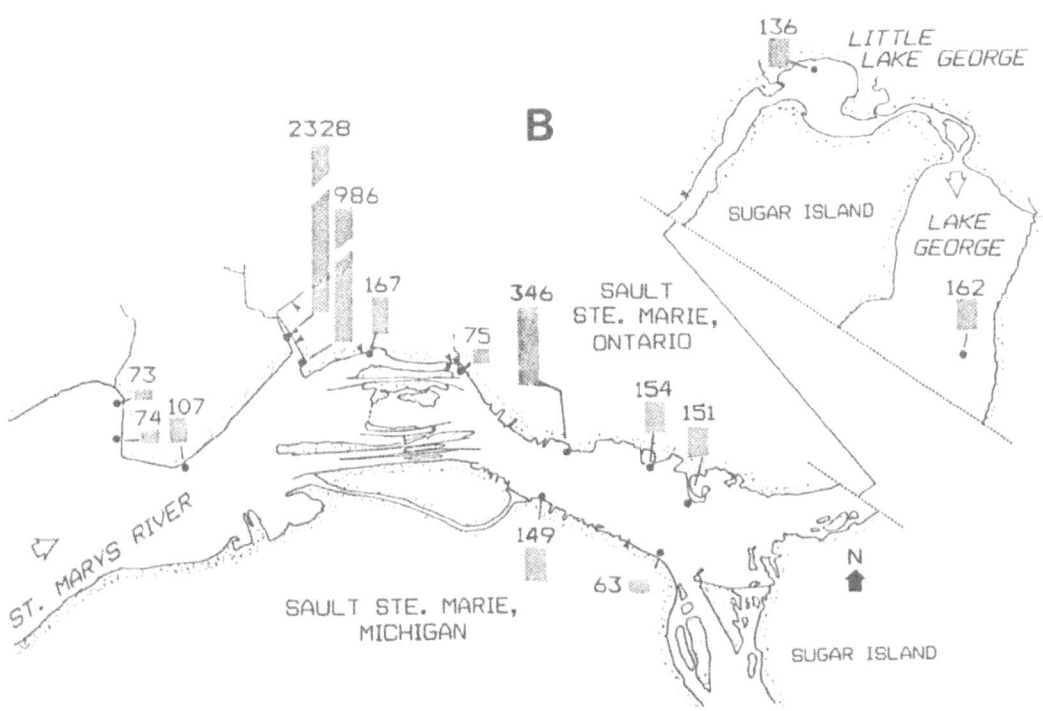

*Fig. 21.* Concentrations ($\mu$g kg$^{-1}$, wet wt) of phenanthrene (A) and total polycyclic aromatic hydrocarbons (B) in caged mussels (*Elliptio complanata*) after 3 week's exposure at different locations in the St. Marys River in 1984 and 1985, respectively. Sources: A – Kauss (1986); B – Kauss & Hamdy (1991).

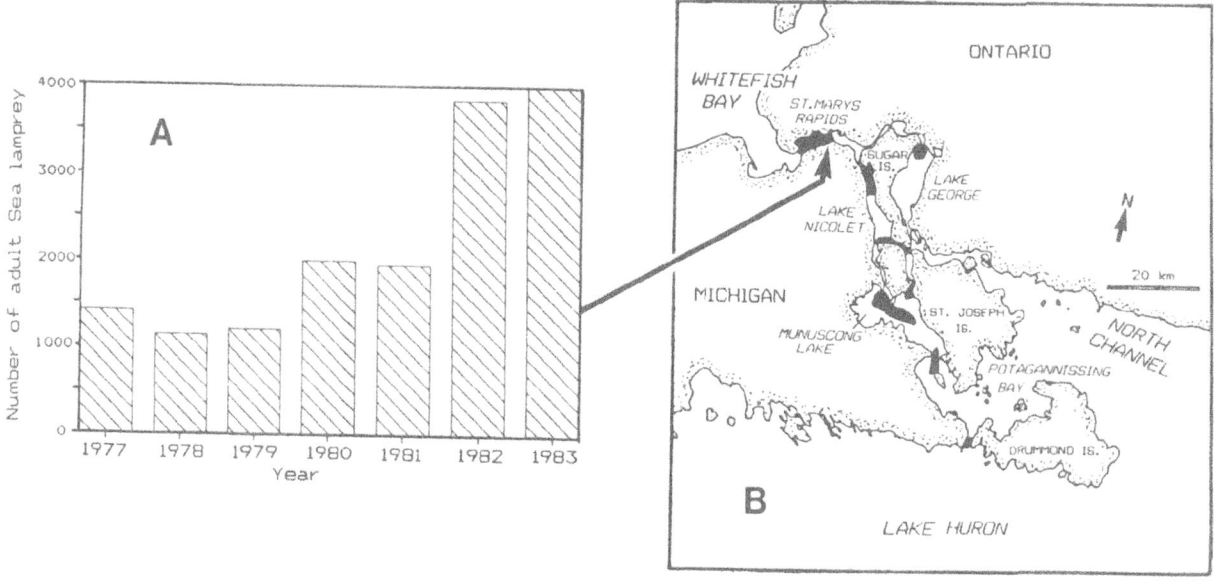

*Fig. 22.* Abundance of adult sea lamprey below the U.S. Army Corps of Engineers hydroelectric plant at Sault Ste. Marie, Michigan (A) and the spatial distribution of sea lamprey ammocoetes (shaded areas) in the St. Marys River (B). Sources: Daugherty & Purvis (1985); Daugherty *et al.* (1984).

*Table 7.* Predominant fish species in the primary habitats of the St. Marys River.

| Scientific name (common name) | Habitat | | | |
|---|---|---|---|---|
| | Open-water & embayments | Emergent wetlands | Sand & gravel beaches | St. Marys Rapids |
| Amiidae | | | | |
| *Amia calva* (bowfin) | | L, J | | |
| Acipenseridae | | | | |
| *Acipenser fulvescens* (lake sturgeon) | A | | | |
| Catostomidae | | | | |
| *Catostomus catostomus* (longnose sucker) | | | | A |
| *Catostomus commersoni* (white sucker) | A, L, J | | | A |
| Centrarchidae | | | | |
| *Lepomis macrochirus* (bluegill) | | L, J | | |
| *Micropterus dolomieui* (smallmouth bass) | A | A, L, J | | |
| *Pomoxis nigromaculatus* (black crappie) | A | | | |
| Clupeidae | | | | |
| *Dorosoma cepedianum* (gizzard shad) | | A | | |
| Cottidae | | | | |
| *Cottus bairdi* (mottled sculpin) | A | | | |
| *Cottus cognatus* (slimy sculpin) | | | | A |

*Table 7* (Continued). Predominant fish species in the primary habitats of the St. Marys River.

| Scientific name (common name) | Habitat | | | |
|---|---|---|---|---|
| | Open-water & embayments | Emergent wetlands | Sand & gravel beaches | St. Marys Rapids |
| **Cyprinidae** | | | | |
| *Cyprinus carpio* (carp) | | L | | |
| *Notropis atherinoides* (emerald shiner) | | A, L | A | |
| *Notropis cornutus* (common shiner) | | A, L | A | |
| *Notropis hudsonius* (spottail shiner) | A | A, L | A | |
| *Notropis volucellus* (mimic shiner) | A | A, L | A | |
| *Pimephales notatus* (bluntnose minnow) | | A, L | | |
| *Rhinichthys cataractae* (longnose dace) | | | | A |
| **Esocidae** | | | | |
| *Esox lucius* (northern pike) | A | A, L, J | | |
| **Gladidae** | | | | |
| *Lota lota* (burbot) | J | | | |
| **Gasterosteidae** | | | | |
| *Pungitius pungitius* (ninespine stickleback) | A | | | |
| **Ictaluridae** | | | | |
| *Ictalurus nebulosus* (brown bullhead) | | J | | |
| **Lepisosteidae** | | | | |
| *Lepisosteus osseus* (longnose gar) | | J | | |
| **Osmeridae** | | | | |
| *Osmerus mordax* (rainbow smelt) | A, L | | | |
| **Percidae** | | | | |
| *Etheostoma nigrum* (johnny darter) | A | | | |
| *Perca flavescens* (yellow perch) | J | A, J, L | | |
| *Stizostedion vitreum vitreum* (walleye) | A | A | J | A |
| **Percopsidae** | | | | |
| *Percopsis omiscomaycus* (trout-perch) | A | | A | |
| **Petromyzontidae** | | | | |
| *Petromyzon marinus* (sea lamprey) | | | | A, L |
| **Salmonidae** | | | | |
| *\*Coregonus artedii* (lake herring) | A | | | |
| *Coregonus clupeaformis* (lake whitefish) | A | | | A |
| *Salmo gairdneri* (rainbow trout) | | | | A |
| *Salmo trutta* (brown trout) | | | | A |
| *Salvelinus fontinalis* (brook trout) | | | | A |
| *Salvelinus namaycush* (lake trout) | | | | A |
| *Oncorhynchus gorbuscha* (pink salmon) | A | | | |
| *Oncorhynchus tshawytscha* (chinook salmon) | A | | | A |
| **Umbridae** | | | | |
| *Umbra limi* (central mud minnow) | | L | | |

Sources: Liston *et al.* as cited in Duffy *et al.* (1987).

Notes:  '*'  = on Michigan threatened species list.

'A'  = adult.

'J'  = juvenile.

'L'  = larva.

the river since 1937 (Goodyear *et al.*, 1982). Changed waterflow at the St. Marys Rapids below the U.S. government hydroelectric plant resulted in a near doubling in the numbers of adults caught after 1981 (Fig. 22). Of concern is the fact that the large population in the river is contributing to increased fish mortality in Lake Huron. The large size of the river precludes chemical treatment to control the sea lamprey population that spawns in the river (Daugherty *et al.*, 1984; Daugherty & Purvis, 1985).

The four primary fish habitats of the river and the most abundant associated species are listed in Table 7. However, it is important to note that many species are associated with more than one habitat, either as they mature from larvae to adults or as a result of diel or seasonal migration.

Emergent wetlands serve as spawning, nursery, and feeding areas for 44 species. Depending on the site and time of year, these wetlands may be susceptible to adverse impacts resulting from the passage of large vessels and the associated changes in hydrologic patterns. Although laboratory studies simulating the impact of vessel-induced drawdown or dewatering of littoral habitat have shown adverse effects on the survival of larval walleye and northern pike (Holland & Sylvester, 1983), studies to confirm these results have not yet been made in the St. Marys River.

Embayments as well as tributaries are also important nursery areas (Goodyear *et al.*, 1982) and many of the fish species native to the river spawn in tributaries (e.g., Fig. 23).

Open-water areas of the river provide a wide

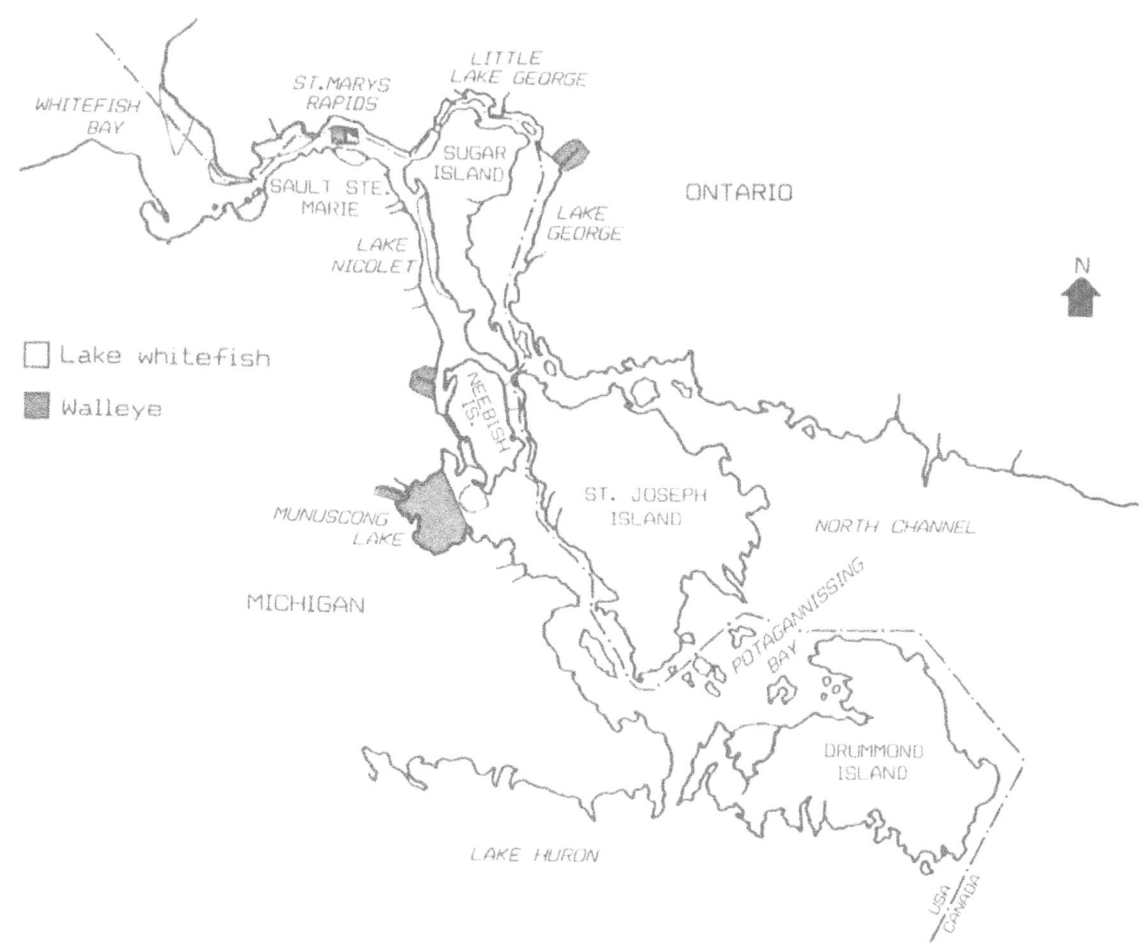

*Fig. 23.* Spawning areas of lake whitefish and walleye in the St. Marys River. Source: Goodyear *et al.* (1982).

variety of habitats, mainly for adult bottom-dwelling (demersal) and open-water species (Table 7).

Beach habitats in the middle of the river are frequented by fish common in emergent wetlands, as well as small demersal species common to open waters (Table 7). Furthermore, species such as lake whitefish, suckers, walleye, trout-perch, and cyprinids may spawn in gravelly areas (Duffy *et al.*, 1987).

The fish community of the St. Marys Rapids is distinct from that of others in the river (Table 7) and provides spawning habitat for such important species as walleye, lake whitefish (Fig. 23), rainbow trout, lake sturgeon, brook trout, slimy sculpin, and possibly also for the introduced pink and chinook salmon (Goodyear *et al.*, 1982). The rapids and other gravelly areas of the river and its tributaries also provide suitable habitat for spawning and nursery habitat for the introduced sea lamprey (Fig. 22). Unfortunately, the rapids have also been subject to dewatering, with attendant adverse impacts on fish habitat, fish fry, and benthic macroinvertebrates. As a consequence, a berm was constructed to maintain water levels along the Canadian side of the rapids during periods of reduced flow (Table 2).

Fish production in the Lake Nicolet reach of the river amounted to 1.2 g AFDW $m^{-2} y^{-1}$, or 29 tonnes AFDW $y^{-1}$ (Edwards *et al.*, 1989). The latter was 0.5 and 6.3 percent, respectively of the annual primary and secondary production in Lake Nicolet.

Commercial fishing for lake trout, lake herring, lake whitefish, and walleye existed in the St. Marys River from the mid-1600s until it was stopped in the early 1900s. However, it is still permitted for lake whitefish and walleye in the North Channel of Lake Huron which borders Potagannissing Bay. The effect of this commercial harvest on these two species that seasonally migrate in and out of the river is presently unknown (Duffy *et al.*, 1987). The St. Marys Rapids area has long been the focus of fishing

*Table 8.* Organochlorine contaminants in juvenile fish from Ontario waters of the St. Marys River (1983).

| Contaminant | Concentration (mg kg$^{-1}$, wet wt) | |
|---|---|---|
| | Agreement objective* | Range of means |
| PCBs (total) | 0.100 | <0.020–0.025 |
| Chlordanes ($\alpha$, $\gamma$) | NA | ND–0.0003 |
| BHC ($\alpha$, $\beta$, $\gamma$) | NA | ND–0.005 |
| DDT and metabolites | 1.000 | 0.005–0.010 |
| Mirex | <detection | ND |
| Chlorinated benzenes (tri-, tetra-, penta-, hexa-; trichlorotoluene, octachlorostyrene) | NA | ND |
| Chlorinated aliphatics (hexachloroethane, hexachlorobutadiene) | NA | ND |
| Chlorinated phenols (tri-, tetra-, penta-) | NA | ND |

Source: Suns *et al.* (1985).
Notes: Range of means for young-of-the-year yellow perch and spottail shiners from Sault Ste. Marie, Ont. and Little Lake George, respectively.
'*' = specific objectives for the protection of birds and animals which consume fish (IJC, 1978).
'NA' = not available.
'ND' = not detected.

activity. Archeological evidence indicates that lake whitefish and lake sturgeon were an important food for native peoples as long as 5 000 years ago (Duffy *et al.*, 1987). Subsistence fishing is still permitted on a limited basis by the Sault Band of the Chippewa Tribe (Duffy *et al.*, 1987). In addition to lake whitefish, trout, salmon, lake herring, and walleye are also taken by subsistence fishermen (Edwards *et al.*, 1989).

Sport fishing is a major recreational activity in the St. Marys River, with about 145 000 angler-days being recorded each year in 1981 and 1982 (Limno-Tech, 1985). Principal species caught are: walleye, yellow perch, and northern pike, with seasonal additions of rainbow smelt and white sucker in spring, lake herring in summer, and chinook and pink salmon in fall (Duffy *et al.*,

1987). Notably, catch per unit effort in the river (excluding the rapids) has declined roughly 3-fold from the 1930s through the 1970s to a present average of about 0.5 fish angler-h$^{-1}$ (Duffy *et al.*, 1987). It has been shown that anglers may harvest a significant percentage of such species as smallmouth bass, northern pike, and muskellunge (Liston *et al.*, 1986).

In the rapids area, the main species caught are rainbow trout and lake whitefish; however, average catch per unit effort is only about 20 percent of that in the remainder of the river (Koshinsky & Edwards, 1983). Ice fishing, mainly for walleye, accounted for about 6,500 angler-days during the winter months of 1984–85 (USACE, 1988).

Thus far, major impacts of pollution, such as those observed for benthic macroinvertebrates,

*Table 9.* Contaminants in adult fish from the St. Marys River and neighbouring Lakes Superior and Huron.

| Contaminant | Concentration (mg kg$^{-1}$, wet wt) | | | |
|---|---|---|---|---|
| | Agreement objective* | L. Superior | St. Marys R. | L. Huron |
| | | a | a | b |
| PCBs (total) | 0.100 | 0.009 | 0.146–1.488 | > 1.000 |
| BHC ($\alpha$) | NA | ND | ND–0.001 | ND–0.019 |
| Chlordanes ($\alpha$, $\gamma$) | NA | 0.0001 | 0.001–0.005 | 0.009–0.110 |
| Nonachlors | NA | 0.0004 | 0.003–0.008 | 0.012–0.120 |
| Dieldrin | NA | 0.0002 | 0.002–0.003 | ND–0.010 |
| DDT and metabolites | 1.000 | 0.002 | 0.010–0.143 | 0.145–0.730 |
| Heptachlor epoxide | NA | ND | ND | ND |
| Mirex | <detection | NA | NA | ND–0.069 |
| Dichlorobenzene | NA | NA | NA | 0.610–1.800 |
| 1, 2. 3, 4-Tetra-chlorobenzene | NA | NA | ND | NA |
| 1, 2, 4, 5-Tetra-chlorobenzene | NA | ND | ND | NA |
| Pentachlorobenzene | NA | ND | ND–0.001 | NA |
| Hexachlorobenzene | NA | ND–0.0001 | 0.0006–0.001 | ND–0.017 |
| Octachlorostyrene | NA | ND | ND–0.017 | ND–0.006 |
| PAHs (total) | NA | NA | NA | 0.045–0.119 |

Sources: a – Jaffe *et al.* (1985), semi-quantitative 1983 data for white sucker and carp from Michigan waters of the St. Marys River and for white sucker from the Tahquamenon River, Whitefish Bay;

b – Zenon (1985), 1983 data for white sucker and brown bullhead from Ontario waters of the North Channel, Lake Huron.

Notes: '*' – specific objectives for the protection of birds and animals which consume fish (IJC, 1978).
'NA' – not available.
'ND' – not detected.

have not been demonstrated for fish (Edsall *et al.*, 1988). Analyses of juvenile (young-of-the-year) yellow perch and spottail shiners collected in 1983 from Ontario waters at Sault Ste. Marie and in Little Lake George showed that concentrations of organochlorine contaminants in whole fish, when detected, were low and below Great Lakes Water Quality Agreement objectives (IJC, 1978) for the protection of higher trophic levels (Table 8).

Data on organochlorine contaminants in whole adult fish homogenates (Table 9) shows that, in general, concentrations in St. Marys River fish were intermediate between those in fish from an upstream tributary to Lake Superior and from downstream Lake Huron. However, in view of the small sample sizes (1 to 5 fish) and the semi-quantitative nature of some of the data, these differences may be only approximate. Also, lack of data on some contaminants precludes comparisons between locations and interpretations (e.g., mirex, chlorinated benzenes, PAHs). With the exception of total PCBs, concentrations in St. Marys River fish were below available objectives for the protection of birds and animals which consume fish. Nevertheless, the lack of agency or Great Lakes Water Quality Agreement objectives for the majority of the contaminants detected makes it difficult to determine their importance to higher trophic levels.

Regular monitoring of the edible (dorsal fillet) portions of sport fish for contaminants has shown that only concentrations of mercury in the larger sizes of certain species are above the Canadian guideline for unlimited consumption by humans (Table 10). Advisories for restricted consumption by anglers apply to larger sizes of longnose sucker, white sucker, walleye, northern pike, and lake trout. However, it should be noted that similar advisories are in effect for walleye caught in Goulais Bay, Lake Superior (OME & OMNR, 1986), suggesting that higher mercury levels are at least partly due to background inputs to the river. Mean mercury concentrations in rainbow trout from the St. Marys Rapids area declined almost 60 percent between 1978 (0.39 mg kg$^{-1}$ wet wt.) and 1985 (0.16 mg kg$^{-1}$); however, more data is required for statistical analysis of trends (OME, unpubl. data).

*Table 10.* Contaminants in dorsal fillets of adult sport fish from Ontario waters of the St. Marys River.

| Contaminant | Concentration (mg kg$^{-1}$, wet wt) | |
|---|---|---|
| | Consumption guideline* | Range |
| Mercury | 0.5 | 0.04–1.30 |
| PCBs (total) | 2.0 | ND–1.260 |
| Chlordanes ($\propto$, $\gamma$) | NA | ND–0.045 |
| Dieldrin | 0.3 | ND |
| DDT and metabolites | 5.0 | ND–0.486 |
| Heptachlor | 0.3 | ND |
| Endrin | 0.3 | NA |
| Lindane ($\gamma$–BHC) | 0.3 | ND–0.001 |
| Mirex | 0.1 | ND |
| Hexachlorobenzene | NA | ND–0.011 |
| Octachlorostyrene | NA | ND–0.009 |

Source: Ontario Ministry of the Environment data (A. Johnson, pers. comm.) for dorsal fillets of various species from St. Marys R. (below Rapids; Lake George; St. Joseph Channel; and St. Joseph Island).

Notes: '*' = Health and Welfare Canada guidelines and/or Great Lakes Water Quality Agreement specific objectives for the protection of human consumers of fish.

'NA' = not available.

'ND' = not detected.

## 4.6. Birds

About 172 species of waterfowl, colonial waterbirds, shorebirds, passerines, and raptors are associated with the riparian areas of the river, either as residents or as temporary inhabitants. The river is an important staging area and migration corridor for dabbling ducks, diving ducks, and geese (Fig. 24). Ice-free areas in the St. Marys Rapids, along the Sault Ste. Marie, Ontario shoreline, and near the outflow area of the Edison Sault Canal in Michigan are also important to over-wintering mallards, black ducks, Canada geese, common goldeneye, common mergansers, and greater and lesser scaup (Robinson & Jensen, 1980; Weise, 1985).

The river is a breeding and rearing area for mallards, common mergansers, black ducks, Canada geese, common goldeneye, blue-winged

29

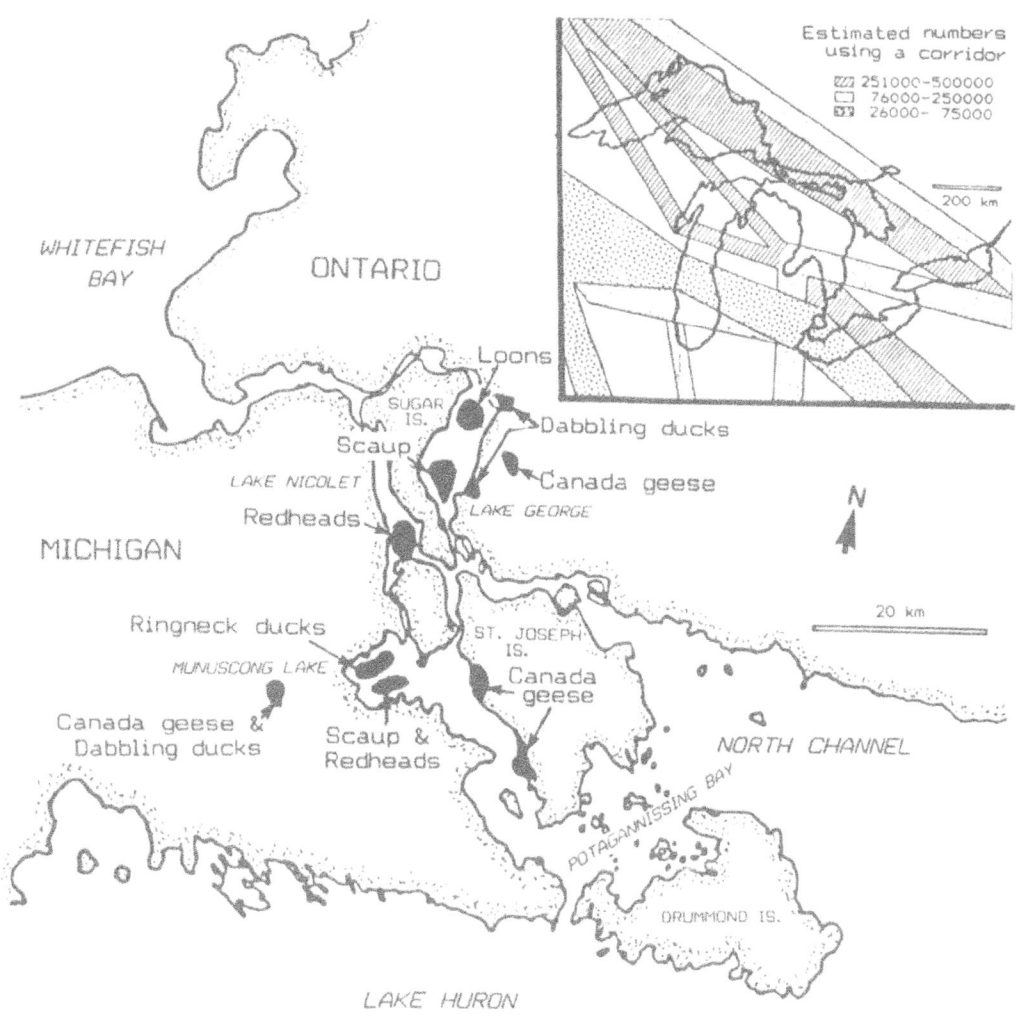

*Fig. 24.* Areas of waterfowl congregation in the St. Marys River (shaded areas) during the fall and spring. Sources: Thomas, Ontario Ministry of Natural Resources, unpubl. manuscr.; figure from Duffy *et al.* (1987). Inset shows major migration corridors of diving ducks in the Great Lakes region. Sources: Bellrose (1968); figure from Duffy *et al.* (1987).

teals, black ducks, American wigeons, American coots, northern pintails, common loons, and ring-necked ducks. Colonial waterbirds nesting on the many islands and in the marshes of the river include ring-billed gulls, common terns, double-crested cormorants, great blue herons, black terns, herring gulls, and black-crowned night herons (Duffy *et al.*, 1987; also Fig. 25). Ring-billed gull populations have been increasing dramatically in the Great Lakes (Fig. 25) and double-crested cormorants are staging a significant recovery (Edsall *et al.*, 1988).

Raptors including northern bald eagles, osprey, snowy owls, great gray owls, gyrfalcons, peregrine falcons, and burrowing owls are attracted to the river by the diversity of habitats (Duffy *et al.*, 1987). For example, northern bald eagles presently nest in two locations on Sugar Island and the number of active nests has remained at one or two from 1974 to 1985. The number of active nests of osprey increased dramatically between 1977 and 1982 and has stabilized at 15 or 16 nests (Fig. 26).

Production data for birds in the river is limited; however, the number of young produced by com-

30

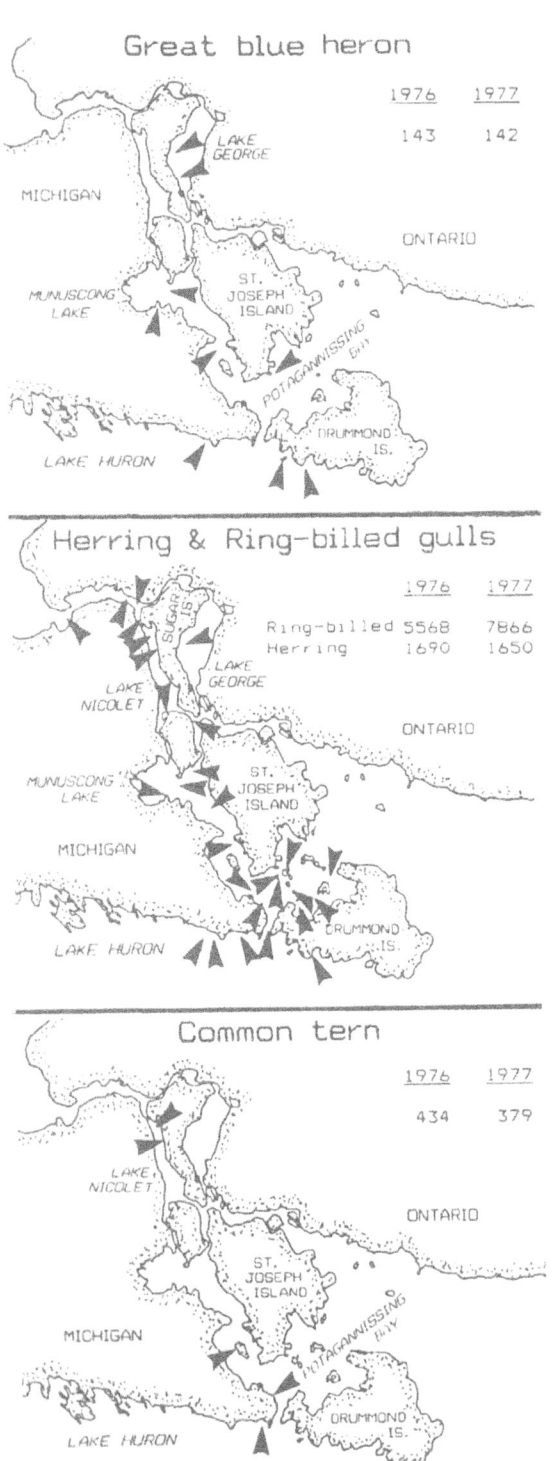

**Great blue heron**

| | 1976 | 1977 |
|---|---|---|
| | 143 | 142 |

LAKE GEORGE

MICHIGAN

ONTARIO

MUNUSCONG LAKE

ST. JOSEPH ISLAND

POTAGANNISSING BAY

DRUMMOND IS.

LAKE HURON

**Herring & Ring-billed gulls**

| | 1976 | 1977 |
|---|---|---|
| Ring-billed | 5568 | 7866 |
| Herring | 1690 | 1650 |

SUGAR IS.

LAKE GEORGE

LAKE NICOLET

ONTARIO

MUNUSCONG LAKE

ST. JOSEPH ISLAND

MICHIGAN

DRUMMOND IS.

LAKE HURON

**Common tern**

| | 1976 | 1977 |
|---|---|---|
| | 434 | 379 |

LAKE NICOLET

ONTARIO

ST. JOSEPH ISLAND

MICHIGAN

POTAGANNISSING BAY

DRUMMOND IS.

LAKE HURON

*Fig. 25.* Nesting sites of selected colonial waterbirds in the St. Marys River (indicated by arrows) and their estimated numbers in 1976 and 1977. Sources: Scharf (1978); figures from Duffy *et al.* (1987).

*Table 11.* Waterfowl, colonial waterbirds and raptors which are or could be used as important water quality indicators in the St. Marys River.

| Scientific name | Common name |
|---|---|
| **ANSERIFORMES** | |
| Anatidae: | |
| *Anas acuta* | Northern pintail |
| *Anas discors* | Blue-winged teal |
| *Anas platyrhynchos* | Mallard |
| *Anas rubripes* | American black duck |
| *Anas strepera* | Gadwall |
| *Aythya affinis* | Lesser scaup |
| *Aythya americana* | Redhead |
| *Aythya collaris* | Ring-necked duck |
| *Aythya marila* | Greater scaup |
| *Aythya valisineria* | Canvasback |
| *Branta canadensis* | Canada goose |
| *Bucephala clangula* | Common goldeneye |
| *Lophodytes cucullatus* | Hooded merganser |
| *Mareca americana* | American wigeon |
| *Mergus merganser* | Common merganser |
| *Mergus serrator* | Red-breasted merganser |
| **CHARADRIIFORMES** | |
| Laridae: | |
| *Chlidonias niger* | Black tern |
| *Larus argentatus* | Herring gull |
| *Larus delawarensis* | Ring-billed gull |
| *Sterna hirundo* | Common tern |
| **CICONIIFORMES** | |
| Ardeidae: | |
| *Ardea herodias* | Great blue heron |
| **FALCONIFORMES** | |
| Accipitridae: | |
| *Buteo lineatus* | Red-shouldered hawk |
| *Circus cyaneus* | Northern harrier |
| *Haliaeetus leucocephalus* | Bald eagle |
| Pandionidae: | |
| *Pandion haliaetus* | Osprey |
| **GAVIIFORMES** | WATERFOWL |
| Gaviidae: | |
| *Gavia immer* | Common loon |
| **PELECANIFORMES** | |
| Phalacrocoracidae: | |
| *Phalacrocorax auritus* | Double-crested cormorant |
| **STRIGIFORMES** | |
| Strigidae: | |
| *Bubo virginianus* | Great horned owl |

Source: Edsall *et al.* (1988).

mon terns in 1984 varied from 2.2 per nest on Raber Island, to 0.43 on Steamboat Island, to none on Lime Island (Smith & Heinz, 1984). This decreased productivity in areas that are close to the navigation channel, such as Lime Island, may be related to high water levels combined with increased erosion from natural and ship-induced wave action, as well as to competition from increasing numbers of ring-billed gulls, which nest earlier than the terns (Scharf, 1978; Edsall *et al.*, 1988).

For both osprey and eagles, there has been a marked resurgence in the number of young produced per year since 1981 (Fig. 26) and this may be due to a general decrease in the concentrations of contaminants in their food, which is composed mainly of other birds and fish.

Due to their habits, some 28 bird species are or could be used as biological monitors in the St. Marys River (Table 11). Present monitoring is largely focused on such indicators as population stability, fledgling deformities and success, and on eggs, due to the susceptibility of embryos to organochlorine contaminants (Gilbertson, 1974)

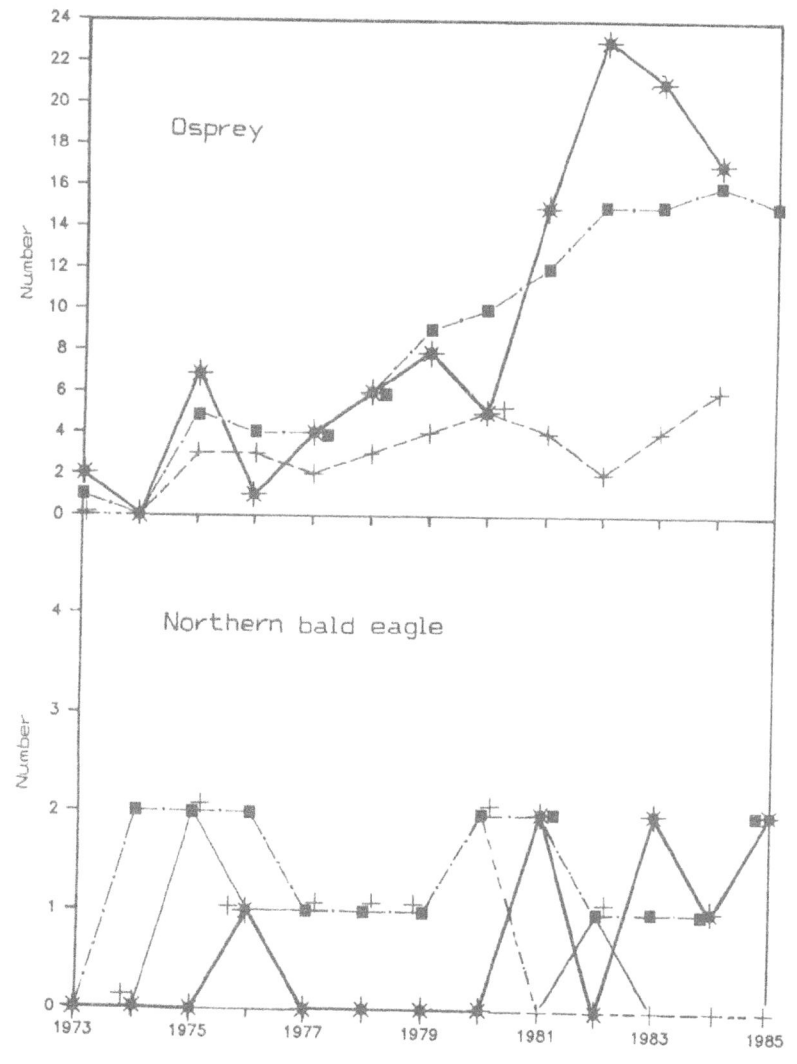

*Fig. 26.* Number of active (■) and failed (+) osprey and northern bald eagle nests and of young (*) produced in the St. Marys River during the period 1973–85. Source: Weise (1985).

32

*Table 12*. Organochlorine contaminants in eggs of piscivorous birds in Lake Superior and the St. Marys River, 1984–1986.

| Species | Location | Contaminant | Mean Concentration (mg kg$^{-1}$, wet wt) | Source |
|---|---|---|---|---|
| Herring gull | Agawa Rock (Lake Superior) | p, p′–DDE | 3.0/3.1 | a |
| | | total PCBs | 12.0/14.0 | a |
| | | 1, 2, 3, 4–TeCB | 0.007/NA | a |
| | Pumpkin Point, Lake George (St. Marys River) | p, p′–DDE | 4.0/3.2 | a |
| | | total PCBs | 22.0/14.0 | a |
| | | 1, 2, 3, 4–TeCB | 0.030/NA | a |
| | | 2, 3, 7, 8–TCDD | $4 \times 10^{-6}/16 \times 10^{-6}$ | a |
| Common tern | Lime Island (lower St. Marys River) | p, p′–DDE | 1.8 | b |
| | | total PCBs | 3.9 | b |

Sources: a – Canada Wildlife Service (unpubl. data for 1985/1986, C.V. Weseloh, pers. comm. Apr., 1988);
    b – U.S. Fish and Wildlife Service (unpubl. data for 1984 in Edsall *et al.*, 1988).

Notes: 'NA'  = not available.
    'TeCB' = tetrachlorobenzene.
    'TCDD' = tetrachlorodibenzo-p-dioxin.

and the effects of these substances on shell thickness (Wiemeyer *et al.*, 1988). Table 12 summarizes 1984–86 data on organochlorine contaminants in the eggs of herring gulls and common terns, two piscivorous species that have been routinely used for monitoring in the Great Lakes Basin. It should be noted that these species are not permanent residents of the river, and therefore contaminant levels in their eggs can reflect exposure of the adult female elsewhere.

Mean concentrations of PCBs, p,p′-DDE, and 2,3,7,8-tetrachlorodibenzo-p-dioxin in herring gull eggs from the St. Marys River colony, while elevated, are typical of other areas of the Great Lakes (i.e., upstream Lake Superior) or in the background range. However, 1,2,3,4-tetrachlorobenzene in eggs from the St. Marys River were the highest of any Great Lakes site in 1985.

Mean levels of total PCBs and p,p′-DDE were less in common tern eggs from the lower river than in herring gull eggs from Lake George. However, the highest individual PCB concentration in the former (7.3 mg kg$^{-1}$) was within the range that could produce harmful effects in eggs (Edsall *et al.*, 1988). These contaminants levels may also pose a threat to higher trophic levels (i.e., ospreys and bald eagles), particularly in the case of gulls, which are increasing in numbers (Fig. 25).

## 5. Conclusions and recommendations

1. Overall, the St. Marys River can be characterized as oligotrophic, with good water quality, a thermal regime largely reflective of Lake Superior surface waters, and the potential to support coldwater fish for most of the year. Although sediment, and in particular water quality, are improving along the Canadian shoreline, municipal and industrial pollution from Sault Ste. Marie, Ontario has caused widespread contamination of river sediments.

2. Aquatic plants (phytoplankton, periphyton, macrophytes, and macroalgae) are probably the basis for most of the secondary (invertebrate) production in the river. However, further research is necessary, particularly on the smaller size classes of phytoplankton that contribute significantly to food web dynamics elsewhere in the Great Lakes (Munawar & Munawar, 1978) and which are also very sensitive to contaminants (Munawar *et al.*, 1985).

Additional, basic knowledge is also required to determine the degree to which the zooplankton of the river may be affected by contaminants.

3. The contamination of sediments along the Canadian shoreline has severely impacted the benthic macroinvertebrate community. There is a need for further reductions in contaminants loadings from the Sault Ste. Marie, Ontario area to accelerate recovery of sediment quality and the associated benthos.

4. No effects of contaminants on the health of the fish community of the river have been discovered as yet, and, with the exception of mercury, contaminants in edible portions of adult sports fish are below consumption guidelines to protect human health. However, additional information is required on the trends of mercury in fish over time and also on the levels and significance of other contaminants, such as PAHs, detected in whole fish and accumulated by caged mussels.

5. The increasing abundance of gulls and the presence of low levels of PCBs, p,p'-DDE and other organochlorine contaminants in their eggs may adversely affect their young or their predators. The relative significance of PCBs and other organochlorine contaminants, physical factors, and interspecific competition in the decline of common terns should be investigated. Production and contaminant levels in other common species should also be monitored.

6. Increases in the productivity of osprey during the 1980s suggests a reduction of organochlorine contaminants in their diet; however, the population should be closely monitored to ensure their continued breeding success.

7. Sea lamprey from resident St. Marys River populations continue to adversely impact fisheries of the St. Marys River and Lake Huron, and more cost-effective methods of controlling this predator need to be developed and applied. Catch per unit effort by sport fish anglers has declined from 1930s levels and the relative impact of various fishing activities and sea lamprey on fish stocks should be investigated.

8. Other major impacts by man on the river and its biota include physical alteration of channels by dredging and flow regulation for navigation and hydropower in the St. Marys Rapids area, which have reduced benthic macroinvertebrate and fish habitats. These losses should be better quantified and remediated.

The passage of large vessels, particularly in the lower river, increases current speed, wave action and turbidity, sediment transport, erosion, detritus drift, and dewatering of littoral areas, and can adversely affect emergent wetlands and sensitive stages of associated fauna. The importance of emergent wetlands as habitat, spawning, and nursery areas suggests the need for further study of navigational impacts resulting from both commercial and recreational vessels.

## 6. Acknowledgements

Thanks are due to T.A. Edsall and M. Munawar for their invitation to participate in this symposium and its proceedings. The author also gratefully acknowledges the efforts of: B. Woo and A. Foley for typing the manuscript; and the constructive comments and suggestions of W.G. Duffy, T.A. Edsall and an anonymous reviewer.

The publications by Duffy *et al.* (1987) and Edsall *et al.* (1988) formed the basis for much of the biological and background material in this paper.

## References

Bellrose, F. C., 1968. Waterfowl migration corridors east of the Rocky Mountains in the United States. Illinois Ntrl. Hist. Surv., Biol. Notes No. 61, 24 pp.

Botts, L. & B. Krushelnicki, 1987. The Great Lakes: An environmental atlas and resource book. Produced by: Env. Canada, U.S. Env. Protect. Agency, Brock Univ. (Ontario) and Northwestern Univ. (Illinois). Chicago, Ill. and Toronto, Ont., 44 pp.

Burt, A. J., P. M. McKee & D. R. Hart, 1988. Benthic invertebrate survey of the St. Marys River, 1985. Volume 1 – Main Report. Prep. for Ontario Min. Env., Toronto, Ont. by Beak Consultants Ltd., Brampton, Ont. (Contract Nos. A89538 & A98444), 88 pp. + app.

34

Burt, A. J., P. M. McKee, D. R. Hart & P. B. Kauss, 1991. Effects of pollution on benthic invertebrate communities of the St. Marys River, 1985. Hydrobiologia 219: 63–81.

CCGL (Coordinating Committee on Great Lakes Basic Hydraulic and Hydrologic Data), 1977. Coordinated Great Lakes physical data.

Daugherty, W. E. & H. A. Purvis, 1985. Sea lamprey control in the United States. Great Lakes Fish. Comm. Ann. Rep. Ann Arbor, Mi.

Daugherty, W. E., H. H. Moore, J. J. Tibbles, S. M. Dustin & B. G. H. Johnson, 1984. Sea lamprey management in the Great Lakes. Great Lakes Fish. Comm. Ann. Rep. Ann Arbor, Mi.

Duffy, W. G., 1985. The population ecology of the damselfly, *Lestes disjunctus disjunctus* in the St. Marys River, Michigan. Ph.D. Thesis, Michigan State Univ. East Lansing, Mi., 119 pp.

Duffy, W. G., T. R. Batterson & C. D. McNabb, 1987. The St. Marys River, Michigan: An ecological profile. U.S. Dept. Inter., Fish Wildl. Serv. Biol. Rep. 85(7.10). Washington, D.C., 138 pp.

Edsall, T. A., P. B. Kauss, D. Kenaga, T. Kubiak, J. Leach, M. Munawar, T. Nalepa & S. Thornley, 1988. St. Marys River biota and their habitats: A geographic area report of the Biota Work Group, Upper Great Lakes Connecting Channels Study. Unpubl. Rep. for Activities Integration Committee. 73 pp. + app.

Edsall, T. A., B. A. Manny, D. W. Schloesser, S. J. Nichols & A. M. Frank, 1991. Production of *Hexagenia limbata* nymphs in contaminated sediments in the Upper Great Lakes Connecting Channels. Hydrobiologia 219: 353–361.

Edwards, C. J., P. L. Hudson, W. G. Duffy, S. J. Nepszy, C. D. McNabb, R. C. Hass, C. R. Liston, B. Manny & W.-D. N. Busch, 1989. Hydrological, morphometrical, and biological characteristics of the connecting rivers of the international Great Lakes: a review. In; D. P. Dodge (ed.) Proc. Internat. Large River Symp., Can. Spec. Publ. Fish. Aquat. Sci. 106: (in press).

Gilbertson, M., 1974. Pollutants in breeding herring gulls in the lower Great Lakes. Can. Field-Nat. 88: 273–280.

Goodyear, C. S., T. A. Edsall, D. M. Ormsby-Dempsey, G. D. Moss & P. E. Polanksi, 1982. Atlas of the spawning and nursery areas of Great Lakes fishes. Volume 3: St. Marys River. U.S. Fish Wildl. Serv. Publ. FWS/OBS-82/52. Washington, D.C., 22 pp.

Hamdy, Y., J. D. Kinkead & M. Griffiths, 1978. St. Marys River water quality investigations, 1973–74. Ont. Min. Env., Wat. Resources Br. Internal Rep. Toronto, Ont., 53 pp.

Hiltunen, J. K. & D. W. Schloesser, 1983. The occurrence of oil and the distribution of *Hexagenia* (Ephemeroptera: Ephemeridae) nymphs in the St. Marys River, Michigan and Ontario. Freshwat. Invert. Biol. 2: 199–203.

Holland, L. E. & J. R. Sylvester, 1983. Evaluation of simulated drawdown due to navigation traffic on eggs and larvae of two fish species of the upper Mississippi River. U.S. Army Corps Engin. Rep. (Contract No. NCR-LO-83-C9). Rock Island, Ill.

Hopkins, G. J., 1986. The trophic status of nearshore waters in Lake Superior at three Ontario water supply intakes, 1979–1984. Ont. Min. Env. Rep. Toronto, Ont., 21 pp. + app.

IJC (International Joint Commission), 1914. Progress report of the International Joint Commission on the reference by the United States and Canada in re the pollution of boundary waters. Including report of the sanitary experts, 384 pp. + app.

IJC, 1951. Report of the International Joint Commission on the pollution of boundary waters. Washington, D.C. and Ottawa, Ont., 312 pp.

IJC, 1978. Great Lakes water quality agreement of 1978. Windsor, Ont., 52 pp.

IJC, 1985. 1985 report on Great Lakes water quality. Great Lakes Water Quality Board Rep. to the IJC. Kingston, Ont., 212 pp.

Jaffe, R., E. A. Stemmler, B. D. Eitzer & R. A. Hites, 1985. Anthropogenic, polyhalogenated, organic compounds in sedentary fish from Lake Huron and Lake Superior tributaries and embayments. J. Great Lakes Res. 11: 156–162.

Jude, D. J., M. Winnell, M. S. Evans, F. J. Tesar & R. Futyma, 1986. Drift of zooplankton, benthos and larval fish and distribution of macrophytes and larval fish during winter and summer, 1985. U.S. Army Corps Engin. Rep. DACW-35-15-C-0005. Detroit, Mi., 173 pp. + app.

Kauss, P. B., 1986. Presentation to citizens hearing (Great Lakes United) on St. Marys River water pollution, August 7, 1986, Sault Ste. Marie, Michigan. Ont. Min. Env., Wat. Resources Br. File Rep. Toronto, Ont., 9 pp. + tables & figs.

Kauss, P. B. & Y. S. Hamdy, 1991. Polycyclic aromatic hydrocarbons in surficial sediments and caged mussels of the St. Marys River, 1985. Hydrobiologia 219: 37–62.

Kenaga, D., 1979. A biological and chemical survey of the St. Marys River in the vicinity of Sault Ste. Marie, June 19–20, 1978. Michigan Dept. Ntrl. Resources, Wat. Qual. Div. Rep. Lansing, Mi., 12 pp.

Kenaga, D., 1980. Chromium in the St. Marys River in the vicinity of the old North Western Leather Company at Sault Ste. Marie, Michigan, June 27 & 28, 1979. Michigan Dept. Ntrl. Resources, Wat. Qual. Div. Rep. Lansing, Mi., 9 pp.

Koshinsky, G. D. & C. J. Edwards, 1983. The fish and fisheries of the St. Marys Rapids: an analysis of status with reference to water discharge, and with particular reference to 'condition 1.(b)'. Rep. to Internat. Joint Commiss., 164 pp. + app.

Kreis, R. G., Jr., T. B. Ludewski & E. F. Stoermer, 1983. Influence of the St. Marys River plume on northern Lake Huron phytoplankton assemblage. J. Great Lakes Res. 9: 40–51.

Larson, J. W., 1981. Essays on: a history of the Detroit District United States Army Corps of Engineers. U.S. Army Corps Engin. Detroit, Mi., 215 pp.

Limno-Tech., 1985. Summary of the existing status of the Upper Great Lakes Connecting Channels data. Rep. prep. for Upper Great Lakes Connecting Channels Study by Limno-Tech, Inc. Ann Arbor, Mi., 156 pp. + app.

Liston, C. R., C. D. McNabb, D. Brazo, J. Bohr, J. Craig, W.

Duffy, G. Fleischer, G. Knoecklein, F. Koehler, R. Ligman, R. O'Neal, M. Siami & P. Roettger, 1986. Limnological and fisheries studies in relation to proposed extension of the navigation season, 1982 and 1983. U.S. Fish Wildl. Serv. Publ. FWS/OBS-80/62.3, 764 pp. + app.

Manny, B. A., D. W. Schloesser, S. J. Nichols & T. A. Edsall, 1987. Drifting submersed macrophytes in the upper Great Lakes channels. U.S. Fish Wildl. Serv., Unpubl. MS. (cited in Edsall *et al.*, 1988).

Manny, B. A., D. W. Schloesser, S. J. Nichols & T. A. Edsall, 1990. Drift of submersed macrophytes in the Upper Great Lakes Connecting Channels, 1986.

McNabb, C. D., T. R. Batterson, J. R. Craig, P. Roettger & M. Siami, 1986. Ship-passage effects on emergent wetlands. Rep. Michigan State Univ., Dept. Fish. Wildl. East Lansing, Mi., 78 pp.

Munawar, M., 1987. Bioavailability and toxicity of in-place pollutants in the Upper Great Lakes Connecting Channels. Unpubl. Rep. Fish. Oceans Can. Burlington, Ont., 11 pp. + table & figs.

Munawar, M. & I. F. Munawar, 1978. Phytoplankton of Lake Superior 1973. J. Great Lakes Res. 4: 415–422.

Munawar, M., R. L. Thomas, W. Norwood & A. Mudroch, 1985. Toxicity of Detroit River sediment-bound contaminants to ultraplankton. J. Great Lakes Res., 11: 264–274.

OME (Ontario Ministry of the Environment), 1984. Water management. Goals, policies, objectives and implementation procedures of the Ministry of the Environment. November, 1978, revised May, 1984. Toronto, Ont., 70 pp.

OME & OMNR (Ontario Ministry of the Environment & Ontario Ministry of Natural Resources), 1986. Guide to eating Ontario sport fish. Toronto, Ont., 280 pp.

Osborne, B. S. & D. Swainson, 1986. The Sault Ste. Marie canal. A chapter in the history of Great Lakes transport. Parks Canada. Ottawa, Ont., 148 pp.

Persaud, D. & W. D. Wilkins, 1976. Evaluating construction activities impacting on water resources. Ont. Min. Env. Rep. Toronto, Ont.

Persaud, D., T. D. Lomas & A. Hayton, 1987. The in-place pollutants program. Vol. III. Phase I studies. Ont. Min. Env. Rep. Toronto, Ont., 94 pp.

Poe, T. P. & T. A. Edsall, 1982. Effects of vessel-induced waves on the composition and amount of drift in an ice environment in the St. Marys River. U.S. Fish Wildl. Serv., Great Lakes Fish. Lab. Admin. Rep. 82-6. Ann Arbor, Mi., 45 pp.

Quinn, F. H. & R. N. Kelley, 1983. Great Lakes monthly hydrologic data. Great Lakes Env. Res. Lab. Ntnl. Oceanogr. Atmos. Admin. Data Rep. ERL GLERL-26. Ann Arbor, Mi., 79 pp.

Robinson, W. L. & R. W. Jensen, 1980. Effects of winter navigation on waterfowl and raptors in the St. Marys River area. U.S. Army Corps Engin. Rep. DACW-35-30-X-0194. Detroit, Mi., 102 pp.

Scharf, W. C., 1978. Colonial birds nesting on man-made and natural sites in the U.S. Great Lakes. U.S. Army Corps Engin. Waterways Exper. Stn. Tech. Rep. D-78-10. Vicksburg, Miss., 165 pp.

Schirripa, M. J., 1983. Colonization and production estimates of rock basket samplers in the St. Marys River, 1983. Unpubl. MS., Michigan State Univ. East Lansing, Mi.

Schloesser, D. W., T. A. Edsall, B. A. Manny & S. J. Nichols, 1991. Distribution of *Hexagenia* nymphs and visible oil in sediments of the Upper Great Lakes Connecting Channels. Hydrobiologia 219: 345–352.

Selgeby, J. H., 1975. Life histories and abundance of crustacean zooplankton in the outlet of Lake Superior, 1971–1972. J. Fish. Res. Bd Can. 32: 461–470.

Smith, G. J. & G. H. Heinz, 1984. Effects of industrial contaminants on common terns in the Great Lakes. U.S. Fish Wildl. Serv. Draft Rep. Study Plan 889.01.01. Laurel, Maryland (cited in Duffy *et al.*, 1987).

Suns, K., G. E. Crawford, D. D. Russell & R. E. Clement, 1985. Temporal trends and spatial distribution of organochlorine and mercury residues in Great Lakes spottail shiners (1975–1983). Ont. Min. Env. Rep. Toronto, Ont., 43 pp.

Thomas, M. & C. R. Liston, 1986. Zooplankton composition and density in a navigation channel and near a littoral site of the lower St. Marys River, Michigan. Michigan Academician, 18: 365–373.

Upchurch, S. B., 1976. Lake basin physiography. In; Great Lakes basin framework study. App. 4: Limnology of lakes and embayments. Great Lakes Basin Comm. Ann Arbor, Mi.: 17–26.

USACE (United States Army Corps of Engineers), 1988. Draft environmental impact statement. Supplement II to the final environmental impact statement. Operations, maintenance, and minor improvements of the federal facilities at Sault Ste. Marie, Michigan (July 1977). Operation of the lock facilities to 31 January ± 2 weeks. Detroit, Mi.

UGLCCS (Upper Great Lakes Connecting Channels Study Management Committee), 1989. Volume II – Final report of the Upper Great Lakes Connecting Channels study. Env. Can., U.S. Env. Protect. Agency, Michigan Dept. Ntrl. Resources, Ont. Min. Env., Ntnl. Oceanogr. Atmos. Admin., U.S. Fish Wildl. Serv., U.S. Army Corps Engin., and Detroit Water and Sewerage Dept. December, 1988. 591 pp. + app.

Veal, D. M., 1968. Biological survey of the St. Marys River. Ont. Wat. Resources Comm. Toronto, Ont., 23 pp. + app.

Weise, T. F., 1985. Waterfowl, raptor and colonial bird records for the St. Marys River. Unpubl. Rep. Michigan Dept. Ntrl. Resources. Sault Ste. Marie, Mi.

Wiemeyer, S. N., C. M. Bunck & A. J. Krynitsky, 1988. Organochlorine pesticides, polychlorinated biphenyls and mercury in osprey eggs – 1970–79 – and their relationships to shell thinning and productivity. Arch. Env. Contam. Toxicol. 17: 767–787.

Zenon, 1985. To devise and implement a revised monitoring scheme for persistent and toxic organics in Great Lakes sport fish. Rep. prep. for Ont. Min. Env., Toronto, Ont. by Zenon Environmental Inc., Burlington, Ont.

*Hydrobiologia* **219**: 37–62, 1991.
*M. Munawar & T. Edsall (eds),*
*Environmental Assessment and Habitat Evaluation of the Upper Great Lakes Connecting Channels.*
© *1991 Kluwer Academic Publishers.*

# Polycyclic aromatic hydrocarbons in surficial sediments and caged mussels of the St. Marys River, 1985

P. B. Kauss & Y. S. Hamdy
*Ontario Ministry of the Environment, Water Resources Branch, 1 St. Clair Ave. W., Toronto, Ontario, M4V 1K6, Canada*

*Key words:* bioindicators, freshwater mussels, sediments, pollution

## Abstract

During 1985, a biological monitoring study was conducted in the St. Marys River to determine the availability and source areas of polycyclic aromatic hydrocarbons (PAHs) and the relative importance of water and sediment to tissue contaminant concentrations. Clean unionid mussels (*Elliptio complanata*) were exposed in cages for three weeks at 14 stations and surficial sediments were obtained at 12 of these. The highest concentrations of total PAHs as well as many individual compounds were found in sediments and mussels from the Algoma Steel Slip above St. Marys Falls and in sediments below the discharge of the Edison Sault Electric Co. Canal. Although concentrations were lower below the Falls, elevated levels persisted downstream along the Ontario shore into Lake George. PAHs were also accumulated by mussels exposed along the Michigan shoreline, but at lower levels than along the Ontario shore. From 23 to 63 percent of the PAHs in sediments and 6 to 27 percent of those in mussels were comprised of compounds with mutagenic and/or carcinogenic properties. The predominance of higher molecular weight PAHs in both sediments and mussels as well as lower molecular weight PAHs and nitrogen- and sulfur-containing PAHs in certain areas, indicates that high temperature combustion of fossil fuels is the major source of PAHs, augmented by localized spills of fossil fuels and coke oven by-products.

Mussels contained fewer PAHs (up to 18 vs. 25) at lower concentrations ($\mu g\ kg^{-1}$ vs. $mg\ kg^{-1}$) than the associated surficial sediments. Also, mussels suspended at mid-depth in the water column accumulated similar concentrations to those exposed to sediments, indicating that the major exposure pathway is via the water filtered by these organisms, and hence, is due primarily to external industrial inputs.

## 1. Introduction

The St. Marys River, connecting Lake Superior with Lake Huron, is a large river supporting a diverse aquatic flora and fauna (Kauss, 1991). At the same time it provides drinking and process water for, and receives wastes from, a sizeable urban and industrial area centred at the twin cities of Sault Ste. Marie, Ontario and Michigan, in the upper part of the river (Fig. 1).

The major Sault Ste. Marie point source dischargers (and their 1985 average daily discharge rates) include the Michigan Publicly Owned Treatment Works ($\sim 11 \times 10^3\ m^3\ d^{-1}$). On the Ontario side of the river, sources include: the east end Water Pollution Control Plant ($50.3 \times 10^3\ m^3\ d^{-1}$); St. Marys Paper Inc. ($28.2 \times 10^3\ m^3\ d^{-1}$), a producer of groundwood specialties (e.g., coated newsprint) from stoneground hardwood and purchased Kraft pulp; and Algoma Steel

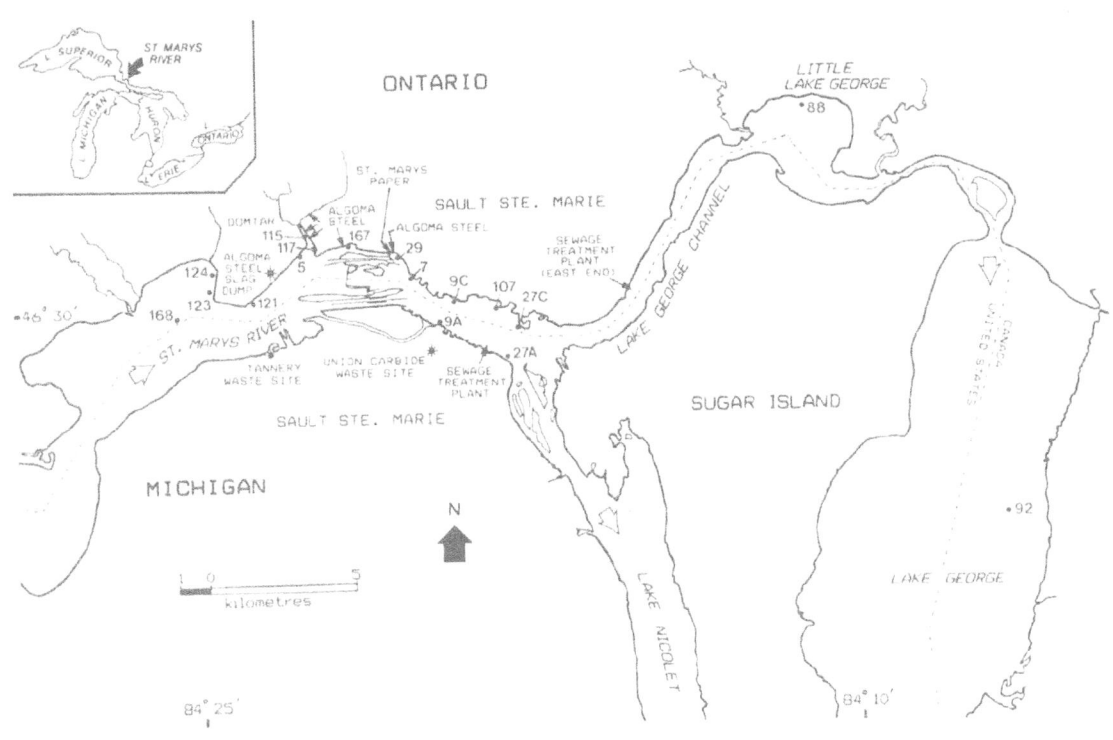

*Fig. 1.* 1985 St. Marys River biomonitoring and sediment sampling stations. Arrows indicate location of major industrial and municipal discharges.

Corp. Ltd. ($506.3 \times 10^3$ m$^3$ d$^{-1}$), which manufactures iron and steel from iron concentrates/pellets, limestone and coke (the latter produced on-site from coal) (OME, 1986; 1987a). Cumulatively, the discharge from these facilities was about 0.3 percent of the average daily flow rate of the St. Marys River during 1985 ($21,440 \times 10^3$ m$^3$ d$^{-1}$).

Inputs of contaminants from the above sources and from storm sewers, combined sewer overflows, shipping traffic, and power generation have resulted in degraded water and sediment quality, mainly along the Ontario shoreline. This has led to classification of the St. Marys River as an 'Area of Concern' by the International Joint Commission (IJC, 1985), i.e. as an area where desirable water uses are restricted due to exceedences of Great Lakes Water Quality Agreement (IJC, 1978), and provincial (OME, 1984) objectives for such contaminants as ammonia, cyanide, phenols, iron, zinc, oil and grease, and bacteria.

On November 4, 1986, Algoma Steel was served with an Amending Control Order by the Ontario Ministry of the Environment requiring measures to ensure that emissions to air and water do not adversely affect the environment. Examples of such reductions and their effective dates are: decreases of daily loadings of phenols to 22.7 kg d$^{-1}$ by 1989 and of oil and grease to 1 023 kg d$^{-1}$ by 1990. For comparison, 1985 average daily loadings of phenols and oil and grease were 102 kg d$^{-1}$ and 1 880 kg d$^{-1}$, respectively (OME, 1987a).

Monitoring of effluent, water, and sediment quality by the OME and other agencies during the recent binational Upper Great Lakes Connecting Channels Study of 1985/86 and the Municipal-Industrial Strategy for Abatement (MISA) pilot site study of Algoma Steel during 1986/87 focused on many of the above-listed contaminants in order to assess current impacts and temporal trends. Samples of potable water from the Sault

Ste. Marie, Ontario municipal water treatment plant contained the highest total polycyclic aromatic hydrocarbon (PAH) concentrations of any Ontario city tested in 1980 (Williams *et al.*, 1982). (The intake at that time was located in the power plant headrace downstream of some Algoma Steel discharges, but has since been moved to Gros Cap in Lake Superior.) Consequently, the above studies also emphasized determining the concentrations, distribution, and biological availability of PAHs in the river, as well as the magnitude of source inputs.

PAHs are a diverse and large group of hydrocarbons containing two or more fused benzene rings, with or without substituent groups. Parent and substituted PAHs are present in fossil fuels (coal, crude oil) and some petroleum products such as used crankcase oil. Consequently, they can reach the environment through spills or as a result of the incomplete and sometimes uncontrolled combustion of organic matter. Other potential sources of PAHs include: gasification and liquifaction of coal; incineration of wastes; burning of wood; and the production of coke,

*Table 1.* Polycyclic aromatic hydrocarbons analyzed for in sediments and mussel tissues.

| Compound | Abbreviation | Quantification ion (secondary ion) | Mutagenicity | Carcinogencity |
|---|---|---|---|---|
| Naphthalene | N | 128 | − | − |
| *Acenaphthylene | Acy | 152 | − | − |
| *Acenaphthene | Ac | 154 | + | − |
| *Fluorene | F | 166 (166) | − | − |
| *Phenanthrene | Ph | 178 | − | − |
| *Anthracene | An | 178 | − | − |
| *Fluoranthene | Ft | 202 (101) | | − |
| *Pyrene | P | 202 (101) | − | − |
| *Chrysene | C | 228 (114) | + | ± |
| *Benzo(a)anthracene | B(a)An | 228 (114) | + | + |
| *Benzo(b)fluoranthene | B(b)Ft | 252 (126) | | + + |
| *Benzo(k)fluoranthene | B(k)Ft | 252 (126) | | − |
| Benzo(j)fluoranthene | B(j)Ft | 252 (126) | | + + |
| Benzo(e)pyrene | B(e)P | 252 (126) | + | ± |
| *Perylene | Per | 252 (126) | | − |
| Dimethylbenz(a)anthracene | DMB(a)An | 256 (241) | | + + + + |
| *Benzo(a)pyrene | B(a)P | 252 (126) | + | + + + |
| *Indeno(1,2,3-c,d)pyrene | IP | 276 (138) | | + |
| *Dibenzo(a,h)anthracene | DB(ah)An | 278 (139) | + | + + + |
| *Benzo(g,h,i)perylene | B(ghi)Per | 276 (138) | | − |
| Anthanthrene | Ant | 276 (138) | | |
| Benzo(b)chrysene | B(b)C | 278 (139) | | |
| Coronene | Co | 300 (150) | | − |
| Quinoline | Q | 129 (102) | + | + |
| Carbazole | Car | 167 (166) | | − |
| Acridine | Acr | 179 (89) | + | − |
| Benz(a)acridine | B(a)Acr | 229 (114) | + | |
| Benzothiophene | Bt | 134 (89) | | |
| Dibenzothiophene | DBt | 184 (92) | | |

Notes: 1) Compounds prefixed by '*' are on USEPA Priority Pollutants List (Keith & Telliard, 1979).

2) Information on mutagenic and carcinogenic properties is from Verschueren (1983); Oehme (1985); and from U.S. National Academy of Sciences, reported in National Research Council (1983).

3) Carcinogencity ranking is: '−', not carcinogenic; '±' uncertain or weakly carcinogenic; '+', carcinogenic; '+ +', '+ + +', '+ + + +', strongly carcinogenic.

carbon black, coal tar pitch, asphalt, and creosote. Through atmospheric transport, PAHs are widely distributed in the environment (Lunde & Bjorseth, 1977), and are found at low levels even in remote areas (Hites & Gschwend, 1982). However, concentrations of these compounds are usually elevated near urban and industrial centres, where PAHs can reach surface waters via direct discharges (industrial and municipal), urban run-off, spills, and seepage. For example, in the Great Lakes basin, elevated sediment concentrations of such PAHs as fluoranthene ($> 10$ mg kg$^{-1}$) and benzo(a)pyrene ($> 1$ mg kg$^{-1}$) have been found in localized areas of: Lake Superior (near Duluth); western Lake Erie (Black River and Cleveland Harbour); eastern Lake Erie (Buffalo River); and western Lake Ontario (Hamilton Harbour). It is noteworthy that all of these areas have coke manufacturing facilities (Baumann & Whittle, 1988).

The low water solubility of many PAHs results in their tendency to partition out of the aqueous phase and to adsorb onto particulate matter, with eventual deposition in the sediments of aquatic systems. The resulting elevated concentrations can mean increased exposure risk to resident benthic biota. This high escaping tendency (or fugacity) from the aqueous phase also results in the potential for PAHs to accumulate in the lipids of aquatic organisms, particularly if these compounds are only slowly depurated and not readily metabolized (Mackay, 1982; Gobas et al., 1987). PAHs can be metabolized by fish and some invertebrates that possess inducible enzyme systems (Frank et al., 1986; Maccubbin et al., 1988; Vandermeulen & Penrose, 1978), thereby increasing the rate of elimination and decreasing the bioconcentration factor (BCF) of the parent compound. This may make the analysis of parent PAH levels in fish a poor indicator of their environmental exposure (Eadie et al., 1982a).

Several PAHs are of environmental concern because they are mutagenic or their metabolites are carcinogenic (see Table 1). PAHs and their reduced derivatives have been recommended for further research and possible regulatory attention (Passino & Smith, 1987) as well as for effluent monitoring (OME, 1987b). In fact, the detection of tumors (neoplasms) in some resident bottom-feeding fish species has been correlated with high concentrations of such PAHs in sediments (Fabacher et al., 1988; Baumann et al., 1982), high total body burdens of PAHs (Baumann et al., 1982), high PAH concentrations in stomach contents (Maccubbin et al., 1985) or elevated concentrations of aromatic compound metabolites in their bile (Krahn et al., 1986).

The objectives of the study reported here were to:

i) Determine the distribution, biological availability and major source area(s) of PAHs in the St. Marys River.

ii) Determine the relative importance of PAH-contaminated sediments and water on tissue contaminant levels.

The organism used was the freshwater mussel, *Elliptio complanata* Lightfoot (Fam. Unionidae; Subf. Ambleminae). This bivalve filter-feeding mussel has been used as a biomonitor of organochlorine contaminants by OME in Ontario lakes (Suns et al., 1980) and rivers (Curry, 1977/78) as well as in Great Lakes Connecting Channels such as the Niagara River (Niagara River Toxics Committee, 1984) and the St. Clair and Detroit Rivers (Kauss & Hamdy, 1985; EC-OME, 1986). Studies in New York State (Heit et al., 1980) and in the St. Marys River (Kauss, 1986) showed that *E. complanata* was also suitable for monitoring the biological availability of PAHs.

## 2. Materials and methods

### 2.1. Field

Sampling stations were selected on the basis of past water and sediment quality data, their proximity to major industrial outfalls, and to permit both cross-channel and upstream-downstream comparisons (Fig. 1). Stations were located in the nearshore, away from major shipping channels, in 0.7 m to 10 m of water (Table 2).

Table 2. Physical and chemical characteristics of St. Marys River waters, 1985.

| Station | Water depth m | Mid(M) or bottom(B) | Current speed m s⁻¹ | Temperature °C | Dissolved oxygen mg l⁻¹ | pH | Conductivity μS cm⁻¹ @25°C | Suspended solids mg l⁻¹ | Calcium mg l⁻¹ | Magnesium mg l⁻¹ | Hardness mg l⁻¹ |
|---|---|---|---|---|---|---|---|---|---|---|---|
| 124 | 0.7 | B | 0.00/0.00 | 15/12 | 10.1*/10.8* | 8.1/8.3 | 115/106 | 5.06/1.90T | 15.6/15.3 | 2.96/2.76 | 51/50 |
| 123 | 1.5 | B | 0.00/0.02 | 16/12 | 10.0*/10.8* | 7.8/7.8 | 123/112 | 1.11T/0.09T | 16.6/16.7 | 2.54/2.71 | 52/53 |
| 121 | 1.7 | B | 0.00/0.03 | 15/12 | 10.2*/10.8* | 7.8/7.8 | 91/91 | 1.62T/1.63T | 13.4/14.3 | 2.53/2.67 | 44/46 |
| 115 | 5.0 | M | 0.07/0.05 | 20/16 | 6.0/6.8 | – | – | – | – | – | – |
|  |  | B | 0.03/0.03 | 20/16 | 0.0/0.0 | 7.6/8.2 | 152/188 | 7.58/6.44 | 22.0/20.5 | 3.12/3.13 | 68/64 |
| 117 | 6.0 | M | 0.23/0.19 | 19/17 | 7.2/8.0 | – | – | – | – | – | – |
|  |  | B | 0.09/0.08 | 16/13 | 4.7/5.3 | 6.7/7.3 | 115/141 | 11.49/5.89 | 18.5/18.0 | 3.67/3.92 | 62/61 |
| 167 | 3.0 | M | 0.05/0.24 | 17/14 | 9.7*/10.4* | – | – | – | – | – | – |
|  |  | B | 0.04/0.11 | 17/14 | 9.4/10.1 | 8.3/8.4 | 129/117 | 2.09/1.60T | 17.3/15.1 | 3.05/2.84 | 56/50 |
| 29 | 2.0 | B | 0.04/0.08 | 15/12 | 9.9/10.5 | 7.7/7.7 | 91/91 | 0.91T/0.77T | 13.1/13.8 | 2.64/2.70 | 44/46 |
| 9C | 2.0 | B | 0.00/0.04 | 15/12 | 9.5/10.3 | 7.7/7.7 | 92/92 | 0.62T/0.27T | 13.3/13.7 | 2.66/2.74 | 44/46 |
| 107 | 2.0 | M | 0.04/0.04 | 15/12 | 10.2*/10.4 | – | – | – | – | – | – |
|  |  | B | 0.04/0.04 | 15/12 | 7.7/8.6 | 7.7/7.7 | 91/91 | 1.13T/3.90 | 13.6/14.1 | 2.64/2.80 | 44/47 |
| 27C | 1.4 | B | 0.07/0.13 | 15/12 | 9.7/10.0 | 7.3/7.8 | 92/89 | 0.50T/0.95T | 13.4/13.8 | 2.68/2.68 | 44/46 |
| 9A | 10.0 | B | 0.04/0.07 | 15/12 | 6.8/8.2 | 7.8/7.8 | 91/88 | 1.73T/0.90T | 13.2/13.6 | 2.68/2.66 | 44/45 |
| 27A | 2.0 | B | 0.04/0.06 | 15/12 | 9.4/10.0 | 7.6/7.6 | 91/89 | 5.05/0.48T | 13.2/13.6 | 2.72/2.73 | 44/45 |
| 88 | 2.0 | B | 0.05/0.04 | 20/12 | 9.0/10.5 | 7.6/7.6 | 88/91 | 1.48T/1.03T | 13.2/13.5 | 2.68/2.70 | 44/45 |
| 92 | 2.0 | B | 0.00/0.02 | 20/12 | 9.0/10.5 | 7.6/7.3 | 85/91 | 29.5/5.36 | 12.9/13.7 | 2.47/2.71 | 42/46 |
| Balsam Lake | 1.5 | B | 0.00/0.00 | 23/14 | 9.2/5.2 | 8.4/5.6 | 130/123 | 2.51/0.75 | 20.2/14.9 | 2.56/2.08 | 61/46 |

Notes: 1) Station numbers correspond to locations on Fig. 1.
2) St. Marys River data is from two samplings in 1985: Oct. 3/Oct. 24;
Balsam L. data are maximum/minimum of five samplings spanning May 29–Sept. 10, 1984.
3) '–' indicates no sample or no analytical result.
'T' indicates tentative value.
'*' indicates saturated oxygen level.

### 2.1.1. Caged mussels

*E. complanata* specimens of a restricted size class (6.5 to 7.2 cm, maximum shell length) were diver-collected from Balsam Lake in central Ontario. Mussels were transported to the study area in bags of lake water within 48 h of collection.

At each of 17 stations, five mussels were placed in a hexane-rinsed wire cage (about 30 cm × 36 cm × 10 cm) fabricated from 1.5 cm open-mesh galvanized wire. Cages were anchored to the bottom with a weight or tethered to a shore-line structure so that mussels were in contact with the sediments. In addition, mussels were also suspended at mid-depth at four of these stations, numbers 115, 117, 167 and 107 (Fig. 1).

After exposure for three weeks (October 3–24, 1985), mussels were retrieved from 14 of the stations and immediately put on ice. [Cages could not be recovered from three of the stations (see Fig. 1), due either to strong currents (Stn. 168), burial by rubble (Stn. 5) or suspected vandalism (Stn. 7). Mortality of mussels was very low, with 98 percent of recovered individuals being alive after their exposure. Three died, one at Station 123 and two at Station 9A.] Within 24 h, mussels were shucked, rinsed, and the drained soft tissues wrapped in hexane(glass-distilled)-rinsed aluminum foil and frozen. Soft tissues were weighed and growth rings of the shells counted to determine age. A random sub-sample of mussels from Balsam Lake was also processed in this manner immediately after collection to provide data on contaminant levels prior to exposure in the St. Marys River. Mussel tissues were kept frozen ($-20\,°C$) until analyzed.

During deployment and retrieval of the caged mussels, field measurements were obtained of conductivity, water temperature, dissolved oxygen, pH, and current speed. Water samples (unfiltered) were also collected for laboratory analysis of hardness, calcium, magnesium, and suspended solids, using a 1-l glass bottle (OME, 1985).

### 2.1.2. Bottom sediments

Bottom sediments were also collected in 1985 during the period September 24 to October 4, as part of a larger study of sediment quality and benthic invertebrates (Burt *et al.*, 1988). However, samples could not be obtained at 2 of the above 14 stations (Stns. 167 and 29), due to the fast current and unavailability of sediment (Fig. 1).

At each of the 12 sampled stations, the top 3 cm of at least three Shipek grabs (each of 0.04 $m^2$ surface area) were composited and homogenized. A subsample of this composite was frozen ($-20\,°C$) in a 0.5-l wide-mouth solvent-rinsed amber jar with foil-lined cap until analyzed.

### 2.2. Analytical

Calcium (as $Ca^{2+}$) and magnesium (as $Mg^{2+}$) in water samples were determined by automated atomic absorption spectroscopy in the presence of lanthanum chloride. Hardness (as calcium carbonate) was determined using a semi-automated titrimetric procedure employing EDTA and Erichrome Black T indicator. Suspended solids concentration was obtained gravimetrically after drying the filtrate at 103 $°C$ (OME, 1983).

The surficial sediment sample and the entire soft tissues of each of three mussels from each station were analyzed for 29 PAHs (Table 1). This list includes the 17 PAHs on the U.S. Environmental Protection Agency's (USEPA) Priority Pollutants List.

Wet mussel tissue (6.5 to 10.8 g) was spiked with two deuterated surrogate standards [$d_{10}$-anthracene and $d_{12}$-benzo(a)pyrene] and then refluxed for 3 h with 5 percent potassium hydroxide in methanol. The cooled sample was diluted with organic-free water, acidified, and then extracted with three portions of hexane. The combined hexane extracts were dried over sodium sulphate ($Na_2SO_4$) and then concentrated by rotary evaporation, with solvent exchange into iso-octane. After concentration by evaporation, half of the sample was submitted for clean-up (alumina column) and eluted with benzene. The eluate was again concentrated by evaporation and spiked with an internal standard ($d_{10}$-phenanthrene) prior to gas chromatography (GC)/ mass spectrometry-selected ion monitoring (MS–SIM) analysis using a Finnegan 4500 system. GC analysis employed a 30 m × 0.25 mm ID DB5 column, helium carrier gas @20 cm s$^{-1}$, with an oven temperature profile of: 60 °C for 2 min; 60 °C–210 °C @ 10 °C min$^{-1}$; 210 °C–315 °C @ 16 °C min$^{-1}$; hold for 20 min. On-column injection was used. MS–SIM analysis in electron impact mode included an ionization voltage of 70eV, 0.5 A filament emission, 1500 V electron multipler and stepped-ion MID scan.

For sediments, a weighed portion (10 to 20 g) of wet sample was spiked with the above surrogate standards and then Soxhlet extracted for 16 h with benzene. Na$_2$SO$_4$ and concentrated hydrochloric acid were added to ensure good solvent-sample contact. The extract was then concentrated by rotary evaporation followed by solvent exchange into iso-octane. The final volume was submitted for clean-up as described above for mussel tissue and spiked with the internal standard. GC analysis employed the same column and carrier gas as utilized for mussel extracts. However, the oven temperature profile was; 80 °C for 2 min; 80 °C–210 °C @ 10 °C min$^{-1}$; 210 °C–290 °C @ 16 °C min$^{-1}$; hold for 10 min. A splitless injection mode was used. MS conditions were identical to those described above, but the electron multiplier was operated at 2000 V. This was followed by analysis under the following conditions to provide additional sensitivity for the higher molecular weight PAHs: GC using the above-noted column and carrier gas and a temperature profile of 120 °C for 2 min, followed by 120 °C–220 °C @ 10 °C min$^{-1}$, 220 °C–310 °C @ 16 °C min$^{-1}$ and hold for 12 min; MS was as described above for mussels.

A direct couple GC/MS interface was used with a transfer area of 250 °C.

In addition to the isotopically-labelled surrogate standards added to both mussel tissue and sediment matrices, quality control measures included method blanks and analysis of the U.S. National Bureau of Standards Urban Dust Reference Material (SRM-1649).

Confirmation of the identity of individual PAHs required: the presence of appropriate secondary ions (Table 1) in the mass spectrum; a signal to noise ratio of at least 3 to 1; and retention time within 2 percent of reference standard.

Quantification of PAH compounds was achieved by comparing MS responses of selected ions (Table 1) to those of high purity external standards. [Pure crystals of benzo(j)fluoranthene, benzo(b)chrysene, and anthanthrene were not readily available, so the available solution was used only for retention time confirmation and an isomeric compound was used for quantification.] Calculations were based on the sample dry weight, and no correction was made for deuterated surrogates' spike recoveries.

## 3. Results and discussion

### 3.1. Water quality

Measurements of basic physical conditions and water chemistry were restricted to the beginning and end of the mussels' 3 week exposure period. Nevertheless, it is evident from the range of values (Table 2) that, with the exception of the Algoma Slip, the spatial and temporal variability of factors such as dissolved oxygen (6.8–10.8 mg O$_2$ l$^{-1}$), conductivity (85–129 μS cm$^{-1}$), calcium (12.9–17.3 mg l$^{-1}$) and hardness (44–56 mg l$^{-1}$) was not great. Also, these concentrations as well as the range of pH (6.7–8.3) were within applicable provincial (OME, 1984) and/or Agreement (IJC, 1978) objectives for the protection of aquatic life.

In contrast, somewhat higher temperatures (up to 20 °C) were noted in waters of the Algoma Slip, which receive effluents from Algoma Steel (Fig. 1) and in the shallow areas of Little Lake George and Lake George (Stns. 88 and 92). Algoma Slip Stations 115 and 117 also tended to have somewhat higher conductivities (115–188 μS cm$^{-1}$), calcium (18.0–22.0 mg l$^{-1}$), magnesium (3.12–3.92 mg l$^{-1}$), hardness (61–68 mg l$^{-1}$) and lower dissolved oxygen levels (0.0–8.0 mg l$^{-1}$), particularly near the bottom.

Current speed varied spatially and temporally, ranging from the still waters of Leigh Bay (0.0 m s$^{-1}$) to the more rapid flow (0.24 m s$^{-1}$) downstream of an Algoma Steel discharge in the Great Lakes Power Corp. headrace (Stn. 167).

Suspended solids concentrations were quite changeable, varying up to 20-fold at the same station (e.g., 27A). However, the majority (60 percent) of values were below 2.0 mg l$^{-1}$, reflecting

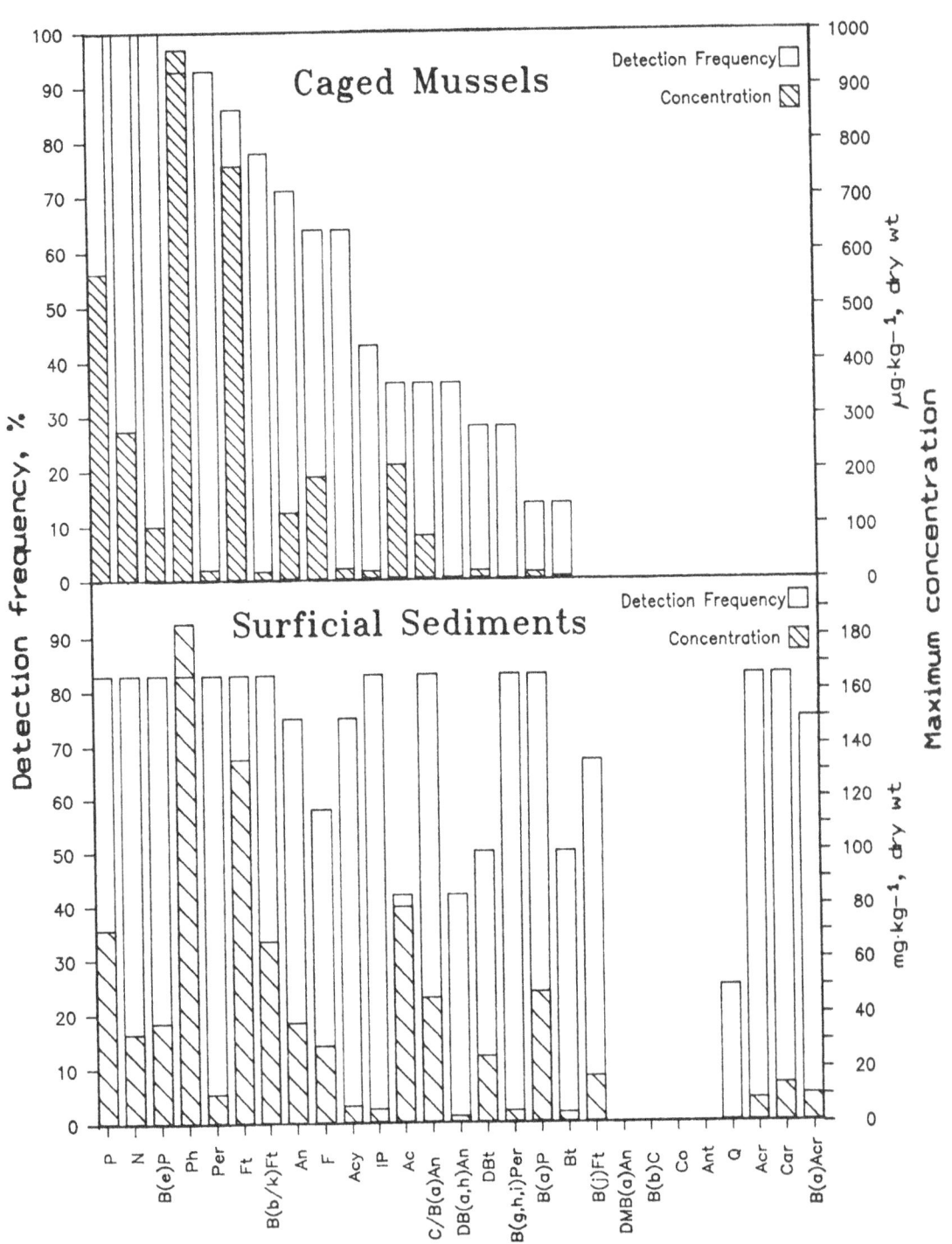

*Fig. 2.* Polycyclic aromatic hydrocarbons detected in caged mussels and surficial sediments of the St. Marys River in 1985. Abbreviations as per Table 1.

the low levels of particulates (0.5–1.0 mg l$^{-1}$) in the river's major source, Lake Superior (Eadie, 1984). The above-noted spatial and temporal variability is unlikely to have had adverse effects on the biomonitor, however, since both the actual concentrations as well as their ranges were similar to those experienced by the mussels in Balsam Lake, their native environment (Table 2).

## 3.2. *PAHs detected in sediments and mussels*

For the sake of convenience, individual PAH compounds in this and following sections are referred to by their abbreviations (see Table 1).

Of the 29 PAHs analyzed for, more compounds were detected in St. Marys River surficial sediments than in mussel tissues (Fig. 2). The largest number, 24 to 25 [B(b)Ft and B(k)Ft isomers could not be resolved in sediment samples], was found at Station 115 in the Algoma Slip, and 17 to 18 of these PAHs were also identified at >50 percent of the sediment sampling locations. These included, in decreasing order of frequency: [P = N = B(e)P = Ph = Per = Ft = B(b/k)Ft = IP = C = B(a)An = B(g,h,i)Per = B(a)P = Acr = Car] > [An = Acy] > B(j)Ft > F. In contrast, a maximum of 16 to 18 compounds [B(b)Ft and B(k)Ft isomers as well as C and B(a)An could not be resolved in mussel tissues] were found in *E. complanata* after three weeks' exposure at Station 115. Of these, 10 to 11 were also detected at >50 percent of the biomonitoring stations. In decreasing order of detection frequency, these were: [P = N = B(e)P] > [Ph = Per] > Ft > B(b/k)Ft > An > [F = Acy].

DMB(a)An, B(b)C, Co, and Ant were not detected in either matrix at the method detection levels employed in this study (Tables 3 and 5), indicating that there were no recent inputs of these compounds to the river.

In contrast, the nitrogen heterocycles Q, Acr, Car, and B(a)Acr were not accumulated by mussels, although the latter three were present in most of the sediment samples.

For those PAHs common to both sediments and mussels, there was generally a similar pattern of relative maximum concentrations (Fig. 2). At the majority of sampling stations, both sediments and mussel tissues were dominated by the 3- and 4-ring compounds Ph, Ft and P (Tables 3 and 5), suggesting that high temperature combustion of fossil fuels (e.g., burning of coal or wood, coke manufacturing) was the major contributor of PAH's to the local aquatic environment (Gschwend & Hites, 1981; Helfrich & Armstrong, 1986). This same predominance was observed in urban runoff (Marsalek & Ng, 1987) and snowpack samples (Boom & Marsalek, 1988) collected in Sault St. Marie, Ontario during 1986 and 1987, respectively. The relatively high concentrations of the lower molecular weight 2- and 3-ring compounds N, Acy, Ac, and F relative to the high molecular weight C, B(a)An, B(e)P, B(a)P, and IP suggest that fossil fuel spills (e.g., coal, petroleum) are also important sources of anthropogenic PAHs in this area (Helfrich & Armstrong, 1986). This is particularly evident for the sediments at Stations 115 and 117 in the Algoma Slip and at Station 9A at Sault Ste. Marie, Michigan. Sediments of all three stations generally contained elevated levels of nitrogen heterocycles [Q,Car,Acr, B(a)Acr] and sulfur heterocycles (Bt, DBt) (Table 3). Both Bt and DBt are components of petroleum (Lake & Hershner, 1977) and DBt and Acr (in addition to unsubstituted and substituted PAHs) have been identified in certain coals (Barrick *et al.*, 1984; Bender *et al.*, 1987; Tripp *et al.*, 1981). Q and Acr are also found in wastewaters from high temperature industrial processes such as coal coking and conversion (Cassidy *et al.*, 1988). It is worth noting that coal pellets or dust were observed in the sediments of these three stations during their collection (Burt *et al.*, 1988). Spills are also possible sources of the heterocyclic PAHs detected in the Algoma Slip. DBt and Car were detected in coal tar, and these compounds as well as Bt, Q and Acr were identified in creosote manufactured by a nearby plant (Domtar) from Algoma Steel coal tar (OME, 1988, unpublished data).

*Table 3.* Concentration of polycyclic aromatic hydrocarbons, oil and grease and organic carbon and percent fines in St. Marys River surficial sediments, 1985.

| Compound | MDL | Station | | | | | | | | | | | |
| --- | --- | --- | --- | --- | --- | --- | --- | --- | --- | --- | --- | --- | --- |
| | | 124 | 123 | 121 | 115 | 117 | 9C | 9A | 107 | 27C | 27A | 88 | 92 |
| Naphthalene | 0.02 | * | * | 0.06 | 29.0 | 3.3 | 5.90 | 0.69 | 6.60 | 1.70 | 0.03 | 0.13 | 0.13 |
| Acenaphthylene | 0.02 | * | * | 0.08 | 4.80 | 0.42 | 0.55 | 4.60 | 0.52 | 0.55 | * | 0.18 | 0.07 |
| Acenaphthene | 0.02 | * | * | * | 19.5 | 0.93 | 0.18 | 0.62 | 0.17 | * | * | * | * |
| Fluorene | 0.02 | * | * | * | 28.5 | 1.90 | 0.47 | 3.40 | 0.36 | 0.17 | 0.09 | * | 0.03 |
| Phenanthrene | 0.02 | * | * | 0.86 | 185 | 14.0 | 3.90 | 25.0 | 2.20 | 1.10 | 0.03 | 0.40 | 0.25 |
| Anthracene | 0.02 | * | * | * | 37.0 | 3.50 | 1.40 | 10.0 | 1.00 | 0.52 | 0.03 | 0.18 | 0.07 |
| Fluoranthene | 0.02 | * | * | 4.00 | 135 | 13.0 | 6.40 | 30.0 | 3.70 | 1.70 | 0.12 | 0.59 | 0.43 |
| Pyrene | 0.02 | * | * | 3.00 | 71.5 | 7.40 | 3.60 | 17.0 | 2.20 | 1.10 | 0.07 | 0.36 | 0.27 |
| Chrysene | 0.02 | * | * | 3.40 | 9.30 | 1.50 | 1.00 | 9.90 | 1.00 | 0.45 | 0.02 | 0.14 | 0.06 |
| Benzo(a)anthracene | 0.02 | * | * | 18.0 | 32.0 | 5.50 | 4.60 | 36.0 | 4.20 | 2.10 | 0.09 | 0.55 | 0.28 |
| Benzo(b/k)fluoranthene | 0.04 | * | * | 16.0 | 32.5 | 4.30 | 6.20 | 67.0 | 11.0 | 2.80 | * | 0.97 | 0.33 |
| Benzo(j)fluoranthene | 0.02 | * | * | 1.10 | 5.50 | 0.64 | 1.10 | 17.0 | 3.00 | * | 0.08 | 0.16 | 0.05 |
| Benzo(e)pyrene | 0.02 | * | * | 7.40 | 16.0 | 2.10 | 3.00 | 37.0 | 7.00 | 2.00 | 0.03 | 0.56 | 0.18 |
| Perylene | 0.02 | * | * | 2.20 | 7.30 | 0.79 | 1.20 | 11.0 | 2.30 | 0.55 | * | 0.33 | 0.11 |
| Dimethylbenz(a)anthracene | 0.02 | * | * | * | * | * | * | * | * | 1.90 | 0.11 | 0.71 | 0.21 |
| Benzo(a)pyrene | 0.02 | * | * | 9.20 | 24.5 | 2.80 | 4.40 | 48.0 | 8.40 | * | 0.04 | 0.34 | 0.07 |
| Indeno(1,2,3-c,d)pyrene | 0.02 | * | * | 3.60 | 5.30 | 0.82 | 1.10 | 4.90 | 0.69 | * | * | 0.11 | 0.03 |
| Dibenzo(a,h)anthracene | 0.02 | * | * | 1.50 | 2.30 | 0.35 | 0.46 | 2.50 | * | 0.04 | 0.04 | 0.31 | 0.07 |
| Benzo(g,h,i)perylene | 0.02 | * | * | 3.50 | 4.50 | 0.82 | 0.88 | 1.70 | 0.21 | * | * | * | * |
| Anthanthrene | 0.02 | * | * | * | * | * | * | * | * | * | * | * | * |
| Benzo(b)chrysene | 0.04 | * | * | * | * | * | * | * | * | * | * | * | * |
| Coronene | 0.02 | * | * | * | * | * | * | * | * | * | * | * | 0.03 |
| Quinoline | 0.02 | * | * | * | 0.46 | 0.12 | 0.81 | 2.20 | 0.30 | 0.10 | 0.04 | 0.26 | 0.19 |
| Carbazole | 0.02 | * | * | 0.28 | 14.0 | 1.70 | 0.61 | 1.20 | 0.22 | 0.16 | 0.08 | 0.25 | 0.20 |
| Acridine | 0.02 | * | * | 0.58 | 8.70 | 1.30 | 0.84 | 5.70 | 0.49 | 0.31 | * | 0.13 | 0.19 |
| Benz(a)acridine | 0.02 | * | * | 2.70 | 10.2 | 1.50 | 0.31 | 0.06 | 0.26 | * | * | * | 0.06 |
| Benzothiophene | 0.02 | * | * | * | 3.80 | 0.39 | 0.59 | * | * | * | * | * | * |
| Dibenzothiophene | 0.02 | * | * | * | 24.5 | 1.70 | * | 2.30 | * | 0.14 | * | * | 0.15 |
| Oil & Grease | – | 117 | 123 | 237 | 4722 | 2276 | 9612 | 1216 | 4651 | 4053 | 505 | 2258 | 2178 |
| Total Organic Carbon (g·kg⁻¹) | – | 5.0 | 5.0 | 10.0 | 335 | 140 | 100 | 22.0 | 57.0 | 36.0 | 22.6 | 22.0 | 18.0 |
| <62 μm diameter (%) | – | 3.3 | 47.6 | 36.0 | 37.7 | 35.1 | 74.4 | 12.0 | 26.0 | 27.8 | 8.5 | 75.8 | 81.1 |

Notes: 1) Station numbers correspond to locations on Fig. 1.
2) Concentration units are mg kg⁻¹ (dry weight) except as noted.
3) '*' indicates not detected at method detection limit (MDL) shown.

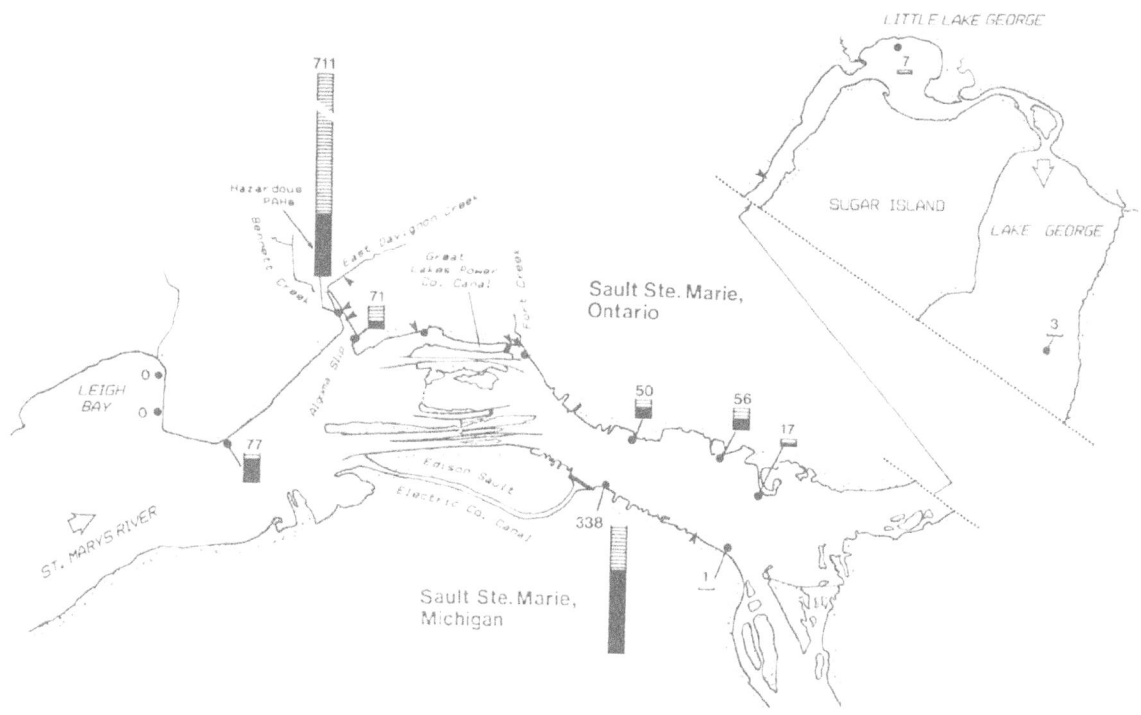

*Fig. 3.* Total polycyclic aromatic hydrocarbons (mg kg$^{-1}$, dry wt) in surficial sediments of the St. Marys River in 1985. Bar height is proportional to concentration; shaded portion of bars indicates the proportion of the total comprised of mutagenic and/or carcinogenic compounds.

## 3.3. *PAH concentrations in sediments*

The sum of individual PAHs at each station ('total PAHs' in Fig. 3) was highest in surficial sediments collected at Station 115 in the Algoma Slip (711 mg kg$^{-1}$ dry wt). This station also had the highest concentrations of the following individual PAHs: N, Acy, Ac, F, Ph, An, Ft, P, IP, B(g,h,i)Per, Q, Car, Acr, B(a)Acr, Bt, and DBt. Concentrations of both total and individual PAHs generally decreased with increasing distance downstream (Table 3 and Fig. 3). The second highest concentration of total PAHs (338 mg kg$^{-1}$) was found along the U.S. shore immediately downstream of the Edison Sault Electric Co. Canal discharge at Station 9A. Maximum survey concentrations of C, B(a)An, B(b/k)Ft, B(a)P, DMB(a)An, B(e)P, B(j)Ft, and Per were also recorded at this location (Table 3).

Figure 3 also shows the relative contribution of 'hazardous PAHs' to the total PAH levels at each station. This sub-group was comprised of 13 compounds considered to be mutagenic and/or carcinogenic (see Table 1): Ac, C, B(a)An, B(b)Ft, B(j)Ft, B(e)P, DMB(a)An, B(a)P, IP, DB(a,h)An, Q, Acr, and B(a)Acr. The percentages of these hazardous PAHs varied greatly between stations, from a low of 23 to 30 percent in the Algoma Slip to a maximum of 80 percent upstream of Algoma Steel discharges (Stn. 121). At the remaining downstream stations, percentages ranged between 45 and 63 percent. The major contributors to these percentages were: B(a)An, B(b/k)Ft, B(e)P, and B(a)P (Table 3).

Table 4 compares the ranges of concentrations (minimum to maximum) of individual PAHs in St. Marys River sediments with those from five other areas in the Great Lakes basin that also receive inputs from steel and/or coke manufacturing facilities, in addition to urban inputs.

Table 4. Comparison of polycyclic aromatic hydrocarbon concentrations (mg kg⁻¹, dry wt) in Great Lakes and tributary sediments.

| Compound | Lake Superior 1978 (a) | Colin Scott Lake, Ont. (b) | St. Marys River 1985 (c) | Detroit River 1982 (d) | Detroit River 1985 (e) | Black River 1980 (f) | Black River 1984 (g) | Lackawanna & Union Canals 1981 (h) | Buffalo River 1981 (h) | Buffalo River 1983 (i) | Hamilton Harbour 1982 (j) |
|---|---|---|---|---|---|---|---|---|---|---|---|
| *Naphthalene | | | ND–29.0 | 0.12–16.8 | ND–4.40 | 31.0 | 14.0 | ND–23.0 | ND–239.1 | | |
| *Acenaphthylene | | | ND–4.80 | ND–1.34 | ND–1.00 | 40.0 | 17.0 | ND–2.5 | ND | | |
| *Acenaphthene | | | ND–19.5 | ND–42.3 | ND–2.60 | 36.0 | 2.50 | ND–49.0 | ND–80.0 | ND–0.64 | |
| *Fluorene | 0.027 | | ND–28.5 | ND–5.10 | ND–12.6 | | 16.0 | ND–2.20 | ND–5.10 | 0.67–9.60 | |
| *Phenanthrene | 0.003 | | ND–185 | 0.10–55.2 | ND–85.6 | 390 | 52.0 | 4.2–16.7 | ND–16.3 | 0.20–3.40 | |
| *Anthracene | 0.091 | 0.038 | ND–37.0 | ND–31.4 | ND–55.0 | 220 | 15.0 | 5.5–63.9 | ND–35.6 | 2.45–10.1 | 1.9–4.3 |
| *Fluoranthene | 0.055 | 0.023 | ND–135 | ND–31.5 | ND–119 | 140 | 33.0 | 5.5–49.6 | 1.0–26.7 | 1.40–21.9 | |
| *Pyrene | 0.091 | 0.023 | ND–71.5 | ND–31.5 | ND–86.1 | 51.0 | 24.0 | } 3.5–66.4 | } ND–60.2 | 0.24–1.77 | |
| *Chrysene | 0.027 | 0.007 | ND–9.90 | } ND–35.6 | } ND–81.9 | 51.0 | 10.0 | | | 0.46–2.60 | |
| *Benzo(a)anthracene | 0.025 | 0.025 | ND–36.0 | | | } 75.0 | } 15.0 | ND–63.0 | ND–96.9 | ND–5.50 | 1.1–9.0 |
| *Benzo(b)fluoranthene | 0.025 | 0.025 | } ND–67.0 | } ND–57.9 | } ND–63.5 | | | | | 0.32–1.15 | |
| *Benzo(k)fluoranthene | 0.008 | 0.008 | | | | | | | | | |
| *Benzo(j)fluoranthene | | | ND–17.0 | | | | | | | | |
| Benzo(e)pyrene | 0.082 | 0.028 | ND–37.0 | | | 28.0 | 6.00 | | ND | ND–4.51 | 1.2–9.7 |
| Perylene | | 0.020 | ND–11.0 | | | 12.0 | 3.60 | | | 2.10–13.7 | |
| Dimethylbenz(a)anthracene | | | ND | | | | | | | | |
| *Benzo(a)pyrene | 0.036 | 0.013 | ND–48.0 | ND–23.7 | ND–67.5 | 43.0 | 8.80 | ND–106.5 | ND–72.5 | 0.21–2.54 | 1.2–11.1 |
| *Indeno(1,2,3-c,d)pyrene | | | ND–5.30 | ND–8.15 | ND–13.8 | 26.0 | 6.40 | ND ~ 10.0 | ND | 0.62–2.80 | 1.1–9.7 |
| *Dibenzo(a,h)anthracene | | 0.028 | ND–2.50 | ND–12.6 | ND–14.3 | 9.40 | 1.60 | ND–14.0 | ND | 0.12–6.31 | 1.6–8.6 |
| *Benzo(g,h,i)perylene | | 0.006 | ND–4.50 | | | 24.0 | 5.40 | | | 0.21–2.61 | |
| Coronene | | | ND–0.46 | | | | | | | | |
| Quinoline | | | ND–14.0 | | | | | | | | |
| Carbazole | | | ND–8.70 | | | | ND | | | | |
| Acridine | | | ND–10.2 | | | | 0.67 | | | | |
| Benz(a)acridine | | | ND–3.80 | | | | | | | | |
| Benzothiophene | | | | | | | | | | | |
| Dibenzothiophene | | | ND–24.5 | | | 22.0 | 3.70 | | | | |
| Total PAHs (sum of *) | | | ND–621 | 0.58–254 | 0.60–600 | 1144 | 221 | 9.70–392 | 0.6–285 | 8.74–60.2 | |
| Number of Samples | 1 | 1 | 12 | 36 | 20 | 1 | 1 | 8 | 17 | 10 | 6 |

Data sources: (a) Gschwend & Hites (1981); 1955–78 section of core. (b) Brown & Starnes (1978); surficial, using pipe dredge. (c) This study; surficial (top 3 cm.) section of dredge sample. (d) Pranckevicius (1987); surficial section of dredge or core. (e) Pranckevicius & Kitsuse (1989); unspecified depth of sample. (f) Baumann et al. (1982); unspecified depth of sample. (g) Fabacher et al. (1988); complete dredge sample. (h) Rockwell et al. (1984); unspecified depth of dredge sample. (i) Niagara River Toxics Committee (1984); unspecified depth of sample. (j) Poulton (1987); surficial (top 3 cm.) section of dredge sample.

Notes: 1) The highest concentrations are underlined when more than one data set is available.
2) ND = not detected.
3) 'Total PAHs' range is the minimum and maximum of summed concentrations at individual stations.

Where comparisons are possible, it is evident that the range of concentrations of individual compounds in these other areas was similar to those in the St. Marys River. It is noteworthy that the highest B(a)P concentrations in all six areas (and 58, 56 and 31%, respectively, of stations in the St. Marys River, Detroit River and Detroit River tributaries) exceeded the recommended IJC objective of 1 mg kg$^{-1}$ (IJC, 1983). Unfortunately, there are no available objectives or guidelines for other PAHs in sediments. Nevertheless, the maximum concentrations found in these six urban/industrial areas were from an order of magnitude [B(g,h,i)Per in the Buffalo River] to four orders of magnitude (Ph in the Black River, a tributary of Lake Erie) higher than in 'background' sediment samples from Lake Superior and a wilderness lake (Colin Scott) in Ontario (Table 4), indicating significant localized contamination.

Differences in the number of PAH compounds

analyzed for makes a comparison of 'total PAH' concentrations from different studies difficult. Consequently, concentrations of only the 16 asterisked compounds in Table 4 (representing 94% of the PAHs on the USEPA Priority Pollutants List) were summed for individual stations in the St. Marys River, Detroit River and tributaries, Black River, Lackawanna and Union Canals, and the Buffalo River. The ranges of 'total PAHs' are presented in Table 4. Of the six data sets, sediment collected from the Black River in 1980 contained the highest concentration of total PAHs (1 144 mg kg$^{-1}$), whereas samples from the Buffalo River in 1983 had the lowest (8.74–60.2 mg kg$^{-1}$). The maximum concentration of total PAHs in the St. Marys River (621 mg kg$^{-1}$) was intermediate between these extremes and similar to the maximum total PAH concentration found in sediments of tributaries to the Detroit River in 1985 (600 mg kg$^{-1}$). This similarity is partially a reflection of the low energy

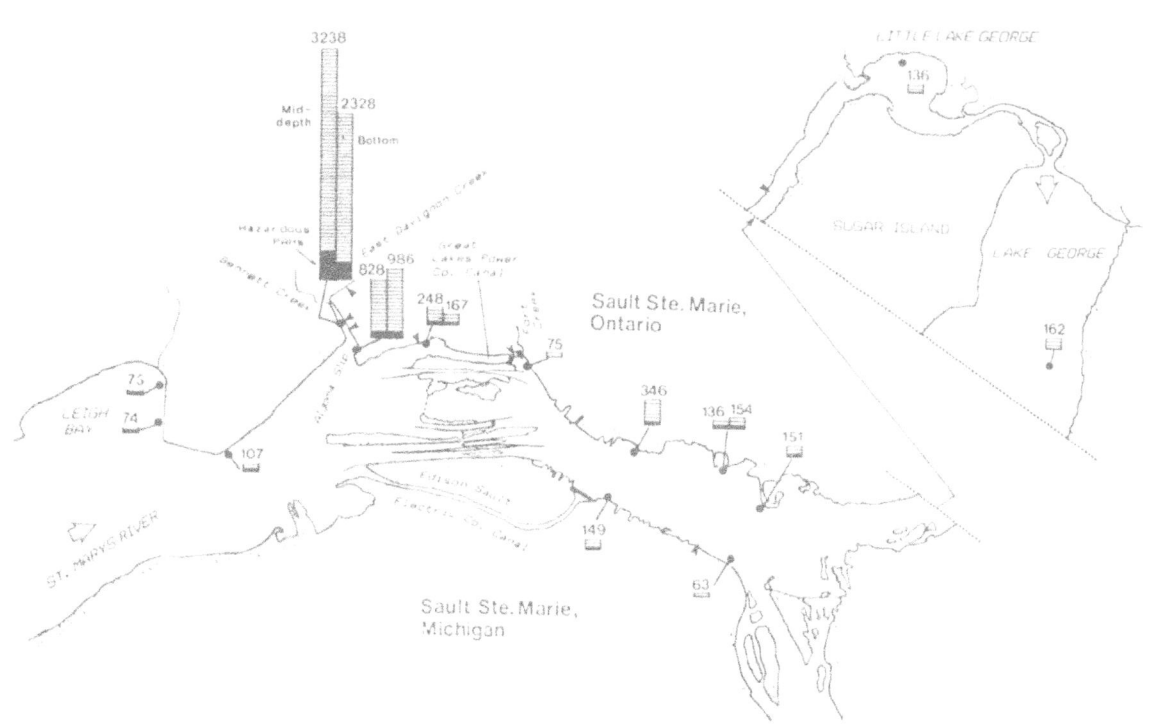

*Fig. 4.* Total polycyclic aromatic hydrocarbons ($\mu$g kg$^{-1}$, wet wt) in caged mussels (*Elliptio complanata*) after 3 weeks' exposure in the St. Marys River in 1985. Bar height is proportional to concentration; shaded portion of bars indicates the proportion of the total comprised of mutagenic and/or carcinogenic compounds.

Table 5. Concentrations of polycyclic aromatic hydrocarbons in mussels (*Elliptio complanata*) after three weeks' exposure in the St. Marys River, 1985.

| Compound | MDL | Station | | | | | | | | | | | | | | | | | | |
|---|---|---|---|---|---|---|---|---|---|---|---|---|---|---|---|---|---|---|---|---|
| | | 124 | 123 | 121 | 115 M | 115 B | 117 M | 117 B | 167 M | 167 B | 29 | 9C | 9A | 107 M | 107 B | 27C | 27A | 88 | 92 | Balsam Lake |
| Naphthalene | 1 | 45 | 17 | 86 | 184 | 116 | 43 | 40 | 37 | 27 | 21 | 275 | 89 | 44 | 41 | 73 | 25 | 30 | 103 | 11 |
| Acenaphthylene | 1 | 0.4 | * | * | 20 | 8 | * | 5 | 8 | 0.9 | * | * | * | 10 | 5 | 4 | 3 | 5 | 1 | * |
| Acenaphthene | 1 | * | 0.9 | * | 208 | 145 | 7 | 40 | 12 | 2 | * | * | * | 1 | 1 | * | * | * | * | 0.3 |
| Fluorene | 1 | 0.4 | 12 | * | 189 | 146 | 19 | 44 | 5 | 2 | 14 | * | * | * | * | 4 | 1 | 1 | 1 | 1 |
| Phenanthrene | 1 | 2 | 8 | 0.8 | 969 | 751 | 163 | 331 | 72 | 45 | * | 22 | * | 11 | 11 | 14 | 6 | 12 | 3 | 1 |
| Anthracene | 1 | 0.8 | * | 2 | 123 | 83 | 29 | 39 | 14 | 1 | 5 | * | 3 | * | 0.6 | 0.6 | 2 | * | * | 2 |
| Fluoranthene | 1 | 13 | 21 | 5 | 756 | 570 | 304 | 356 | 23 | * | * | 13 | 3 | 23 | 36 | 17 | 2 | 36 | 13 | 1 |
| Pyrene | 1 | 9 | 12 | 3 | 563 | 392 | 180 | 30 | 64 | 49 | 6 | 11 | 2 | 15 | 15 | 10 | 1 | 17 | 3 | 2 |
| Chrysene/ Benzo(a)anthracene | 2 | * | * | * | 79 | 57 | 48 | 39 | 2 | * | * | 2 | 49 | 1 | 4 | * | * | 2 | * | * |
| Benzo(b/k)fluoranthene | 2 | 0.8 | 0.6 | 0.6 | 16 | 8 | * | * | * | * | 0.8 | 0.7 | * | 2 | 3 | 2 | 13 | * | 12 | 0.8 |
| Benzo(j)fluoranthene | 1 | * | * | * | * | * | * | * | * | * | * | * | * | * | * | * | * | * | * | * |
| Benzo(e)pyrene | 1 | 0.7 | 1 | 3 | 100 | 42 | 36 | 40 | 9 | 31 | 7 | 9 | 4 | 20 | 30 | 20 | 4 | 18 | 7 | 1 |
| Perylene | 1 | 0.4 | * | 0.9 | 13 | 6 | 8 | 8 | 2 | 8 | 21 | 4 | 2 | 7 | 5 | 3 | 6 | 8 | 5 | * |
| Dimethylbenz(a)anthracene | 1 | * | * | * | * | * | * | * | * | * | * | * | * | * | * | * | * | * | * | * |
| Benzo(a)pyrene | 1 | * | * | * | 13 | 1 | 1 | * | * | 1 | * | 0.2 | * | 1 | 1 | 1 | * | 2 | 14 | * |
| Indeno(1,2,3-c,d)pyrene | 1 | * | * | * | 1 | * | * | * | * | * | * | 4 | * | 1 | 1 | * | * | 1 | 0.2 | 0.2 |
| Dibenzo(a,h)anthracene | 1 | * | * | * | 1 | * | * | * | * | * | * | 4 | * | 1 | 1 | 1 | * | 2 | * | * |
| Benzo(g,h,i)perylene | 1 | * | * | * | 1 | 0.7 | * | * | * | * | * | 1 | * | 0.2 | 0.2 | * | * | * | * | * |
| Anthanthrene | 1 | * | * | * | * | * | * | * | * | * | * | * | * | * | * | * | * | * | * | * |
| Benzo(b)chrysene | 1 | * | * | * | * | * | * | * | * | * | * | * | * | * | * | * | * | * | * | * |
| Coronene | 1 | * | * | * | * | * | * | * | * | * | * | * | * | * | * | * | * | * | * | * |
| Quinoline | 1 | * | * | * | * | * | * | * | * | * | * | * | * | * | * | * | * | * | * | * |
| Carbazole | 1 | * | * | * | * | * | * | * | * | * | * | * | * | * | * | * | * | * | * | * |
| Acridine | 1 | * | * | * | * | * | * | * | * | * | * | * | * | * | * | * | * | * | * | * |
| Benz(a)acridine | 1 | * | * | * | * | * | * | * | * | * | * | * | * | * | * | * | * | * | * | * |
| Benzothiophene | 1 | * | * | 5 | 2 | 0.2 | * | 14 | * | * | * | * | * | 0.2 | * | * | * | * | * | * |
| Dibenzothiophene | 1 | * | * | * | * | 2 | 2 | * | * | * | * | 0.4 | * | * | * | * | * | 2 | * | 0.3 |
| Tissue wet weight(g) | – | 8.1 | 8.2 | 8.2 | 9.1 | 10.1 | 8.2 | 8.7 | 8.6 | 8.3 | 8.7 | 8.6 | 10.9 | 9.4 | 8.9 | 8.5 | 8.6 | 9.3 | 9.3 | 8.8 |
| Age (yr) | – | 8 | 8 | 9 | 11 | 8 | 9 | 9 | 9 | 9 | 10 | 9 | 8 | 14 | 9 | 9 | 10 | 10 | 8 | 9 |

M = mid-depth; B = bottom.

Notes: 1) Station numbers correspond to locations on Fig. 1.
2) Concentrations are $\mu g\,kg^{-1}$ wet weight and are the geometric means of 3 replicates; other parameters are arithmetic means.
3) '*' indicates not detected at method detection limit (MDL) shown.

nature of these highly contaminated areas (i.e., Algoma Slip and slow-flowing tributaries) and the resultant deposition of contaminated materials. However, since other factors such as field methodology (depth of sample collected, proximity to source-see Table 4), sediment physical and chemical characteristics [%fine ($< 62 \mu$m diam.) particles, total organic carbon content] and even analytical methodology can be important determinants in the final 'bulk' concentrations of PAHs reported, the magnitude of differences (or similarities) between areas may not be as significant as they at first appear.

### 3.4. PAH concentrations in mussels

The presence of detectable PAH concentrations in *E. complanata* from some stations after three weeks' exposure indicates a rapid accumulation of these compounds. Pittinger *et al.* (1985) observed that this occurred within as short a period as three days in the oyster *Crassostrea virginica*. As with surficial sediments, total PAHs in mussel tissues were highest at Station 115 in the Algoma Slip (2328–3238 $\mu$g kg$^{-1}$ wet wt). The mean concentrations of the following compounds were also highest at this and at nearby Station 117: Acy, Ac, F, Ph, An, Ft, P, C/B(a)An, B(e)P, and B(a)P (Fig. 4 and Table 5). Downstream, total PAH concentrations decreased rapidly to 151 $\mu$g kg$^{-1}$ at Station 27C. However, similar concentrations were still observed in Little Lake George (136 $\mu$g kg$^{-1}$) and Lake George (162 $\mu$g kg$^{-1}$), some 13 km and 20 km distant, respectively, and along the Michigan shoreline at Station 9A (149 $\mu$g kg$^{-1}$). This suggests inputs or biological availability of PAHs in the Sault Ste. Marie, Ontario and Michigan areas, with additional input along the Lake George Channel.

The proportion of total PAHs represented by 'hazardous PAHs' varied considerably in mussels (Fig. 4), but generally was less than that observed in sediments and did not correspond with the areas of maximum and minimum values noted for the latter. Percentages of this sub-group more than doubled in the Algoma Slip (11–13%) from the low levels present in mussels prior to their exposure (5%) and at upstream stations (2–3%). At the remaining stations, percentages ranged from 6 percent (Stn. 9C) to a maximum of 36 percent (Stn. 9A), with the remainder being around 20 percent. Compounds contributing most to these percentages differed between Ontario and Michigan stations in the Sault Ste. Marie area. The former were dominated by B(e)P and additionally by Ac and C/B(a)An in the Algoma Slip, whereas at the latter, B(b/k)Ft was the major contributor (Table 5). It is noteworthy that even the maximum concentration of B(a)P in mussels (13 $\mu$g kg$^{-1}$) was about two orders of magnitude lower than the recommended objective of 1000 $\mu$g kg$^{-1}$ for organisms serving as a food source for fish (IJC, 1983).

Statistical analysis on log$_e$-transformed replicate data (one-way ANOVA; Tukeys Multiple Range Test) was used to determine those stations or station groups with significantly different ($\alpha = 0.05$) concentrations of individual PAHs in *E. complanata* exposed to the sediments. Although mean concentrations of many PAHs were highest in the Algoma Slip, the high variability between replicates for some compounds resulted in no significant station-to-station differences for seven of the compounds: N, Acy, B(b/k)Ft, B(a)P, DB(a,h)An, B(g,h,i)Per, and DBt. [Similar large variability among individual replicates has been observed for this organism (Kauss & Hamdy, 1985) and for the marine bivalve *Rangia cuneata* (Neff & Anderson, 1975). The latter felt that this was typical of molluscs that can close their valves and remain isolated from the external environment for variable time periods.] The remaining 11 compounds exhibited varying degrees of difference or overlap between stations and the areas with significantly higher concentrations were compound-specific (Fig. 5). For example, mussels exposed at one or both stations in the Algoma Slip contained significantly higher mean concentrations of the following: Ft(356–570 $\mu$g kg$^{-1}$), Ph(331–751 $\mu$g kg$^{-1}$), Ac(40–145 $\mu$g kg$^{-1}$), F(146 $\mu$g kg$^{-1}$), An(39–93 $\mu$g kg$^{-1}$), B(e)P(40–42 $\mu$g kg$^{-1}$), P(392 $\mu$g kg$^{-1}$), and C/B(a)An (39–57 $\mu$g kg$^{-1}$).

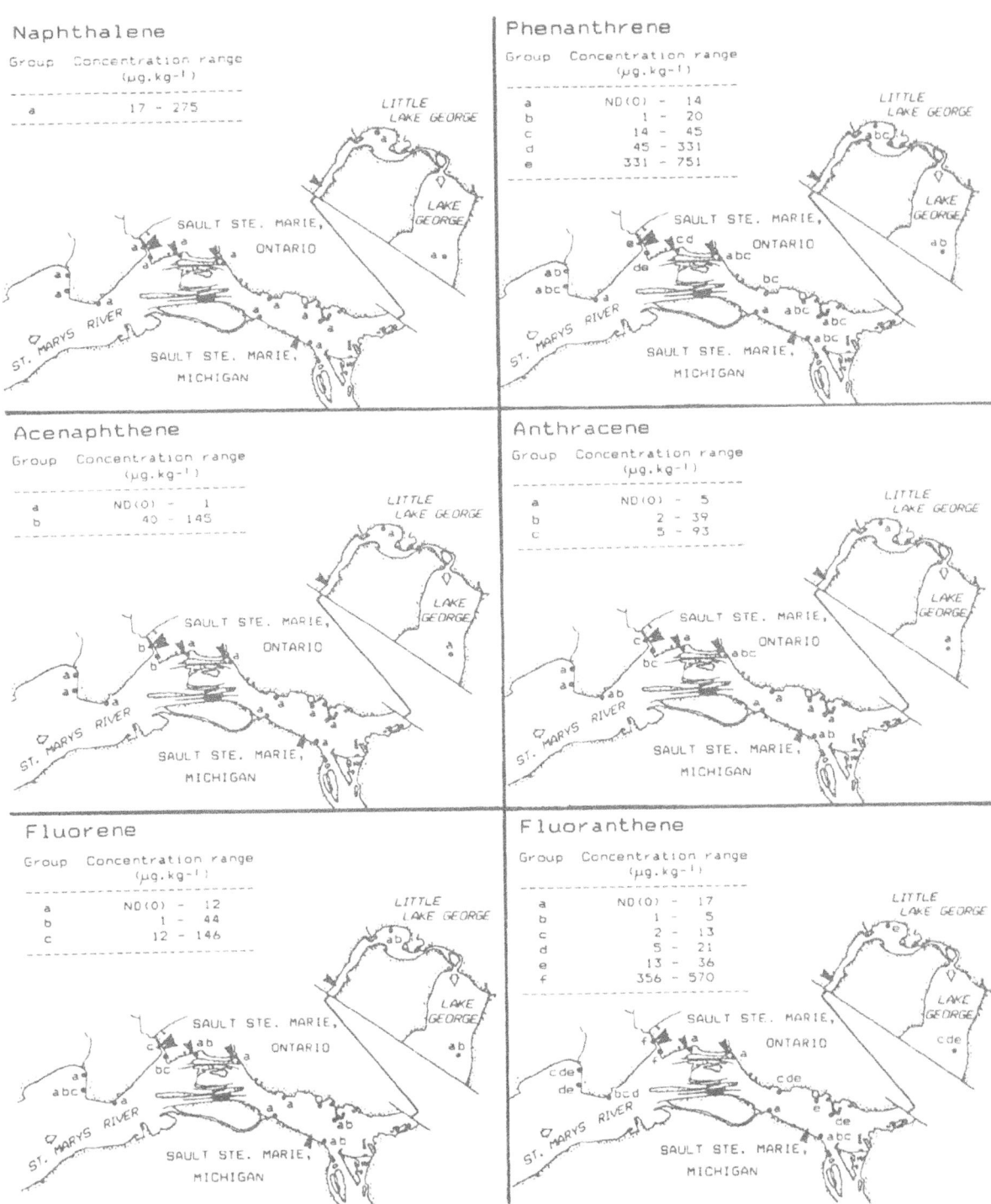

*Fig. 5.* Distribution of naphthalene, acenaphthene, fluorene, phenanthrene, anthracene, fluoranthene, pyrene, chrysene/ benzo(a)anthracene, benzo(e)pyrene, perylene, indeno(1,2,3-cd)pyrene and benzothiophene concentrations in caged mussels (*Eliptio complanata*) after 3 weeks' exposure in the St. Marys River in 1985. Letters delineate groups of significantly different ($\alpha = 0.05$) stations.

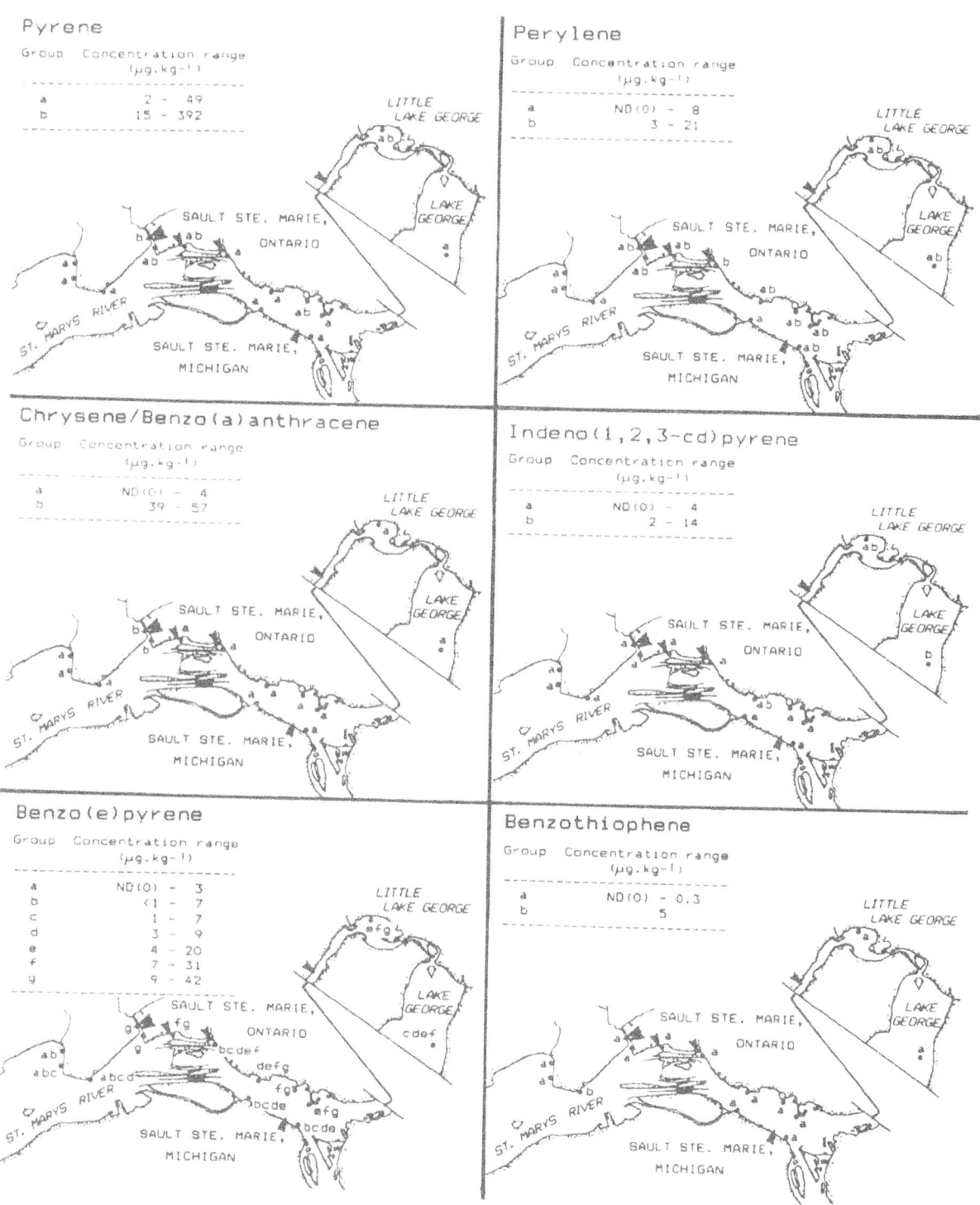

*Fig. 5.* (Continued).

Table 6. Comparison of polycyclic aromatic hydrocarbon concentrations ($\mu g\ kg^{-1}$, wet wt) in clams and mussels from various studies.

| Compound | Freshwater | | Marine | | | | | | | | | | |
|---|---|---|---|---|---|---|---|---|---|---|---|---|---|
| | Elliptio complanata | Elliptio complanata & Lampsilis radiata | Crassostrea galloprovincialis | Crassostrea virginica | | | Macoma inquinata | Mytilus californianus | Mytilus edulis | | | Rangea cuneata | |
| | (a) | (b) | (c) | (d) | (e) | (f) | (g) | (h) | (h) | (i) | (j) | (d) | (e) |
| Naphthalene | ND–275 | | | 14700 | 35 | 12000 | | | | | | 3800 | 51–120 |
| Acenaphthylene | ND–20 | | | | 36 | | | | | | | | 34–130 |
| Acenaphthene | ND–208 | | 7–8 | | 46 | | | | | | | 300 | 24–46 |
| Fluorene | ND–189 | <0.2–59 | 2 | 1000 | 21 | | | | | | | 500 | ND |
| Phenanthrene | ND–969 | <0.2–9.6 | 8–9 | 1900 | 220 | | 38–96 | | | | | | 72 |
| Anthracene | ND–123 | | 8–9 | | 44 | 2500 | | | | | | | 36–43 |
| Fluoranthene | ND–756 | | 7 | | 80 | 4000 | | | | | | | ND |
| Pyrene | 1–563 | <0.2–72 | 4 | | 200 | | | | | | | | ND |
| Chrysene | }ND–79 | <1.6 | ≤8–≤18 | | 58 | 1800 | 297–308 | | | | | | ND |
| Benzo(a)anthracene | | | 2–4 | | 9.9 | | | | | | | | ND |
| Benzo(b)fluoranthene | }ND–49 | | | | 12 | | | | | | | | ND |
| Benzo(k)fluoranthene | | | | | | | | | | | | | |
| Perylene | ND–21 | <1.6–48 | 1–14 | | | | 297–856 | | | | | 7200 | |
| Benzo(a)pyrene | ND–13 | <1.6 | 2 | | | 300 | 37–59 | <0.1–2.3 | 0.4–8.2 | 17–215 | | | |
| Dimethylbenz(a)-anthracene | ND–4 | | ≤1–≤2 | | | | | | | | | | |
| Benzo(g,h,i) perylene | ND–2 | | 4–5 | | | | | | | | | | |
| Coronene | ND | | 3–4 | 300 | | | | | | | 34–809 | | 100 |
| Dibenzothiophene | ND–14 | <0.2–0.6 | | | | | | | | | | | |

Data sources: (a) This study, caged mussels exposed for 3 weeks; range of samples. (b) Heit et al. (1980); native species homogenate, range of samples; dry weight data converted to wet weight basis by dividing by 6.25. (c) Iosifidou et al. (1982); Chrysene includes Triphenylene; Dimethylbenz(a,h)anthracene includes (a,c) isomer; range of means for native species from two locations. (d) Neff et al. (1976); after 24 h exposure to No. 2 fuel oil (1%) or to 0.0305 g m$^{-3}$ B(a)P. (e) McFall et al. (1985); native species. (f) Lee et al. (1978); after 8 days' exposure to Prudhoe Bay crude oil dispersion enriched with PAHs. (g) Roesijadi et al. (1978); after 7 days' exposure to detritus containing 2000 mg kg$^{-1}$ Prudhoe Bay crude oil and spiked with $^{14}$C-labelled PAH compounds; range of means. (h) Dunn & Young (1976); homogenates of native species from various locations. (i) Dunn & Stich (1976); homogenates of native species from various locations. (j) Kira et al. (1983); homogenates of native species from various locations.

Note: ND = not detected.

A decreasing downstream gradient was most evident for Ac, F, An, B(e)P, Ph, and P, while elevated levels of C/B(a)An and Ac were restricted to the Slip. In contrast, the highest concentrations of Bt(5 $\mu$g kg$^{-1}$), Per(21 $\mu$g kg$^{-1}$), and IP(14 $\mu$g kg$^{-1}$) were found at the eastern end of the Algoma Slag Dump, immediately downstream of the Algoma Steel Terminal Basins discharge and in Lake George, respectively.

Table 6 compares the range of PAH concentrations detected in *E. complanta* after exposure in the St. Marys River with levels found in various freshwater and marine bivalves. Differences between studies with respect to duration and type of exposure (short-term vs. life-long; field vs. laboratory), route of major exposure (water vs. sediment), analytical methodology, and species (filter-feeder vs. detritovore; lipid content, etc.) make such comparisons difficult at best. However, it is evident that the ranges of concentrations of individual PAHs detected in our study are similar to those reported by other researchers, particularly for native species.

With respect to *E. complanata*, the presence of high concentrations of certain PAHs in tissues after only three weeks' exposure indicates an ability to accumulate these compounds quite rapidly from its environment. It also suggests that this species lacks the necessary mixed function oxidase (MFO) enzymes to metabolize such contaminants. While this would seem to be the case for most bivalves, the issue is not yet completely resolved. For example, Vandermeulen & Penrose (1978) could detect no induction of aryl hydrocarbon hydroxylase (AHH) or N-dimethylase activity in the marine species *Mya arenaria*, *Mytilus edulis* or *Ostrea edulis* exposed to oils for both short (4 day) or long (6 years) periods in the field. Similarly, Livingstone *et al.* (1985) found no increase in the activity of B(a)P hydroxylase (an AHH) in *M. edulis* exposed to diesel oil for eight days, but detected elevated activity of the cytochrome P-450 monooxygenase system. In contrast, B(a)P hydroxylase activity was found in *Mercenaria mercenaria* exposed to B(a)P, the greatest activity being in the digestive gland (Giam *et al.*, 1987). Preliminary tests have not detected the presence of AHH in *Elliptio complanata* (D. Rokosh, OME, pers. comm., Dec., 1988). Although lack of such enzymes in an organism may protect it from exposure to carcinogenic PAH metabolites, it may have adverse effects on higher trophic levels possessing such toxicogenic/detoxification systems, since they would be exposed to higher concentrations of the procarcinogenic parent compounds if they are not metabolized by their prey.

### 3.5. *Sources of PAHs to mussels*

PAHs such as N, F, Ph, An, Ft, P, B(e)P, and Per were present at low levels in mussels exposed at stations upstream of dischargers as well as in mussels from Balsam Lake prior to their exposure (Table 5). This is suggestive of a general background source of a more diffuse nature. In fact, PAHs have been detected in both air (vapour and particulate phases) and precipitation samples from the Great Lakes basin (Eisenreich *et al.*, 1981) and atmospheric input appears to be a major source of PAHs to the Great Lakes (Eadie, 1984). Eisenreich *et al.* (1981) estimated the atmospheric fluxes of Ph, An, P, B(a)An, B(a)P, and Per to Lake Superior at 4.8, 4.8, 8.3, 4.1, 7.9, and 4.8 tonnes y$^{-1}$, respectively.

Nevertheless, the higher total PAH concentrations in both sediments and mussels along the Canadian shoreline and the significantly higher concentrations of certain PAH compounds near Algoma Steel discharges suggest the presence of additional ongoing sources, or of greater biological availability in these areas of the St. Marys River.

Sediments can be an important source of contaminants to benthic organisms. However, the magnitude of this contribution not only varies with the species and its life habits (Eadie *et al.*, 1983), but also depends on the degree of sediment contamination, sediment particle size distribution, and organic content (Eadie *et al.*, 1982b) as well as the particular compound in question (Eadie *et al.*, 1983; Roesijadi *et al.*, 1978).

Statistical analysis (one-way ANOVA; Tukeys

*Table 7.* Correlation matrix of interrelationships between polycyclic aromatic hydrocarbon concentrations in sediments and mussels.

| Sediment | Mussel | | | | | | | | | | | | | | | | | |
|---|---|---|---|---|---|---|---|---|---|---|---|---|---|---|---|---|---|---|
| | N | Acy | Ac | F | Ph | An | Ft | P | C/B(a)An | B(b/k)Ft | B(a)P | IP | DB(a,h)An | B(g,h,i)P | B(e)P | Per | Bt | DBt |
| N | .260 | .552 | .553 | .157 | .635 | .156 | .579 | .575 | .786 | .012 | .464 | .211 | .333 | .274 | .903 | .667 | .092 | .637 |
| Acy | .374 | .293 | .273 | -.041 | .377 | -.052 | .356 | .363 | .519 | .200 | .567 | .138 | .326 | .327 | .657 | .377 | .222 | .422 |
| Ac | .187 | .282 | .626 | .123 | .452 | .219 | .437 | .476 | .713 | .110 | .593 | -.174 | -.096 | -.088 | .602 | .460 | .168 | .622 |
| F | .326 | .281 | .488 | .111 | .421 | .074 | .355 | .348 | .566 | .287 | .552 | .016 | .105 | -.004 | .654 | .428 | .084 | .466 |
| Ph | .350 | .301 | .415 | .000 | .406 | .120 | .354 | .382 | .625 | .158 | .533 | -.016 | .117 | .121 | .691 | .487 | .291 | .542 |
| An | .176 | .423 | .436 | .133 | .514 | .018 | .422 | .416 | .668 | .223 | .536 | .094 | .188 | .184 | .755 | .588 | .011 | .591 |
| Ft | .420 | .202 | .367 | -.046 | .329 | .159 | .284 | .333 | .578 | .116 | .533 | -.086 | .016 | .038 | .600 | .431 | .431 | .542 |
| P | .420 | .202 | .367 | -.046 | .329 | .159 | .284 | .333 | .578 | .116 | .533 | -.086 | .016 | .038 | .600 | .431 | .431 | .542 |
| C/B(a)An | .434 | .087 | .252 | -.175 | .133 | .123 | .137 | .193 | .430 | .158 | .404 | -.141 | .047 | .017 | .460 | .305 | .512 | .388 |
| B(b/k)Ft | .420 | .116 | .236 | -.297 | .063 | .045 | .109 | .165 | .414 | .250 | .404 | -.062 | .047 | .058 | .446 | .263 | .512 | .263 |
| B(a)P | .420 | .116 | .236 | -.297 | .063 | .045 | .109 | .165 | .414 | .250 | .404 | -.062 | -.047 | -.058 | .446 | .263 | .512 | .263 |
| IP | .416 | .072 | .281 | -.223 | .113 | .086 | .127 | .211 | .510 | .152 | .536 | -.102 | -.134 | -.084 | .381 | .354 | .574 | .465 |
| DB(ah)An | .471 | -.140 | .073 | -.159 | .007 | -.030 | .036 | .109 | .307 | .066 | .479 | -.121 | -.202 | -.129 | .196 | .188 | .530 | .470 |
| B(ghi)Per | .437 | .065 | .235 | -.194 | .127 | .111 | .141 | .232 | .503 | .053 | .053 | .536 | -.094 | -.059 | .360 | .352 | .644 | .511 |
| B(e)P | .420 | .116 | .236 | -.297 | .063 | .045 | .109 | .165 | .414 | .250 | .404 | -.062 | .047 | .058 | .446 | .263 | .512 | .263 |
| Per | .371 | .177 | .281 | -.297 | .091 | .020 | .151 | .200 | .461 | .274 | .404 | -.023 | .094 | .096 | .495 | .298 | .442 | .263 |
| Bt | .329 | .352 | .630 | .126 | .512 | .130 | .471 | .490 | .746 | .113 | .632 | .125 | .092 | -.044 | .682 | .531 | .124 | .612 |
| DBt | .440 | .106 | .302 | .249 | .384 | .059 | .224 | .205 | .379 | .273 | .568 | -.012 | -.029 | -.151 | .497 | .370 | .123 | .504 |

Underlined values are significantly different at the 95% confidence level ($\alpha = 0.05$; $r = 0.671$, 10 d.f.) Abbreviations as per Table 1.

Multiple Range Test) showed that overall, there was no significant difference ($\alpha = 0.05$) between the concentrations of individual PAHs in mussels exposed to sediments or only to water at stations near or downstream of point sources (Stns. 115, 117, 167 and 107). There was significant interaction ($\alpha = 0.05$) for F and B(e)P and this seemed related to Station 167. This may be due to incomplete vertical mixing at this station which is located below an Algoma Steel discharge to the power plant headrace.

Correlation analysis performed on $log_e$-transformed data for the PAH compounds common to both sediments and mussels (bottom exposure) at the 12 coincident stations showed that there were no significant correlations ($\alpha = 0.05$) between concentrations of the same compound in

the two media. However, there were significant correlations between: C/B(a)An in mussels and N and Bt in sediments; and between B(e)P in mussels and N, Ph, An and Bt in sediments (Table 7).

Mussel-sediment concentration factors (dry weight basis) were usually $< 1$ for all compounds at all stations (range: $< 0.0001–0.62$, Table 8). A similarly low range of values was obtained for Ph, C, B(a)P, and DMB(a)An in *Macoma inquinata* exposed to detritus (Roesijadi *et al.*, 1978), for N, Ph, Ft, P, C, and B(a)An in caged *Crassostrea virginica* relative to creosote-contaminated sediments (Elder & Dresler, 1988) and generally also for Ph, F, P, B(a)An, B(e)P in *E. complanata/L. radiata* relative to sediment from northern Lake George in New York State (Heit

*Table 8.* Comparison of bivalve-sediment concentration factors and partition coefficients.

| Compound | Concentration factor (dry wt) | | | Partition coefficient | |
|---|---|---|---|---|---|
| | (a) | (b) | (c) | $log_{10}K_{OC}$ | $log_{10}K_{OW}$ |
| Naphthalene | 0.02   −8.96 | | | $2.81–2.82^h$ | $3.35^f, 3.36^e, 3.59^d$ |
| Acenaphthylene | $<0.007 - \geq 1.88$ | | | $3.31^h$ | $3.96^g$ |
| Acenaphthene | $<0.005 -0.27$ | | | $3.32^h$ | $3.92^f$ |
| Fluorene | $<0.0009–0.21$ | | | $3.48–3.49^h$ | $4.18^f$ |
| Phenanthrene | $<0.0001–0.42$ | $<0.007-0.4$ | 0.20 | $3.62^h$ | $4.46^d, 4.57^f$ |
| Anthracene | $<0.003 -0.42$ | | | $4.38–4.45^h$ | $4.34^i, 4.45^e, 4.54^f$ |
| Fluoranthene | 0.008 −0.38 | $<0.006-2.3$ | | $4.02^h$ | $4.30^g, 5.22^f$ |
| Pyrene | } 0.0007–0.30 | $<0.007-3.0$ | | $4.15^h$ | $4.88^i, 4.90^d, 5.18^f$ |
| Chrysene | } | | 0.04 | $5.25–5.28^h$ | $5.48^g, 5.79^f$ |
| Benzo(a)anthracene | $<0.0001–0.03$ | $<0.33$ | | $4.75^h$ | $5.61^g$ |
| Benzo(b/k)fluoranthene | $<0.0002–0.62$ | | | $5.19–5.34^h$ | $5.35^g$ |
| Benzo(j)fluoranthene | 0(ND) | | | $5.07^h$ | $5.35^g$ |
| Benzo(e)pyrene | 0.0007–0.31 | $<0.62 ~19$ | | $4.96^h$ | $5.90^g$ |
| Perylene | 0.001 −1.25 | | | $5.51^h$ | $5.27^e$ |
| Dimethylbenz(a) anthracene | 0(ND) | | 0.06 | $4.33^h$ | $6.49^g, 6.50^f$ |
| Benzo(a)pyrene | $<0.0006–1.25$ | $<0.33$ | 0.09 | $5.18^h$ | $5.98^g, 6.06^f$ |
| Indeno(1,2,3-c,d)–pyrene | $<0.0006–1.25$ | | | $5.69^h$ | $5.70^g$ |
| Dibenzo(a,h)–anthracene | $<0.001 -0.05$ | | | $5.10^h$ | $6.50^e, 7.19^f$ |
| Benzo(g,h,i)perylene | $<0.001 - \geq 0.62$ | | | $5.61^h$ | $6.25^g$ |
| Quinoline | $0(ND)-<0.10$ | | | $1.01^h$ | $2.03–2.06^l$ |
| Carbazole | $<0.0002-<0.08$ | | | | $3.29^l$ |
| Acridine | $<0.0003-<0.04$ | | | $2.19–3.41^l, 4.22^j$ | $3.30^l, 3.40^m$ |
| Benzo(a)acridine | $0(ND)-<0.02$ | | | | $4.45^l$ |
| Benzothiophene | $0.0003- \geq 3.12$ | | | $2.51^h$ | $3.10^k$ |
| Dibenzothiophene | $0(ND)- \geq 1.25$ | | | $2.92–4.17^j, 4.05^j$ | $4.40^k$ |

Data sources: (a) This study: caged *Elliptio complanata*; mussel wet weight concentration data multiplied by 6.25 (Kauss & Hamdy, 1985) to convert to dry weight basis. (b) Heit *et al.* (1980); homogenate of native *E. complanata* & *L. radiata*. (c) Roesijadi *et al.* (1978); *Macoma inquinata* exposed for 7 days. (d) Abernethy & Mackay (1987). (e) Freitag *et al.* (1985). (f) Gobas *et al.* (1987). (g) Kamlet *et al.* (1988). (h) Kenaga & Goring (1980); using formula $log_{10}K_{OC} = 3.64 - 0.55 log_{10}S$ and solubility (S) data in Billington *et al.* (1988); Das (1983); Eadie *et al.* (1982a); Mackay & Shiu (1975); Vasilaros *et al.* (1982); Verschueren (1983); and Zachara *et al.* (1987). (i) Mackay (1982). (j) Sabljic (1987). (k) Vasilaros *et al.* (1982). (l) Verschueren (1983). (m) Zachara *et al.* (1987).

Notes: 1) Concentration factors prefixed by '$\geq$' or '$<$' indicate that there were no detectable concentrations in sediments or mussels, respectively and half the detection limits (see Tables 3 and 4) were used.
   2) ND = not detected.

*et al.*, 1980). In our study, concentration factors for N ranged up to almost 9 and for Bt $\geq 3$ at Station 121. This may be due to the greater water solubility of these compounds, a decreased tendency to associate with organic matter in sediment (see 'log $K_{oc}$' in Table 8) and hence, greater biological availability. PAH concentrations in sediment pore water and on the fine fraction of sediments ($<62 \, \mu$m dia.) can be much higher than in overlying water and in bulk sediments (Eadie *et al.*, 1982b) and may therefore be important sources of PAHs to some benthic organisms. However, for *E. complanata*, such exposure routes would tend to further decrease the concentration factors. Consequently, our data indicate that sediment-associated PAHs had only limited bioavailability to *E. complanata* and that the major exposure route was via the water filtered by these organisms and hence, from discharges to the river.

Due to the low water solubility of higher molecular weight PAHs and their correspondingly high organic carbon partition coefficients ($K_{oc}$), they will tend to partition out onto fine particulate matter (solids) suspended in the water column. It might therefore be argued that particulate-associated PAHs are the major source of the elevated concentrations in mussels at certain stations. However, concentrations of suspended solids during this study were mostly below 10 mg $l^{-1}$ (Table 2), even near point sources. At these concentrations, most (70–100%) of the PAHs in the water column would be in the dissolved phase (Eadie, 1984) and would therefore accumulate in the lipids of aquatic organisms through equilibrium partitioning as a result of their high octanol-water partition coefficients (see log $K_{ow}$ in Table 8).

Similarly, leaching of PAHs from the coal particles or dust detected at Stations 115, 117 and 9A is not likely to have contributed to mussel body burdens. Bender *et al.* (1987) were unable to detect any elevation of PAHs above background in estuarine water after contact with fine coal dust. However, since the mussels in our study were not allowed to clear their guts prior to analysis, coal dust in the water filtered by these organisms may

have contributed to elevated tissue levels, but only if present at high concentrations, i.e., $> 1 \, \text{mg} \, l^{-1}$ (Bender *et al.*, 1987).

The higher concentrations of many PAHs in mussels exposed in the Algoma Slip are not necessarily reflective of higher loadings relative to other areas, but are probably a function of limited initial dilution and hence, higher water concentrations of these contaminants. Flow rates of the two creeks and two Algoma Steel blast furnace sewers discharging to the slip average a total of $\sim 3.1 \, \text{m}^3 \, \text{s}^{-1}$ (OME, 1986 unpubl. data). In contrast, lower concentrations of many PAHs were found in mussels at Station 29, just downstream of the major discharge of Algoma Steel, the Terminal Basins (see Fig. 1). This unexpected result may be due to two phenomena: (i) a high initial nearshore dilution of this discharge (averaging $3.9 \, \text{m}^3 \, \text{s}^{-1}$ in 1985) by cleaner water from the Great Lakes Power Corp. tailrace ($\sim 1000 \, \text{m}^3 \, \text{s}^{-1}$) and incomplete lateral mixing (McCorquodale & Yuen, 1987; McCorquodale, pers. comm., Dec., 1988); and (ii) incomplete vertical mixing of the thermally buoyant plume until it moves further downstream (Abdel-Gawad, 1985). The higher concentrations observed at downstream Stations 9C, 107 and 27C suggests that these mussels were exposed to a more homogeneous plume, or that additional PAH inputs to the waterfront, such as urban runoff (Marsalek & Ng, 1987), resulting from local atmospheric inputs (Boom & Marsalek, 1988) are also important. However, there were insufficient monitoring stations in our study to permit an evaluation of the contribution of individual storm sewers to the Sault Ste. Marie, Ontario nearshore.

## 4. Summary

Sampling in the St. Marys River during 1985 identified the presence of up to 25 PAH compounds in surficial sediments. As many as 18 of these PAHs were also accumulated by caged mussels during a three-week exposure period at the same locations. The highest concentrations of many of these compounds were detected in both media in the Sault Ste. Marie area along the Canadian shore in the Algoma Slip, which receives effluent from Algoma Steel Corp., and in sediments along the U.S. shore, immediately downstream of the Edison Sault Electric Co. Canal discharge. At and downstream of major discharges, from 23 to 63 percent of the total PAHs concentrations in sediments and from 6 to 27 percent in mussels were comprised of compounds with mutagenic and/or carcinogenic properties. Sediments from 58 percent of the stations sampled contained concentrations of B(a)P above the proposed IJC objective of 1 mg $kg^{-1}$. However, levels of this compound in mussels were well below the proposed IJC objective of $1\,000\ \mu g\ kg^{-1}$ for the protection of higher trophic levels.

The predominance of high molecular weight compounds such as Ph, Ft, and P in both sediments and mussel tissues at most stations indicates that the major source of PAHs in the St. Marys River is the high temperature combustion of fossil fuels, such as from coal coking. However, the presence of lower molecular weight PAHs (as well as N- and S-containing PAHs in sediments) in the Algoma Slip and at the outlet of the Edison Sault Canal also indicates spills of fossil fuels such as coal and additionally, coke oven by-products in the slip. Therefore, in the Algoma Slip, elevated PAH concentrations likely result from a combination of: (i) contamination of cooling water sewers discharging to the slip with process water from blast furnace operations; (ii) previous spills to the two creeks discharging to the Slip; and iii) input from coal piles in the vicinity of the Slip, either via wind or run-off.

The ability of the mussel, *Elliptio complanta*, to accumulate significantly elevated concentrations of compounds in three weeks indicates that it is a useful biomonitor for detecting spatial differences in the biological availability of PAHs in freshwater aquatic ecosystems. Present evidence suggests that this is due to the absence of inducible metabolic enzymes in this organism.

Although mussel tissues indicated the presence of some PAHs even upstream of point sources (i.e., at background levels), the significantly higher concentrations in the Sault Ste. Marie, Ontario area suggest ongoing inputs from Algoma Steel, or greater biological availability in these areas as a result of past losses. However, statistical analysis showed that PAH concentrations in sediments had no significant effect, or correlation with, levels in mussels, indicating that exposure is primarily via water for this filter-feeding organism and therefore due to ongoing inputs.

Although *E. complanata* was used as an introduced (caged) organism in this study, the fact that it indicated the presence of PAH compounds in the river that can potentially be accumulated by native benthic species poses some concern and merits further work to determine the present distribution, concentration and effect of these compounds in indigenous biota. It also indicates the necessity to reduce inputs of PAHs to the St. Marys River ecosystem.

In this regard, as part of MISA, an Effluent Monitoring Regulation for the Ontario iron and steel sector was served in draft form by OME for a 21-day public review on February 6, 1989. This draft regulation is a step towards the virtual elimination of persistent toxic chemicals such as PAHs from discharges. The detailed monitoring results from this regulation will be used to set stringent discharge limits for the iron and steel mills.

### Acknowledgements

The authors are grateful to the following: staff of the OME Laboratory Services Branch (LSB), Water Quality Section, for analysis of water samples; V. Taguchi and D. Wang of the LSB Mass Spectrometry Unit for evaluation of the PAH analyses; the Great Lakes Section (GLS) field crew (E. Law, O. Moore, R. Savage) for collection of sediment samples; D. Rokosh (OME) and J. McCorquodale, (Univ. of Windsor) for helpful discussions; S. Agnew and M. Kirby of the GLS for statistical analyses; A. Foley for typing the manuscript; and M. Kirby and two anonymous reviewers for helpful comments on the manuscript.

Field work with mussels was carried out under

contract to OME by Integrated Explorations (Guelph, Ontario).

Analysis of sediment and mussel samples for PAHs was performed by Zenon Environmental Inc. (Burlington, Ont.) under contract to OME.

This study was partially funded by Environment Canada under terms of the Canada-Ontario Agreement respecting Great Lakes Waters Quality.

# References

Abdel-Gawad, S. T., 1985. Mixing and decay of pollutants from shore-based outfalls discharging into cross-flowing streams. Ph.D. Thesis, Univ. of Windsor. Windsor, Ont.

Abernethy, S. & D. Mackay, 1987. A discussion of correlations for narcosis in aquatic species. In; K. L. E. Kaiser (ed), QSAR in Environmental Toxicology – II. D. Reidel Publ. Co., Boston, Mass.: 1–16.

Barrick, R. C., E. T. Furlong & R. Carpenter, 1984. Hydrocarbon and azaarene markers of coal transport to aquatic sediments. Env. Sci. Technol. 18: 846–854.

Baumann, P. C. & D. M. Whittle, 1988. The status of selected organics in the Laurentian Great Lakes: an overview of DDT, PCBs, dioxins, furans, and aromatic hydrocarbons. Aquat. Toxicol. 11: 241–257.

Baumann, P. C., W. D. Smith & M. Ribick, 1982. Hepatic tumor rates and polynuclear aromatic hydrocarbon levels in two populations of brown bullhead (*Ictalurus nebulosus*). In; M. Cooke, A. J. Dennis & G. L. Fisher (eds.), Polynuclear Aromatic Hydrocarbons: 6th Int'l. Symp., Physical and Biological Chemistry. Batelle Press, Columbus, Oh./Springer-Verlag, New York, N.Y.: 93–102.

Bender, M. E., M. H. Roberts, Jr. & P. O. DeFur, 1987. Unavailability of polynuclear aromatic hydrocarbons from coal particles to the eastern oyster. Env. Pollut. 44: 243–260.

Billington, J. W., G.-L. Huang, F. Szeto, W.-Y. Shiu & D. Mackay, 1988. Preparation of aqueous solutions of sparingly soluble organic substances: I. Single component systems. Env. Toxicol. Chem. 7: 117–124.

Boom, A. & J. Marsalek, 1988. Accumulation of polycyclic aromatic hydrocarbons (PAHs) in an urban snowpack. Sci. Total Env. 74: 133–148.

Brown, R. A. & P. K. Starnes, 1978. Hydrocarbons in the water and sediment of Wilderness Lake II. Mar. Pollut. Bull. 9: 162–165.

Burt, A. J., P. M. McKee & D. R. Hart, 1988. Benthic invertebrate survey of the St. Marys River, 1985. Volume I – Main REPORT (Contract Nos. A89538 & A98444) Prep. for Ont. Min. Env. Toronto, Ont., by Beak Consultants Ltd., Brampton, Ont., 88 pp + app.

Cassidy, R. A., W. J. Birge & J. A. Black, 1988. Biodegradation of three azaarene congeners in river water. Env. Toxicol. Chem. 7: 99–105.

Curry, C. A., 1977/1978. The freshwater clam (*Elliptio complanata*), a practical tool for monitoring water quality. Wat. Pollut. Res. Can. 13: 45–52.

Das, B. S., 1983. Applications of HPLC to the analysis of polycyclic aromatic hydrocarbons in environmental samples. In; J. F. Lawrence, (ed), Liquid Chromatography in Environmental Analysis. The Humana Press Inc., Clifton, N.J.: 19–75.

Dunn, B. P. & H. F. Stich, 1976. Monitoring procedures for chemical carcinogens in coastal waters. J. Fish. Res. Bd Can. 33: 2040–2046.

Dunn, B. P. & D. R. Young, 1976. Baseline levels of benzo(a)pyrene in southern California mussels. Mar. Pollut. Bull. 12: 231–234.

Eadie, B. J., 1984. Distribution of polycyclic aromatic hydrocarbons in the Great Lakes. In; J. O. Nriagu & M. S. Simmons (eds.), Toxic Contaminants in the Great Lakes. John Wiley & Sons, Inc., New York, N.Y.: 195–211.

Eadie, B. J., P. F. Landrum & W. F. Faust, 1982a. Polycyclic aromatic hydrocarbons in sediments, pore water and the amphipod *Pontoporeia hoyi* from Lake Michigan. Chemosphere 9: 847–858.

Eadie, B. J., W. R. Faust, P. F. Landrum, N. R. Morehead, W. S. Gardner & T. Nalepa, 1982b. Bioconcentration of PAH by some benthic organisms. In; M. W. Cooke & A. J. Dennis (eds.), Polynuclear Aromatic Hydrocarbons: 7th Internat. Symp. on Formation, Metabolism and Measurement. Battelle Press, Columbus, Oh.: 437–449.

Eadie, B. J., W. R. Faust, P. F. Landrum & N. R. Morehead, 1983. Factors affecting bioconcentration of PAH by the dominant benthic organisms of the Great Lakes. In; M. W. Cooke & A. J. Dennis (eds.), Polynuclear Aromatic Hydrocarbons: 8th Internat. Symp. on Mechanisms, Methods and Metabolism. Battelle Press, Columbus, Oh.: 363–377.

Eisenreich, S. J., B. B. Looney & J. D. Thornton, 1981. Airborne organic contaminants in the Great Lakes ecosystem. Env. Sci. Technol. 15: 30–38.

Elder, J. F. & P. V. Dresler, 1988. Accumulation and bioconcentration of polycyclic aromatic hydrocarbons in a nearshore estuarine environment near a Pensacola (Florida) creosote contamination site. Env. Pollut. 49: 117–132.

EC-OME (Environment Canada-Ontario Ministry of the Environment), 1986. St. Clair River pollution investigation (Sarnia area). Toronto, Ont., 135 pp + app.

Fabacher, D. L., C. J. Schmitt & J. M. Besser, 1988. Chemical characterization and mutagenic properties of polycyclic aromatic compounds in sediment from tributaries of the Great Lakes. Env. Toxicol. Chem. 7: 529–543.

Frank, A. P., P. F. Landrum & B. J. Eadie, 1986. Polycyclic aromatic hydrocarbon rates of uptake, depuration, and biotransformation by Lake Michigan *Stylodrilus heringianus*. Chemosphere 15: 317–330.

Freitag, D., L. Ballhorn, H. Geyer & F. Korte, 1985. Environmental hazard profile of organic chemicals. An experimental method for the assessment of the behaviour of organic chemicals in the ecosphere by means of simple laboratory tests with $^{14}$C labelled chemicals. Chemosphere 14: 1589–1616.

Giam, C. S., L. E. Ray, R. S. Anderson, C. R. Fries, R. Lee, J. M. Neff, J. J. Stegeman, P. Thomas & M. R. Trip, 1987. Pollutant responses in marine animals: the program. In; C. S. Giam & L. E. Ray (eds.), Pollutant Studies in Marine Animals. CRC Press Inc., Boca Raton, Fl.: 1–21.

Gschwend, P. M. & R. A. Hites, 1981. Fluxes of PAHs to marine and lacustrine sediments in the north-eastern United States. Geochim. Cosmochim. Acta 45: 2359–2367.

Gobas, F. A. P. C., W.-Y. Shiu & D. Mackay, 1987. Factors determining partitioning of hydrophobic organic chemicals in aquatic organisms. In; K. L. E. Kaiser (ed), QSAR in Environmental Toxicology – II. D. Reidel Publ. Co., Boston, Mass.: 107–123.

Heit, M., C. S. Klusek & K. M. Miller, 1980. Trace element, radionuclide, and polynuclear aromatic hydrocarbon-concentrations in Unionidae mussels from northern Lake George. Env. Sci. Technol. 14: 465–468.

Helfrich, J. & D. E. Armstrong, 1986. Polycylic aromatic hydrocarbons in sediments of the southern basin of Lake Michigan. J. Great Lakes Res. 12: 192–199.

Hites, R. A. & P. M. Gschwend, 1982. The ultimate fates of polycyclic aromatic hydrocarbons in marine and lacustrine sediments. In; M. Cook, A. J. Dennis & G. L. Fisher (eds.), Polynuclear Aromatic Hydrocarbons: 6th Internat. Symp. Physical and Biological Chemistry. Battelle Press, Columbus, Oh./Springer-Verlag, New York, N.Y.: 357–365.

IJC (International Joint Commission), 1978. Great Lakes Water Quality Agreement of 1978. Windsor, Ont.

IJC, 1983. 1983 annual report. Report of the Aquatic Ecosystems Objectives Committee. Windsor, Ont.

IJC, 1985. 1985 report on Great Lakes water quality. Great Lakes Water Quality Board report to the International Joint Commission. Windsor, Ont.

Iosifidou, H. G., S. D. Kilikidis & A. P. Kamarianos, 1982. Analysis for polycyclic aromatic hydrocarbons in mussels (Mytilus galloprovincialis) from the Thermaikos Gulf, Greece. Bull. Env. Contam. Toxicol. 28: 535–541.

Kamlet, M. J., R. M. Doherty, P. W. Carr, D. Mackay, M. H. Abraham & R. W. Taft, 1988. Linear solvation energy relationships. 44. Parameter estimation rules that allow accurate prediction of octanol/water partition coefficients and other solubility and toxicity properties of polychlorinated biphenyls and polycyclic aromatic hydrocarbons. Env. Sci. Technol. 22: 503–509.

Kauss, P. B., 1986. Presentation to citizens hearing (Great Lakes United) on St. Marys River water pollution, Aug. 7, 1986, Sault Ste. Marie, Michigan. Ont. Min. Envir. file rept. Toronto, Ont., 9 pp. + tables & figs.

Kauss, P. B., 1991. Biota of the St. Marys River: habitat evaluation and environmental assessment. Hydrobiologia 219: 1–35.

Kauss, P. B. & Y. S. Hamdy, 1985. Biological monitoring of organochlorine contaminants in the St. Clair and Detroit Rivers using introduced clams, Elliptio complanatus. J. Great Lakes Res. 11: 247–263.

Keith, L. H. & W. A. Telliard, 1979. Priority pollutants. I - A perspective view. Env. Sci. Technol. 13: 416–423.

Kenaga, E. E. & C. A. I. Goring, 1980. Relationship between water solubility, soil sorption, octanol-water partitioning and concentration of chemicals in biota. In; J. G. Eaton, P. R. Parrish & A. C. Hendricks (eds.), Aquatic Toxicology. Amer. Soc. for Testing and Materials. Philadelphia, Penn.: 78–115.

Kira, S., T. Izumi & M. Ogata, 1983. Detection of dibenzothiophene in mussel, Mytilus edulis, as a marker of pollution by organosulfur compounds in a marine environment. Bull. Env. Contam. Toxicol. 31: 518–525.

Krahn, M. M., L. D. Rhodes, M. S. Myers, L. K. Moore, W. D. MacLeod, Jr. & D. C. Malins, 1986. Associations between metabolites of aromatic compounds in bile and the occurrence of hepatic lesions in English sole (Parophyrys vetulus) from Puget Sound, Washington. Arch. Env. Contam. Toxicol. 15: 61–67.

Lake, J. L. & C. Hershner, 1977. Petroleum sulfur-containing compounds and aromatic hydrocarbons in the marine mollusks Modiolus demissus and Crassostrea virginica. In; Proc. Joint Conf. on Prevention and Control of Oil Spills. American Petroleum Institute/United States Coast Guard: 627–632.

Lee, R. F., W. S. Gardner, J. W. Anderson, J. W. Blaylock & J. Barwell-Clarke, 1978. Fate of polycyclic aromatic hydrocarbons in controlled ecosystem enclosures. Env. Sci. Technol. 12: 832–838.

Livingstone, D. R., M. N. Moore, D. M. Lowe, C. Nasci & S. V. Farrar, 1985. Responses of the cytochrome P-450 monooxygenase system to diesel oil in the common mussel, Mytilus edulis L., and the periwinkle, Littorina littorea L. Aquat. Toxicol. 7: 79–91.

Lunde, G. & A. Bjorseth, 1977. Polycyclic hydrocarbons in long-range transported aerosols. Nature 268: 518–519.

Maccubbin, A. E., P. Black, L. Trzeciak & J. J. Black, 1985. Evidence for polynuclear aromatic hydrocarbons in the diet of bottom-feeding fish. Bull. Env. Contam. Toxicol. 34: 876–882.

Maccubbin, A. E., S. Chidambaram & J. J. Black, 1988. Metabolites of aromatic hydrocarbons in the bile of brown bullheads (Ictalurus nebulosus). J. Great Lakes Res. 14: 101–108.

Mackay, D., 1982. Correlation of bioconcentration factors. Env. Sci. Technol. 16: 274–278.

Mackay, D. & W.-Y. Shin, 1975. The aqueous solubility and air-water exchange characteristics of hydrocarbons under environmental conditions. In; Proc. Symp. on Chemistry and Physics of Aqueous Gas Solutions. The Electrochemical Society: 93–110.

Marsalek, J. & J. Y. F. Ng, 1987. Contaminants in urban runoff in the Upper Great Lakes Connecting Channels area. Envir. Can. Ntnl. Wat. Res. Inst. Contrib. No. 87–112. June, 1987, Burlington, Ont., 54 pp + app.

McCorquodale, J. A. & E. M. Yuen, 1987. Report on St. Marys River hydrodynamic and dispersion study. Univ. of Windsor, Ind. Res. Inst. Rept. IRI 18–61 to Ont. Min. Envir. Windsor, Ont., 90 pp + app.

McFall, J. A., S. R. Antoine & I. R. DeLeon, 1985. Base-neutral extractable organic pollutants in biota and sediments from Lake Pontchartrain. Chemosphere 14: 1561–1569.

National Research Council of Canada, 1983. Polycyclic aromatic hydrocarbons in the aquatic environment: formation, sources, fate and effects on aquatic biota. NRCC No. 18981. Ottawa, Ont., 209 pp.

Neff, J. M. & J. W. Anderson, 1975. Accumulation, release and distribution of benzo(a)pyrene-$C^{14}$ in the clam *Rangia cuneata*. In; Proc. Conf. on Prevention and Control of Oil Pollution. American Petroleum Institute/United States Environmental Protection Agency/United States Coast Guard: 469–471.

Neff, J. M., B. A. Cox, D. Dixit & J. W. Anderson, 1976. Accumulation and release of petroleum-derived aromatic hydrocarbons by four species of marine animals. Mar. Biol. 38: 279–289.

Niagara River Toxics Committee, 1984. Ambient river monitoring. In; Report of the Niagara River Toxics Committee. U.S. Env. Protect. Agency/Env. Can./New York State Dep. Env. Conserv./Ont. Min. Env. Toronto, Ont.

Oehme, M., 1985. Negative ion chemical ionization mass spectrometry – a useful technique for the selective detection of polar substituted polycyclic aromatic hydrocarbons with mutagenic properties. Chemosphere 14: 1285–1297.

OME (Ontario Ministry of the Environment), 1983. Handbook of analytical methods for environmental samples. Lab. Serv. Appl. Res. Br. Toronto, Ont.

OME, 1984. Water management. Goals, policies, objectives and implementation procedures of the Ministry of Environment. November, 1978, revised May, 1984. Toronto, Ont., 70 pp.

OME, 1985. A guide to the collection and submission of samples for laboratory analysis. 2nd edition. Lab. Serv. Br. Toronto, Ont., 94 pp.

OME, 1986. Report on the 1985 discharges from municipal wastewater treatment facilities in Ontario. Toronto, Ont., 241 pp.

OME, 1987a. Report on the 1985 industrial discharges in Ontario. December 22, 1986 (revised February, 1987). Toronto, Ont., 47 pp. + app.

OME, 1987b. Development of an Ontario effluent monitoring priority pollutants list. A guidance document for review. OME rep., Hazardous Contaminants Branch for the MISA Priority Pollutants Task Force. Toronto, Ont., 66 pp. + app.

Passino, D. R. M. & S. B. Smith, 1987. Acute bioassays and hazard evaluation of representative contaminants detected in Great Lakes fish. Env. Toxicol. Chem. 6: 901–907.

Pittinger, C. A., A. L. Buikema, Jr., S. G. Hornor & R. W. Young, 1985. Variation in tissue burdens of polycyclic aromatic hydrocarbons in indigenous and relocated oysters. Env. Toxicol. Chem. 4: 379–387.

Poulton, D. J., 1987. Trace contaminant status of Hamilton Harbour. J. Great Lakes Res. 13: 193–201.

Pranckevicius, P. E., 1987. 1982 Detroit Michigan area sediment survey. U.S. Env. Protect. Agency, Great Lakes Ntnl. Program Off. Rep. EPA 905/4-87-003. Chicago, Ill., 43 pp. + app.

Pranckevicius, P. E. & B. Kitsuse, 1989. Upper Great Lakes Connecting Channels tributary sediments, 1985. U.S. Env. Protect. Agency, Great Lakes Ntnl. Program Off. Draft rept. in prep. Chicago, Ill.

Rockwell, D. C., R. E. Claff & D. W. Kuehl, 1984. 1981 Buffalo, New York, area sediment survey (BASS). U.S. Env. Protect. Agency, Great Lakes Ntnl. Program Off. Rep. EPA 905/3-84-001. Chicago, Ill., 31 pp. + app.

Roesijadi, G., J. W. Anderson & J. W. Blaylock, 1978. Uptake of hydrocarbons from marine sediments contaminated with Prudhoe Bay crude oil: influence of feeding type, of test species and availability of polycyclic aromatic hydrocarbons. J. Fish. Res. Bd Can. 35: 608–614.

Sabljic, A., 1987. On the prediction of soil sorption coefficients of organic pollutants from molecular structure: application of molecular topology model. Env. Sci. Technol. 21: 358–366.

Suns, K., C. Curry, D. Wilkins & G. Crawford, 1980. The effects of road-oiling on PCB accumulation in aquatic life. In; Proc. Technology Transfer Conf. No. 1. Ont. Min. Env., Res. Advisory Committee. Toronto, Ont., 19–36.

Tripp, B. W., J. W. Farrington & J. M. Teal, 1981. Unburned coal as a source of hydrocarbons in surface sediments. Mar. Pollut. Bull. 12: 122–126.

Vandermeulen, J. H. & W. R. Penrose, 1978. Absence of aryl hydrocarbon hydroxylase (AHH) in three marine bivalves. J. Fish. Res. Bd Can. 35: 643–647.

Vasillaros, D. L., D. A. Eastmond, W. R. West, G. M. Booth & M. L. Lee, 1982. Determination of bioconcentration of polycyclic aromatic sulfur heterocycles in aquatic biota. In; M. Cooke, A. J. Dennis & G. S. Fisher (eds.), Polynuclear Aromatic Hydrocarbons: Physical and Biological Chemistry. Battelle Press, Columbus, Oh./Springer-Verlag, New York, N.Y.: 845–857.

Verschueren, K., 1983. Handbook of Environmental Data on Organic Chemicals. 2nd edition. Van Nostrand Reinhold Co., Inc., New York, N.Y. 1310 pp.

Williams, D. T., E. R. Nestman, G. L. LeBel, F. M. Benoit & R. Otson, 1982. Determination of mutagenic potential and organic contaminants of Great Lakes drinking water. Chemosphere 11: 263–276.

Zachara, J. M., C. C. Ainsworth, C. E. Cowan & B. L. Thomas, 1987. Sorption of binary mixtures of aromatic nitrogen heterocyclic compounds on subsurface materials. Env. Sci. Technol. 21: 397–402.

*Hydrobiologia* **219**: 63–81, 1991.
*M. Munawar & T. Edsall (eds),*
*Environmental Assessment and Habitat Evaluation of the Upper Great Lakes Connecting Channels.*
© *1991 Kluwer Academic Publishers.*

63

# Effects of pollution on benthic invertebrate communities of the St. Marys River, 1985

A. J. Burt, P. M. McKee, D. R. Hart & P. B. Kauss[1]
*Beak Consultants Limited, 14 Abacus Road, Brampton, Ontario, L6T 5B7 Canada;* [1]*Ontario Ministry of the Environment, 1 St. Clair Avenue West, Toronto, Ontario, M4V 1K6, Canada*

*Key words:* St. Marys River, sediment quality, benthic communities, pollution

## Abstract

A survey was undertaken in 1985 to assess spatial and temporal trends in the benthic community structure in relation to sediment contamination and wastewater sources at 70 stations between Whitefish Bay and lower Lake George in the St. Marys River. Cluster analysis identified seven benthic communities. Three were identified as pollution impacted, based on a preponderance of tubificids and nematodes, usually at high densities (up to 259000 m$^{-2}$), but sometimes at low densities ($< 100$ m$^{-2}$) at individual stations. Impacted communities occurred downstream of industrial and municipal sources and in depositional areas, and were confined mainly to Canadian waters. Unimpacted communities had greater numbers of taxa, and occurred upstream of point sources, along the U.S. shoreline, and in most areas of downstream lakes. Impacted and unimpacted communities were separated along particle size and contaminant gradients in river sediments. Despite recent reductions in pollutant loadings and improvements in sediment quality, no major changes were apparent in the status of the benthic community from earlier surveys.

## 1. Introduction

The St. Marys River, which flows from Lake Superior to Lake Huron, has been known to be polluted by steel and paper mill discharges since the 1940's, as evidenced by impairment of water quality, sediments, and the zoobenthos. Surveys of the impacted benthic communities in 1967 (Veal, 1968) and 1973 (Hamdy *et al.*, 1978) indicated zones of severe impairment downstream of discharges from Algoma Steel, Abitibi-Price Paper and the Sault Ste. Marie Sewage Treatment Plant (STP). On the basis of the latter survey, further reductions in waste loading from the contributing industries were recommended by the Ontario Ministry of the Environment (OME) and International Joint Commission.

A follow-up survey was conducted in 1983 (McKee *et al.*, 1984) to examine benthos and sediment quality relationships in light of ongoing pollution abatement programs at Algoma Steel and changing hydrological conditions brought about by the Great Lakes Power Corporation generating station. This survey indicated that the benthic communities of the St. Marys River were similar to those found in 1967 (Veal, 1968) and 1973 (Hamdy *et al.*, 1978), with zones of severe or moderate impact adjacent to and downstream of the industrial and municipal discharges. A severely impacted community characterized earlier by high densities of *Limnodrilus/Tubifex* was less pronounced in 1983, and was confined to a region downstream of St. Marys Falls. This has been identified as a region of heavy metal and

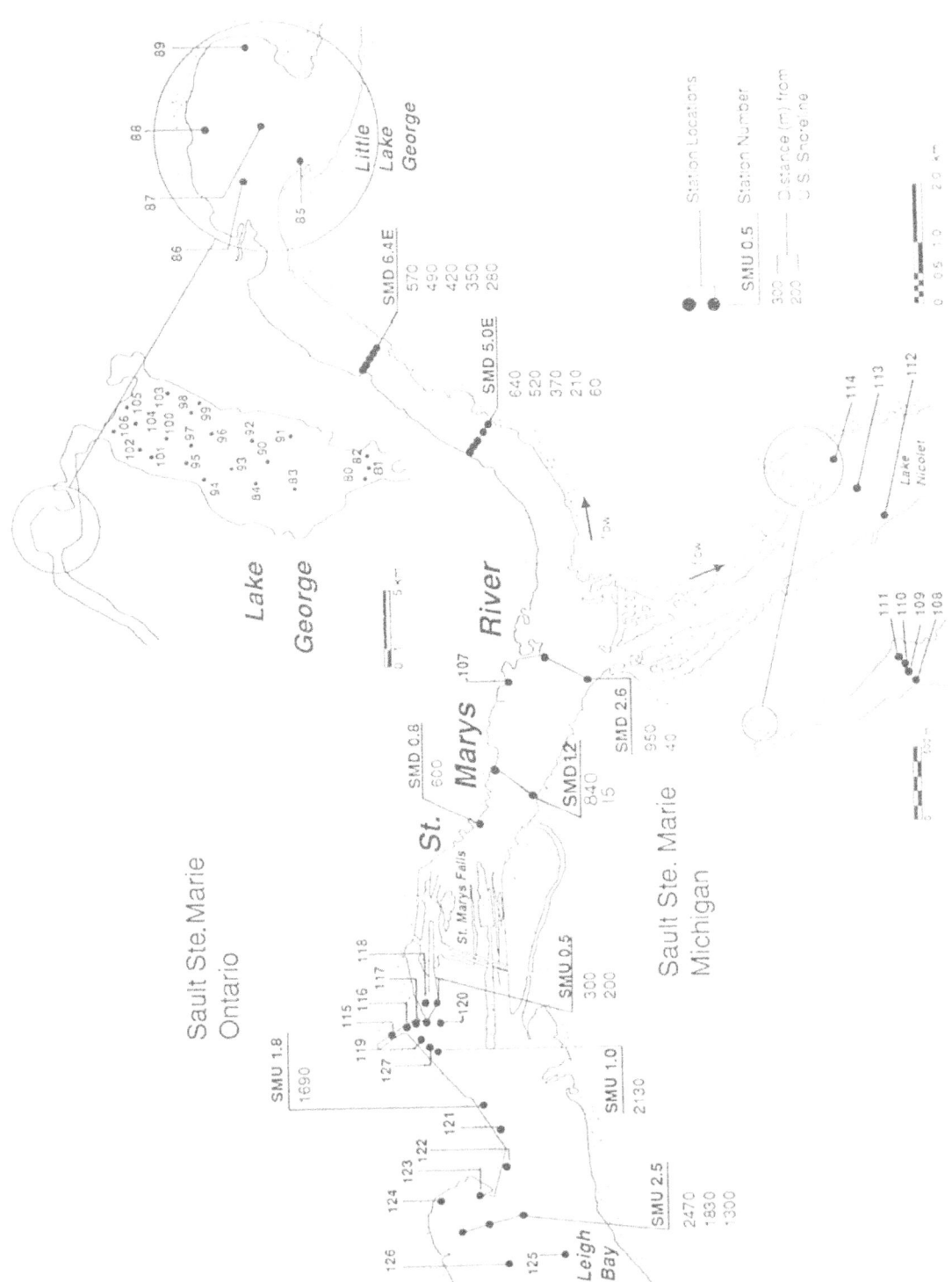

*Fig. 1.* Sediment and benthic fauna sampling station locations in 1985.

oil and grease contamination. Reductions in pollutant loadings by industry prior to 1983 appeared to have contributed to the minor community changes.

The 1983 survey was less extensive than those in earlier years, with little or no sampling downstream of the industrial region in Lake George, Little Lake George or Lake Nicolet, or upstream of Algoma Steel to Whitefish Bay (Fig. 1). Since benthic community effects were still apparent throughout the industrial region in 1983, additional sampling further upstream and downstream was identified as necessary to define their full extent. On the basis of a survey by Hiltunen & Schloesser (1983) in 1974–75, the benthic community was believed to be impacted to a point at least 30 km downstream of Sault Ste. Marie. The present survey was undertaken to augment the 1983 results by McKee *et al.* (1984) and verify Hiltunen & Schloesser (1983). Emphasis was placed on the lower and upper river areas not sampled in 1983. This paper documents the results of the 1985 benthic survey, compares the 1985 data with those from earlier studies, and evaluates relationships between benthic community structure and sediment quality.

## 2. Methods

### 2.1. Field survey

The 1985 survey was carried out in September–October. A total of 70 stations were sampled (Fig. 1), with 11 stations corresponding to 1983 locations. Station transects were denoted as either upstream (SMU) or downstream (SMD) distances from the St. Marys Falls and from the U.S. shore. For example, Station SMD 6.4E–490 indicates a location of 6.4 miles downstream of the falls in the eastern channel and 490 m from the U.S. shore.

Benthic samples were collected primarily by Ponar grab, but two Stations (SMU 0.5–200 and SMD 0.8–600) required airlift sampling (described by Barton & Hynes, 1978) due to substrate characteristics. The Ponar grab sampled a

bottom area of $0.052 \, \text{m}^2$. The diver-operated airlift sampler was equipped with $200 \, \mu m$ mesh Nitex collecting bags, and was used to sample $0.075 \, \text{m}^2$ quadrats on the river bed. All benthic samples were washed through $200 \, \mu m$ mesh and were preserved in 5 to 10 percent formalin.

A survey completed in 1983 (McKee *et al.*, 1984) indicated that three replicate samples at each sampling station were sufficient for estimation of log-transformed total organism density with a standard error = 20 percent of the mean density. On this basis, three replicate samples were collected at each 1985 sampling station.

Surficial sediments (top 3 cm) were collected at each station for physical/chemical analysis using a Shipek sampler. Substrate characteristics, water depth, and perceived current strength were also recorded. Sediments from at least 3 grabs at each station were homogenized in a clean, solvent-rinsed stainless steel tray using stainless steel utensils. Sediments were then partitioned into subsamples for analysis of trace organic and inorganic constituents by the OME laboratories, using documented procedures (OME, 1983). Results of these analyses were used to evaluate sediment quality-benthic community relationships.

### 2.2. Benthic analyses

Benthic organisms were sorted from sediment and debris using a dissecting microscope (10 × power) and grouped by major taxa. After sorting, a subsample of the tubificids (all of sample up to 100 individuals) and chironomids ( > = 10 percent subsample of each major chironomid group sorted) in each sample was cleared and mounted in 'CMCP-10' mounting medium on microscope slides. Chironomids were generally decapitated to facilitate clearing of the head capsule. Benthic organisms were identified using various taxonomic keys – Klemm (1985) for Oligochaeta; Oliver & Roussel (1983) for Chironomidae; Mackie *et al.*, (1980) and Clarke (1973) for Mollusca; Wiggins (1977) for Trichoptera; and Pennak (1978) for other groups.

## 2.3. Statistical analyses

The 1985 benthic communities were described on the basis of species compositional similarities between stations. Squared Euclidean distance was used as an inverse similarity measure for this purpose. Biologically similar stations are characterized by a small interstation distance. The distance formula for Stations $j$ and $k$ (Norusis, 1986) is as follows:

$$D^2(j, k) = \sum_{h=1}^{n} (X_{hj} - X_{hk})^2$$

where:

$X_{hj}$ = log transformed density of species $h$ at Station $j$.

The distance matrix contains an interstation distance for each pair of sampling stations. This information can be used to group the stations into clusters of biologically similar stations. The grouping procedure, or Q-mode cluster analysis, used in this study was a hierarchical agglomerative technique (Ward's method; Norusis, 1986). At each stage, one station is combined either with another station or with an existing cluster of stations, according to its affinities, and the affinities of the newly formed cluster are recalculated. The process continues until all stations are accounted for. Each cluster so defined can be interpreted as a distinct benthic community.

Characterization of station clusters was based on comparison of total organism density, species richness, tubificid dominance, and species composition between clusters. This information was used to identify station clusters, and from the spatial distribution of these clusters to delimit impacted zones. Spatial patterns and impacted zones in 1985 were related to known past and present pollution inputs, and compared to spatial patterns and impacted zones during previous surveys.

Several independent methods were used to examine the sediment quality characteristics of the St. Marys River benthic communities. First, discriminant analysis was performed on the sediment quality data in order to identify environmental gradients in sediment quality which tend to distinguish between the stations in different benthic faunal clusters. This method has been described by Green (1979), and other authors (e.g., Hutchinson, 1978), as a means of representing the ecological niche space occupied by different communities.

A second method of examining sediment quality-benthic faunal relationships is by derivation of a quantitative benthic community index, with a value at each station, to be used as a dependent variable in regression analysis of the sediment data. Thus, the sediment characteristics which are the best predictors of the community index can be identified. This approach differs from discriminant analysis primarily in the nature of the dependent variable. In discriminant analysis, the dependent variable reflects cluster membership, based on overall species composition, while a community index can reflect more specific attributes of species composition in a quantitative manner.

The community index used in this approach was a measure of dominance by a particular group of taxa (or guild). Guilds were defined by R-mode cluster analysis, which groups species based on similar spatial distributions (i.e., similar densities at each station). Densities were log transformed, as in the Q-mode cluster analysis. An index of dominance by each guild was calculated at each station as follows:

$$I_g = \frac{\sum_{i=1}^{S_g} \log(X_i + 1)}{\sum_{i=1}^{S} \log(X_i + 1)}$$

where:

$S_g$ = number of taxa in the guild,
$S$ = total number of taxa present, and
$X_i$ = density of species i.

The dominance index ranges from zero to one in magnitude.

Some guilds consisted of recognized pollution tolerant taxa. Their index values were plotted on study area maps to illustrate association with known point sources. Bivariate correlations with sediment variables were determined, and stepwise multiple regression analysis was performed to find the best set of sediment variables for prediction of the community index.

Some sediment variables were transformed prior to use in discriminant, correlation or multiple regression analysis, in order to improve their statistical distributions. Trace metals, pesticides, oil and grease, total organic carbon (TOC), total Kjeldahl nitrogen (TKN), and phosphorus were log-transformed. An angular transformation was used for particle size variables, and a square root transformation for loss on ignition (LOI). The pH and Eh distributions were not transformed. Water depth was included as a predictor variable, with log transformation.

## 3. Results and discussion

### 3.1. Indicator species in the St. Marys River

In the following discussion, reference is frequently made to the relative tolerance or intolerance of benthic species found in the St. Marys River (Table 1) to polluted conditions. Clean-water indicators are reported as more prevalent at locations well removed from effluent discharges, while tolerant forms prevail in the apparently impacted areas found downstream of the major point sources. It would be expected that station locations remote from pollution sources in the St. Marys River would reflect a healthy benthic community structure. The source of the St. Marys River is Lake Superior (Whitefish Bay, Fig. 1) and therefore, the water quality is very good. The pollution tolerance of benthic invertebrates, as outlined below, typically refers to tolerance of organic pollution and associated low dissolved oxygen conditions, rather than to tolerance of toxic metals and organic compounds.

Invertebrate forms found in the study area which are generally considered to be intolerant included mayflies such as *Hexagenia, Ephemera, Baetis* and *Caenis* (Roback, 1974). *Hexagenia* has

been reported as rare or absent in sediments contaminated by oily substances in the St. Marys River (Hiltunen & Schloesser, 1983). The lumbriculid *Stylodrilus heringianus* may also be considered a clean-water indicator. This species prefers sandy to gravelly substrates such as found in much of the study area (i.e. near SMD 6.4E, SMD 1.2), but is typically absent or reduced in numbers in disturbed areas such as near urban centres (Nalepa & Thomas, 1976).

Other species such as the tubificids *Aulodrilus* spp., *Potamothrix moldaviensis* and *Spirosperma ferox* are mesotrophic indicators, and *Rhyacodrilus* spp. is usually found in oligotrophic waters (Cook & Johnson, 1974); thus, these species are probably intolerant of heavy organic pollution. Nematodes apparently reach their highest densities in mesotrophic regions, with a low tolerance for highly polluted habitats (Golini, 1979). The polychaete *Manayunkia speciosa* is also considered indicative of moderate organic pollution (Poe & Stefan, 1975), but intolerant of severe organic pollution (Mackie & Qadri, 1971). Other mesotrophic indicators include the chironomids *Polypedilum, Nanocladius, Psectrocladius, Rheotanytarsus, Dicrotendipes* and *Microtendipes*. Gill-breathing gastropods such as *Valvata* and *Amnicola* also appear to be relatively intolerant to heavy organic pollution (Freitag *et al.*, 1973).

Among the pollution-tolerant forms, tubificids and chironomids are most noteworthy. High densities of the tubificids *Limnodrilus* spp. (including immatures without capilliform chaetae), *Tubifex tubifex, Quistadrilus multisetosus* and *Ilyodrilus templetoni* (both including immatures with capilliform chaetae) are characteristic of areas showing organic enrichment throughout the Great Lakes (Cook & Johnson, 1974; Brinkhurst & Cook, 1974). Similarly, the chironomids *Procladius, Chironomus* and *Cryptochironomus* are common in polluted conditions in the Great Lakes (Cook & Johnson, 1974).

### 3.2. Cluster analysis of spatial patterns in 1985

Seven clusters were defined in 1985 using Q-mode (station) clustering (Fig. 2, Table 2) with three

*Table 1.* Benthic invertebrates collected from the St. Marys River in October 1989 and their frequency of occurrence

| Taxa | Status | Taxa | Status | Taxa | Status |
|------|--------|------|--------|------|--------|
| | | Immature with hair setae | C | *Microtendipes sp.* | R |
| *COELENTERATA* | | Immature without hair setae | L(C) | *Pagastiella sp.* | L(R) |
| *Hydra sp.* | L(R) | *Dina sp.* | R | *Parachironomus sp.* | R |
| *PLATYHELMINTHES* | | *Glossiphonia complanata* | R | *Paralauterborniella sp.* | R |
| *Planaria sp.* | R | *Helobdella sp.* | R | *Paratendipes sp.* | R |
| *NEMERTEA* | | *Helobdella stagnalis* | R | *Phaenopsectra sp.* | R |
| *Prostoma rubrum* | R | *Piscicola punctata* | R | *Polypedilum sp.* | C |
| *NEMATODA* | A | | | *Pseudochironomus sp.* | R |
| *ANNELIDA* | | *ANTHROPODA* | | *Rheotanytarsus sp.* | L(R) |
| *Manayunkia speciosa* | L(C) | *Asellus sp.* | R | *Stempellina sp.* | R |
| *Branchiobdellidae* | R | *Lirceus sp.* | L(R) | *Stictochironomus sp.* | R |
| *Enchytraeidae* | L | *Crangonyx sp.* | R | *Tanytarsus sp.* | R |
| *Glossoscolecidae* | R | *Gammarus sp.* | R | *Xenochironomus sp.* | R |
| *Lumbriculus variegatus* | R | *Hyallela azteca* | R | *Diamesa sp.* | R |
| *Stylodrilus heringianus* | R | *Orconectes sp.* | R | *Monodiamesa sp.* | R |
| *Amphichaeta americana* | R | *Hydracarina* | C | *Odontomesa sp.* | R |
| *Chaetogaster diaphanus* | R | *Collembola* | R | *Potthastia sp.* | R |
| *C. diastrophus* | R | *Chloroperlidae* | R | *Corynoneura sp.* | R |
| *C. setosus* | R | *Baetis sp.* | R | *Cricotopus sp.* | R |
| *Nais barbata* | R | *Ephemera sp.* | R | *Epiococladius sp.* | R |
| *N. bretscheri* | R | *Hexagenia sp.* | C | *Heterotrissocladius sp.* | R |
| *N. communis* | R | *Ephemerella sp.* | R | *Nanocladius sp.* | R |
| *N. pardalis* | R | *Paraleptophlebia sp.* | R | *Orthocladius sp.* | R |
| *N. simplex* | R | *Brachycerus sp.* | R | *Paracladius sp.* | R |
| *N. variabilis* | R | *Caenis sp.* | R | *Psectrocladius sp.* | R |
| *Ophidonais serpentina* | R | *Stenonema sp.* | R | *Pseudosmittia sp.* | R |
| *Piquetiella michigansis* | R | *Tetragoneura sp.* | R | *Synorthocladius sp.* | R |
| *Pristina foreli* | R | *Notonecta sp.* | R | *Thienemanniella sp.* | R |
| *P. osborni* | R | *Sialis sp.* | R | *Ablabesmyia sp.* | R |
| *Ripistes parasita* | R | *Hydropsche sp.* | R | *Clinotanypus sp.* | R |
| *Slavina appendiculata* | R | *Agraylea sp.* | R | *Larsia sp.* | R |
| *Specaria josinae* | R | *Hydroptrila sp.* | R | *Nilotanypus sp.* | R |
| *Stylaria fossularis* | R | *Oxyethira sp.* | R | *Procladius sp.* | C |
| *S. lacustris* | R | *Lepidostoma sp.* | R | *Tanypus sp.* | R |
| *Uncinais uncinata* | R | *Ceraclea sp.* | R | *Thienemannimyia sp.* | L(C) |
| *Vejdovskyella comata* | R | *Mystacides sp.* | R | *Bezzia complex* | C |
| *V. intermedia* | R | *Nectopsyche sp.* | R | *Hemerodromia sp.* | R |
| *Aulodrilus americana* | C | *Oecetis sp.* | R | *MOLLUSCA* | |
| *A. limnobius* | R | *Triaenodes sp.* | R | *Elliptio sp.* | R |
| *A. piqueti* | R | *Molanna sp.* | R | *Sphaerium sp.* | R |
| *A. pluriseta* | C | *Phryganea sp.* | R | *S. simile* | R |
| *Ilyodrilus templetoni* | R | *Ptilostoma sp.* | R | *S. striatium* | R |
| *Isochaetides curvisetosus* | R | *Phylocentropus sp.* | R | *Pisidium sp.* | C |
| *I. freyi* | R | *Polycentropus sp.* | R | *Ferrissia sp.* | R |
| *Limnodrilus claparedianus* | R | *Chironomus sp.* | R | *Amnicola limosa* | C |
| *L. hoffmeisteri* | R | *Cladopelma sp.* | R | *Lymnaea sp.* | R |
| *L. udekemianus* | R | *Cladotanytarsus sp.* | R | *Helisoma sp.* | R |
| *Potamothrix moldaviensis* | R | *Cryptochironomus sp.* | C | *Gyraulus parvus* | R |
| *P. vejdovskyi* | R | *Cryptotendipes sp.* | C | *Physa gyrina* | R |
| *Quistadrilus multisetosus* | R | *Demicryptochironomus sp.* | R | *Valvata sincera* | L(C) |
| *Rhyacodrilus coccineus* | R | *Dicrotendipes sp.* | R | *V. tricarinata* | R |
| *Spirosperma ferox* | C | *Glyptotendipes sp.* | R | *Campeloma decisum* | R |
| *Tubifex tubifex* | R | *Harnischia sp.* | R | | |

Presence codes:
A – Abundant (> = 20% community at > = 50% of the stations)
L – Locally abundant (> = 20% community at any station)
C – Common (<20% of the community at >30 of the stations)
R – Rare (<20% of the community at < = 30 of the stations)

*Table 2.* Characteristics of benthic community clusters in the St. Marys River – 1985

| Parameter | Cluster number | | | | | | |
|---|---|---|---|---|---|---|---|
| | 1 | 2 | 3 | 4 | 5 | 6 | 7 |
| Number of stations | 16 | 7 | 9 | 11 | 8 | 14 | 5 |
| Dominant groups (Abundant) | Nematoda (33%) Manayunkia sp. (23%) Chironomidae (18%) Tubificidae[1] (14%) | Manayunkia sp. (43%)* Nematoda (25%) Tubificidae[1,2] (16%) | Nematoda (45%) Tubificidae[1] (39%) Chironomidae (6%) | Nematoda (37%) Tubificidae[1] (33%) Limnodrilus hoffmeisteri (26%) Chironomidae (16%) | Chironomidae (54%) Nematoda (23%) | Chironomidae (32%) Nematoda (20%) Isopoda (19%) Tubificidae[1] (11%) | Chironomidae (38%) Nematoda (28%) Tubificidae[1] (10%) |
| Common taxa (Present at all Stations) | Procladius sp. Bezzia complex Spirosperma ferox | Nemertea Stylodrilus heringianus Spirosperma ferox Pisidium sp. | Nais variabilis | | Tubificidae[1] Procladius sp. Cricotopus sp. Bezzia complex | Manayunkia sp. Pisidium sp. Amnicola sp. Valvata sincera Procladius sp. Polypedilum sp. | Helisoma sp. Procladius sp. Chironomus sp. Polypedilum sp. Hyallela sp. Hydracarina |
| Mean no. taxa | 27 | 23 | 15 | 12 | 39 | 40 | 32 |
| Mean total density (No/m$^2$) | 128,000 | 192,000 | 259,000 | 71,000 | 56,000 | 165,000 | 201,000 |
| Substrate | Brown silt over sand/clay | Coarse sand variable silt | Organic silt | Silt | Variable silty-coarse sand | Silt over sand | Brown silt over sand/clay |
| Water depth (m) | 2-4 | 3-14 | 1-16 | 2-9 | 1-6 | 2-11 | 3-12 |
| Visible oil | Absent to slight (10)[3] | Absent to abundant (5)[3] | Slight to abundant (9)[3] | Absent to abundant (10)[3] | Absent to slight (2)[3] | Absent to slight (1)[3] | Absent to abundant (4)[3] |
| Current | None to slight | Slight to moderate | None | None to strong | None to moderate | None to slight | None to slight |

*( ) Mean percent composition of major taxonomic groups within a cluster.
1 Immature Tubificidae without hair chaetae – i.e., Limnodrilus spp.
2 Immature Tubificidae with hair chaetae – i.e., Tubifex tubifex.
3 Number of stations where visible oil present.

Squared Euclidean Distance (Rescaled)

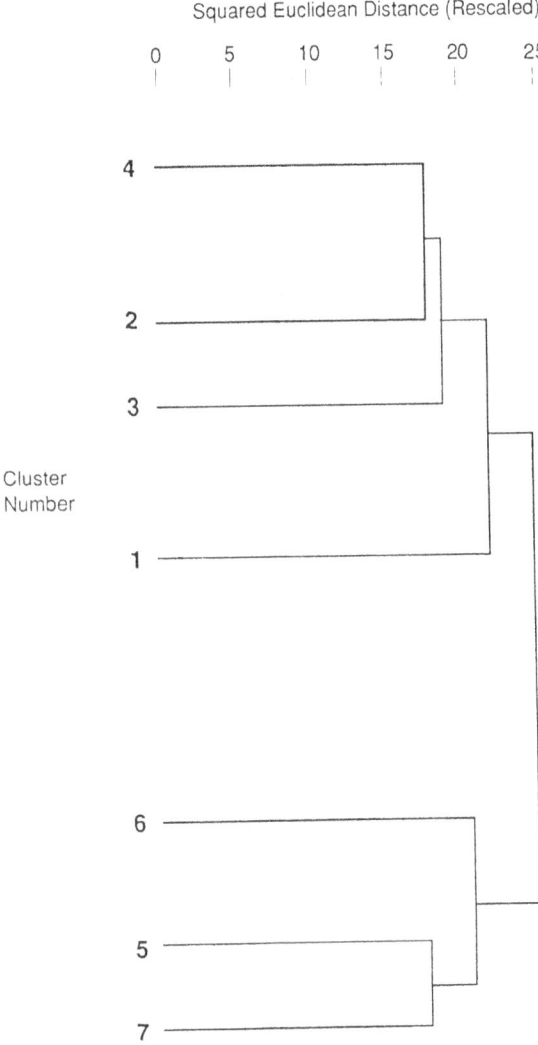

*Fig. 2.* Relationships between 1985 benthic station clusters derived by Ward's method.

clusters (2 to 4) identified as including pollution-impacted stations. The stations in these three clusters were dominated by facultative (including mesotrophic) to severe pollution-tolerant taxa. As the degree of organic enrichment increased, total community density also increased, together with a decline in the total number of taxa present. In situations where toxicity was apparent, very low densities and number of taxa were reported. These groups were associated with known industrial outfalls or downstream depositional areas. The remaining four clusters (1, 5, 6, and 7) generally describe stations that were upstream of

effluent discharges, near the U.S. shoreline or well downstream of discharges.

Cluster 2 consisted of five stations downstream of the Sault Ste. Marie STP, a station at the north end of Lake Nicolet, and one near the Algoma slip (Fig. 3). The mean number of taxa per station was relatively low (23), while the mean total density was relatively high (192 000 m$^{-2}$) in comparison with the unimpacted group of station clusters (1, 5, 6, and 7) (Table 2). Most of the common species among Cluster 2 stations were indicative of mesotrophic conditions, with more sensitive groups such as gastropods, isopods, ephemeropterans, and trichopterans either substantially reduced in density or absent. Consequently, stations included in Cluster 2 were considered to be slightly impaired.

Stations included in Cluster 3 were situated near the Canadian shore of the river, downstream of the main industrial outfalls; downstream of the Sault Ste. Marie STP; in Little Lake George; and in upper Lake George (Fig. 3). These stations are in depositional areas characterized by a silt substrate, a weak current, and oil accumulation in the sediment. Characteristic organisms at stations within Cluster 3 consisted of nematodes, unidentified immature tubificids without capilliform chaetae (likely *Limnodrilus* spp.), and the naidid oligochaete *Nais variabilis*, as well as other organisms that tend to be pollution-tolerant (Table 2). The relative abundance of the tubificid fraction was substantially greater than that observed at Cluster 2 stations (39% vs 16%, respectively). Of particular note was the low density or absence of the polychaete *M. speciosa* at Cluster 3 stations, in contrast to its abundance at stations in the closely related Cluster 2 (Fig. 2), suggesting greater impairment of water and sediment quality than occurred at stations in Cluster 2. This greater degree of impairment is reflected in a much lower mean number of taxa (15) and a much greater mean total density (259 000 m$^{-2}$) relative to Cluster 2 (Table 2). Both of these features are typical of organic enrichment. On this basis, as well as the species assemblage, Cluster 3 stations were considered moderately impaired.

71

*Fig. 3.* Distribution and zones of impairment of benthic fauna in 1985.

72

Cluster 4 included eight stations in the immediate vicinity of the Algoma Steel slip, two stations near the Canadian shore below the St. Marys rapids, and one station immediately downstream of the STP (Fig. 3). Physical environmental conditions described at stations within this cluster group were variable and the substrate was frequently heavily contaminated by oily residue (Table 2). Cluster 4 was defined biologically by the presence of unidentified immature tubificids without capilliform chaetae (likely *Limnodrilus* spp.) and nematodes were also present at a majority of stations. Other common species included the tubificid *Q. multisetosus* and the chironomids *Procladius* and *Cryptochironomus*. All of these species are tolerant of organic pollution. A few other pollution-tolerant taxa were also common (Table 2). The mean number of taxa per station (12) in Cluster 4 was the lowest of all clusters (Table 2). The variation in total densities at stations within Cluster 4 was very large, with very high densities occurring at organically enriched areas, and very low densities at some locations where effluent discharges may be exerting a toxic effect. For example, the mean total density of organisms at Station 116 in the Algoma slip was only 88 m$^{-2}$, suggesting a toxic inhibition of benthic community development by steel mill effluents, perhaps in combination with poor physical habitat conditions. At Station SMD 5.0E–520 near the STP, severe organic enrichment is implied by the extremely high mean total density of 345 000 m$^{-2}$. Biologically unusual stations within Cluster 4 included Station 127 where nematode densities reached 220 000 organisms m$^{-2}$ (70% of the population) and SMD 0.8–600 where coelenterates (*Hydra* sp.) accounted for 51 percent of the community. The very low number of taxa and preponderance of pollution-tolerant forms in Cluster 4 are indicative of severe environmental impairment.

### 3.3. Unimpacted areas

The remaining four clusters (1, 5, 6 and 7) included stations which were either upstream of the discharges, near the U.S. shore or too far downstream to be substantially affected (Fig. 3). As such, they were characterized by benthic associations which were low in tubificid oligochaetes, high in average numbers of taxa (27 to 40 per station), and are considered to be relatively unimpaired.

Cluster 1, the largest cluster in terms of numbers of member stations (16), occurred primarily in Lake George on silty substrate. Other environmental factors were variable, although oil tended to be either visibly absent or present only as slight traces (Table 2). Cluster 1 was biologically characterized by nematodes, polychaetes, the chironomid *Procladius*, immature tubificids without capilliform chaetae (likely *Limnodrilus* spp.), and the ceratopogonid *Bezzia* (Table 2). Several other species were also common to the majority of the stations in this cluster. The presence of non-tolerant organisms at most of the stations and relatively high mean numbers of taxa suggest that Cluster 1 stations were unimpaired overall, with some degree of organic enrichment.

Stations in Cluster 5 were generally situated upstream of pollution sources in the St. Marys River. Environmental characteristics were variable for all recorded parameters (Table 2). Characteristic taxa were Nematoda, immature Tubificidae, *Bezzia*, *Procladius*, and *Cricotopus*. Other frequently occurring taxa included non-tolerant forms such as ephemeropterans and trichopterans. Tubificids tended to be a minor portion of this community, with chironomids forming the majority of the assemblage. The high mean number of taxa (39), the low mean total density (56 000 m$^{-2}$), and the location of these stations either upstream or well downstream of pollution sources suggest that this cluster represents a relatively unimpaired community.

Cluster 6 was the second largest cluster (16 stations), with stations located in Leigh Bay, near the U.S. shore below the St. Marys Falls, in Lake Nicolette, and in Lake George. Essentially all locations were either upstream or across the river from point sources, or were well removed downstream (Lake George) in depositional areas. Characteristics of these locations included a substrate of silty sand and a local macrophyte community (Table 2). Characteristic taxa included the ubiquitous nematodes and immature tubificids (without capilliform chaetae), chironomids, polychaetes, pea clams, and snails (Table 2). Other commonly occurring taxa included the isopods *Asellus* and *Lirceus*, the amphipod *Hyalella azteca*, and the intolerant groups such as ephemeropterans. This assemblage is a mixture of facultative and non-tolerant organisms. The mean number of taxa at each station in this cluster (40) was the highest of all seven (Table 2). As in Cluster 5, chironomids were dominant (32% of the community). This cluster is classified as relatively unimpaired, based on the high number of taxa, the presence of intolerant forms, and remote location of these stations from sources of pollution.

Cluster 7 is the smallest of all the clusters (5 stations), and included only stations located upstream of the St. Marys Falls, downstream of the STP, and in Little Lake George. Stations in this cluster were characterized by a brown silty substrate and a general presence of aquatic macrophytes. Taxa common to all stations in Cluster 7 included primarily chironomids and nematodes with immature tubificids, Hydracarina, amphipods and snails also present (Table 2). Most of these taxa are facultative or tolerant forms. In general, stations within this cluster may be classified as unimpaired by virtue of the relatively high mean number of taxa (32), notwithstanding the fact that some stations (SMD 6.4E–280 and Station 89) have a high visible oil content in the sediments.

## 3.4. Sediment quality relationships

Discriminant analysis revealed a possible sediment quality basis for separation of the seven station groups defined by cluster analysis. Stations in the same cluster (with the same numeric label in Fig. 2) have similar overall species compositions. Clusters 2, 3, and 4 were considered to represent impacted groups, based on the recognized pollution tolerance of their characteristic species.

Clusters 3 and 4 were distinguished by high sediment concentrations of iron and zinc in relation to particle size (DF1 in Fig. 4). Cluster 2 sediments were distinguished on the same discriminant function by a greater fraction of Particle Size 2 (45 to 1 000 $\mu$m) but lower concentrations of these metals. Relationships between bulk sediment chemistry as examined here, and benthic community response, are poorly understood, and may be strongly influenced by the fractions of sediment contaminants that are loosely associated with the sediment and therefore are more bioavailable. Nonetheless, geometric mean iron concentrations of 3.1 percent and 6.3 percent and zinc concentrations of 174 $\mu$g g$^{-1}$ and 257 $\mu$g g$^{-1}$ at stations defined by Clusters 3 and 4 (Table 3), respectively, are considerably higher than the OME dredge disposal guidelines of 1 percent for iron and 100 $\mu$g g$^{-1}$ for zinc (Persaud & Wilkens, 1976), indicating that these key elements in DF1 could adversely affect benthic organisms and influence benthic community structure. Cluster 3 sediments were distinguished from those of Cluster 4 by higher concentrations of PCBs and a reduced fraction of finer particulate material (DF2) (Table 3). The first three discriminant functions were statistically significant ($p < 0.05$), with the first two collectively accounting for 72 percent of the variance in cluster membership.

Station Clusters 1, 5, 6, and 7 were not readily distinguishable from each other on the basis of the first two discriminant functions, but were reasonably well distinguished as a group from each of the three impacted station clusters. With the other functions, primarily DF3 (a particle size gradient), some discrimination within this group was achieved. Cluster 1 had the most fine material ($<45$ $\mu$m), while Cluster 5 had the least. Three stations within this four-cluster group were misclassified by the full set of discriminant functions (i.e., predicted on the basis of sediment characteristics to be in the wrong cluster). The overall percentage of stations correctly classified was 95.7 percent.

74

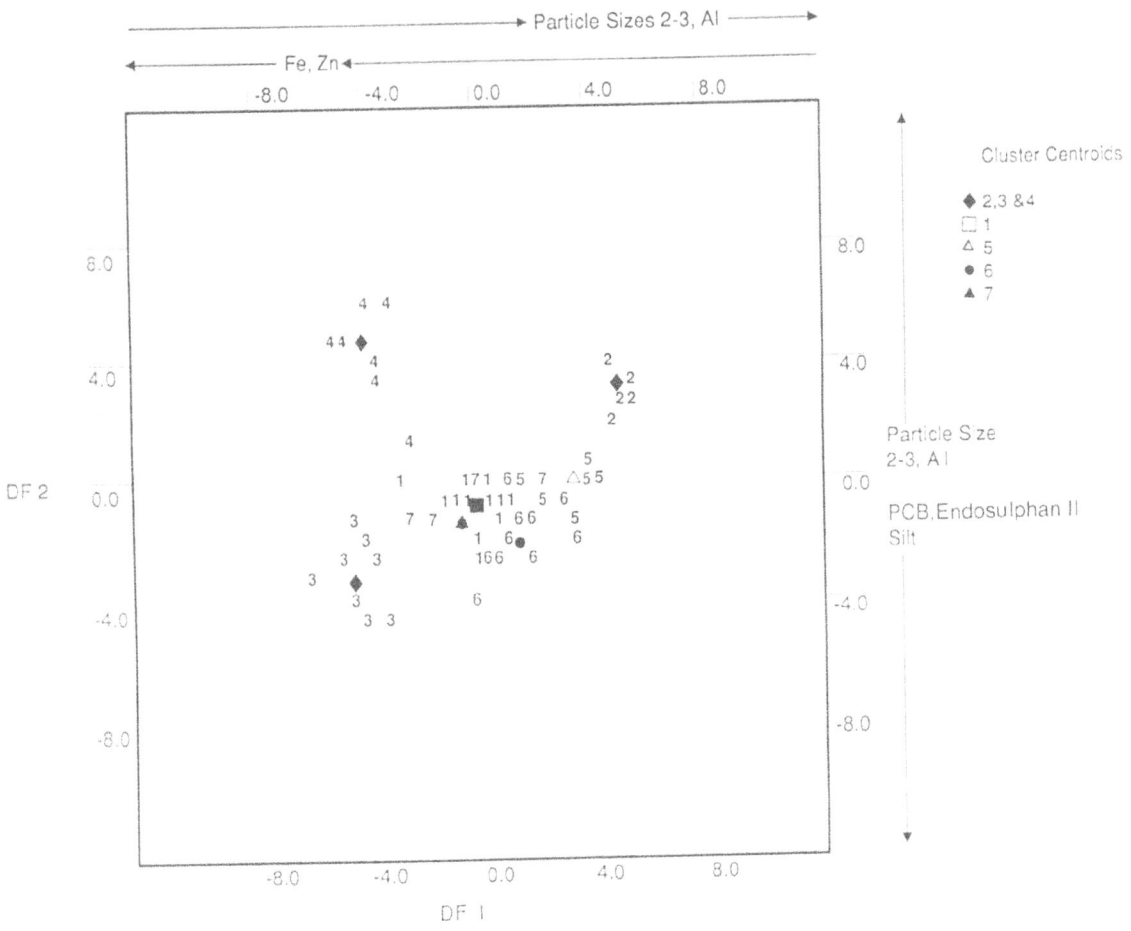

*Fig. 4.* Station clusters plotted on first two discriminant functions.

*Table 3.* Station cluster means of selected sediment characteristics

| Sediment Variable[1] | Station cluster mean | | | | | | |
|---|---|---|---|---|---|---|---|
| | 1 | 2 | 3 | 4 | 5 | 6 | 7 |
| Iron ($\mu g\,g^{-1}$) | 24,000 | 7,940 | 30,900 | 63,100 | 9,120 | 8,130 | 22,390 |
| Zinc ($\mu g\,g^{-1}$) | 123.0 | 30.2 | 173.8 | 257.0 | 22.4 | 29.5 | 97.7 |
| Particle Size 2 (45-1000 $\mu$m) (%) | 37.6 | 88.8 | 42.5 | 64.3 | 83.3 | 77.0 | 47.5 |
| Particle Size 3 (<45 $\mu$m) (%) | 62.3 | 8.7 | 56.4 | 31.9 | 12.4 | 22.2 | 51.5 |
| Silt (%) | 63.3 | 9.9 | 58.4 | 24.7 | 17.4 | 29.1 | 56.4 |
| Aluminium ($\mu g\,g^{-1}$) | 11,500 | 3,020 | 9,120 | 7,760 | 3,720 | 4,370 | 8,710 |
| PCBs ($ng\,g^{-1}$) | 21.4 | 20.0 | 37.2 | 25.1 | 20.0 | 24.0 | 22.9 |
| Endosulphan II ($ng\,g^{-1}$) | 4.1 | 4.0 | 4.5 | 5.9 | 4.0 | 4.0 | 4.0 |

[1] Sediment variables shown are the main contributors to discriminant functions 1 and 2, with standardized coefficients exceeding 1.9 in absolute value.

### 3.5. Temporal comparisons of spatial patterns

Results of the 1985 cluster analysis can be summarized in terms of the following pollution impairment zones:

1. severe:
   a) extreme tubificid dominance (i.e., *L. hoffmeisteri* and immatures without capilliform chaetae), tolerant chironomids, low numbers of taxa but high total densities (Cluster 4), or
   b) communities with either very low total densities and low numbers of taxa, and/or high densities of nematodes with few other taxa (Cluster 4);
2. moderate: tubificid dominance with high densities of nematodes and the presence of various facultative chironomids, absence of polychaete worms, reduced numbers of taxa, and high total densities (Cluster 3);
3. slight: nematode and polychaete dominance with moderate densities of tubificids and some non-tolerant groups present (Cluster 2); and
4. unimpaired: communities tending towards chironomid dominance, with several non-tolerant groups (e.g., cphemeropterans and trichopterans) present, low tubificid densities, and high numbers of taxa (Clusters 1, 5, 6 and 7).

The spatial distribution of these four zones, based on assignment of each cluster of stations to one of the zones (Fig. 3), was used to infer any major changes in benthic community status between years.

In comparisons among surveys, methodological differences must be considered. Increased sorting efficiency (due in part to smaller mesh size in 1983–85 than in 1968–73, and the use of a microscope in 1985 but not in previous years) contributed to higher total densities in 1985 than in earlier surveys. Total organism densities in 1985 ranged from 88 to 591 000 organisms m$^{-2}$, compared to a range of 104 to 40 000 m$^{-2}$ in 1983, a difference of up to one order of magnitude (Burt *et al.*, 1988).

In addition, numerous faunal shifts between 1983 and 1985 were indicated (Burt *et al.*, 1988). Two particularly notable shifts involved nematodes and polychaetes which were rare or absent in all previous OME surveys, and abundant and often dominant in 1985. These organisms are small enough that their enumeration is influenced by mesh size or sorting efficiency (Golini, 1979). Other small organisms, which occurred in higher densities in 1985, included naidid oligochaetes and early instar chironomids. However, other shifts in abundances such as the increased densities in 1985 of *Hexagenia*, a comparatively large mayfly nymph, were probably not related to mesh size or sorting technique.

Total organism densities reported at some locations in 1985 are very high ($> 100 000$ m$^{-2}$) and comparable to the maximum densities reported in the most heavily polluted harbours of the Great Lakes (Cook & Johnson, 1974). These high densities can, to some extent, be accounted for by the large numbers of very small organisms retained by a 200 $\mu$m mesh. In contrast, most benthic surveys of the Great Lakes have been carried out using a U.S. #30 (about 500 $\mu$m) mesh, which would retain fewer organisms.

In spite of the differences between surveys, the inferred zones of impairment in 1985 were similar to those reported in 1983. Both surveys indicated a zone of severe impairment below the St. Marys Falls near the Canadian shore at SMD 0.8. In 1985, however, this zone did not include SMD 1.2–840 (Fig. 1 and 3), suggesting some improvement since 1983. This change was primarily due to the presence in 1985 of large numbers of the nemertean, *Prostoma rubrum*, which is typically found in well-oxygenated, littoral standing water and cannot tolerate low oxygen conditions (Pennak, 1978).

The 1985 survey indicated severe impairment, possible resulting from toxicity, in the vicinity of the Algoma slip. In 1983, this area was identified as moderately impaired, based on 4 stations located outside the slip. The extra stations in closer proximity to the slip accounts for the change in status of this area in 1985 relative to 1983.

Both the 1983 and 1985 surveys suggested a

76

zone of moderate impairment which extends downstream from St. Marys Falls, close to the Canadian shore. Another zone of moderate impairment was associated with the Sault Ste. Marie STP, extending to within 370 m of the U.S. shore. Differences from the 1983 pattern included the identification of SMD 5.0E–520 and Station 107 as severely impaired in 1985. Analysis of the detailed species lists from both years substantiates the differentiation of SMD 5.0E–520 from adjacent stations on the basis of a more complete tubificid dominance (primarily *L. hoffmeisteri*) and the occurrence of only three other taxa. In 1985, the differentiation of Station 107 from nearby stations was related to a lower total density and fewer taxa.

In 1983, a zone of slight impairment was evident within the Lake George Channel discharging into Little Lake George. In 1985, this zone extended from SMD 5.0E–210 downstream of the STP outfall past transect SMD 6.4E (except SMD 6.4E–210 near the U.S. shore), into Little Lake George (Fig. 3). In the earlier OME surveys (1968 and 1973), this section was designated as moderately impacted. The improvement here is based on the presence of taxa such as *S. heringianus* and *P. rubrum*, as well as a slight increase in taxa and reduction in densities relative to the moderate impact zones.

A zone of moderate impairment at Stations 86, 87, and 88 (in Little Lake George) was identified in 1985, not sampled in 1983 but species assemblages were similar to those reported in 1973 and were characterized by high tubificid densities (primarily *L. hoffmeisteri*) and low numbers of taxa. The extent of this zone is similar to that indicated in 1973 (Hamdy *et al.*, 1978). A zone of moderate impairment was also identified in Lake George in 1985, at the two deep water stations.

Hiltunen & Schloesser (1983) and others (Edsall *et al.*, 1990; Schloesser *et al.*, 1990) have related the distribution of the Ephemeroptera *Hexagenia* to the presence or absence of visible oily residues in sediments of the St. Marys River system. Substrates collected from Lake George in 1985 were contaminated by oily substances at most stations between the STP and upper Lake

*Fig. 5.* Distribution of hexagenia nymphs and visible oil in the St. Marys River sediments in 1985.

George (Fig. 5), similar to the pattern indicated by Hiltunen & Schloesser (1983). The 1985 results suggest that the extent of sediment contamination by oil in the Lake George Channel has not changed from that observed in 1975. Oil was very limited in distribution in the more extensive study of the Lake Nicolet Channel in 1975 but was absent in stations sampled in this area in 1985 (Fig. 5). The distribution of *Hexagenia* in 1985 has also remained similar to that observed in 1975. In 1985, *Hexagenia* was absent from sediments in 27 of 32 stations analyzed in the St. Marys River, Lake Nicolet, Lake George Channel, Little Lake George, and Lake George where even slight amounts of visible oil were present (Fig. 5). In the remaining 19 stations where visible oil was absent, 16 had *Hexagenia* populations present. At two of the three remaining stations where *Hexagenia* was absent, the bottom substrate was primarily sand which deters *Hexagenia* inhabitation (Edmunds *et al.*, 1976).

In general, reductions in pollutant loadings

from the Algoma Steel terminal basin and St. Marys Paper operations since 1973 have contributed to minor community changes (Burt *et al.*, 1988). However, these changes have not been reflected in major improvements in pollution status of the benthic community from 1968 or 1973.

### 3.6. Community index analysis

R-mode cluster analysis (species grouping) separated the benthic species into five clusters or guilds (Fig. 6). Similarities between guilds with respect to their spatial distributions are indicated by the rescaled distance values at which they combine. Distance is an inverse measure of similarity. Five guilds, labelled A to E on the dendrogram,

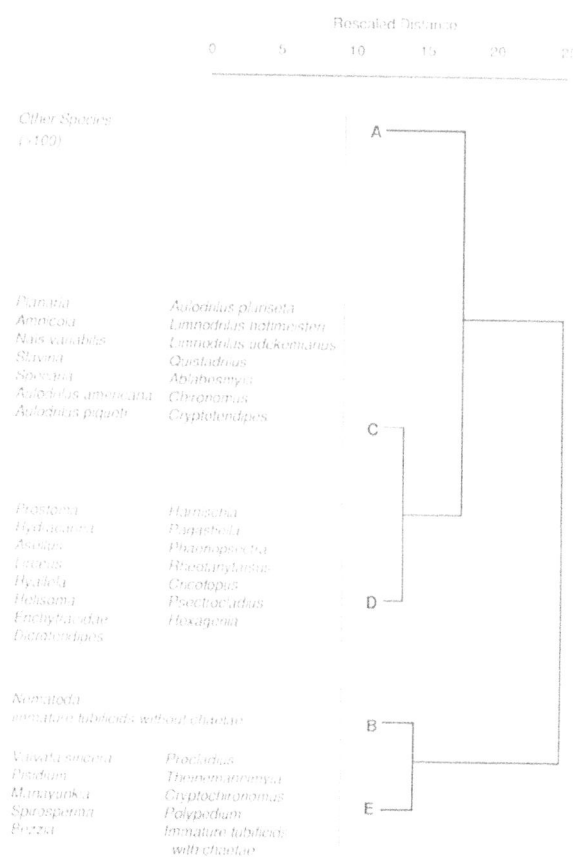

*Fig. 6.* Species clusters (guilds) based on concordance of spatial distributions.

are evident at an interpretation level of 13.5 on the distance scale.

Guilds B and C were comprised primarily of pollution tolerant taxa (Fig. 6). Guild B consisted of Nematoda and immature Tubificidae without capilliform chaetae. The immatures are most likely *L. hoffmeisteri* based on the relative abundance of adults. Station clusters 2, 3, and 4 were all characterized by dominance of Guild B taxa as indicated by the dominance index Ig, and were identified as pollution-impacted. Guild C included both pollution-tolerant species, such as *Limnodrilus* spp., *Q. multisetosus*, and *Chironomus*, and some species that are less tolerant of organic pollution, such as *Aulodrilus* spp., suggesting that Guild C occurs in areas that are less impaired than areas inhabited by Guild B.

Species Guilds D and E included some relatively intolerant and mesotrophic taxa, such as *Hexagenia* (Guild D) and *Polypedilum* (Guild E), with relatively few eutrophic indicators. Guild D was probably more indicative of unimpacted conditions than Guild E, based on the occurrence of the intolerant *Hexagenia* in the former, and greater representation of forms tolerant of organic enrichment in the latter (e.g., Tubificidae, *Procladius, Manayunkia*). High index values for these guilds, particularly for Guild D, tend to occur at stations assigned to Station Clusters 1, 5, 6, and 7, which are described as relatively unimpaired.

Species Guild A included a large number of species (> 100), most of which were uncommon at most locations. Because most of the dominant forms and many of the key indicator species occurred in Guild B to E, interpretation of the pollution indicator value of Guild A was not attempted.

Guild B and Guild C dominance indexes ($I_g$) were each positively correlated with various heavy metals, solvent extractables, and TOC ($p < 0.01$). Guild B dominance was also positively correlated with several pesticides (Table 4), while Guild C dominance was also correlated with fine particulates, phosphorus, LOI, and TKN. Guild B dominance was negatively correlated with Eh, indicating association with reducing environments.

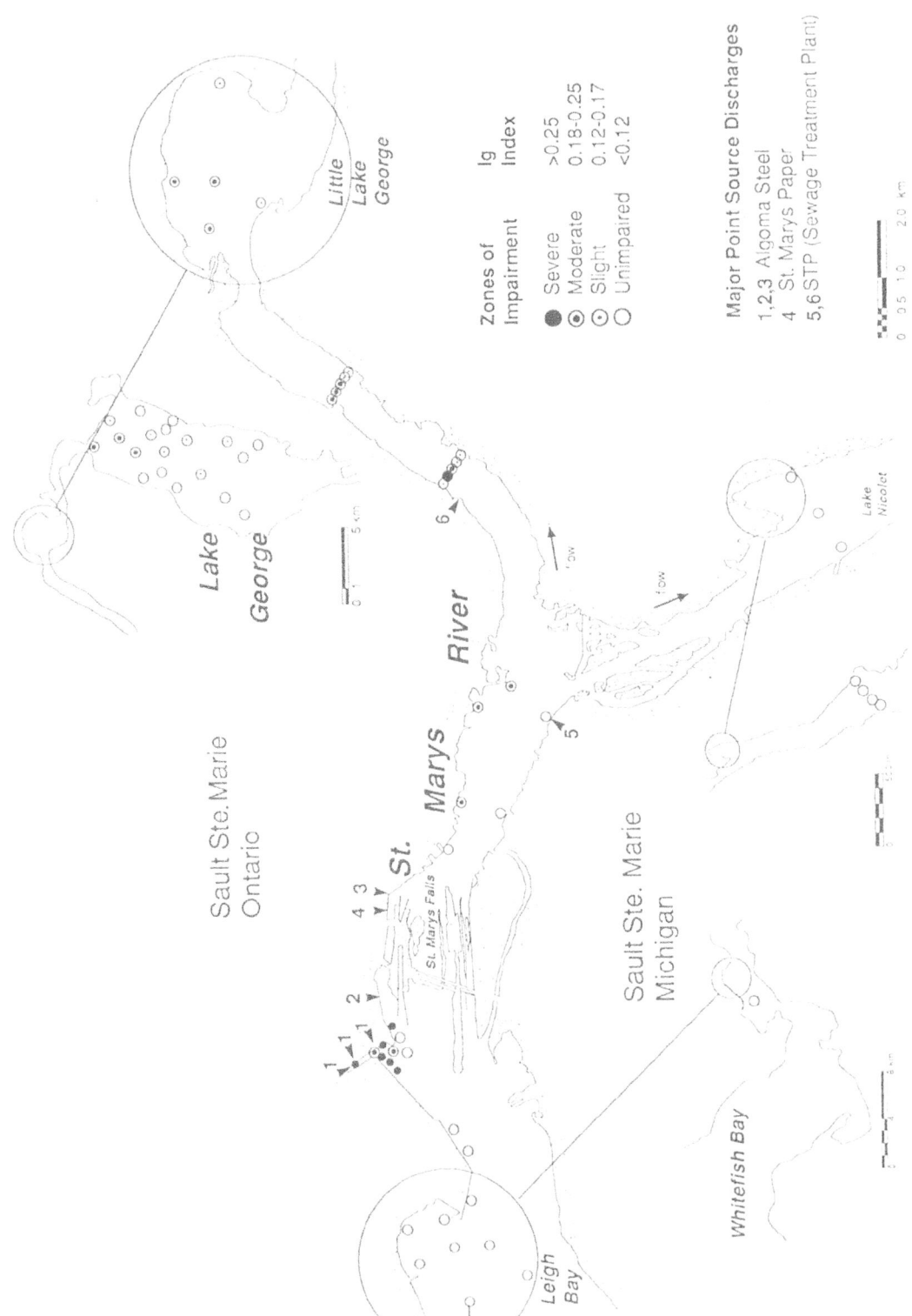

*Fig. 7.* Distribution of dominance index (guild *B*) in 1985.

*Table 4.* Sediment characteristics associated with the guild B dominance index (Ig)

| Sediment Variable | Mean sediment characteristic | | | | | |
|---|---|---|---|---|---|---|
| | Ig Index ranges | | | | Standard partial regression coefficient | Significance level ($p$) |
| | <0.12 | 0.12-0.179 | 0.18-0.249 | ≥0.25 | | |
| Phosphorus (%) | 0.3 | 0.5 | 0.4 | 0.36 | -0.177 | 0.411 |
| Coarse Particulates (>1,000 $\mu m$) (%) | 0.81 | 0.36 | 1.2 | 1.0 | 0.105 | 0.268 |
| Eh (meV) | 478 | 458 | 407 | 262 | -0.147 | 0.142 |
| Loss on Ignition (LOI) (%) | 2.1 | 2.5 | 2.6 | 2.9 | -0.845 | >0.001 |
| Clay Fraction (%) | 4.0 | 7.6 | 7.6 | 6.1 | 0.474 | 0.003 |
| Total Kjeldahl Nitrogen (TKN) (%) | 0.55 | 0.91 | 0.66 | 0.55 | -0.255 | 0.095 |
| Heptachlor Epoxide (ng g$^{-1}$) | 0.00 | 1.1 | 0.00 | 1.9 | 0.254 | 0.008 |
| Arsenic ($\mu g\ g^{-1}$) | 3.1 | 5.5 | 6.5 | 17 | -0.630 | 0.020 |
| Mercury ($\mu g\ g^{-1}$) | 0.02 | 0.08 | 0.07 | 0.10 | 0.653 | >0.001 |
| Chromium ($\mu g\ g^{-1}$) | 22 | 31 | 40 | 41 | -0.308 | 0.040 |
| Zinc ($\mu g\ g^{-1}$) | 40 | 107 | 126 | 316 | -0.331 | 0.159 |
| Copper ($\mu g\ g^{-1}$) | 14 | 28 | 30 | 34 | 0.393 | 0.091 |
| Total Organic Carbon (TOC) (%) | 10 | 18 | 22 | 81 | 1.187 | 0.000 |
| Aluminium ($\mu g\ g^{-1}$) | 5,370 | 8,320 | 7,760 | 7,240 | -0.410 | 0.097 |
| Iron ($\mu g\ g^{-1}$) | 12,000 | 20,900 | 25,700 | 60,300 | 0.570 | 0.071 |

These correlations support the contention that dominance by Guilds B and C is representative of impaired conditions.

Guild E dominance was positively correlated only with depth ($p < 0.01$), and negatively correlated only with p,p-DDE, suggesting that these taxa are not strongly indicative of pollution status. Guild D dominance was negatively associated with various heavy metals, endrin, solvent extractables, phosphorus, loss-on-ignition (LOI), total organic carbon (TOC), and total Kjeldahl nitrogen (TKN). Thus, this group represents a relatively clean-water faunal assemblage, both on the basis of biological indicators and negative correlations with sediment-associated contaminants and organic matter.

Based on correlations with sediment characteristics (Table 4), Guild B dominance appears to be the best general indicator of toxic contaminants (both metals and pesticides). This guild is also defined by two of the most abundant indicator taxa (Nematoda and *Limnodrilus* spp.), further facilitating interpretation of spatial trends in environmental quality. The highest Guild B dominance index values are in the vicinity of the Algoma slip and the Sault Ste. Marie STP (Fig. 7). Values greater than 0.25 for Guild B in these areas were considered indicative of severe impact. Zones of moderate impact (index 0.18 to 0.25) were also apparent in a depositional area downstream of St. Marys Falls on the Canadian shore, downstream of the Sault Ste. Marie STP to Little Lake George, and in a deep water depositional area of Lake George near the inlet.

Impact zones defined on the basis of Guild B dominance (Fig. 7) closely resemble the zones previously defined in terms of station clusters 2, 3, and 4 (Fig. 3). However, they suggest a slightly greater degree and extent of impact of Little Lake George and Lake George than discerned by station cluster distributions.

The best multivariate equation for prediction of Guild B dominance from sediment characteristics includes significant ($p < 0.05$) positive contributions from mercury, heptachlor epoxide, clay, and TOC (Table 3). High concentrations of these constituents tend to increase the predicted dominance index. The equation includes significant negative contributions from arsenic, chromium, and LOI. The complete equation explains 72 percent of variation in the dominance index (adjusted $R^2 = 0.72$, $p < 0.001$).

The sediment characteristics identified by the multiple regression analysis as the best quantitative predictors of Guild B dominance differ from those identified by discriminant analysis as the key contributors to qualitative discrimination between station clusters defined on the basis of overall species composition. This is not surprising in view of the different dependent variables involved (indicator species dominance vs. station cluster membership). Nevertheless, the concordance of impacted zones derived from the dominance index and station cluster patterns strengthens the finding of the continued existence of pollution tolerant benthic communities downstream of the Algoma Steel, St. Marys Paper and Sault Ste. Marie STP discharges, extending downstream along the Canadian shoreline as far as upper Lake George.

## 4. Acknowledgements

We wish to acknowledge the contributions of Mr. Y. Hamdy, Ms. Mary Kirby, and Mr. B. Bowman of the Ontario Ministry of the Environment in the experimental and statistical design of this programme. The captain and crew of the 'Monitor V' assisted with the field survey. Mr. Dennis Farara sorted and analyzed the benthic samples for the oligochaetes while Ms. Marion Brinkman analyzed the Chironomidae and remaining taxa. Mr. P. Mark Green drafted the figures.

## References

Barton, D. R. & H. B. N. Hynes, 1978. Wave-zone macrobenthos of the exposed Canadian shores of the St. Lawrence Great Lakes. J. Great Lakes Res. 4: 27–45.

Brinkhurst, R. O. & D. G. Cook, 1974. Aquatic earthworms. In; C. W. Hart & S. L. H. Fuller (eds.). Pollution Ecology of Freshwater Invertebrates. pp. 143–156. Academic Press, New York.

Burt, A. J., P. M. McKee & D. R. Hart, 1988. Benthic invertebrate survey of the St. Marys River, 1985. Rep. Prep. for Ont. Min. Env., Avail. from Wat. Resources Br., Great Lakes Sect., 1 St. Clear Avenue West, Toronto, Ontario, Canada M4V 1K6.

Clarke, A. H., 1973. The freshwater molluscs of the Canadian Interior Basin. Malacologia 13: 1.

Cook, D. G. & M. G. Johnson, 1974. Benthic macroinvertebrates of the St. Lawrence Great Lakes. J. Fish. Res. Bd Can. 3: 763–782.

Edmunds, G. F., Jr., S. L. Jensen & L. Berner, 1976. The Mayflies of North and Central America. Univ. Minnesota Press, Minneapolis. 330 pp. ISBN 0–8166–0759–1.

Edsall, T. A., B. A. Manny, D. W. Schloesser, S. J. Nichols & A. M. Frank, 1991. Production of *Hexagenia limbata* nymphs in contained sediments in the Upper Great Lakes Connecting Channels. Hydrobiologia 219: 353–361.

Freitag, R., P. Fung, J. S. Mothersill & G. K. Prouty, 1973. Geographical distribution and spatial relationships of macroinvertebrates of Nipigon Bay, Lake Superior, Ontario. Lakehead University M.S. Rep. 1–29.

Golini, V. I., 1979. Benthic macro- and microinvertebrates of Lake Ontario, 1977: distribution and relative abundance. Env. Can., Fish. Mar. Serv., MS Rep. 1519.

Green, R. H., 1979. Sampling Design and Statistical Methods for Environmental Biologists. Wiley & Sons, New York. 257 pp. ISBN 0–471–03901–2.

Hamdy, Y., J. D. Kinkead & M. Griffiths, 1978. St. Marys River water quality investigations 1973–74. Great Lakes Surv. Unit, Wat. Res. Branch, Ont. Min. Env. Avail. from Wat. Res. Br., Great Lakes Sect., 1 St. Clair Avenue West, Toronto, Ontario, Canada M4V 1K6.

Hiltunen, J. K. & D. W. Schloesser, 1983. The occurrence of oil and the distribution of *Hexagenia* (Ephemeroptera: Ephemeridae) nymphs in the St. Marys River, Michigan and Ontario. Freshwat. Invertebr. Biol. 1983, 2 (4): 199–203.

Hutchinson, G. E., 1978. An Introduction to Population Ecology. Yale University Press, New Haven & London. 260 pp. ISBN 0–300–02155–0.

Klemm, D. J. (ed), 1985. A Guide to the Freshwater Annelida (Polychaeta, Naidid and Tubificid Oligochaeta and Hirudinea) of North America. Kendall/Hunt Publishing Co., Dubuque Iowa, 198 pp. ISBN 0–8403–3577–6.

Mackie, G. L. & S. U. Qadri, 1971. A polychaete, *Manayunkia speciosa*, from the Ottawa River and its North American distribution. Can. J. Zool. 49: 780–782.

Mackie, G. L., D. S. White & T. W. Zdeba, 1980. A guide to the freshwater molluscs of the Laurentian Great Lakes with species emphasis on the genus *Pisidium*. U.S. Env. Protect. Agency EPA–600/3–80–068.

McKee, P. M., A. J. Burt & D. R. Hart, 1984. Benthic invertebrate and sediment survey of the St. Marys River, 1983. Rep. Prep. Ont. Min. Env., Wat. Res. Br., Great Lakes Sect., 1 St. Clair Avenue West, Toronto, Ontario, Canada M4V 1K6.

Nalepa, T. F. & N. A. Thomas, 1976. Distribution of macrobenthic species in Lake Ontario in relation to sources of pollution and sediment parameters. J. Great Lakes Res. 2: 150–163.

Norusis, J. J., 1986. Advanced Statistics SPSS/PC + for the IMB PC/XT/AT. SPSS Inc., Chicago, IL.

Oliver, D. R. & M. E. Roussel, 1983. The genera of larval midges of Canada. In; The Insects and Arachnids of Canada, Part 2. Biosystematics Res. Inst., Ottawa, Canada.

OME (Ontario Ministry of the Environment), 1983. Handbook of Analytical Methods for Environmental Samples. Lab. Serv. Appl. Res. Br., Rexdale, Ontario, Canada, M9W 5L1.

Pennak, R. W., 1978. Freshwater Invertebrates of the United States, 2nd Edition. John Wiley and Sons. 803 pp. ISBN 0–471–04249–8.

Persaud, D. & W. D. Wilkens, 1976. Evaluating construction activities impacting on water resources. Ont. Min. Env., Wat. Resources Br., 1 St. Clair Avenue West, Toronto, Ontario, Canada M4V 1K6.

Poe, T. P. & D. C. Stefan, 1975. Several environmental factors influencing the distribution of the freshwater polychaete, *Manayunkia speciosa* Seidy. Chesapeake Sci. 15: 235–237.

Roback, S. S., 1974. Insects: In; C. W. Hart & S. L. H. Fuller (eds.) Pollution Ecology of Freshwater Invertebrates. Academic Press, New York: 314–376.

Schloesser, D. W., T. A. Edsall, B. A. Manny & S. J. Nichols, 1991. Distribution of *Hexagenia* nymphs and visible oil in sediments of the Upper Great Lakes Connecting Channels. Hydrobiologia 219: 345–352.

Veal, D. M., 1968. Biological survey of the St. Marys River 1968. Biol. Br., Ont. Wat. Resources Commiss.

Wiggins, G. B., 1977. Larvae of the North American Caddisfly Genera (Trichoptera). University of Toronto Press. 401 pp. ISBN 0–8020–5344–0.

*Hydrobiologia* **219**: 83–95, 1991.
*M. Munawar & T. Edsall (eds),*
*Environmental Assessment and Habitat Evaluation of the Upper Great Lakes Connecting Channels.*
© 1991 *Kluwer Academic Publishers.*

# Use of a geographic information system data base to measure and evaluate wetland changes in the St. Marys River, Michigan

Donald C. Williams & John G. Lyon
*Detroit District, Corps of Engineers, P.O. Box 1027, Detroit, MI 48231, USA; Department of Civil Engineering, Ohio State University, Columbus, OH 43210, USA*

*Key words:* wetland, St. Marys River, water level, succession, geographic information system, photo interpretation

## Abstract

A digital data base was constructed by photo interpretation, mapping, and digitizing seven dates of aerial photography on the St. Marys River, Michigan, USA. The data base was used in conjunction with geographic information system software to examine historical changes in wetland area. Total wetland area between 1939 and 1985 ranged from 7 200 to 7 317 ha over a 46-year period of high and low water. There was greatest variation in areas of emergent wetland and scrub-shrub wetland, which appeared to be responding primarily to changes in water level.

## 1. Introduction

The St. Marys River (Fig. 1) is the outlet for the oligotrophic water of Lake Superior to the lower Great Lakes. Because of the amount of wetland along the river, and the relatively low phytoplankton productivity of this water, wetlands have been shown to play a dominant role in the primary productivity of the St. Marys ecosystem (Duffy *et al.*, 1987; Liston *et al.*, 1986). The Detroit District, U.S. Army, Corps of Engineers (USACE) proposed to extend winter season navigation through the locks at Sault Ste. Marie, Michigan (USACE, 1987). The USACE undertook further wetland study in conjunction with environmental impact analysis because of the importance of the wetlands to the ecosystem in this connecting channel.

Local distribution of wetlands, their species composition, productivity, and growth rates had been studied in selected parts of the St. Marys

River (Liston *et al.*, 1980; Jude *et al.*, 1986; Liston *et al.*, 1986). The area of St. Marys River in the USACE study (Fig. 2), primarily Lake Nicolet, contains a diversity of wetlands. To determine and evaluate wetland distribution changes it was necessary to classify them, map their current distribution, and measure their areas. A historical inventory of these wetlands was necessary to determine their changes over time. For ease of analysis, the information from the inventory was placed in a Geographical Information System (GIS). From measurements derived from this system it was possible to address the relative abundance of wetlands, and consider past, present, and potential sources of change in the wetlands.

## 2. Methods

The USACE obtained photographs for the summer and fall seasons of 1939, 1953, 1964, 1978,

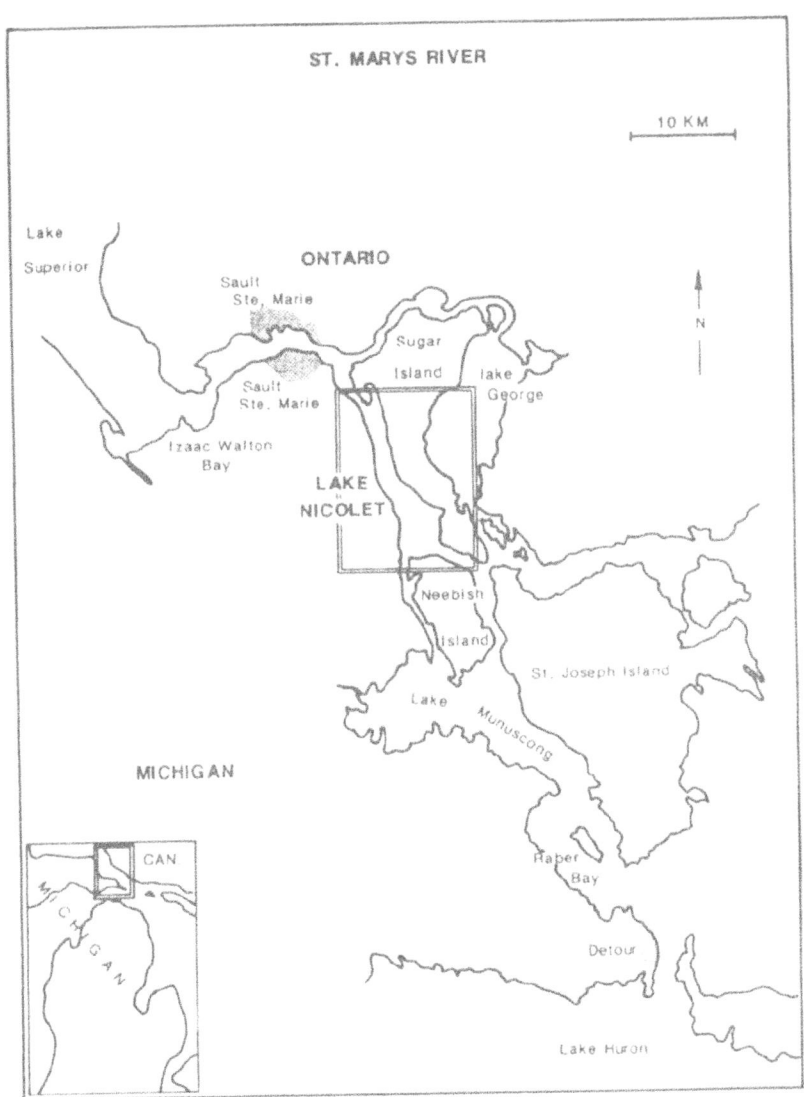

*Fig. 1.* Location of the study area.

1982, 1984 and 1985. The film types included black and white, black and white infrared, color, and color infrared. The scales of the photos ranged from 1:12000 to 1:58000 (Table 1). Wetland types and areas were interpreted from these photos by the National Wetlands Inventory (NWI) of the Fish and Wildlife Service (FWS), which provided wetland maps, tabular data, and digital files for use in the study.

The aerial photos were acquired at seven different water levels. Like the Great Lakes, the connecting channels are subject to a wide range of water levels resulting from large variations in water balance in the Great Lakes basin. Water levels can range plus or minus one meter from the long-term average. The aerial photos obtained for this study documented wetland changes in association with water levels that varied within a range of 1.04 m, from 176.72 to 177.20 m. This allowed an analysis of the influence of St. Marys River water level elevations on the wetlands. The specific dates of the photos and the yearly average

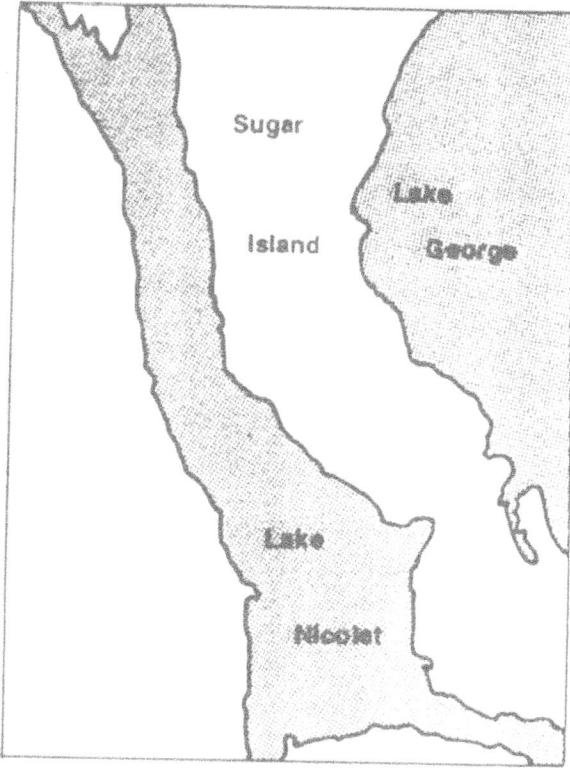

METERS

0        4000

*Fig. 2.* Wetland study site, St. Marys River.

*Fig. 2.* Wetland study site, St. Marys River.

water levels are given in Table 1. The water levels were recorded at the U.S. Slip gauge on the St. Marys River.

*Table 1.* Images for photo interpretation and data base creation and corresponding yearly average water levels.

| Year | Date | Scale | Emulsion | Water level (m) |
|------|------|-------|----------|-----------------|
| 1939 | 7/2 | 1 : 20 000 | black & white | 176.72 |
| 1953 | 7/19 | 1 : 16 000 | black & white infrared | 177.19 |
| 1964 | 6/28 | 1 : 16 000 | black & white | 176.16 |
| 1978 | 6/29 | 1 : 12 000 | black & white | 176.83 |
| 1982 | 10/25 | 1 : 58 000 | color infrared | 176.66 |
| 1984 | 9/18 | 1 : 12 000 | color infrared | 177.00 |
| 1985 | 10/19 | 1 : 24 000 | natural color | 177.20 |

Wetland areas were quantified for each date of aerial photo by a NWI contractor experienced with photo interpretation and identification of wetlands. The project area was also visited by the contractor. Interpretations were according to NWI conventions (USFWS, 1987) which provide specific instructions for applying the FWS classification system (Cowardin *et al.*, 1979). Initial interpretations were completed with an analog stereo plotter using the 1984 images. Wetland boundaries were interpreted in and along the river up to one-half mile inland from the shore. The 1984 study area limits were maintained for the other years. The plotter established geometric control and was used to correct any inaccuracies found in the photos. The interpreter viewed stereo pairs of photos, and outlined the wetland boundaries. The other dates of photography were compared to the 1984 maps and wetland boundary adjustments were made with a zoom transfer scope. The boundaries were plotted on U.S. Geological Survey 7-1/2 minute maps of the project area.

The wetland types were characterized as emergent, aquatic bed, scrub-shrub, forested, unconsolidated shore, and unconsolidated bottom wetland types based on the FWS Classification System (Cowardin *et al.*, 1979), and can be grouped as Riverine and Palustrine wetland ecological systems. However, the unconsolidated bottom type probably includes many areas with submergent vegetation, since much of the bottom of the St. Marys River is vegetated outside the navigation channel (Liston *et al.*, 1986).

Large scale (1 : 24 000) maps were produced by the NWI for the seven 7-1/2 minute U.S. Geological Survey quadrangles covering the study area. A separate map was produced for each year of interpreted aerial photography. Wetlands were mapped on the following quadrangles: Sault Ste. Marie South, Baie De Wasai, Oak Ridge, and small parts of Munuscong and Munuscong NE. The wetland boundaries for each wetland class were converted into area measurements, and reported in tenths of acres (0.04 ha) by the USFWS Wetland Analytical Mapping System (WAMS) software. Area data were summarized in wetland maps and tables of wetland areas for each year and quadrangle.

The maps and tabular summaries provided by the NWI were appropriate for completion of most analyses. However, since the tabular summaries were completed on a quadrangle by quadrangle basis, it was difficult to locate and quantify local changes. For these more detailed analyses, digital files were evaluated.

The digital files of each year for each quadrangle were created by NWI using a geographic information system (GIS). The techniques are those being used in construction of a national georeferenced wetland data base. WAMS was used to digitize the mapped information and place it in a 'common ground' geographic reference system. Digital files from WAMS were imported to the Map Overlay and Statistical System (MOSS). Using MOSS, the files were converted from latitude/longitude to the Universal Transverse Mercator (UTM) projection, rasterized, and con-

verted to ELAS format, which is readable on the Detroit District's GIS system (USFWS, 1988).

Analyses of the digital files were concentrated on one quadrangle, Oak Ridge, because it contained approximately 46 percent of the total wetlands in the study area and a representative cross section of wetland classes. Digital files of four dates of interpreted photos, 1939, 1964, 1978, and 1982, were used for making comparisons.

Comparisons were made between scenes using the ERDAS Geographic Information System software program MATRIX. Wetlands at two dates were compared by creation of an $n \times n$ matrix, where $n$ was equal to the number of classes in the scene. In this case eight classes were used for each scene, the seven FWS classes (unconsolidated bottom, aquatic bed, emergent, unconsolidated shore, scrub-shrub, forested, and upland) plus 'other' areas outside the study area.

These digital analyses allowed exact location of

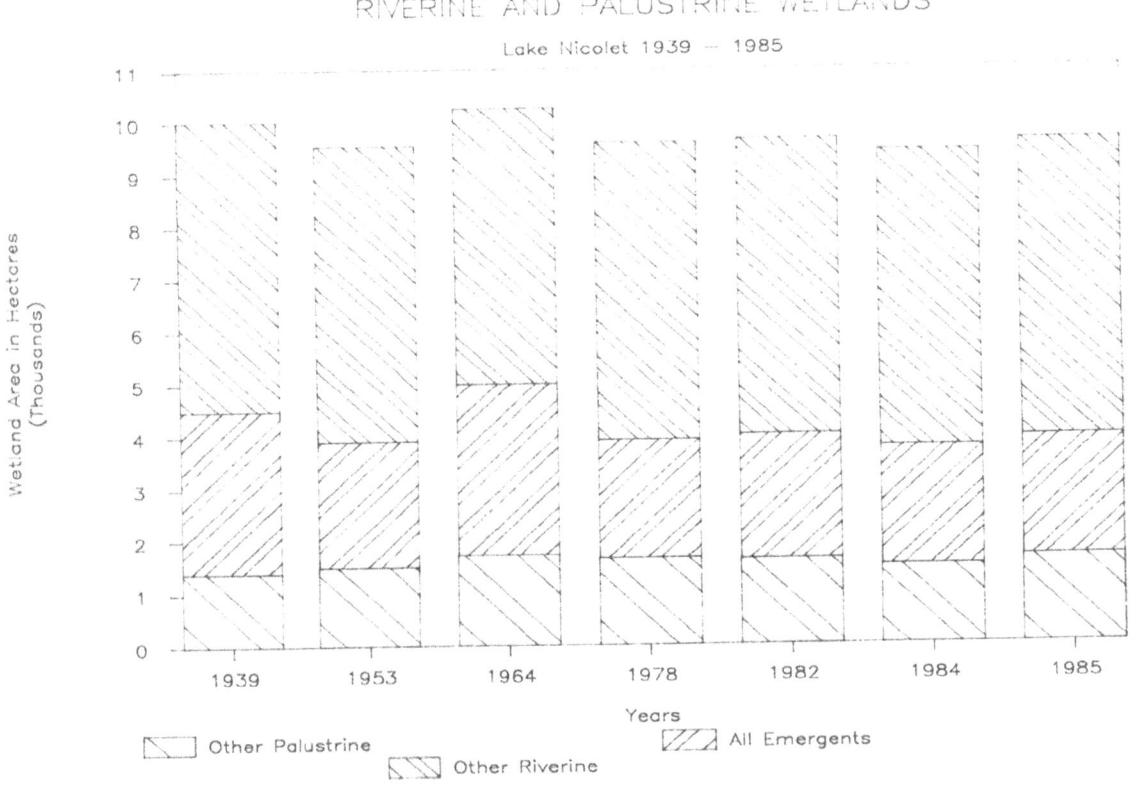

Fig. 3. Emergent and other palustrine and riverine wetland areas in the St. Marys River study area, 1939–1985.

wetland increases and decreases. The MATRIX computation produced an 8 × 8 matrix. The diagonal elements indicated the cells in which no class change occurred between dates, and the off-diagonal elements represented the cells in which there were changes to other wetland and non-wetland classes. The matrix results were then mapped into a digital file which showed where changes occurred.

## 3. Results

### 3.1. Tabular data

Historically, the quantity of total wetlands in the project area has remained nearly the same (Fig. 3). The tabular data from the interpretation of historical aerial photos from the 46-year period indicated that the total area of wetlands (including unconsolidated bottom) ranged from 7200 to 7317 ha. The maximum difference was 1.6 per-

cent. During that period water levels varied by 1.04 m. The total wetland in 1939 was 7317 hectares at water elevation 176.72 m, and in 1985, 7247 ha at water elevation 177.20 m. The wetland areas were similar with the lake levels 0.48 m higher in the recent photos. Comparisons between any two of the four dates of photographs indicated that changes in overall wetland areas were small, ranging between 0.5 and 1.6 percent over the 45-year period.

Related local studies of wetland plant populations have demonstrated little change in quantities of plants from year-to-year (Liston *et al.*, 1986). These year-to-year studies do not explain evidence of past losses at the outer edges of emergent wetland stands. A significant part of these small variations in totals appears to be explained by changes in water levels. This was probably the reason for the small increase in total wetlands in the St. Marys study area between 1953 and 1984. In the latter year, water levels were substantially

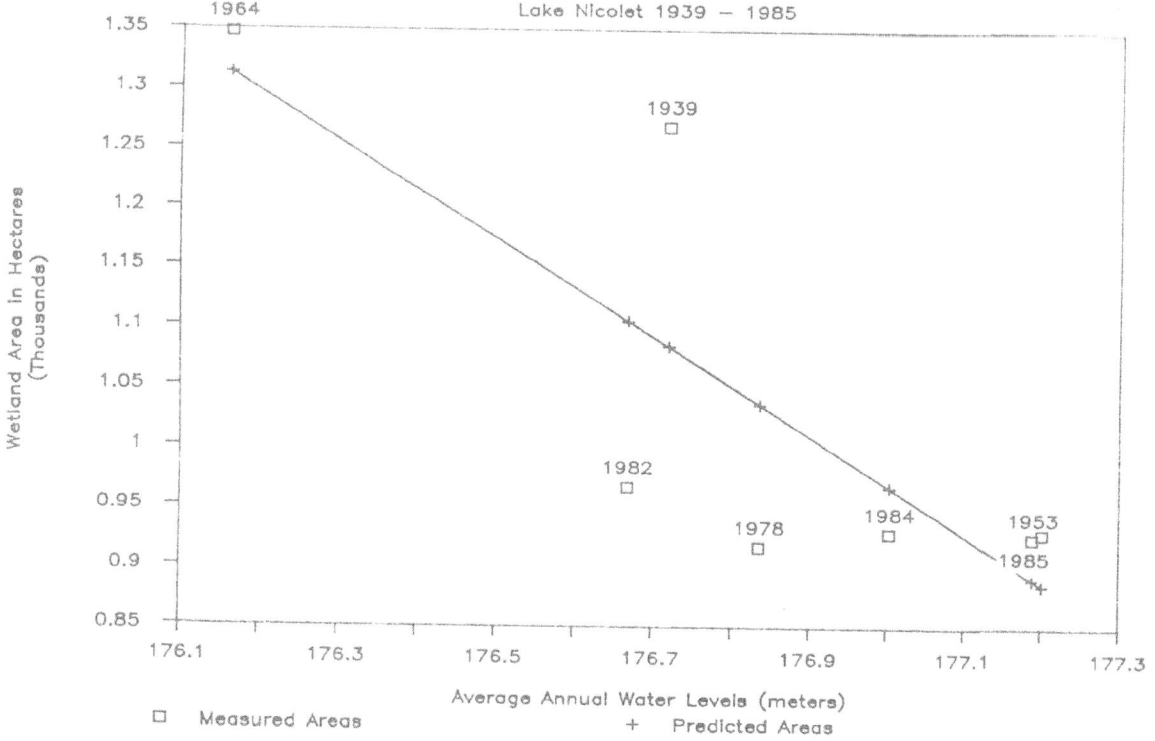

*Fig. 4*. Relationship of emergent wetland area to water levels – 1939–1985.

lower: 1953 – 7200 ha at 177.19 m; 1984 – 7247 ha at 177.00 m.

The most clear relationship to water levels was found in the emergent wetland class. Figure 4 shows strong evidence of a relationship between increasing water levels and the decreasing emergent wetland areas over the course of the study period, between 1939 and 1985. A regression equation was derived relating emergent vegetation areas to average annual water levels (Dixon, 1985). This equation was significant at the 0.05 level. The multiple R-square value for the regression was 0.64. These parameters indicate that there is a strong relationship between the emergent wetlands variability and water level elevation in the St. Marys River. There was approximately a 32 percent change in emergent wetland area between high and low water (1347 vs. 917 ha). The overall results suggest that some emergent

wetland losses can be expected during periods of high water.

There was a parallel relationship between water level and areas of the scrub-shrub wetland class. Areas of scrub-shrub wetland were reduced during periods of high levels and increased during periods of low levels. This relationship is shown in Fig. 5. The regression equation derived for the relationship between average annual water level and area of scrub-shrub wetland was also significant at the 0.05 level, and the multiple R-square value was 0.69. While the relationship was statistically significant, there was only a 16 percent change in area of scrub-shrub wetland from high to low water (390 vs. 328 ha), so that area of scrub-shrub vegetation did not vary as widely under changing water regimes as did emergent wetland.

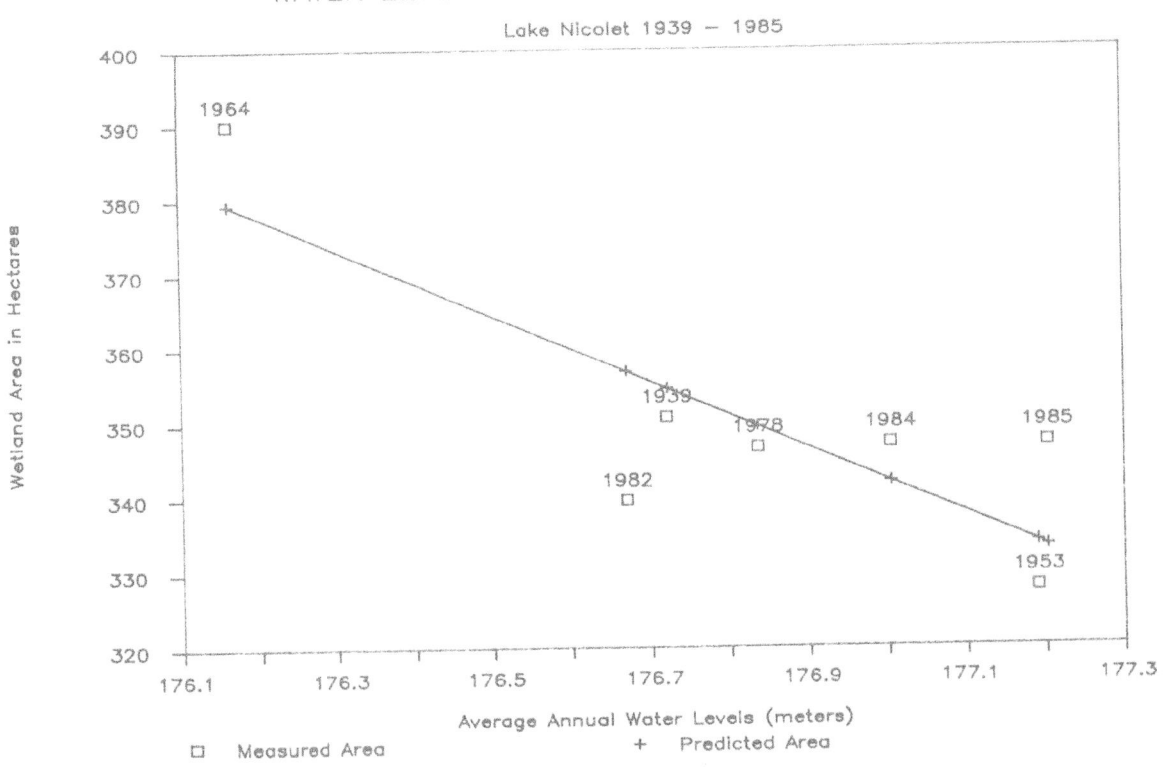

*Fig. 5.* Relationship of scrub-shrub wetland area to water levels – 1939–1985.

## 3.2. Digital data

The digital data analyses of the Oak Ridge wetlands showed summary results similar to the tabular data. In these analyses, unconsolidated bottom was not included as wetland because this class may or may not be vegetated, and includes a large part of the study area. Dates with the largest difference in water levels were compared to investigate changes primarily due to water level. Analysis of the 1964 and 1982 files indicated that 6.2 percent (731 ha) of the quadrangle study area remained wetland between the low water year, 1964, with levels of 176.16 m, and the high water year, 1982, with levels of 176.66 m. Wetlands were lost in 2.39 percent of the area (281 ha), and gained in 0.65 percent of the area (77 ha). The net change was therefore a loss of wetland in 1.7 percent of the study area or a loss of wetland in 204 ha with an increase in water level of 0.50 m.

Additional comparisons were made to deter-mine short term and long term changes at similar water levels. The years 1939 and 1982 were used to indicate changes over the long term. Between 1939 and 1982, 6.0 percent or 710 ha of the Oak Ridge study area remained in wetlands. Wetlands were lost in 1.8 percent of the mapped area (209 ha), and gained in 0.8 percent of the area (97 ha). The net change in total wetlands was therefore 1.0 percent of the quadrangle study area, or 113 ha. This took place over a 43 year period with a decrease of 0.06 m in average annual water level.

Very small changes took place over the short term. Between 1978 and 1982, 5.8 percent or 689 ha of the Oak Ridge study area remained as wetland. There was a loss of 0.6 percent (71 ha) and a gain of 0.7 percent (81 ha). Hence there was no significant net change in wetland area between those two years in the Oak Ridge study area. There was a decrease of 0.17 m in water levels between the two years.

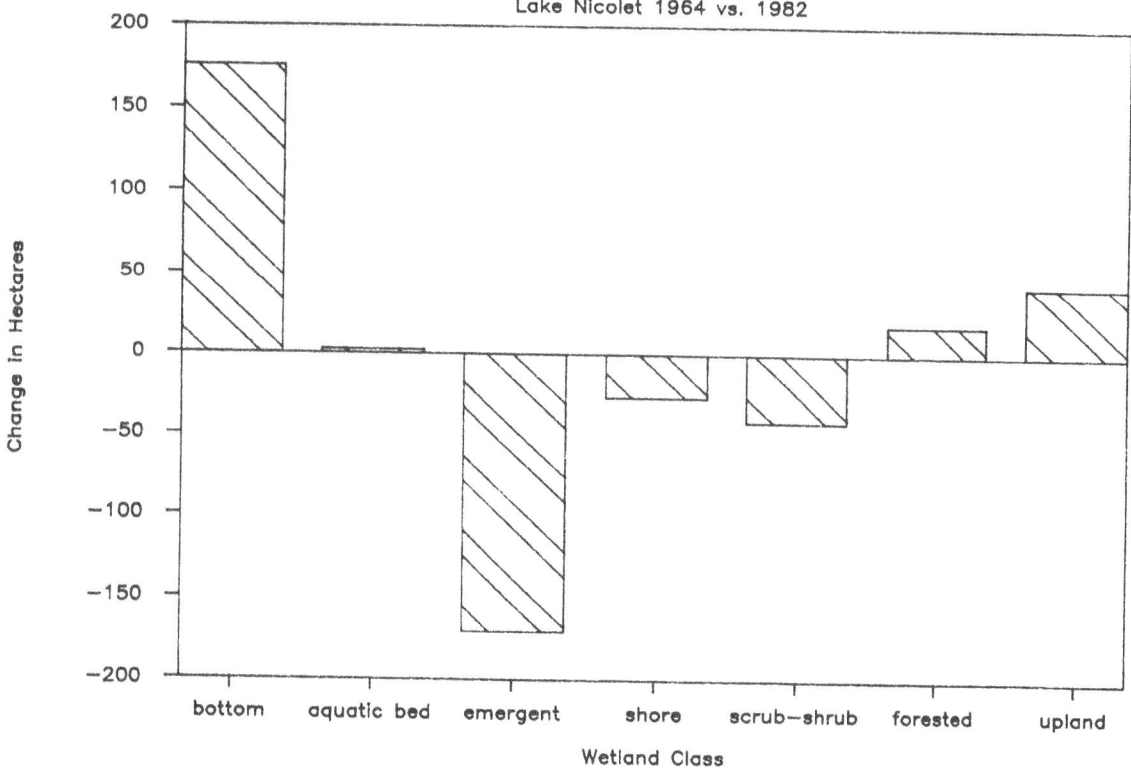

Fig. 6. Changes in wetland class related to changes in water level, 1964 vs. 1982.

# WETLAND COMMUNITY CHANGES WITH TIME

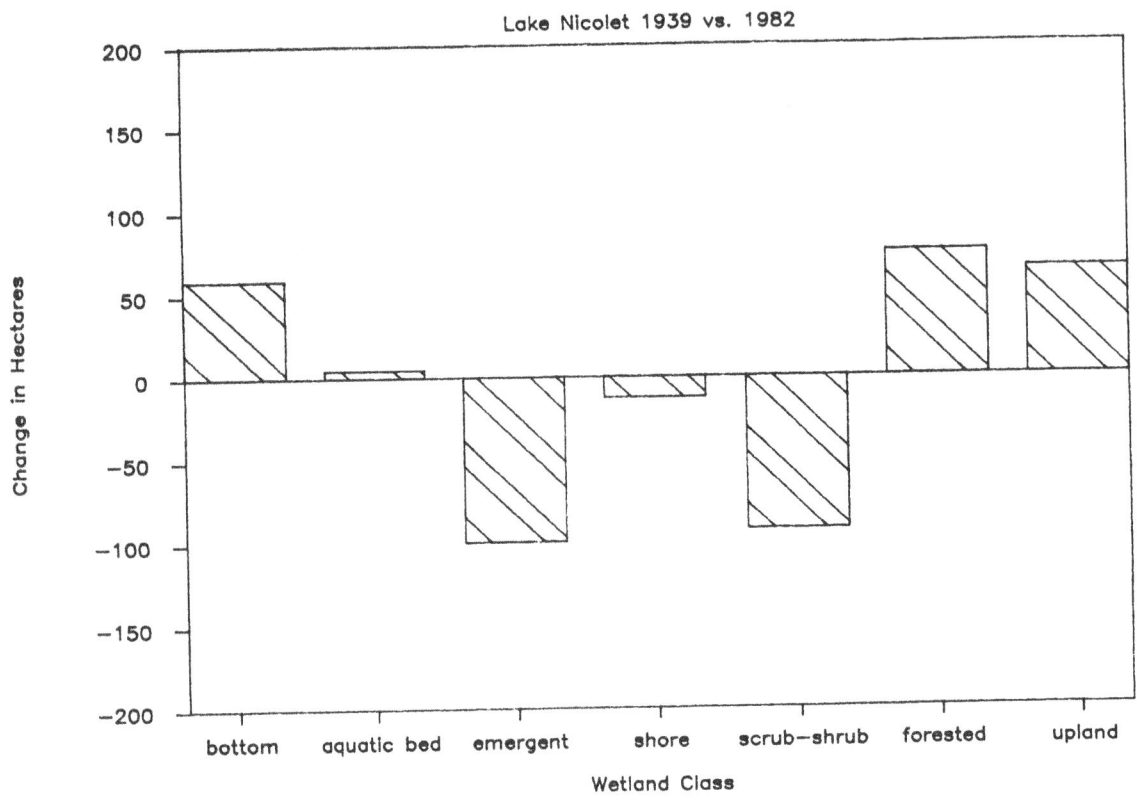

*Fig. 7.* Changes in wetland class from 1939 to 1982.

# RECENT WETLAND COMMUNITY CHANGES

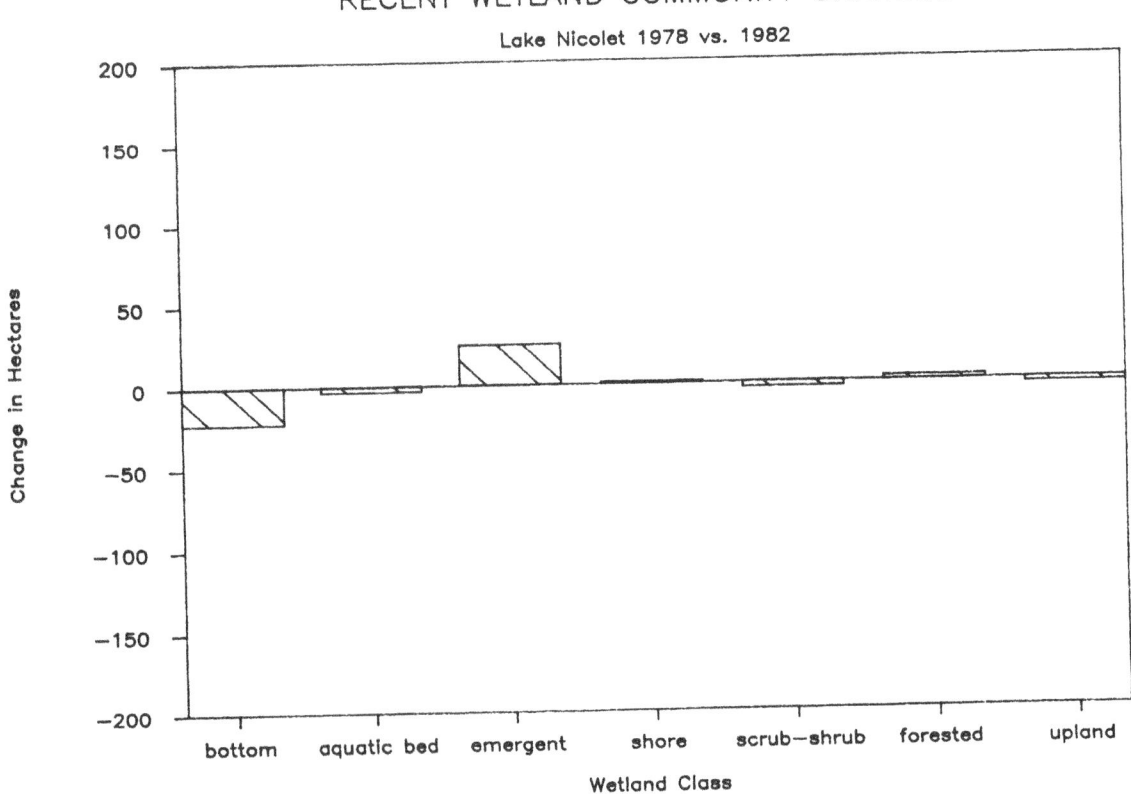

*Fig. 8.* Changes in wetland class from 1978 to 1982.

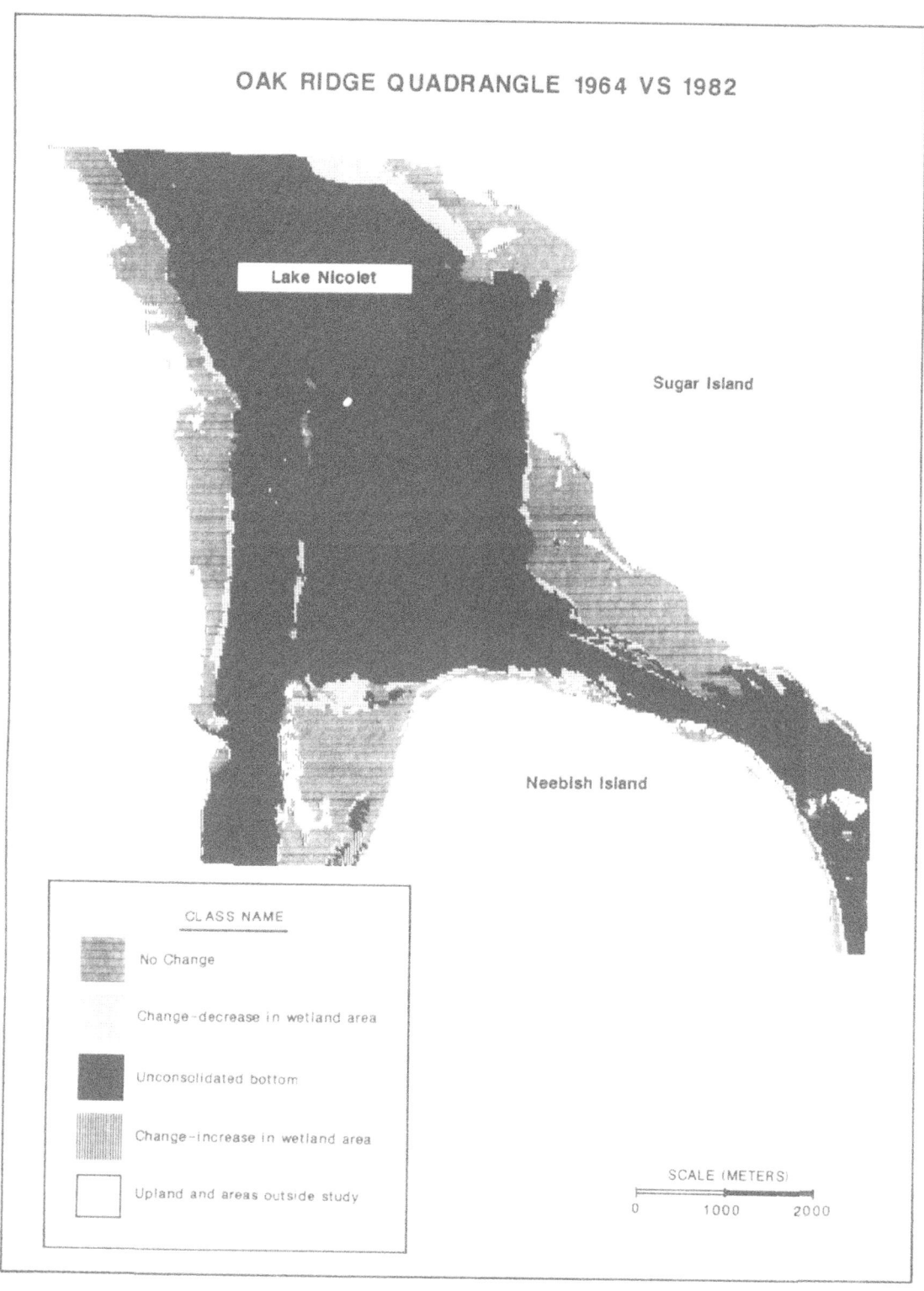

*Fig. 9.* Change in wetland area related to water level changes, 1964 vs. 1982, Oak Ridge quadrangle, Lake Nicolet, St. Marys River. Unconsolidated bottom excluded from wetland area.

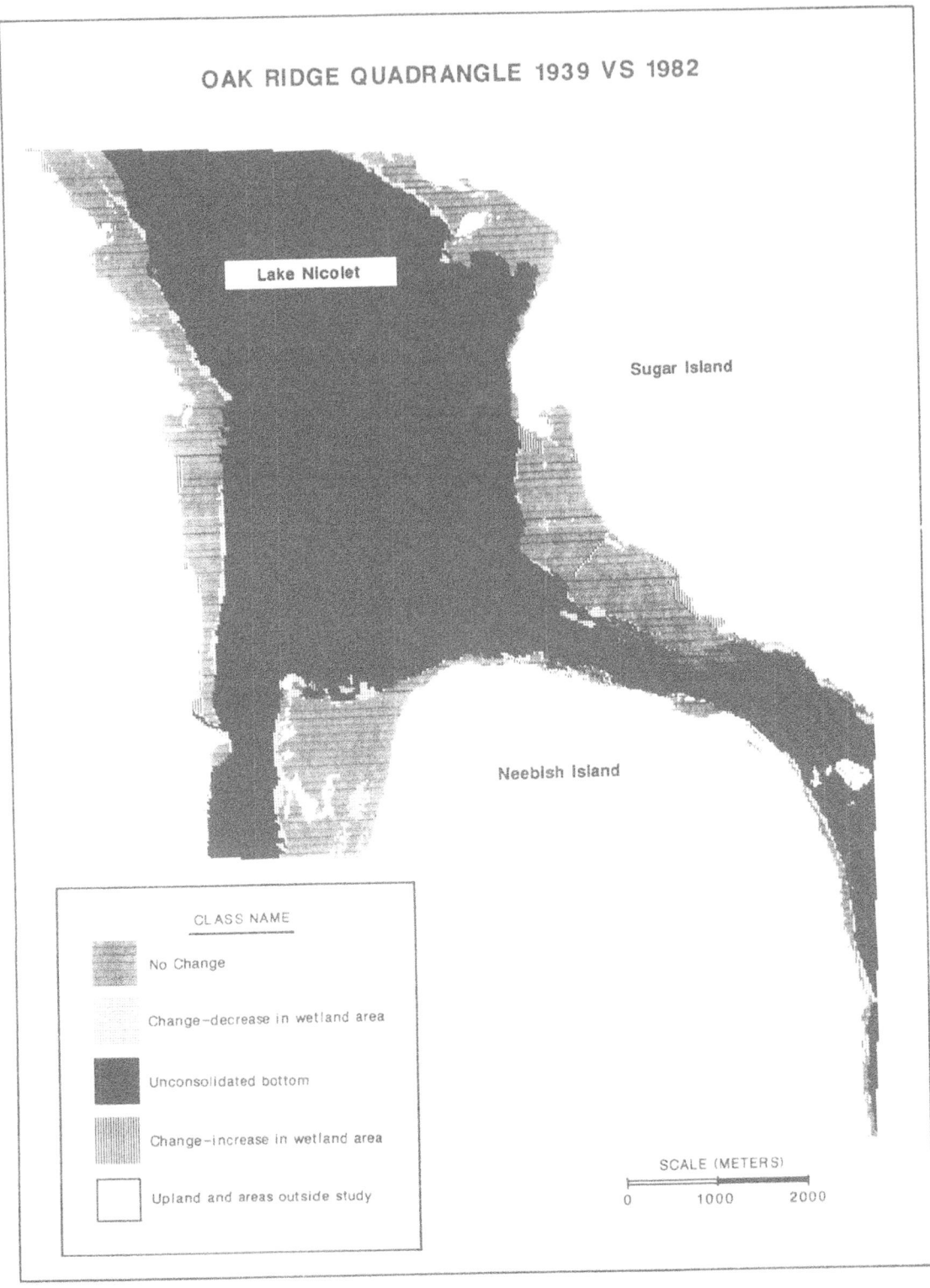

*Fig. 10.* Change in wetland area between 1939 and 1982, Oak Ridge quadrangle, Lake Nicolet, St. Marys River. Unconsolidated bottom excluded from wetland area.

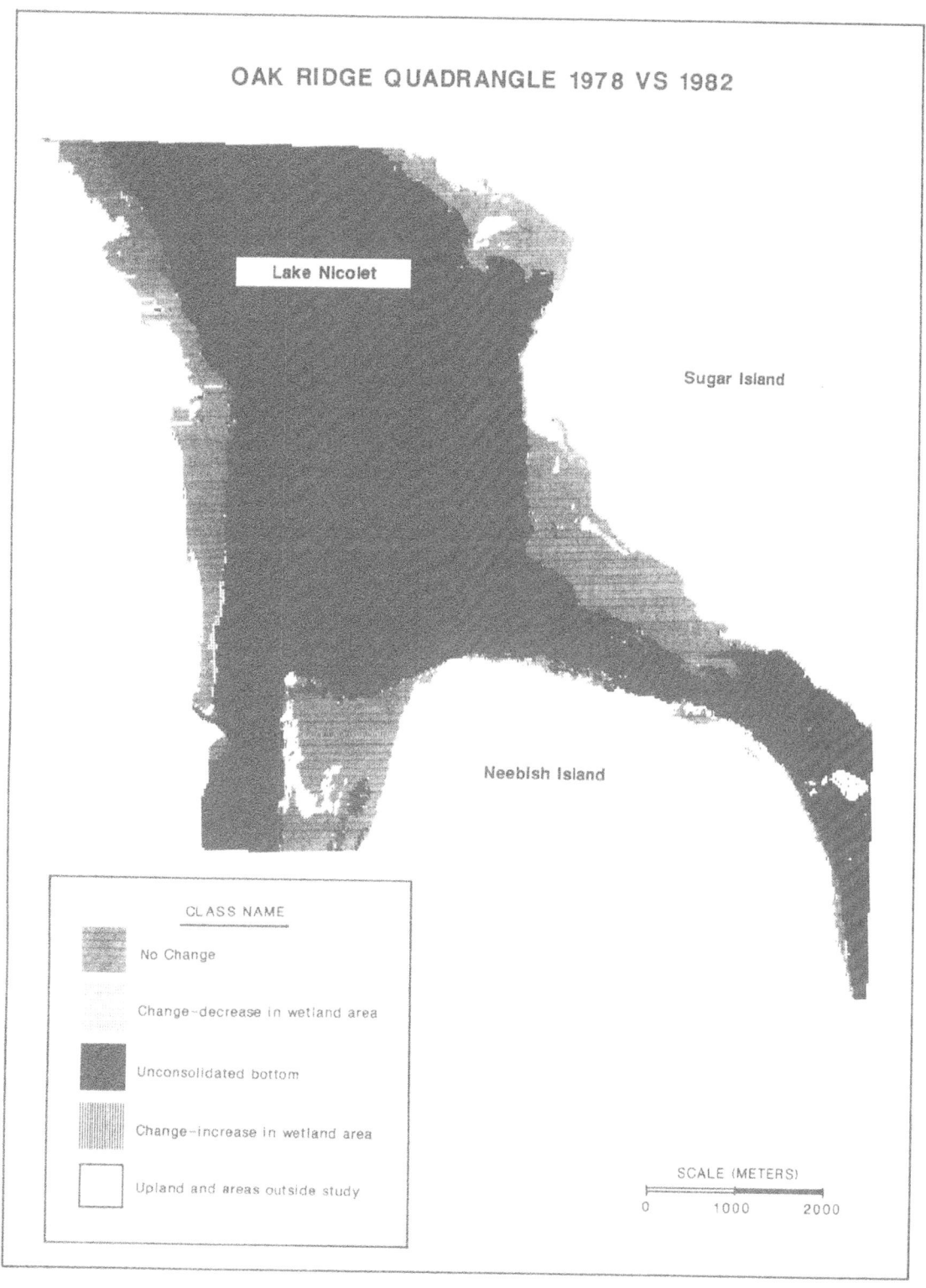

*Fig. 11.* Change in wetland area between 1978 and 1982, Oak Ridge quadrangle, Lake Nicolet, St. Marys River. Unconsolidated bottom excluded from wetland area.

Further analysis of the digital files indicated the types of changes that were occurring in the Oak Ridge wetlands. Inputs and outputs from each wetland type were tabulated for the water level, long term, and short term comparisons described above. These changes included movement between wetland classes, and losses or gains of wetland to upland. The results are given in Fig. 6, 7, and 8. The major changes due to water levels (1964 to 1982) appear to have been (1) losses of emergent vegetation, and (2) gains in unconsolidated bottom. The major long term changes appear to have been (1) losses of emergents, (2) losses of scrub-shrub, and (3) large gains of forested wetlands and unconsolidated bottom. The only changes in the short term (1978–1982) appear to have been small gains in emergent wetlands. These apparent gains may also be due to the early date of the 1978 photography, which was well before the vegetation in the St. Marys had fully emerged (Liston *et al.*, 1986).

The digital analyses allowed mapping of the specific sites of changes in wetland type. The sites of change indicated in the digital analyses are shown on the maps displayed in Fig. 9, 10, and 11. These show the wetland areas in the Oak Ridge quadrangle. Rather than the specific community changes given in graphs above, the maps show areas of stable wetland, and areas where losses and gains have occurred. Maps are given again showing changes associated with water level variation, long term changes, and short term changes.

## 4. Discussion

This analysis indicates that GIS provides a convenient method for documentation and analysis of wetland data. The National Wetland Inventory data are currently processed in a GIS system that provides for quantitative analysis of wetland trends and changes. The availability of digital files on a map-by-map basis allows the use of digital computers so that wetland changes may be tracked and analyzed.

The information developed during this study has provided an opportunity to determine whether the wetlands in the St. Marys River exhibit historical changes like other Great Lakes wetlands. The above analyses have indicated that a significant inverse relationship exists between water level and the extent of emergent wetland. This relationship has been shown to consistently occur in Great Lakes coastal wetlands. Lower water levels result in greater quantities of coastal emergent wetlands (Lyon & Drobney, 1984; Payne, *et al.*, 1985; Lyon, *et al.*, 1986; Jaworski, *et al.*, 1979; Greene, 1987). The relationship may be due to the geometry of the nearshore profile (Bukata, *et al.*, 1988). According to this analysis, the inverse relationship also holds in the case of the wetlands of the Great Lakes connecting channels.

On the other hand, judging from the significance of the regression derived, and the R-square value of the regression, this relationship is less clear than in the case of coastal wetlands of the Great Lakes proper (Table 2). The additional interpretation of wetland trends in the St. Marys made possible by the digital files and the matrix analysis may provide some insight as to differences in response to water levels.

It is interesting to note that the long-term trend of gains in unconsolidated bottom, losses in emergent wetland, and stability or gains in forested wetland found in the St. Marys River, is consistent with long-term trends found in New England wetlands (Larson & Golet, 1982). There is general acceptance of the concept of 'pulse stabilization' in Great Lakes coastal wetlands. That is, these wetlands do not undergo long-term succession because of periodic disturbance from changing water levels (Lyon, 1981; Lyon, *et al.*, 1986; Harris, 1977). The GIS-derived data suggest that

*Table 2.* Coefficients of determination ($R^2$) and significance levels (p) of regression equations; water levels on Great Lakes wetland areas.

| Wetland location | Author(s) | $R^2$ | p |
|---|---|---|---|
| Lake Michigan | Lyon (1981) | 0.93 | <0.001 |
| Lake Huron | Payne *et al.* (1985) | 0.75 | <0.025 |
| Lake Erie | Greene (1987) | 0.87 | <0.005 |
| St. Marys River | – | 0.64 | <0.05 |

in the case of the St. Marys River wetlands, this may not be entirely the case. Possibly because of their more limited exposure than the coastal wetlands, the wetlands in the St. Marys River may be undergoing slight successional changes.

The major conclusions of the wetlands analysis are: (1) that there have been no significant changes in the total amounts of wetland found in the project study area; (2) that the changes that have occurred, especially in the emergent wetland class, appear to be primarily related to changes in water levels, although ice, recreational and commercial vessels, currents, and other factors probably have some effects, and (3) quantitative evaluations using digital files provide some indication that there are long-term successional trends which are not directly evident in Great Lakes coastal wetlands.

## 5. Acknowledgements

Mr. Ross Lunetta, now of the U.S. Environmental Protection Agency, Las Vegas, Nevada, Laboratory, and Ms. Robin Gebhard, Chief Cartographer, National Wetland Inventory, St. Petersburg, Florida, made important contributions to the successful use of NWI digital files in this study. Mr. Les Weigum, Chief of the Environmental Analysis Branch, Detroit District Corps of Engineers, provided generous support toward completion of this project.

## References

Bukata, R. P., J. E. Bruton, J. H. Jerome & W. S. Haras, 1988. A mathematical description of the effects of prolonged water level fluctuations on the areal extent of marshlands. Inl. Wat. Direct., Ntnl. Wat. Res. Inst., Can. Inl. Wat., Burlington, Ontario. Sci. Ser. 166.

Cowardin, L. W., V. Carter, F. C. Golet & E. T. LaRoe, 1979. Classification of wetlands and deepwater habitats of the United States. U.S. Dep. Inter., Fish Wild. Serv., Report No. FWS/OBS-79/31. Washington D.C. 103 pp.

Dixon, W. J. (ed.), 1985. BMDP statistical software. Univ. California Press, Berkeley, California. 734 pp. ISBN 0-520-04408-8.

Duffy, W. G., T. R. Batterson & C. D. McNabb, 1987. The St. Marys River, Michigan: an ecological profile. U.S. Fish Wild. Serv. Biol. Rep. 85(7.10). 138 pp.

Greene, R. G., 1987. Effects of Lake Erie water levels on wetlands as measured from aerial photographs: Pointe Mouillee, Michigan, M.S. Thesis. Ohio State Univ., Dept. Civil Engin. 70 pp.

Harris, H. J., T. R. Bosley & F. D. Roznik, 1977. Green Bay's coastal wetlands: a picture of dynamic change. In; C. B. Dewitt & E. Solway (eds), Wetlands Ecology, Values and Impacts. pp. 337–358. Proc. Waubesa Conf. Wetlands, Madison, Wisconsin, June 2–5, 1977.

Jaworski, E., C. N. Raphael, P. J. Mansfield & B. B. Williamson, 1979. Impact of Great Lakes water level fluctuations on coastal wetlands. Off. Res. Devel., Inst. Res., Michigan State Univ., East Lansing, Michigan. 351 pp.

Jude, D. J., M. Winnell, M. S. Evans, F. J. Tesar & R. Futyma, 1986. Drift of zooplankton, benthos, and larval fish, and distribution of macrophytes and larval fish during winter and summer, 1985. Subm. to Detroit District, Corps of Engineers. 174 pp + appendices.

Larson, J. S. & F. C. Golet, 1982. Models of freshwater wetland change in southeastern New England. In; B. Gopal, (ed.) Wetlands, Ecology and Management. Ntnl. Inst. Ecol. and Internat. Sci. Publ., Jaipur, India. 512 pp.

Liston, C. R. & C. D. McNabb (principal investigators) with D. Brazo, J. Bohr, J. Craig, W. Duffy, G. Fleischer, G. Knoecklein, F. Koehler, R. Ligman, R. O'Neal, M. Siami & P. Roettger, 1986. Limnological and fisheries studies of the St. Marys River, Michigan, in relation to proposed extension of the navigation season, 1982 and 1983. Sub. to Detroit District, Corps of Engineers, Detroit, Michigan and U.S. Fish Wildl. Serv., Twin Cities, Minnesota. Issued as U.S. Fish Wildl. Serv., Off. Biol. Serv. Rep. OBS/85(2). 764 pp. + appendices.

Lyon, J. G., 1981. The influence of Lake Michigan water levels on wetland soils and distribution of plants in the Straits of Mackinac, Michigan. Doc. diss., Univ. Michigan, Ann Arbor, Michigan.

Lyon, J. G. & R. D. Drobney, 1984. Lake level effect as measured from aerial photos. ASCE J. Surv. Engin., 110: 103–111.

Lyon, J. G., R. D. Drobney & C. E. Olson, Jr., 1986. Effects of Lake Michigan water levels on wetland soil chemistry and distribution of plants in the Straits of Mackinac. J. Great Lakes Res. 12(3): 175–183.

Payne, F. C., J. L. Schuette, J. E. Schaeffer, J. B. Lisiecki, D. P. Regalbuto & P. S. Rogers, 1985. Evaluation of marsh losses: Maisou Island complex. Prep. for: Wildl. Div., Michigan Dep. Ntrl. Resources. 67 pp + appendix.

USACE (U.S. Army, Corps of Engineers, 1987.) Draft Environmental Impact Statement: Supplement II to the Final Environmental Impact Statement for Operations, Maintenance, and Minor Improvements of the Federal Facilities at Sault Ste. Marie, Michigan (July 1977). Operation of the Lock Facilities to 31 January ± 2 Weeks. Detroit District, Corps of Engineers. 176 pp + appendices.

USFWS (U.S. Fish & Wildlife Service) 1987. Photointerpretation conventions for the National Wetlands Inventory. May 1, 1987. 51 pp.

USFWS (U.S. Fish & Wildlife Service) 1988. St. Marys River, Michigan, geographic information system applications. Final report, January 19, 1988. 5 pp.

*Hydrobiologia* **219**: 97–123, 1991.
*M. Munawar & T. Edsall (eds),*
*Environmental Assessment and Habitat Evaluation of the Upper Great Lakes Connecting Channels.*
© 1991 *Kluwer Academic Publishers.*

# Limnological aspects of the St. Clair River

Ronald W. Griffiths, Stewart Thornley & Thomas A. Edsall
*Ministry of the Environment, Southwestern Region, Water Resources Assessment Unit, London, Ontario N6E 1V3; U.S. Fish and Wildlife Service, National Fisheries Research Centre – Great Lakes Ann Arbor, Michigan 48105, USA*

*Key words:* Environmental impact, macroinvertebrates, macrophytes, fishes, water quality, Great Lakes system

## Abstract

The St. Clair River is a major navigable waterway transporting water southwards for 63 km from Lake Huron to Lake St. Clair at an average flow of $5\,100\ \text{m}^3\ \text{s}^{-1}$. Water entering the river is low in suspended solids, organic carbon, phosphorus and nitrates, typical of clear, oligotrophic waters. In contrast to many large rivers, dissolved and colloidal solids account for 90 to 95 percent of the total solids load transported by the river, giving the river a turquoise colour common of glacial meltwater streams.

The river supports a diverse floral and faunal community that includes 20 taxa of submergent macroflora, at least 300 benthic macroinvertebrates and 83 fishes. A number of exotic (European) species, including 3 plants, 4 molluscs and 11 fishes, occur in the river with the macroalga, *Nitellopsis obtusa*, zebra mussel (*Dreissena polymorphora*), Asian clam (*Corbicula fluminea*), and white perch (*Morone americana*) being the most recent invaders. Production is estimated to be $200\ \text{g m}^{-2}\ \text{a}^{-1}$ ash-free dry mass for submergent macrophytes and periphyton, 7 g for macroinvertebrates and 5 g for fishes.

The river also supports a variety of water-oriented recreational activities, is a source of municipal and industrial water, a receiver of municipal and industrial wastes, and a shipping corridor. Industrial discharges have adversely affected aquatic life, particularly in the nearshore areas along the Canadian shoreline south of Sarnia, Ontario. In addition, channel dredging and shoreline modifications (bulk-heading and backfilling) have destroyed large areas of valuable habitat in the main channel and along the shoreline. Improvements in the nearshore benthic macroinvertebrate community of the river over the past 20 years show that the river will respond to reductions in contaminants loadings.

## Introduction

The St. Clair River (Fig. 1) is a major international waterway that separates Canada (Ontario) and the United States (Michigan). It has a limited tributary drainage system and is essentially a strait carrying about $5\,100\ \text{m}^3\ \text{s}^{-1}$ of water from Lake Huron to Lake St. Clair. Because Lake Huron is a source of high quality water, the river supports a relatively diverse and productive flora and fauna, including the more than 80 species of fish that are permanent or seasonal residents of the river or use it as a migratory pathway between Lake Huron and Lake St. Clair (Edsall *et al.*, 1988).

The river provides a variety of water-oriented recreational activities, including fishing, waterfowl hunting, and pleasure boating. It is also used, however, as a source of municipal and industrial water, a receiver for municipal and industrial

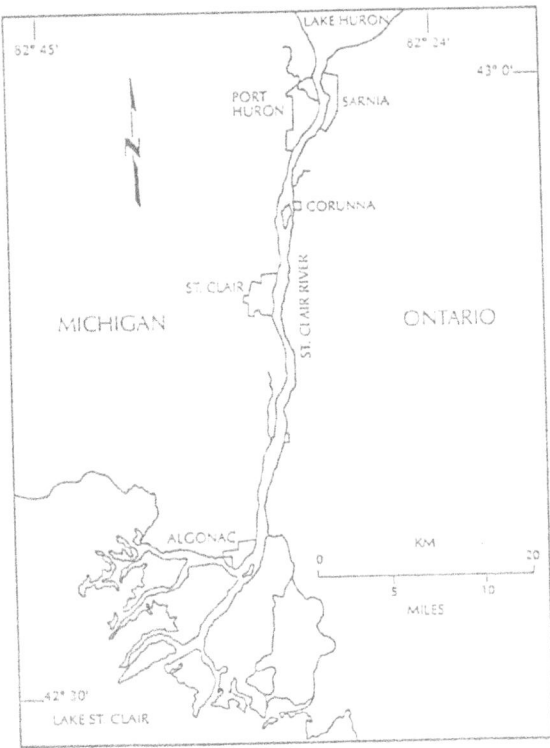

*Fig. 1.* Map of the St. Clair River.

wastes, and as a commercial shipping corridor. Millions of cubic metres of waste-water, carrying heavy metals, chlorinated organics and other petrochemical compounds, are discharged daily to the river from industries located mainly in the Chemical Valley, south of Sarnia, Ontario. These wastes have impaired sediment and water quality and contaminated the biota of the river (EC & US EPA, 1989). In addition, channel dredging, bulk-heading, and backfilling required to accommodate commercial vessel traffic and the development and maintenance of attendant industrial and commercial shoreline docking facilities have destroyed large areas of valuable fish and wildlife habitat in the river channel and along the shore-line. Because of these problems, the river has been designated as an 'Area of Concern' by the IJC (1987).

Our purpose here was to assemble the most recent information, most of it published in the grey literature, and present an integrated overview of the current limnological condition of this large

waterway. A better understanding of the present ecological structure and functioning of the entire St. Clair River system is necessary if an eco-system approach (Vallentyne & Hamilton, 1987) for managing the resources of the river is to be implemented to restore and maintain all beneficial uses.

## Physical and chemical characteristics

The St. Clair River transports water some 63 km in a southerly direction from Lake Huron to Lake St. Clair (Fig. 1). The river consists of a channel 45 km long that broadens into an extensive birds-foot delta, covering about 100 km$^2$ along the north shore of Lake St. Clair. Three major channels through the delta average 18 km in length and transport 92 percent of the flow.

The St. Clair River can be divided into three distinct hydraulic reaches (Derecki, 1984). The steep upper reach from Lake Huron to the Black River falls 0.3 m over only 5 km. The channel width is less than 450 m, the mid-channel depth varies from 9 to 21 m, and the mean surface velocity is about 1.5 m s$^{-1}$ (5.4 km h$^{-1}$). The more shallow middle reach from the Black River to Russell Island, in contrast, falls 1.1 m over a 40 km stretch. The channel is generally 600 to 900 m wide except near Stag and Fawn Islands and the St. Clair middle ground shoal where the width increases up to 1 200 m. Mid-channel depths range from 8 to 15 m and the mean surface velocity is about 1.0 m s$^{-1}$ (3.6 km h$^{-1}$). The flat, lower reach encompasses the delta region of the river where the vertical fall is less than 0.2 m over 18 km. The three main channels vary from 200 to 900 m in width, mid-channel depth varies greatly but averages 12 m and mean surface velocity is about 0.7 m s$^{-1}$ (2.5 km h$^{-1}$). Mean travel time from Lake Huron to Lake St. Clair thus requires about 21 h, including 8.5 h to travel through the delta (Derecki, 1983). The historical mean discharge of the river is approximately 5 100 m$^3$ s$^{-1}$, with a winter low of 4 200 m$^3$ s$^{-1}$ and a summer high of 5 500 m$^3$ s$^{-1}$ (Quinn & Kelly, 1983). During the mid-1980's, however, mean flow has averaged $6\,000\ m^3\,s^{-1}$ (Hudson *et al.*, 1986).

The riverbed is incised into unconsolidated glacial deposits consisting of hard, stony glaciolacustrine clay (Black Shale Till) in the mid-channel and finer glaciolacustrine silt-clay (St. Joseph Till) in the nearshore area (Fitzgerald *et al.*, 1979). The poor erodibility of these deposits (Ritter, 1986) is probably an important factor accounting for the relative straightness of the river's channel and the lack of floodplain features typical in many alluvial valleys. It also explains the high water clarity of the river, even though its bed is composed of fine sediments. Turbidity throughout the river is typically less than 6 Formazin Turbidity Units and colour less than 5 True Colour Units (Ont. Min. Env., unpubl. data).

Little sediment accumulation occurs in the upper river because of the high current velocities. Radionuclide measurements of a sediment core near Sarnia, Ontario, indicated that the top 3 cm of sediment was less than nine months old, suggesting that the sediments are transitory with a river retention time of less than one year (Oliver, 1988). Bed materials of the river thus reflect the general composition of the tills and their erosional products. In the central channel, hard clay is covered with a layer of well-rounded boulders, cobbles, and gravels all infiltrated with sand. The hard and irregular bottom is characteristic of an actively eroding or dredged surface. In nearshore areas, i.e. those with depths less than 7 or 8 m, the bottom is more regular, has a steeper slope, and a convex profile. Bottom sediments are composed of sands and gravels over glacial clay. Sediment thickness is greatest at the shoreline and decreases in an offshore direction over about 100 m. Organic content of these sediments is low ($<1.5\%$).

The characteristic turquoise colour of the river, common of glacial meltwater streams, is a result of the high dissolved solids (particles $<2\ \mu m$) content (120 to $150\ mg\,l^{-1}$) of the water, of which very fine colloidal clays, eroded from glacial clay deposits, may account for 20 to $30\ mg\,l^{-1}$. In contrast, the concentration of suspended solids transported by the river typically ranges from just 2 to $8\ mg\,l^{-1}$. Duane (1967) showed that the mean diameter of the suspended river sediment is 0.2 to 3.2 mm (i.e. fine sand to gravel), while that of the bed load is 2.0 to 2.6 mm (i.e. coarse sand to gravel). The principle source of these sediments is probably the southern basin of Lake Huron, although the slow erosion of the glacial clay riverbed would also be a source of sands. Artificial fill used to stabilize and extend shore banks would be an additional source of gravels.

Lau *et al.* (1989) estimated that the suspended solids load transported in the upper river was in the order of $2\,500$ to $3\,000\ t\,d^{-1}$ while the bed load was in the order of 1 to $4\ t\,d^{-1}$. These loads, however, are small in comparison with the dissolved solids load of the river, estimated in the order of $50\,000\ t\,d^{-1}$, and colloidal solids load, estimated in the order of $10\,000\ t\,d^{-1}$. In contrast to other large rivers of the world (Whitton, 1975; Davis & Walker, 1986), the dissolved and colloidal sediment load of the St. Clair River is 90 to 95 percent of the total load.

Water temperatures generally vary from a summer high of 18.5 to 21.5 °C to a winter low of 0.5 to 1.0 °C. Ice rarely forms over the entire river; however, ice flows periodically enter the river in large quantities from Lake Huron and have formed huge ice jams throughout the river in 1901, 1920, 1942, and 1984, greatly reducing the flow of water (Edsall *et al.*, 1988).

Water entering the river from Lake Huron is consistently of high quality (Table 1); total phosphorus, generally below $10\ \mu g\,l^{-1}$, and nitrate, generally below $400\ \mu g\,l^{-1}$, are typical of oligotrophic waters (Wetzel, 1975) and metal concentrations are low. Dissolved oxygen concentrations are always near saturation and the pH is alkaline, generally varying between 8.0 and 8.3.

## Biological communities and habitats

The aquatic macrophyte and macroalgal flora, and the benthic macroinvertebrate and fish fauna have been the most intensively studied biotic com-

*Table 1.* Concentrations of chemicals in water entering the St. Clair River. Analysis of monthy samples of raw water from Lambton Water Treatment Plant, 1988. All units are $\mu g\,l^{-1}$ except where noted.

| Constituent | Concentrations |
|---|---|
| Alkalinity (mg l$^{-1}$) | 80.3–88.0 |
| Chloride (mg l$^{-1}$) | 5.7–7.5 |
| Ammonia | 7–34 |
| Kjeldahl nitrogen | 120–375 |
| Nitrate | 245–350 |
| Phosphorus | 2–17 |
| Aluminum | 4.3–114 |
| Chromium | 0.2–5.6 |
| Copper | 0.4–52 |
| Iron | <4.0–210 |
| Lead | <0.1–0.5 |
| Mercury | <0.1 |
| Nickel | <0.1–1.0 |
| Zinc | 0.3–5.8 |

ponents of the St. Clair River. Although most of these studies were conducted to document the specific effects of pollution, dredging, shipping, and other anthropogenic stresses on the aquatic biota of the river (e.g. OME, 1979; Hudson *et al.*, 1986; Griffiths, 1989; Schloesser & Manny, 1989), they collectively also provide a reasonably good understanding of the biological organization and dynamics of this large lotic system.

*Phytoplankton*

The phytoplankton of the St. Clair River has received little study, but resembles that of southern Lake Huron, consisting mainly of the diatoms *Cyclotella, Fragilaria, Melosira, Stephanodiscus, Synedra,* and *Tabellaria* (ENCOTEC, 1974; Vollenweider *et al.*, 1974). Mean annual phytoplankton density at the head of the river is low, about 500 A.S.U. ml$^{-1}$ from 1976 to 1981, with peak seasonal abundances occurring each spring and autumn. Similarly, chlorophyll *a* concentrations are low, averaging 1.2 to 1.4 $\mu g\,l^{-1}$ from May to October throughout the river (Ont. Min. Env., unpubl. data).

*Macroflora*

Twenty submergent macrofloral taxa, composed of 3 macroalgal taxa, 10 macrophyte taxa (Table 2) and the filamentous green alga *Cladophora*, occur in the St. Clair River. The Characeae (*Chara* and *Nitella*) and *Cladophora* are the most widely distributed macrofloral taxa (Ont. Min. Env., unpubl. data), while among macrophytes, *Potamogeton richardsonii* (Benn.) Rydb., *Elodea canadensis* Michx., *Myriophyllum spicatum* L. (Eurasian milfoil, an exotic species), and *Vallisneria americana* Michx. are the most common species.

The most extensive growths of submergent vegetation occur in Sarnia Bay, along the Canadian shoreline near Stag and Fawn Islands, around Stag and Fawn Islands, and in the north and middle channels of the delta (Schloesser & Manny, 1982; OME, 1979). Aquatic macrophytes are generally found at depths <4.5 m (maximum depth of 7.9 m for *E. canadensis*), Characeae commonly occur at depths <5.5 m (maximum depth of 6.7 m), and *Cladophora* attaches to rocks and other solid objects and is most abundant at depths <8 m (maximum depth of 10 m) (Ont. Min. Env., unpubl. data). Macroflora are rarely observed at depths greater than 10 m (Schloesser & Manny, 1982; Ont. Min. Env., unpubl. data). The maximum depths colonized by macrophytes and Characeae in the St. Clair River are considerably greater than would be predicted from Secchi disc measurements of 1.6 to 2.1 m for the river using the relation established by Chambers & Kalff (1985) for lakes. The macrophyte habitat of the river, between the shoreline and the 4.5 m depth contour, presently includes approximately 2000 ha, and probably has been reduced significantly by dredging and filling to accommodate navigation and shoreline development.

The growing season for aquatic macrophytes generally extends from June through October (Schloesser *et al.*, 1985), when water temperatures are above 10 °C, the threshold temperature for growth (Kunii, 1981; Haag, 1979). Maximum biomass, which is achieved between July and

*Table 2.* Occurrence (%) and biomass (g m$^{-2}$) of submergent macrophytes and macroalgae in the St. Clair River.

| Taxa | Common name | Occurrence[1] | Biomass[2] | | |
|---|---|---|---|---|---|
| | | | June | July-August | September |
| Macroalgae: | | | | | |
| *Chara* | Stonewort | 68 | 44 | 116 | 149 |
| *Nitella* | | | 43 | 73 | 14 |
| *Nitellopsis obtusa** | | 0 | 0 | 0 | 2 |
| Macrophytes: | | | | | |
| *Elodea canadensis* | Waterweed | 36 | 34 | 41 | 27 |
| *Heteranthera dubia* | Water stargrass | <1 | 0 | 0 | 0 |
| *Myriophyllum exalbescens* | Water milfoil | <1 | 0 | 0 | T |
| *Myriophyllum spicatum** | Eurasian milfoil | 28 | 30 | 31 | 61 |
| *Najas flexilis* | Bushy pondweed | <1 | 0 | 0 | T |
| *Potamogeton crispus** | Curley pondweed | 2 | 29 | 31 | 6 |
| *Potamogeton gramineus* | Variable pondweed | 11 | 12 | 25 | 62 |
| *Potamogeton natans* | Floating-leaf pondweed | <1 | 0 | 0 | 0 |
| *Potamogeton nodosus* | Long-leaf pondweed | 2 | 0 | 0 | 0 |
| *Potamogeton richardsonii* | Clasping-leaf pondweed | 49 | 11 | 44 | 69 |
| *Potamogeton zosteriformis* | Flatstem pondweed | <1 | 0 | 5 | 15 |
| *Potamogeton kk* | Narrow-leaf pondweeds | 24 | 26 | 40 | 74 |
| *Rununculus* | Buttercup | 2 | 0 | 0 | 0 |
| *Vallisneria americana* | Wild celery | 28 | 0 | 4 | 22 |
| *Zannichellia palustris* | Horned pondweed | 0 | 0 | T | T |

[1] data from Schloesser & Manny (1982); samples collected at 217 sites from August-October of 1978.
[2] data from Hudson *et al.* (1986); biomass values represent mean dry mass found at Stag Island in 1983 and 1984.
T trace mass
\* non-native species
kk *pectinatus* & *filiformis*

October, is about 130 to 270 g m$^{-2}$ dry mass (Hudson *et al.*, 1986; Schloesser *et al.*, 1985), and is slightly higher than the lower end of the range observed for temperate rivers (Westlake, 1963; Edwards & Owens, 1960). Most taxa grow continuously throughout the summer reaching peak biomass in early autumn (Table 2). However, a few taxa such as *Potamogeton crispus* L. (an exotic), *E. canadensis*, and *Nitella* reach a peak biomass in mid-summer. While the Characeae overwinter as green plants, the macrophytes die back and decompose during the autumn, contributing to the detrital food-base of the river.

Macrophytes and their associated periphyton probably are a significant source of the organic matter available for secondary production. Water entering from Lake Huron is oligotrophic, low in total organic carbon (2 mg l$^{-1}$; OME, 1979),

phytoplankton (1.2 to 1.4 mg l$^{-1}$; Ont. Min. Env., unpubl. data) and zooplankton (Watson & Carpenter, 1974). Other allochthonous inputs of organic matter are chiefly contributed by the bank vegetation, but these are <1 percent of the total budget (Edwards *et al.*, 1990) because the amount of shoreline is small relative to the volume of flow. Net production of submerged vegetation and associated periphyton in the St. Clair River is estimated to be 200 g m$^{-2}$ a$^{-1}$ ash-free dry mass (Edwards *et al.*, 1990). Manny *et al.* (1988) estimated that about 900 t ash-free dry mass of living plant material drifted out of the upper river at Algonac during April–October of 1986. Although production of emergent macrophytes is estimated to be about four times that from submergent vegetation and periphyton (Edwards *et al.*, 1990), emergent macrophyte stands are primarily concentrated in the lower deltic part of the river (Edsall *et al.*, 1988). Thus, in the upper St. Clair River submergent vegetation and periphyton form most of the biological production in the river.

Macrophytes not only provide detritus and fresh plant material that is fed upon by invertebrates, fish, and waterfowl (Schloesser, 1986), they also add important physical structure and increase habitat diversity to an environment that has been simplified by man through navigation-related dredging, removal of snags and debris, filling, and bulkheading. The species composition and distribution of weed beds had been shown to influence the phytomacroinvertebrate (Brown *et al.*, 1988) and fish (French 3rd, 1988; Poe *et al.*, 1986) communities. Macrophyte beds are used by a number of fishes as spawning habitat (Schloesser, 1986) and by juveniles and adult fish as cover from avian and piscine predators (Hunt, 1963; Werner *et al.*, 1983). They also increase the deposition of organic matter by reducing flow velocities up to 80 percent (Hudson *et al.*, 1986), thereby increasing the food supply for epifaunal and infaunal invertebrates. The retention of organic matter in the river for further biological processing increases productivity by shortening the spiralling length of nutrients and the turnover length of organic carbon (Elwood *et al.*, 1983). Although most of the macrophyte canopy is absent during the winter and spring, the Characeae canopy remains and can perform this function throughout the year.

The stable flows of the St. Clair River promote long-term stability in weed beds in the river. Scouring of beds by high flows or ice action was not observed by Hudson *et al.* (1986) during an ice jam in 1984, but ship traffic can affect the composition and stand density of the macrofloral community (Schloesser & Manny, 1989).

### Zooplankton

A study conducted with a Tucker trawl (333 $\mu$m mesh) from May to December 1974 at the electrical generating plant of Detroit Edison Co. near Marine City, Michigan, revealed that the limnetic

zooplankton in the river consisted of 11 cladocerans (*Bosmina longirostris* O.F. Muller, 7 species of *Daphnia*, *Holopedium gibberum* Zaddach, *Leptodora kindti* Foche, *Polyphemus pediculus* Linne), 10 calanoid copepods, and 7 cyclopoid copepods (Texas Instruments, 1975). Densities increased exponentially from May (50 individuals m$^{-3}$) through August (4000 individuals m$^{-3}$) then dropped in September and remained similar through December (300 to 500 m$^{-3}$). *Daphnia ambigua* Scourfield and *P. pediculus* occurred during the summer, while no seasonal preference was observed for *Diaptomus oregonensis* Lilljeborg, *Diaptomus sicilis* S.A. Forbes, *Diacyclops thomasi* S.A. Forbes, *Mesocyclops edax* S.A. Forbes, *B. longirostris*, and *H. gibberum*. A seasonal succession was observed with *D. thomasi* dominating in May, *Daphnia retrocurva* Forbes in June, *D. retrocurva* and *H. gibberum* codominating in August, *H. gibberum* dominating in September, *Daphnia galeata* Sars *mendotae* Birge in October, and *D. sicilis* from November through December.

The zooplankton in the river, like the phytoplankton, largely represent production exported from Lake Huron. This accounts for the close resemblance of the zooplankton from the St. Clair River and Lake Huron (Watson & Carpenter, 1974). It is likely, therefore, that the European spiny water flea *Bythotrephes cederstroemi* Schoedler, that was recently introduced to Lake Huron (Burr *et al.*, 1986) is now a member of the plankton community of the river.

### Macrozoobenthos

Presently, at least 179 macrobenthic invertebrate taxa, including insects, crayfishes, amphipods, isopods, molluscs, annelids, flatworms, nematodes and nemerteans occur in the St. Clair River (Table 3). Twenty-four taxa commonly occur in the near-shore area (depth <8 m) of the river (Table 4). Assuredly, the list in Table 3 underestimates the species richness of the river because the sieve aperture size used in the listed studies was large (0.6 mm) and suitable keys to identify

*Table 3.* Occurrence of benthic macroinvertebrates in the St. Clair River.

| | 1976 Oct | 1977 Mar | May (1) | Oct | 1977 May | Jul (2) | 1983 & 84 May | Oct (3) | 1985 May (4) |
|---|---|---|---|---|---|---|---|---|---|
| **INSECTA:** | | | | | | | | | |
| **AQUATIC MOTHS:** | | | | | | | | | |
| Pyralidae | P | P | P | P | P | P | P | P | P |
| **BEETLES:** | | | | | | | | | |
| Elmidae: | | | | | | | | | |
| *Dubiraphia* | P | P | P | P | P | P | P | P | P |
| **BUGS:** | | | | | | | | | |
| Corixidae: | P | | P | | P | P | P | | |
| *Sigara lineata* | | | | | | | | | P |
| **CADDISFLIES:** | | | | | | | | | |
| Brachycentridae: | | | | | | | | | |
| *Brachycentrus* | P | P | P | P | P | P | P | P | |
| *Micrasema* | | | | | | | | P | |
| Hydropsychidae: | | | | | | | | | |
| *Cheumatopsyche* | P | P | P | P | P | P | P | P | P |
| *Hydropsyche* | P | P | P | P | P | P | P | P | P |
| Hydroptilidae: | | | P | | | | P | | |
| *Agraylia* | | | | | P | P | | | |
| *Hydroptila* | | | | | | | | P | |
| *Ochrotrichia* | | | | | | | P | P | |
| Glossosomatidae: | | | | | | | | | |
| *Protoptila maculata* | P | P | | | | | P | P | P |
| Lepidostomatidae: | | | | | | | | | |
| *Lepidostoma* | P | | P | | | | | | P |
| Leptoceridae: | P | | | | | | | | |
| *Ceraclea* | P | P | P | P | P | P | P | P | P |
| *Mystacides* | P | P | P | P | | | P | P | P |
| *Nectopsyche* | P | P | | | | | P | P | |
| *Oecetis* | P | P | P | P | P | P | P | P | P |
| *Setodes* | | | | | | | P | P | P |
| *Triaenodes* | P | P | P | P | | | P | P | P |
| Limnephilidae: | | | | | | | | | |
| *Limnephilus* | P | | P | | | | | | |
| *Pyncopsyche* | | | | | | P | P | | |
| Molannidae: | | | | | | | | | |
| *Molanna* | | P | | | | | | | |
| Phryganeidae: | | | | | | | | | |
| *Phryganea* | P | | | | | | | P | |
| Polycentropodidae: | | | | | | | | | |
| *Neureclipsis* | P | P | P | P | | | P | P | P |
| *Phylocentropus* | P | P | P | P | | | | P | P |
| *Polycentropus* | P | P | | P | P | P | P | P | |
| **DAMSELFLIES:** | | | | | | | | | |
| Coenagrionidae: | P | P | P | P | | | P | P | |
| *Argia* | | | | | | | | | P |
| *Coenagrion* | | | | | | | | | P |

104

*Table 3.* (Continued).

| | 1976 Oct | 1977 Mar | May (1) | Oct | 1977 May | Jul (2) | 1983 & 84 May | Oct (3) | 1985 May (4) |
|---|---|---|---|---|---|---|---|---|---|
| DRAGONFLIES: | | | P | | | | | | |
| Gomphidae: | P | P | | P | | | | | |
| *Dromogomphus* | | P | | | | | | | |
| *Gomphurus* | | | | | | | | | P |
| *Gomphus* | P | | | P | | | P | P | |
| *Stylurus* | | | | | | | P | | P |
| MAYFLIES: | | | | | | | | | |
| Baetidae: | | | | | | | | | |
| *Baetis* | | | | | | P | | P | |
| Baetiscidae: | | | | | | | | | |
| *Baetisca* | P | P | P | P | P | P | P | P | P |
| Caenidae: | | | | | | | | | |
| *Caenis* | P | P | P | P | P | P | P | P | P |
| Ephemerellidae: | | | | | | | | | |
| *Ephemerella* | | P | P | P | P | P | P | P | P |
| *Serratella* | | | | | | | | | P |
| Ephemeridae: | | | | | | | | | |
| *Ephemera simulans* | | P | | | | | | | P |
| *Hexagenia* | P | P | P | P | P | P | P | P | P |
| Heptageniidae: | | | | | | | | | |
| *Stenonema* | P | P | P | P | P | P | P | P | P |
| Leptophlebiidae: | | | | | | | | | |
| *Paraleptophlebia* | | | | | | | | | P |
| Tricorythidae: | | | | | | | | P | |
| *Tricorythodes* | | | | | | | P | P | |
| STONEFLIES: | P | | | | | | | | |
| Perlidae: | | | | | P | P | | | P |
| *Perlesta placida* | | | | | | | P | | |
| Perlodidae: | | | | | | | | P | |
| *Isogenoides* | | | | | | | | | P |
| *Isoperla bilineata* | | | | | | | | | |
| TRUE FLIES: | | | | | | | | | |
| Athericidae: | | | | | P | P | | | |
| *Atherix* | | | | | P | P | | | |
| Ceratopogonidae | P | P | P | P | P | P | P | P | P |
| Chaoboridae: | P | | | | | | | | |
| *Chaoborus* | P | P | P | P | | | | | |
| Chironomidae: | | | | | | | | | |
| *Chernovskiia* | | | | | | | ? | ? | P |
| *Chironomus* | | | | | P | P | ? | ? | P |
| *Cladotanytarsus* | | | | | | | ? | ? | |
| *Cryptochironomus* | P | | | | P | P | ? | ? | P |
| *Demicryptochironomus* | | | | | | | | | P |
| *Dicrotendipes* | | | | | P | P | ? | ? | P |
| *Glyptochironomus* | | | | | | P | | | |
| *Harnischia* | | | | | | | | | P |
| *Microtendipes* | | | | | P | P | | | P |
| *Parachironomus* | | | | | P | P | | | P |
| *Paracladopelma* | | | | | | P | | | P |

*Table 3.* (Continued).

| | 1976 Oct | 1977 Mar (1) | May (1) | Oct (1) | 1977 May (2) | Jul (2) | 1983 & 84 May (3) | Oct (3) | 1985 May (4) |
|---|---|---|---|---|---|---|---|---|---|
| *Paralauterbomiella* | | | | | P | | | | |
| *Paratanytarsus* | | | | | | | | | P |
| *Paratendipes* | | | | | P | P | | | P |
| *Phaenopsectra* | | | | | P | P | | | P |
| *Polypedilum* | P | | | | P | P | ? | ? | P |
| *Pseudochironomus* | | | | | | | ? | ? | P |
| *Rheotanytarsus* | | | | | | | ? | ? | P |
| *Robackia* | | | | | | | ? | ? | |
| *Saetheria* | | | | | | | | | P |
| *Stempellina* | | P | | | | | ? | ? | |
| *Stenochironomus* | | | | | P | | | | P |
| *Stictochironomus* | | | | | P | P | | | P |
| *Tanytarsus* | | | | | P | | | | P |
| *Tribelos* | P | | | | P | P | | | P |
| *Xenochironomus ( = Axarus)* | | | | | | | ? | ? | P |
| *Potthastia* | | | | | P | P | ? | ? | P |
| *Corynoneura* | | | | | P | P | ? | ? | |
| *Cricotopus* | | | | | P | P | ? | ? | |
| *Epiococladius* | | | | | | | ? | ? | P |
| *Heterotrissocladius* | | | | | P | P | | | P |
| *Hydrobaenus* | | | | | | | ? | ? | P |
| *Lopescladius* | | | | | | | ? | ? | |
| *Nanocladius* | | | | | | | ? | ? | |
| *Orthocladius* | | | | | | | ? | ? | P |
| *Parakiefferiella* | | | | | | | ? | ? | |
| *Psectrocladius* | | | | | P | P | | | P |
| *Monodiamesa* | | | | | P | P | ? | ? | P |
| *Ablabesmyia* | | | | | | | ? | ? | P |
| *Coelotanypus* | | | | | | | ? | ? | |
| *Conchapelopia* | | | | | P | P | ? | ? | P |
| *Djalmabatista* | | | | | | | | | P |
| *Pentaneura* | | | | | | | ? | ? | |
| *Procladius* | | | | | P | P | ? | ? | P |
| Culicidae | | | | | P | P | | | |
| Dolichopodidae | | | P | | | | | | |
| Empididae | P | P | P | P | | | P | P | P |
| Psychodidae | | P | | | | | P | P | P |
| Tipulidae | | | | | | | P | | |
| CRUSTACEA: | | | | | | | | | |
| AMPHIPODS: | | | | | | | | | |
| Gammaridae: | | | | | | | | | |
| *Crangonyx* | P | P | P | P | | | | | |
| *Gammarus* | P | P | P | P | P | | P | P | |
| *G. fasciatus* | | | | | | P | | | P |
| Haustoriidae: | | | | | | | | | |
| *Pontoporeia hoyi* | P | P | P | P | P | P | P | P | P |

*Table 3.* (Continued).

| | 1976 Oct | 1977 Mar | May (1) | Oct | 1977 May (2) | Jul | 1983 & 84 May (3) | Oct | 1985 May (4) |
|---|---|---|---|---|---|---|---|---|---|
| Taltridae: | | | | | | | | | |
| *Hyalella azteca* | P | P | P | P | P | P | P | P | P |
| CRAYFISHES: | | | | | | | | | |
| Astacidae: | | | | | | | | | |
| *Cambarus diogenes* | | P | | | | | | | |
| *Orconectes* | P | P | | | | | P | | |
| *O. propinquus* | | | P | | P | P | | | |
| ISOPODS: | | | | | | | | | |
| Asellidae: | | | | | | | | | |
| *Asellus* (Caecidotea) | P | P | P | P | P | P | P | P | P |
| *Lirceus* | P | P | P | P | | | P | P | P |
| | | | | | | | | | |
| MOLLUSCA: | | | | | | | | | |
| | | | | | | | | | |
| CLAMS: | | | | | | | | | |
| Sphaeriidae: | | | | | | | | | |
| *Pisidium* | P | P | P | P | P | P | P | P | P |
| *Sphaerium* | | P | | | | P | P | P | P |
| *S. striatinum* | | P | | | | | P | P | |
| Unionidae: | | | | | | | | | |
| *Anodonta grandis* | | P | | | | | | | |
| *Lampsilis radiata* | | | P | | | | P | | |
| *Villosa iris* | | | | | | | | | P |
| SNAILS: | | | | | | | | | |
| Ancylidae: | P | P | P | P | | | | | |
| *Ferrissia* | | | | | P | P | P | P | P |
| Bithyniidae: | | | | | | | | | |
| *Bithynia tentaculata** | | | | | P | P | | | |
| Hydrobiidae: | | | | | | | | | |
| *Amnicola* | P | P | P | P | P | P | P | P | P |
| *Probythinella lacustris* | | | | | | P | P | | P |
| *Somatogyrus subglobosus* | P | P | P | P | P | P | P | P | |
| Lymnaeidae: | | | | | | | | | |
| *Fossaria* | | | | | | | | | P |
| *Lymnaea stagnalis* | | | P | | | | | | |
| *Pseudosuccinea columella* | | | | | | | | | P |
| Physidae: | P | | | | | | | | |
| *Physella gyrina* | | P | P | P | P | P | P | P | P |
| Planorbidae: | | | | | | | | | |
| *Gyraulus* | P | P | P | P | P | P | P | P | P |
| *Promenetus* | | | | | | | | | P |
| Pleuroceridae: | | | | | | | | | |
| *Elimia livescens* | P | P | P | P | P | P | P | P | P |
| *Pleurocera acuta* | | | | | | | | P | P |
| Valvatidae: | | | | | | | | | |
| *Valvata piscinalis** | | | | | P | P | | | P |
| *V. sincera* | P | P | P | P | P | P | P | P | |
| *V. tricarinata* | P | P | P | P | P | P | P | P | P |
| Viviparidae: | | | | | | | | | |
| *Campeloma decisum* | P | | | P | | P | P | | |

Table 3. (Continued).

| | 1976 Oct | 1977 Mar | May (1) | Oct | 1977 May (2) | Jul | 1983 & 84 May (3) | Oct | 1985 May (4) |
|---|---|---|---|---|---|---|---|---|---|
| **ANNELIDA:** | | | | | | | | | |
| **LEECHES:** | | P | | P | P | P | | | |
| Erpobdellidae: | P | P | P | | | | P | | |
| *Erpobdella punctata* | | | | | | | | P | |
| *Mooreobdella fervida* | P | | | | | | | | |
| *M. microstoma* | | | P | | | | | | P |
| Glossiphonidae: | | | | | | | | | |
| *Actinobdella inequiannulata* | | | | | | | | P | P |
| *Alboglossiphonia heteroclita* | P | P | | | | | | P | |
| *Batracobdella phalera* | | P | P | | | | | P | |
| *Glossiphonia complanata* | P | | P | | | | | P | |
| *Helobdella elongulata* | P | | P | | | | P | P | |
| *H. papillata* | | | | | | | | P | |
| *H. stagnalis* | P | | P | | | | P | P | |
| *H. triserialis* | | P | P | | | | P | | |
| *Placobdella* | | | | | | | | P | |
| *P. montifera* | P | P | | | | | | | |
| *P. papillifera* | P | | | | | | P | | |
| Pisicolidae: | | | | | | | | | P |
| *Piscicola milneri* | P | | | | | | P | P | |
| *P. punctata* | | | P | | | | | | |
| **POLYCHAETES:** | | | | | | | | | |
| Sabellidae: | | | | | | | | | |
| *Manayunkia speciosa* | | P | | | | | P | P | |
| **WORMS:** | | | | | | | | | |
| Enchytraeidae | P | | | | | | | | |
| Lumbriculidae: | | | | | | | | | |
| *Stylodrilus heringianus* | | | | | P | P | | | P |
| Naididae: | | | | | | | | | |
| *Amphichaeta* | P | P | | | | | | | |
| *Chaetogaster diaphanus* | P | P | P | P | | | | | |
| *Nais* | | | | | | | | | |
| *N. bretscheri* | | | | | | | P | P | |
| *N. communis* | | | | | | P | | | P |
| *N. pseudobtusa* | | | | | | P | | | P |
| *N. simplex* | | | | | P | P | | | P |
| *N. variabilis* | | | | | | | | | P |
| *Ophidonais serpentina* | P | | | | P | P | | | P |
| *Slavina appendiculata* | P | | | | | | | | |
| *Specaria josinae* | P | | | | | | | | |
| *Stylaria* | | | | | | | P | P | |
| *S. lacustris* | P | | P | P | | P | | | P |
| *Uncinata uncinata* | | | | | | P | | | P |
| *Wapsa mobilis* | P | | | | | | | | P |

*Table 3.* (Continued).

| | 1976 Oct | 1977 Mar | May (1) | Oct | 1977 May | Jul (2) | 1983 & 84 May | Oct (3) | 1985 May (4) |
|---|---|---|---|---|---|---|---|---|---|
| Tubificidae: | P | P | P | P | | | P | P | |
| *Aulodrilus americanus* | | | | | P | P | | | P |
| *A. pigueti* | P | | | | | | | | |
| *A. pluriseta* | P | | | | P | P | | | P |
| *Branchiura sowerbyi* | | | | | | P | | | |
| *Isochaetides freyi* | | | | | P | P | | | P |
| *Ihyodrilus templetoni* | | P | | | P | P | | | P |
| *Limnodrilus angustipenis* | | | | | P | | | | P |
| *L. cervix* | | | | | P | P | | | P |
| *L. claparedianus* | | | | | P | P | | | P |
| *L. hoffmeisteri* | | P | | | P | P | | | P |
| *L. maumeensis* | | | | | P | P | | | P |
| *L. udekemianus* | | P | | | P | P | | | P |
| *Potamothrix moldaviensis* | | | | | P | | | | P |
| *P. vejdovskyi* | P | | | | | | | | P |
| *Quistadrilus multisetosus* | P | | | | P | P | | | P |
| *Spirosperma ferox* | P | | | | P | P | P | P | P |
| *Tubifex ignotus* | | | | | | P | | | P |
| *T. tubifex* | | | | | P | P | | | P |
| NEMERTEA: | | | | | | | | | |
| PROBOSCIS WORMS: | | | | | | | | | |
| *Prostoma* | P | P | P | P | P | P | P | P | P |
| NEMATODA: | | | | | | | | | |
| ROUNDWORMS | P | P | P | P | P | P | P | P | P |
| PLATYHELMINTHES: | | | | | | | | | |
| FLATWORMS: | | | | | | | | | |
| Turbellaria | P | P | P | P | P | P | P | P | P |

1: Hiltunen (1980), chironomids and worms were generally not identified beyond family.
2: Griffiths (1978); OME (1979)
3: Hudson *et al.* (1986), worms were not generally identified beyond family.
4: Griffiths (1989)
* non-native species; P present; ? present but sample month unknown.

immature organisms to the specific level are still lacking. For example, Hudson *et al.* (1986) collected larvae of 18 caddisfly and 27 chironomid taxa, but at the same time collected adults of 28 caddisfly and 121 chironomid species. Furthermore, several species of sphaerid clams, nematodes, flatworms, mites, and benthic cladocerans and copepods likely exist. We, therefore, concur with Hudson *et al.* (1986) that at least 300 benthic macroinvertebrate taxa occur in the river.

Exotic species reported from the river (Table 3) include *Bithynia tentaculata* L. (faucet snail), *Valvata piscinalis* Muller (European valve snail), *Corbicula fluminea* Muller (Asian clam), and *Dreissena polymorpha* Pallas (zebra mussel). *B. tentaculata* was imported to the Great Lakes

*Table 4.* Occurrence of common benthic macroinvertebrate taxa at 78 nearshore sampling sites in the St. Clair River, May 1985. Data from Griffiths (1989).

|  | % occurrence |
|---|---|
| CADDISFLIES: | |
| *Hydropsyche* | 50 |
| *Cheumatopsyche* | 40 |
| MAYFLIES: | |
| *Hexagenia limbata* | 45 |
| *Caenis* | 40 |
| TRUE FLIES: | |
| *Cryptochironomus* | 67 |
| *Polypedilum* | 65 |
| *Procladius* | 65 |
| *Tribelos* | 45 |
| AMPHIPODS: | |
| *Gammarus fasciatus* | 68 |
| SNAILS: | |
| *Amnicola* | 77 |
| *Elimia livenscens* | 68 |
| *Physella gyrina* | 67 |
| *Valvata tricarinata* | 49 |
| *Valvata piscinalis* | 46 |
| *Gyraulus* | 41 |
| CLAMS: | |
| *Pisidium* | 77 |
| WORMS: | |
| *Limnodrilus hoffmeisteri* | 83 |
| *Spirosperma ferox* | 74 |
| *Quistadrilus multisetosus* | 59 |
| *Limnodrilus udekemianus* | 42 |
| *Potamothrix moldaviensis* | 42 |
| *Limnodrilus claparedianus* | 40 |
| *Stylodrilus heringianus* | 40 |
| FLATWORMS: | |
| Turbellaria | 63 |

locations downstream of Marine City (W. Kovalak, Detroit Edison Co., pers. commun.) and at the head of the river in 1990.

The qualitative composition of the macroinvertebrate fauna appears to change little between seasons (Table 3). Most taxa have been collected from May through October and do not seem to be regulated by the presence of macrophyte beds or other seasonally varying factors.

Griffiths (1989) noted that the nearshore macroinvertebrate fauna was organized into four assemblages, occupying four different habitats (Tables 5 and 6; Fig. 2). One assemblage was dominated by the often overlooked, small-bodied and highly specialized chironomids *Chernovskiia* and *Saetheria*. These organisms occurred only at the head of the river where the substrate was composed almost solely of sand and was low in organic carbon and nutrients – a unique habitat known as the 'psammon' (Hynes, 1970). A second assemblage was characterized by the caddisflies *Cheumatopsyche* and *Hydropsyche*, the snail *Elimia* ( = *Goniobasis*) *livescens* Menke, and flatworms. This assemblage of invertebrates occurred in the 'erosional' habitat of the river characterized by swift flows, coarse sediments, and little organic matter. This habitat was concentrated in the upper quarter of the river upstream from Stag Island and also along river bends. A third assemblage was characterized by the mayfly *Caenis*, the amphipod *Gammarus fasciatus* Say, a variety of snails, notably *Amnicola*, and the worms *Limnodrilus hoffmeisteri* Claparede and *Spirosperma ferox* Eisen. This was the most common assemblage of macroinvertebrates observed in the river, occurring from Sarnia to Lake St. Clair. The assemblage occupied the 'run' habitat of the river characterized by intermediate flows and sandy-silt sediments. The fourth assemblage was characterized by the mayfly *Hexagenia limbata* Serville, a variety of chironomids, notably *Procladius*, *Cryptochironomus* and *Polypedilum*, and the worms *Quistradrilus multisetosus* Smith, *Limnodrilus claparedianus* Ratzel, *L. udekemianus* Claparede and *Potamothrix moldaviensis* Vejdovsky & Mrazek. This assemblage was found in the 'depositional' habi-

area from northern Europe in the late 1870's as an aquarium 'pet' (Pennak, 1978), and today can be found throughout the lower Great Lakes region. *V. piscinalis*, first recorded in Lake Ontario in 1913 (Clark, 1981), may have caused the disappearance of the native conspecific *Valvata sincera* Say. In contrast, *Corbicula* and *Dreissena* are recent invaders. *C. fluminea* is restricted to the nearshore area warmed by the heated effluent from the electrical generating plant of Detroit Edison (Fig. 1) located upstream of Marine City (French 3rd & Schloesser, 1991). *D. polymorpha* was observed in 1989 in low numbers at a few

110

*Table 5.* Density of common macroinvertebrates (No. 0.052 m$^{-2}$) in four nearshore benthic habitats reported by Griffiths (1989) from the St. Clair River, May 1985. P denotes a mean density of <1 individual.

| | Habitat | | | |
| --- | --- | --- | --- | --- |
| | Psammon | Erosional | Run | Depositional |
| CADDISFLIES: | | | | |
| Cheumatopsyche | | 7.7 | P | P |
| Hydropsyche | | 1.8 | P | P |
| Protoptila maculata | | 1.0 | | |
| MAYFLIES: | | | | |
| Caenis | | 1.2 | 4.5 | 1.2 |
| Hexagenia limbata | | P | 2.8 | 12.3 |
| Stenonema | | 1.2 | P | |
| TRUE FLIES: | | | | |
| Chernovskiia | 3.4 | | | |
| Chironomus | P | | 1.8 | 13.3 |
| Cryptochironomus | 2.1 | P | 8.8 | 33.0 |
| Harnischia | | | 1.0 | 4.6 |
| Paracladopelma | | P | P | 2.6 |
| Polypedilum | | P | 16.2 | 30.7 |
| Saetheria | 3.0 | | | |
| Stictochironomus | | | 2.7 | 7.9 |
| Tanytarsus | | P | 3.5 | 15.3 |
| Tribelos | | P | 18.0 | 13.7 |
| Monodiamesa | P | P | 1.0 | 4.0 |
| Procladius | | P | 11.3 | 67.1 |
| AMPHIPODS: | | | | |
| Gammarus fasciatus | | 1.4 | 6.2 | 4.4 |
| CLAMS: | | | | |
| Pisidium | P | P | 6.2 | 5.5 |
| SNAILS: | | | | |
| Amnicola | | P | 9.4 | 2.2 |
| Physella gyrina | | 1.0 | 2.7 | P |
| Gyraulus | | P | 1.8 | P |
| Elimia livescens | | 8.4 | 2.6 | P |
| Valvata piscinalis | | P | 3.2 | |
| Valvata tricarinata | | P | 3.6 | P |
| WORMS: | | | | |
| Stylodrilus heringianus | | P | 1.2 | P |
| Limnodrilus claparedianus | | P | 3.7 | 3.9 |
| Limnodrilus hoffmeisteri | | 3.8 | 31.2 | 12.3 |
| Limnodrilus udekemianus | | P | 1.4 | 3.1 |
| Potamothrix moldaviensis | | | 1.1 | 3.4 |
| Quistadrilus multisetosus | | | 2.6 | 8.9 |
| Spirosperma ferox | | P | 17.3 | 9.6 |
| FLATWORMS | | 4.5 | 4.3 | 2.5 |
| Mean richness (No. taxa 0.052 m$^{-2}$) | 2.6 | 8.9 | 15.3 | 17.5 |
| Mean density | 9.9 | 47.3 | 216 | 332 |

*Table 6.* Physicochemical characterization of sediments associated with the four benthic habitats reported by Griffiths (1989) from the St. Clair River, May 1985. Units are $g\,kg^{-1}$, except where specified.

| | Habitat | | | |
| --- | --- | --- | --- | --- |
| | Psammon | Erosional | Run | Depositional |
| Total organic carbon | <5.00 | 8.28 | 7.39 | 13.62 |
| Total phosphorus | 0.10 | 0.22 | 0.21 | 0.29 |
| Total Kjeldahl nitrogen | 0.13 | 0.36 | 0.35 | 0.71 |
| Particle size: % gravels | 0.60 | 67.00 | 0.70 | 0.00 |
| % sands | 98.40 | 32.00 | 79.20 | 51.80 |
| % silts and clays | 1.00 | 1.00 | 20.10 | 48.20 |

tat of the river characterized by quiet flows, fine silty sediments and accumulated organic matter. This habitat occurred primarily in the lower quarter of the river, downstream of Fawn Island and in Sarnia Bay.

The macroinvertebrate assemblages of the river, therefore, reflect the changes in physical conditions which range from areas of swift current with coarse, organically poor sediments in the headwaters to areas of slow current with fine, organically enriched sediments in the tailwaters. Invertebrate species richness and density both increase from the head of the river to its mouth (Table 5), suggesting that biological productivity may also increase along this gradient. Edwards *et al.* (1990) estimated the productivity of the macroinvertebrate fauna in the St. Clair River at $7.4\,g\,m^{-2}\,a^{-1}$ ash-free dry mass.

The change in functional organization of the macroinvertebrate assemblages suggests a downstream shift in the energy dynamics of the river towards internal heterotrophic production. Filter-feeders and scrapers are more prevalent in the headwaters of the river than in downstream areas, whereas gatherers exhibit the opposite trend. Shredders are not a significant component of any assemblage. In the upper part of the river, filter-feeders, mainly *Cheumatopsyche* and *Hydropsyche*, probably utilize the seston production from Lake Huron, while scrapers, mainly *E. livescens*, utilize periphyton growing on the hard substrate. The headwater community thus utilizes organic matter from a number of sources – externally produced seston, internally produced periphyton, internally and externally produced detritus. The downstream community, however, is composed largely of gatherers, particularly worms and chironomids, and probably depends greatly on internally produced detritus from the macrophyte community. The increasing productivity of downstream macroinvertebrate assemblages thus appears to be related to the internal production of organic matter by the macrophyte community.

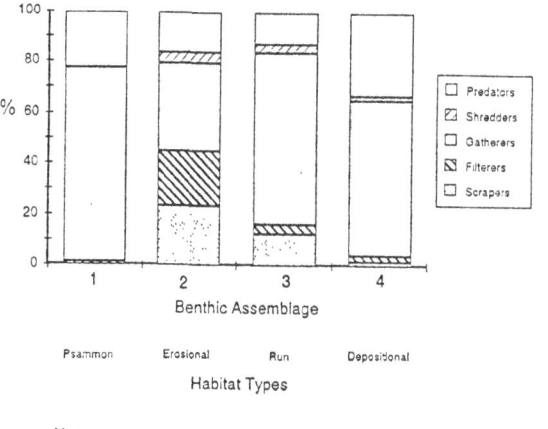

*Fig. 2.* Relative abundance of invertebrates by functional-feeding group in four benthic assemblages noted by Griffiths (1989) from the St. Clair River, May 1985.

*Fish*

The St. Clair River supports a diverse fish fauna. Five recent studies reported a total of 83 sport and forage species (Table 7). Seventeen of these

Table 7. Fish species recorded from the St. Clair River. Numbers refers to studies listed at end of table.

| Common name | Species | Egg | Larva | Juvenile and adult | | | | | | | | | | | |
|---|---|---|---|---|---|---|---|---|---|---|---|---|---|---|---|
| | | | | Jan | Feb | Mar | Apr | May | Jun | Jul | Aug | Sep | Oct | Nov | Dec |
| **LAMPREYS:** | | | | | | | | | | | | | | | |
| Brook lamprey | Ichthyomyzon fossor | | | | | | 2 | | | | | | | | |
| Silver lamprey | Ichthyomyzon unicuspis | | | | | | 2 | 2 | | | | | | | |
| Sea lamprey | Petromyzon marinus* | | 1,2 | | | | | | | | | 3 | | | |
| **STURGEON:** | | | | | | | | | | | | | | | |
| Lake sturgeon | Acipenser fulvescens | | | | | | | 2 | | | | | | | |
| **BOWFIN:** | | | | | | | | | | | | | | | |
| Bowfin | Amia calva | | | | | 4 | 4 | 4 | 4 | 4 | 4 | 4 | | | |
| **HERRINGS:** | | | | | | | | | | | | | | | |
| Alewife | Alosa pseudoharengus* | 1,2 | 1,2 | 2 | | | 2 | 2,4 | 2,4 | 2,4,5 | 2,4 | 2,3,5 | 2 | 2 | |
| Gizzard shad | Dorosoma cepedianum* | 1 | 1,2 | | 2 | 2,4 | 2,4 | 4 | 4 | | | 2,3,4 | 2,4 | 2 | 2 |
| **SALMONS:** | | | | | | | | | | | | | | | |
| Lake herring | Coregonus artedii | | 1 | | | | | | | | | | | | |
| Rainbow trout | Oncorhynchus mykiss* | | | | | | 2,4 | 4 | 2,4 | 2 | | 3 | 2,4 | 2 | 2 |
| Chinook salmon | Oncorhynchus tshawytscha* | | | | | | 4 | | 4 | | | 3 | | | |
| Coho salmon | Oncorhynchus kisutch* | | | | | | 2,4 | | | | | | | | |
| Brown trout | Salmo trutta* | | | 2 | 2 | 4 | | | 2 | | | 3 | 2 | 2 | 2 |
| Lake trout | Salvelinus namaycush | | | | | | | 4 | | | | | | | |
| **SMELT:** | | | | | | | | | | | | | | | |
| Rainbow smelt | Osmerus mordax* | 1,2 | 1,2 | | | 2,4 | 2,4 | 2,4,5 | 2,4,5 | 2 | 2 | 2,3 | 2 | 2 | 2 |
| **MUDMINNOW:** | | | | | | | | | | | | | | | |
| Central mudminnow | Umbra limi | | | | | | | 2 | | | | | | | |
| **MOONEYE:** | | | | | | | | | | | | | | | |
| Mooneye | Hiodon tergisus | | | | | | | 4 | | | | | | | |
| **MINNOWS:** | | | | | | | | | | | | | | | |
| Goldfish | Carassius auratus* | 1,2 | | | 2 | 4 | 4 | 4 | 2,4 | 4 | 2,4 | 2,3,4 | 4 | | |
| Carp | Cyprinus carpio* | 1,2 | 1,2 | | | 4 | 2 | | 2 | 2,4 | 2,4 | 2,3,5 | 2,4,5 | | |
| Silver chub | Hybopsis storeriana | | | 2 | | | | | 5 | | | 2,3,5 | 2,5 | 2 | |
| Hornyhead chub | Nocomis biguttatus | | | | | | | | 5 | 5 | | 2 | 2 | 2 | |
| River chub | Nocomis micropogon | | | | | | | | 4 | | | 3 | | | |
| Golden shiner | Notemigonus crysoleucas | | | | | | | | | 2 | | | | | |
| Satinfin shiner | Notropis analostanus | | | | | | | | | | | | | | |

Table 7. (Continued).

| | Egg | Larva | Juvenile and adult | | | | | | | | | | | |
|---|---|---|---|---|---|---|---|---|---|---|---|---|---|---|
| | | | Jan | Feb | Mar | Apr | May | Jun | Jul | Aug | Sep | Oct | Nov | Dec |
| *Notropis atherinoides* — Emerald shiner | | 1 | 2 | 2 | 2 | 2 | 2 | 2 | 2 | 2 | 2,3 | 2 | 2 | 2 |
| *Notropis bifrenatus* — Bridle shiner | | | | | | | | | | | | | 2 | 2 |
| *Notropis blennius* — River shiner | | | 2 | | | | | | | | | | | 2 |
| *Notropis boops* — Bigeye shiner | | | | | | 2 | 2 | | | | | | | |
| *Notropis chrysocephalus* — Striped shiner | | | | | | | | 5 | | 3 | 3 | 5 | | |
| *Notropis cornutus* — Common shiner | | | | | | 2 | 2 | 2 | | 3 | 3 | | | |
| *Notropis heterolepis* — Blacknose shiner | | | | | | | | | | 2 | 2 | | | |
| *Notropis hudsonius* — Spottail shiner | | 1 | 2 | 2 | 2 | 2 | 2,5 | 2,4,5 | 2,5 | 2 | 2,3,5 | 2,5 | 2 | 2 |
| *Notropis stramineus* — Sand shiner | | 1 | | | | | | | | | | | | |
| *Notropis volucellus* — Mimic shiner | | 1 | 2 | 2 | 2 | 2 | 2 | 2 | 2 | 2 | 2,3 | 2 | 2 | 2 |
| *Opsopoedus emiliae* — Pugnose minnow | | | 2 | 2 | | 2 | 2 | 2 | 2 | 2 | 2 | 2 | 2 | 2 |
| *Pimephales notatus* — Bluntnose minnow | | | 2 | 2 | 2 | 2 | 2 | 2 | | 2 | 2,3 | 2 | 2 | 2 |
| *Pimephales promelas* — Fathead minnow | | | | | | 2 | 2 | 2 | | | | | | |
| *Pimephales vigilax* — Bullhead minnow | | | 2 | | | | 2 | | | 2 | 2 | | | |
| *Rhinichthys atratulus* — Blacknose dace | | | | | | | | | | 2 | 2 | 2 | 2 | 2 |
| *Rhinichthys cataractae* — Longnose dace | | | | | | 2 | | | | 2 | 2 | 2 | 2 | 2 |
| **PIKES:** | | | | | | | | | | | | | | |
| *Esox lucius* — Northern pike | | | 2 | | 4 | 2,4 | 4 | 2,4 | 2,4 | 2,4 | 2,3,4,5 | 2,4 | | 2 |
| *Esox masquinongy* — Muskellunge | | | | | | | 4 | 2,4 | | | 2,4 | 2 | | |
| **SUCKERS:** | | | | | | | | | | | | | | |
| *Carpiodes cyprinus* — Quillback | | | | | 4 | 4 | 4 | 4 | 4 | 4 | 4 | 4 | | |
| *... ... ...* | | | | | 4 | | | | 4 | | | | | |
| *Moxostoma anisurum* — Silver redhorse | | | | | | | | | | – | – | – | | |
| *Moxostoma duquesnei* — Black redhorse | | | | | | | | | | | 5 | | | |
| *Moxostoma erythrurum* — Golden redhorse | | | | | | | | | | | 5 | | | |
| *Moxostoma macrolepidotum* — Shorthead redhorse | | | | | | 2 | | 2 | 2 | 2 | 2 | | | |

Table 7. (Continued).

| | Egg | Larva | Juvenile and adult | | | | | | | | | | | |
| --- | --- | --- | --- | --- | --- | --- | --- | --- | --- | --- | --- | --- | --- | --- |
| | | | Jan | Feb | Mar | Apr | May | Jun | Jul | Aug | Sep | Oct | Nov | Dec |
| **CATFISHES:** | | | | | | | | | | | | | | |
| *Ictalurus melas* Black bullhead | | | | | | 4 | 2 | 4 | 4 | | 4 | 4 | | |
| *Ictalurus nebulosus* Brown bullhead | | 2 | | | | 4 | 2,4 | 4 | 4 | 2,4 | 4 | 4 | | |
| *Ictalurus natalis* Yellow bullhead | | | | | | 4 | 4 | | | | 3 | | 2 | |
| *Ictalurus punctatus* Channel catfish | | | | | | 2 | 4 | 2,4 | 4 | 4 | 2,4 | 4 | | |
| *Noturus flavus* Stonecat | | | | | | | 4 | 4 | 2,4 | | 2,4 | 2 | | |
| *Noturus furiosus* Carolina madtom | | | | | | | | | | | | | | |
| **TROUT-PERCH:** | | | | | | | | | | | | | | |
| *Percopsis omiscomaycus* Trout-perch | 1 | 1,2 | | | 4 | 2 | 2,4 | 2,4,5 | 2 | 2 | 2,3 | | 2 | 2 |
| **BURBOT:** | | | | | | | | | | | | | | |
| *Lota lota* Burbot | | 1 | | | | | 4 | | | | | | | |
| **SILVERSIDE:** | | | | | | | | | | | | | | |
| *Labidesthes sicculus* Brook silverside | | 1 | 2 | | | 2 | | | | 2 | 2,3 | 2 | 2 | 2 |
| **STICKLEBACKS:** | | | | | | | | | | | | | | |
| *Culea inconstans* Brook stickleback | | | | | | 2 | 2 | | | | | | 2 | 2 |
| *Pungitius pungitius* Ninespine stickleback | | | | | | | | | | | | | 2 | 2 |
| **GAR:** | | | | | | | | | | | | | | |
| *Lepisosteus osseus* Long-nose gar | | | | | | | 4 | 2,4 | 2 | | 4 | | | |
| **BASSES:** | | | | | | | | | | | | | | |
| *Morone americana* * White perch | 1 | 1 | | | | | 4 | | 4 | 4 | 3,4 | | | |
| *Morone chrysops* White bass | | 2 | | | | 4 | 4 | 2,4 | 2,4 | 4 | 2,4 | 2,4 | | |
| **SUNFISHES:** | | | | | | | | | | | | | | |
| *Ambloplites rupestris* Rock bass | | 1 | 2 | 2 | 4 | 2,4 | 2,4,5 | 2,4,5 | 2,4,5 | 2,4 | 2,3,4,5 | 2,4,5 | 2 | 2 |
| *Lepomis cyanellus* Green sunfish | | | | | | | 2 | 2 | 4 | 2 | 3 | | | |
| *Lepomis gibbosus* Pumpkinseed | | | | | | 2,4 | 2,4 | 2,4 | 2,4,5 | 2,4 | 2,3,4 | | | |
| *Lepomis macrochirus* Bluegill | | | | | | | 2,4 | 4 | 2,4 | 2,4 | 2,3,4,5 | 4 | | |
| *Micropterus dolomieui* Smallmouth bass | | 2 | | | | | 4 | 2,4 | 4,5 | 2,4 | 2,3,4,5 | 2,4,5 | 2 | |
| *Micropterus salmoides* Largemouth bass | | | | | | | 2,4 | 4 | 4 | 4 | 2,3,4 | 4 | | |
| *Pomoxis annularis* White crappie | | 1 | | | | 4 | | | | | | | | |
| *Pomoxis nigromaculatus* Black crappie | | | | | 4 | 4 | 4 | 2,4 | 2,4 | 2,4 | 2,3,4,5 | 4,5 | | |

Table 7. (Continued).

| | | Egg | Larva | Juvenile and adult | | | | | | | | | | | |
|---|---|---|---|---|---|---|---|---|---|---|---|---|---|---|---|
| | | | | Jan | Feb | Mar | Apr | May | Jun | Jul | Aug | Sep | Oct | Nov | Dec |
| **PERCHES:** | | | | | | | | | | | | | | | |
| *Etheostoma caeruleum* | Rainbow darter | 1 | 1 | 2 | 2 | 2 | 2 | 2 | 2 | | | | 2 | 2 | 2 |
| *Etheostoma nigrum* | Johnny darter | | 1 | | 2 | 2 | 2 | 2 | 2 | | | | 2 | 2 | 2 |
| *Perca flavescens* | Yellow perch | 1,2 | 1,2 | 2 | | 2,4 | 2,4 | 2,4,5 | 2,4,5 | 2,4,5 | 2,4 | 2,3,4,5 | 2,4,5 | 2 | 2 |
| *Percina caprodes* | Logperch | 1 | 1 | | | | 2 | 2 | 2 | 2 | 2 | 2,3 | 2 | | |
| *Percina copelandi* | Channel darter | | | | | | | | | | | | 2 | | |
| *Stizostedion vitreum* | Walleye | 1 | 2 | | | 4 | 4 | 4 | 2,4 | 2,4 | 2,4 | 2,3,4 | 2,4 | 2 | 2 |
| **DRUMS:** | | | | | | | | | | | | | | | |
| *Aplodinotus grunniens* | Freshwater drum | | 1 | | | 4 | 4 | 2,4 | 2,4 | 2,4 | 4 | 3,4 | 4 | | |
| **SCULPINS:** | | | | | | | | | | | | | | | |
| *Cottus bairdi* | Mottled sculpin | 1 | 1,2 | | | | 2 | | 2 | | | 2,3 | 2 | | |
| *Myoxocephalus quadricornis* | Deepwater sculpin | | 1 | | | | | | | | | | 2 | 2 | 2 |

[1] Muth *et al.* (1986); Fish eggs collected with a pump, larval fish with a net towed behind a boat. Samples collected from April to June in 1983 and May to July in 1984.

[2] Texas Instruments (1975); Fish eggs and larvae collected with a Tucker trawl and epibenthic sled. Juvenile and adult fish collected with an Otter trawl, seine, gill net and box trap. Samples collected monthly from May 1974 to April 1975.

[3] Hamilton (1987); Fish collected with a boat equipped with electroshocking equipment in September of 1987.

[4] Haas *et al.* (1983); Fish collected with trap nets from March to September in 1983.

[5] Hudson *et al.* (1986); Fish collected with hoop nets in May, June, July, September and October of 1983 and 1984.

[*] non-native species.

species were rare. Four species, including *Petromyzon marinus* L., the exotic sea lamprey, and *Coregonus artedii* Lesueur, the native lake herring, were only represented by a few larval individuals, while 13 others, including lake sturgeon (*Acipenser fulvescens* Rafinesque) and lake trout (*Salvelinus namaycush* Walbaum), were represented only by one or two juveniles or adults. The most common species were alewife (*Alosa pseudoharengus* Wilson), gizzard shad (*Dorosoma cepedianum* Lesueur), rainbow smelt (*Osmerus mordax* Mitchill), emerald shiner (*Notropis atherinoides* Rafinesque), spottail shiner (*Notropis hudsonius* Clinton), white sucker (*Catostomus commersoni* Lacepede), yellow perch (*Perca flavescens* Mitchill), walleye (*Stizostedion vitreum* Mitchill), and rock bass (*Ambloplites rupestris* Rafinesque). This high species richness reflects a fish community that includes year-round residents, seasonal residents, and also transients that use the river as a migratory route between lakes Huron and St. Clair. Goodyear *et al.* (1982) reported that 28 species spawned in the river system; Table 7 shows that eggs, fish larvae, or both of eight species (spottail, sand and mimic shiners, brook silverside, white crappie, white perch, johnny darter, and deepwater sculpin) not noted by Goodyear *et al.* have also been observed, increasing the number of spawners to at least 36 species. The most abundant eggs and larvae in the river were those of alewife, rainbow smelt, gizzard shad, emerald shiners, logperch (*Percina caprodes* Rafinesque), trout-perch (*Percopsis omiscomaycus* Walbaum) and yellow perch (Muth *et al.*, 1986; Texas Instruments, 1975). The diverse array of habitats ranging from fast flowing, sand or cobble areas in the headwaters to quiet, heavily vegetated, silty areas in the delta thus satisfies the spawning and nursery requirements of a large number of species and contributes to the observed high species richness.

Historically, large runs of lake trout, lake whitefish (*Coregonus clupeaformis* Mitchill) and lake herring entered the St. Clair from lakes Erie and Huron to spawn (Goodyear *et al.*, 1982). These native coldwater species were major seasonal components of the fish fauna in the river; all three occupying the deeper, colder water of the Great Lakes during the warm summer months. Each of these populations disappeared around the turn of the century, probably as a result of overfishing. No lake whitefish and only a few larval lake herring and juvenile lake trout were collected in recent studies (Table 7).

The coldwater fish fauna today is largely composed of exotics: rainbow trout (*Oncorhynchus mykiss* Walbaum), brown trout (*Salmo trutta* L.), chinook salmon (*Oncorhynchus tshawytscha* Walbaum), coho salmon (*Oncorhynchus kisutch* Walbaum), rainbow smelt. They have essentially filled the niche left absent by the native species. Rainbow smelt is a seasonally abundant forage fish which spawns throughout the river. The piscivors use the river for spawning, feeding, and shelter and as a migration corridor between the Great Lakes. They support a small but important recreational fishery (Edsall *et al.*, 1988).

Important members of the coolwater fish fauna are lake sturgeon, northern pike (*Esox lucius* L.), muskellunge (*Esox masquinongy* Mitchill), walleye, and yellow perch. In contrast to coldwater species, coolwater species are present in the river throughout the year. Yellow perch and walleye are well adapted to large river habitats (Kitchell *et al.*, 1977) and support a large component of the recreational fishery (Haas *et al.*, 1983). Tagging studies indicate that the river is an important migration channel for both juveniles and adults (Haas *et al.*, 1983). Yellow perch utilize the river for spawning, feeding, and shelter. They usually occupy shallow waters, feed on benthic invertebrates and small fish, and in turn are prey for walleye, pike, and bass. Walleye and the rare, long-lived lake sturgeon frequent the deeper channels throughout the river, feeding primarily on fish. Both species rarely use the river for spawning. Haas *et al.* (1983) found that the Thames River, a tributary to Lake St. Clair, is probably the major spawning site for walleye in the St. Clair River. Muskellunge and northern pike utilize the channels in the delta as spawning and nursery habitat. Both species move throughout the river, feeding mainly on fish and other animals of the appropriate size.

Important members of the warmwater fish fauna include the suckers, catfishes, basses, sunfishes, and freshwater drum (*Aplodinotus grunniens* Rafinesque). These species are residents of the river throughout the year, migrating only small distances. The sunfishes, particularly rock bass (*Ambloplites rupestris* Rafinesque), and the freshwater drum are important components of the recreational fishery (Haas *et al.*, 1983). The white perch (*Morone americana* Gmelin), one of the most recent exotics to become established in the river, will probably contribute to the fishery in the future.

Overall, the fish community appears well balanced and healthy and able to utilize all energy sources of the ecosystem including benthos, zooplankton, algae, macrophytes, and fish. A large seasonal (e.g. alewife, smelt) and resident (e.g. minnows, trout-perch, darters, silversides) forage base is available to the coldwater, coolwater and warmwater piscivors. Total production of fish was estimated by Edwards *et al.* (1990) to be about $5.2 \, g \, m^{-2} \, a^{-1}$ ash-free dry mass. The effect of the fish community on the ecosystem, however, is unknown; in fact, the trophic structure and food-web dynamics of the St. Clair River still are poorly understood.

*Table 8.* Major effluent discharges along the Ontario shoreline of the St. Clair River.

| | Effluent source | General effluent characteristics [1] | Daily discharge ($10^3 \, m^3$) [2] |
|---|---|---|---|
| Chemical companies | Polysar Ltd. (Sarnia)<br>Dow Chemical Canada Ltd.<br>Ethyl Canada Inc.<br>CIL Inc.<br>Esso Chemical Canada Ltd.<br>DuPont Canada Inc.<br>Novacor Ltd.<br>Chinook Chemical Co. | ammonia<br>phenols<br>oils<br>suspended solids<br>greases<br>chlorinated organics | 1440 |
| Petroleum refineries | Suncor<br>Shell Canada Ltd.<br>Esso Petroleum Canada<br>Polysar Ltd. (Corunna) | ammonia<br>phenols<br>suspended solids<br>sulfide<br>acids & alkalies<br>oils<br>greases | 560 |
| Electrical power generators | Ontario Hydro Lambton Generating Station | heat<br>chlorine | 80000 |
| Cole drain | Scott Road Landfill<br>Fiberglass Canada<br>Cabot Carbon<br>Polysar Styrene II<br>Amaco | cyanide<br>chlorinated organics<br>polycyclic aromatic hydrocarbons<br>oils<br>greases | 147 |
| Waste treatment plants | Sarnia<br>Point Edward | phosphorus<br>metals<br>polychlorinated biphenyls | 56 |

[1] EC & US EPA, 1989
[2] OME, 1988a; 1988b

## Resource-use conflicts and their biological consequences

The St. Clair River flows through an urban and industrial corridor and is readily accessible to approximately four million people from southeastern Michigan and southwestern Ontario. As a result, there are many competing recreational, municipal, and industrial uses of this waterway and its resources. Perhaps of greatest concern from a natural resources standpoint is the use of the river as a source of municipal and industrial water and as a receiver of municipal and industrial wastes. More than $80 \times 10^6$ m³ of water (20% of the total flow of the river) is withdrawn daily from the St. Clair River and then returned to the river as effluent containing a wide variety of contaminants from industries and municipalities along the Ontario shoreline (Table 8).

The river has a history of pollution emanating from the urban centers and from the petrochemical complex on the Ontario shoreline. Abatement programs reduced impacts of some pollutants over the years but more recent concerns over toxic contaminants entering the river resulted in a comprehensive study by Canada and the United States to detail contaminant sources and loadings to the river. That study (EC & US EPA, 1989) revealed most contaminants entered the river as permitted discharges and spills, primarily from sources in the upper river in Ontario. Contaminants of major concern found in water or sediments of the river or both included hexachlorobenzene (HCB), octachlorostyrene (OCS), hexachlorobutadiene, hexachloroethane, benzo-a-pyrene, pentachlorobenzene, benzene, carbon tetrachloride, diphenylether, polychlorinated biphenyls (PCBs), and oil and grease; chlorides and some metals, including mercury and lead, were also contaminants of concern.

The study (EC & US EPA, 1989) also showed that contaminant inputs occur almost exclusively at the shoreline and that the flow regime of the river tends to hold the contaminant plumes against the shoreline as they are swept downstream, mix with the river waters, and are diluted. The concentration of contaminants, therefore,

Table 9. Concentrations and loadings of selected contaminants in St. Clair River water. (EC & US EPA, 1989)

| Contaminant | Concentration (ng l$^{-1}$) | Loadings (kg a$^{-1}$) |
| --- | --- | --- |
| Chloride | $10^6$–$10^7$ | $1.7 \times 10^8$–$1.7 \times 10^9$ |
| Phosphorus | $10^4$–$10^5$ | $1.7 \times 10^6$–$1.7 \times 10^7$ |
| Lead, cobalt, copper | $10^2$–$10^3$ | $1.7 \times 10^4$–$1.7 \times 10^5$ |
| Mercury | $10^1$–$10^2$ | $1.7 \times 10^3$–$1.7 \times 10^4$ |
| PCBs, PAHs, cadmium | $10^0$–$10^1$ | $1.7 \times 10^2$–$1.7 \times 10^3$ |
| HCB, OCS | $10^{-1}$–$10^0$ | $1.7 \times 10^1$–$1.7 \times 10^2$ |

tends to be highest near sources in the nearshore waters. Although the concentration of contaminants in the river is usually quickly reduced by the high volume of clean water entering the river from Lake Huron, the loadings to the system of many contaminants (Table 9) are nevertheless substantial and of considerable concern.

The fate of contaminants released into the river and their impacts on biota and habitats are incompletely known (EC & US EPA, 1989). The more volatile organic contaminants tend to be lost to the atmosphere as they move downstream, but other less volatile organics and the metals may complex with suspended particulate matter and accumulate in riverbed sediments and thus be flushed more slowly from the river. These sediment-bound contaminants are readily available for uptake by benthic organisms and may accumulate in the tissues of fish and birds that feed on these invertebrates, and also in humans that consume the fish and waterfowl from the river. Perhaps the best documented impact of contaminants on aquatic biota and habitat is the impairment of the benthic invertebrate community along a 12 km stretch of Ontario shoreline from the Cole Drain downstream past the southern tip of Stag Island (Fig. 3). This zone of impairment corresponds closely to the distribution of effluent plumes from shore-based discharges. Within the impacted zone' areas exist where the habitat is acutely toxic to the macrozoobenthos as evidenced by the absence of most species: the waterfront along Dow Chemical Canada Inc., for example, is one of these areas (Griffiths, 1989).

The effects of sublethal concentrations of contaminants on the health of aquatic organisms are

taminants act to influence abundance and production in mayflies has not been described.

Presently, concern is greatest for contaminants that bioaccumulate and threaten species at the top of the food chain, including humans (Table 10). Contamination with mercury has necessitated consumption restrictions for walleye, white sucker, yellow perch, and freshwater drum caught in the St. Clair River (OME & OMNR, 1989). In addition, carp (*Cyprinus carpio* L.) and gizzard shad have human health consumption guidelines for PCBs (OME & OMNR, 1989). PCBs in sport fish also exceed the IJC criterion of 100 ng g$^{-1}$ to protect fish-consuming birds and mammals (IJC, 1987). This criterion is also exceeded in young-of-the-year perch and shiners collected from the river (Suns *et al.*, 1985). Studies by the Great Lakes Institute (1987) and Weseloh & Struger (1987) showed that non-migratory ducks accumulated OCS and HCB during residence on the St. Clair River, suggesting that a human health

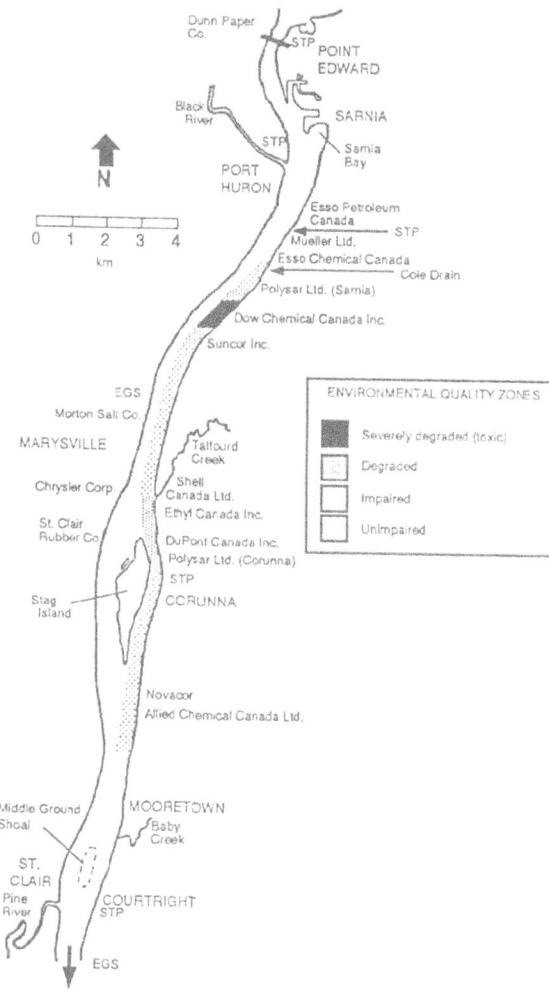

*Fig. 3.* Distribution of environmental quality zones in the St. Clair River, May 1985. The lower river was unimpaired. STP denotes sewage treatment plants. EGS denotes electrical generating stations. From Griffiths (1989).

less well documented or understood than are acute effects. Organic contaminants have been measured in adult mayflies, for example, in the St. Clair River (Ciborowski & Corkum, 1988) and sediments contaminated with oil and grease, other organics, and metals in apparently sublethal concentrations have been shown to limit the abundance and production of mayfly nymphs in the river (Edsall *et al.*, 1991; Schloesser *et al.*, 1991); however, the manner in which these con-

*Table 10.* Contaminants in biota from the St. Clair River.

| Biota | Contaminant | Concentration (ng g$^{-1}$) | Reference |
|---|---|---|---|
| Mussel *Elliptio complanata* | PCBs | 120–190 | 1 |
| | OCS | 140–430 | 1 |
| | HCB | 350–800 | 1 |
| | PAHs | 200–900 | 2 |
| Juvenile shiners *Notropis* | PCBs | 25–175 | 1 |
| | OCS | 25–700 | 1 |
| | HCB | 10–250 | 1 |
| | Mercury | 75–85 | 1 |
| Juvenile yellow perch *Perca flavescens* | Lead | 200–5000 | 1 |
| Adult caddisflies Trichoptera | PCBs | 208–685 | 3 |
| | OCS | 150–415 | 3 |
| | HCB | 40–190 | 3 |
| Alga *Cladophora* | OCS | 5–50 | 2 |
| | HCB | 1–13 | 2 |
| | Mercury | 40–240 | 2 |
| | Lead | 2000–6000 | 2 |

PCBs: polychlorinated biphenyls
OCS: octachlorostyrene
HCB: hexachlorobenzene; PAHs: polycyclic aromatic hydrocarbons
1: EC & OME (1986)
2: Ontario Ministry of Environment, unpublished
3: Ciborowski & Corkum (1988)

consumption guideline might be needed for migratory waterfowl that use the river.

Physical alterations of the river have also adversely affected the aquatic life and their habitats. The mining of river gravel deposits and removal of shoals and littoral areas represent a loss of habitat that could have been used for spawning by the Great Lakes fishes (Edsall et al., 1988). Bulkheading, dredging, and backfilling continue to destroy littoral habitats and wetlands. The paucity of aquatic life, particularly in the deeper channels of the river, results from the passage of commercial ships and associated physical alterations of the river. A shipping channel was dredged to a depth of approximately 8 m into the hard glacial clay of the river bottom. This reduced the physical diversity of the riverbed and increased water velocity and scour of the riverbed in the dredged channel, reducing the suitability of the habitat in the deeper portions of the river. Propeller wash from large commercial vessels that ply the river also contributed to bottom scour and turbulence and reduction of habitat quality in these deep water areas. Vessel passage can also alter the composition of assemblages of submersed macroflora along the river shoulders, causing species that are weakly rooted or less resistant to wave surge and drawdown effects to disappear (Schloesser & Manny, 1989).

The impacts of exotic species on the native biota of the river are not well known but may be substantial. Eurasian milfoil invaded the lower St. Clair River in the mid 1970's and by 1978 was the fourth most common macrophyte in the river (Schloesser & Manny, 1982, 1984) – yet the effect of this species on the native macrophytes, invertebrates or fishes is largely unknown. Proliferation of the zebra mussel will almost certainly affect the abundance of native species. Hundreds of zebra mussels have been found attached to individual native unionid mussels in Lake St. Clair and Lake Erie; in these high densities they appear to interfere with locomotion and feeding by the host unionid and may ultimately result in the death of the host. Zebra mussels may also reduce the abundance of hydropsychid caddisflies and other filter-feeders by monopolizing the sur-

face area of rocks and other firm substrates and by competing successfully for a common food supply. Sea lamprey and carp are clearly undesirable vertebrate exotics; alewife and rainbow smelt are abundant and important forage species but also are generally believed to adversely affect native forage species; and the introduced trout and salmon are highly regarded by sport anglers, but may have significantly altered the native fish community (Edsall et al., 1988).

## Summary and conclusions

Despite its large size, the St. Clair River has demonstrated its sensitivity to a variety of environmental impacts resulting from human activities in the river corridor and the drainage basin. Fortunately, the river has also demonstrated the ability to respond to mitigative measures, as evidenced by improvements in the macrozoobenthos following reductions in contaminant loadings (Griffiths, 1989). The Province of Ontario has begun a Municipal/Industrial Strategy for Abatement (MISA) which calls for the virtual elimination of toxics in discharges (OME, 1986) and which should result in further improvements in habitat quality and the biotic community. The St. Clair River is also one of 42 Areas of Concern identified in the revised Great Lakes Water Quality Agreement signed by the Governments of Canada and the United States of America (IJC, 1988). This agreement recognizes the need for an 'ecosystem approach' to manage the Great Lakes. This new approach gives consideration to the health of the aquatic community and their habitats and thus goes beyond the traditional approach which relied mainly upon numerical water quality standards that were designed to protect human health. A Remedial Action Plan is being developed for each Area of Concern to identify and restore beneficial uses (IJC, 1988). The skill with which these Remedial Action Plans are drafted and the resolve with which they are applied may be major determinants in the future character of the St. Clair River, its habitats, and biota.

## Acknowledgements

The authors thank Patrick L. Hudson and Donald W. Schloesser from the Great Lakes Fishery Laboratory of the U.S. Fish and Wildlife Service and John D. Westwood and Doug Huber of the Ontario Ministry of the Environment, Southwestern Region, for their assistance. Talks with H.B.N. Hynes were most helpful.

## References

Brown, C. L., T. P. Poe, J. R. P. French III & D. W. Schloesser, 1988. Relationships of phytomacrofauna to surface area in naturally occurring macrophyte stands. J. North Am. Benthol. Soc. 7: 129–139.

Bur, M. T., D. M. Klarer & K. A. Krieger, 1986. First records of a European cladoceran, *Bythotrephes cederstroemi*, in lakes Erie and Huron. J. Great Lakes Res. 12: 144–146.

Chambers, P. A. & J. Kalff, 1985. Depth distribution and biomass of submerged aquatic macrophyte communities in relation to Secchi depth. Can. J. Fish. aquat. Sci. 42: 701–709.

Ciborowski, J. J. H. & L. D. Corkum, 1988. Organic contaminants in adult aquatic insects of the St. Clair River and Detroit River, Ontario, Canada. J. Great Lakes Res. 14: 148–156.

Clark, A. H., 1981. The Freshwater Molluscs of Canada. Ntnl. Mus. Ntrl. Sci., Ottawa, Canada. 446 pp.

Davis, B. R. & K. F. Walker (eds.), 1986. The Ecology of River Systems. Dr W. Junk Publishers. Dordrecht, The Netherlands. 574 pp.

Derecki, J. A., 1983. Travel times in the Great Lakes Connecting Channels. Ntnl. Oceanic Atmos. Admin., Great Lakes Envir. Res. Lab., Ann Arbor, Michigan. 12 pp.

Derecki, J. A., 1984. St. Clair River physical and hydraulic characteristics. Ntnl. Oceanic Atmos. Admin., Great Lakes Envir. Res. Lab., Open File Rep. Ann Arbor, Michigan. 8 pp.

Duane, D. B., 1967. Characteristics of the sediment load in the St. Clair River. Proc. Tenth Conf. Great Lakes Res., Internat. Assoc. Great Lakes Res. 115–132.

EC & OME (Environment Canada & Ontario Ministry of the Environment), 1986. St. Clair River pollution investigation (Sarnia area). Toronto, Ontario, Canada. 135 pp.

EC & US EPA (Environment Canada & U.S. Environmental Protection Agency), 1989. Upper Great Lakes Connecting Channels study – Volume 2. Toronto, Ontario, Canada. 626 pp.

Edsall, T. A., B. A. Manny & C. N. Raphael, 1988. The St. Clair River and Lake St. Clair, Michigan: an ecological profile. U.S. Fish Wildl. Serv., Biol. Rep. 85 (7.3). 130 pp.

Edsall, T. A., B. A. Manny, D. W. Schloesser, S. J. Nichols & A. M. Frank, 1991. Production of *Hexagenia limbata* nymphs in contaminated sediments in the Upper Great Lakes Connecting Channels. Hydrobiologia 219: 353–361.

Edwards, C. J., P. L. Hudson, W. G. Duffy, S. J. Nepszy, C. D. McNabb, R. C. Hass, C. R. Liston, B. Manny & W. N. Busch, 1990. Hydrological, morphological and biological characteristics of the connecting rivers of the international Great Lakes: A review. In: D. P. Dodge (ed.). Proceedings of the International Large River Symposium. Can. Spec. Publ. Fish. aquat. Sci. 106: 240–264.

Edwards, R. W. & M. Owens, 1960. The effects of plants on river conditions. I. Summer crops and estimates of net productivity of macrophytes in a chalk stream. J. Ecol. 48: 151–160.

Elwood, J. W., J. D. Newbold, R. V. O'Neill & W. van Winkle, 1983. Resource spiralling: an operational paradigm for analyzing lotic ecosystems. In: T. P. Fontaine & S. M. Bartell (eds.). Dynamics of Lotic Ecosystems. pp. 3–27. Ann Arbor Science, Ann Arbor, Michigan.

ENCOTEC (Environment Control Technology Corporation). 1974. Water pollution investigations: Detroit and St. Clair Rivers. U.S. Envir. Protect. Agency, Enforcement Div., Chicago, Illinois. EPA-905-9-74-013. 361 pp.

Fitzgerald, W. D., E. Janicki & D. J. Storrison, 1979. Sarnia-Bright's Grove area, southern Ontario. Ont. Geol. Surv. Prelim. Map P.2222, Quatern. Geol. Ser. Toronto, Ontario, Canada.

French, J. R. P. III, 1988. Effect of submerged aquatic macrophytes on resource partitioning in yearling rock bass and pumpkinseeds in Lake St. Clair. J. Great Lakes Res. 14: 291–300.

French, J. R. P. III & D. W. Schloesser, 1991. Growth and overwinter survival of the Asiatic clam, *Corbicula fluminea*, in the St. Clair River, Michigan. Hydrobiologia 219: 165–170.

Goodyear, C. D., T. A. Edsall, D. M. O. Dempsey, G. D. Moss & P. E. Polanski, 1982. Atlas of the spawning and nursery areas of Great Lakes fishes. Vol. VI, St. Clair River; Vol. VII, Lake St. Clair; Vol. VIII, Detroit River. U.S. Fish Wildl. Serv. FWS/OBS-82/52. 86 pp.

Great Lakes Institute, 1987. Organochlorinated compounds in duck and muskrat populations of Walpole Island. Walpole Island Band Council, Ontario, Canada. 31 pp.

Griffiths, R. W., 1978. Benthic communities as indicators of water quality in the St. Clair River. B.Sc. Thesis. Univ. Western Ontario, London, Ontario, Canada. 40 pp.

Griffiths, R. W., 1989. Environmental quality of the St. Clair River as reflected by the 1985 distribution of benthic invertebrates. Ont. min. Envir., London, Ontario, Canada. 41 pp.

Haag, R. W., 1979. The ecological significance of dormancy in some rooted aquatic plants. J. Ecol. 67: 727–738.

Haas, R. C., M. G. Galbraith & W. C. Bryant, 1983. Movement and harvest of fish in Lake St. Clair, St. Clair River,

122

and Detroit River. Lake St. Clair Fish. Stat., Fish. Div., Michigan Dept. Ntrl. Resources. 80 pp.

Hamilton, J. G., 1987. Survey of critical fish habitat within International Joint Commission designated Areas of Concern, August–November, 1976. Ont. Min. Ntrl. Resources, Fish. Br., Toronto, Ontario, Canada. 117 pp.

Hiltunen, J. K., 1980. Composition, distribution and density of benthos in the lower St. Clair River, 1976–1977. U.S. Fish Wildl. Serv., Great Lakes Fish. Lab. Admin. Rep. 80-4. Ann Arbor, Michigan. 17 pp.

Hudson, P. L., B. M. Davis, S. J. Nichols & C. M. Tomcko, 1986. Environmental studies of macrozoobenthos, aquatic macrophytes and juvenile fishes in the St. Clair-Detroit River system, 1983–1984. U.S. Fish Wildl. Serv., Great Lakes Fish. Lab. Admin. Rep. 86-7. Ann Arbor, Michigan. 116 pp.

Hunt, G. S., 1963. Wild celery in the lower Detroit River. Ecology 14: 360–370.

Hynes, H. B. N., 1970. The Ecology of Running Waters. Liverpool Univ. Press. Liverpool, England. 555 pp.

IJC (International Joint Commission), 1987. New and revised Great Lakes water quality objectives, Volume I. Windsor, Ontario, Canada.

IJC, 1988. Revised Great Lakes Water Quality Agreement of 1978. Windsor, Ontario, Canada.

Kitchell, J. F., M. G. Johnson, C. K. Minns, K. H. Loftus, L. Creig & C. H. Oliver, 1977. Percid habitat: The river analogy. J. Fish. Res. Bd Can. 34: 1936–1940.

Kunii, H., 1981. Characteristics of winter growth of detached *Elodea nuttallii*, St. John in Japan. Aquat. Bot. 11: 57–66.

Lau, Y. L., B. G. Oliver & B. G. Krishnappan, 1989. Transport of some chlorinated contaminants by the water, suspended sediments and bed sediments in the St. Clair and Detroit Rivers. Envir. Toxicol. Chem. 8: 293–301.

Manny, B. A., D. W. Schloesser, S. J. Nichols & T. A. Edsall, 1988. Drifting submerged macrophytes in the upper Great Lakes channel. U.S. Fish Wildl. Serv., Ntnl. Fish. Res. Centre-Great Lakes. Ann Arbor, Michigan.

Muth, K. M., D. R. Wolfert & M. T. Bur, 1986. Environmental study of fish spawning and nursery area in the St. Clair-Detroit River system. U.S. Fish Wildl. Serv., Sandusky Biol. Stat. Sandusky, Ohio.

Oliver, B. G., 1988. Sediment work group – Geographic area report for the St. Clair River – A level II report for the Upper Great Lakes Connecting Channels Study.

OME (Ontario Ministry of Environment), 1979. St. Clair River organics study, biological surveys 1968 and 1977. Water Resources Assessment Unit, London, Ontario, Canada. 90 pp.

OME, 1986. A policy and program statement of the Government of Ontario on controlling municipal and industrial discharges to surface waters. Communications Branch, Toronto, Ontario, Canada. 53 pp.

OME, 1988a. Report on the 1987 industrial direct discharges in Ontario. Toronto, Ontario, Canada. 250 pp.

OME, 1988b. Report on the 1987 discharges from sewage treatment plants in Ontario. Toronto, Ontario, Canada. 550 pp.

OME & OMNR (Ontario Ministry of Environment & Ontario Ministry of Natural Resources), 1989. Guide to Eating Ontario Sport Fish. Public Information Centre, Toronto, Ontario, Canada. 150 pp.

Pennak, R. W., 1978. Freshwater Invertebrates of the United States. 2nd Edition. John Wiley and Sons, Toronto. 803 pp.

Poe, T. P., C. O. Hatcher, C. L. Brown & D. W. Schloesser, 1986. Comparisons of species composition and richness of fish assemblages in altered and unaltered littoral habitats. J. Freshwat. Ecol. 3: 525–536.

Quinn, F. H. & R. N. Kelly, 1983. Great Lakes monthly hydrologic data. Ntnl. Oceanic Atmos. Admin., Great Lakes Env. Res. Lab., Ann Arbor, Michigan. 79 pp.

Ritter, D. F., 1986. Process Geomorphology. 2nd edition. W.C. Brown, Dubuque, Iowa. 579 pp.

Schloesser, D. W., 1986. A field guide to valuable underwater aquatic plants of the Great Lakes. U.S. Fish Wildl. Serv., Great Lakes Fish. Lab., Ann Arbor, Michigan and Cooperative Extension Serv., Michigan State Univ., East Lansing, Michigan. 32 pp.

Schloesser, D. W. & B. A. Manny, 1982. Distribution and relative abundance of submerged aquatic macrophytes in the St. Clair-Detroit River ecosystem. U.S. Fish Wildl. Serv., Great Lakes Fish. Lab. Admin. Rep. 82-7. Ann Arbor, Michigan. 21 pp.

Schloesser, D. W. & B. A. Manny, 1984. Distribution of eurasian watermilfoil, *Myriophyllum spictatum*, in the St. Clair-Detroit River system in 1978. J. Great Lakes Res. 10: 322–326.

Schloesser, D. W. & B. A. Manny, 1989. Potential effects of shipping on submersed macrophytes in the St. Clair and Detroit Rivers of the Great Lakes. Michigan Academician XXI: 101–108.

Schloesser, D. W., T. A. Edsall & B. A. Manny, 1985. Growth of submerged macrophyte communities in the St. Clair-Detroit River system between Lake Huron and Lake Erie. Can. J. Bot. 63: 1061–1065.

Schloesser, D. W., T. A. Edsall, B. A. Manny & S. J. Nichols, 1991. Distribution of *Hexagenia* nymphs and visible oil in sediments of the Upper Great Lakes Connecting Channels. Hydrobiologia 219: 345–352.

Suns, K., G. E. Crawford, D. D. Russell & R. E. Clement, 1985. Temporal trends and spatial distribution of organochlorine and mercury residues in Great Lakes spottail shiners (1975–1983). Ont. Min. Env., Rexdale, Ontario, Canada.

Texas Instruments, 1975. Report of fish and macrozooplankton studies on the St. Clair River in the vicinity of the proposed Belle River Power Plant. Detroit Edison Co., Detroit, Michigan.

Vallentyne, J. R. & A. L. Hamilton, 1987. Managing human

uses and abuses of aquatic resources in the Canadian ecosystem. In: M. C. Healey & R. R. Wallace (eds.). Canadian Aquatic Resources. Can. Bull. Fish. aquat. Sci. 215: 513–533.

Vollenweider, R. A., M. Munawar & P. Stadelmann, 1974. A comparative review of phytoplankton and primary production in the Laurention Great Lakes. J. Fish. Res. Bd. Can. 31: 739–762.

Watson, N. H. F. & G. F. Carpenter, 1974. Seasonal abundance of crustacean zooplankton and net plankton biomass of Lakes Huron, Erie, and Ontario. J. Fish. Res. Bd Can. 31: 309–317.

Werner, E. E., J. F. Gilliam, D. J. Hall & G. G. Mittlebach, 1983. An experimental test of the effects of predation risk on habitat use in fishes. Ecology 64: 1540–1548.

Weseloh, D. V. & J. Struger, 1987. Contaminants in wildlife in the Upper Great Lakes Connecting Channels. UGLCC Study Report. Can. Wildl. Serv., Burlington, Ontario, Canada. 9 pp.

Westlake, D. F., 1963. Comparisons of plant productivity. Biol. Rev. Camb. Philos. Soc. 38: 385–425.

Wetzel, R. G., 1975. Limnology. W.B. Saunders Co., Philadelphia, Pennsylvania. 743 pp.

Whitton, B. A., 1975. River Ecology. Univ. California Press. Berkeley, California. 725 pp.

*Hydrobiologia* **219**: 125–134, 1991.
*M. Munawar & T. Edsall (eds),*
*Environmental Assessment and Habitat Evaluation of the Upper Great Lakes Connecting Channels.*
© 1991 *Kluwer Academic Publishers.*

# Distribution and abundance of young fish in the St. Clair River and associated waters, Ontario

J.K. Leslie & C.A. Timmins
*Bayfield Laboratory, Great Lakes Laboratory for Fisheries and Aquatic Sciences, 867 Lakeshore Road, Burlington L7R 4A6, Canada*

*Key words:* larval fish, distribution, abundance, St. Clair River

## Abstract

Fish larvae were sampled in 1986 in the St. Clair River, and adjacent waters. Species richness (9 taxa as larvae; 4 others as juveniles) and abundance was lowest in the river, where many larvae (e.g., burbot, rainbow smelt, and yellow perch) were in transit from Lake Huron. The most abundant, and localized, species was gizzard shad, which reached a peak mean density of 4 600 larvae 100 m$^{-3}$ in an agricultural canal. Adjacent waters contribute greatly to the fish communities of the river and adjoining Lakes Huron and Erie, especially in terms of the number and quantity of forage species.

## 1. Introduction

Entrainment studies at power plants during the 1970's and 1980's (Leslie *et al.*, 1979; Leslie, 1986; Texas Instruments, 1975) and general surveys (Hatcher & Nester, 1983; Muth *et al.*, 1986) suggest larvae of only a few species inhabit the main channel of the St. Clair River, because swift currents, low temperatures, and lack of cover provide an environment unsuited to most species. However, the protected shallow areas along the shore of the river, in tributaries, agricultural drainage ditches, and around islands in the river are preferred habitat for many warmwater species that generally are absent from the main channel. This study describes the distribution and abundance of larval fish in these protected habitats on the Canadian side of the river and provides a broader basis for evaluating the impact of pollution and other human activities on the fish community of the river.

## 2. Study area

The St. Clair River is the channel connecting Lakes Huron and St. Clair (Fig. 1). Average monthly discharge is 5 100 m$^3$ s$^{-1}$, with midstream surface velocities ranging from 0.6 to 1.8 m s$^{-1}$ (Derecki, 1984). Because most of the shore in Michigan is bulkheaded and much of the shoreline in Ontario has been developed for industrial and residential use, the amount of available substrate for spawning and for establishment of submersed and emergent macrophytes is limited. Growth of macrophytes is minimal before mid-June (Schloesser *et al.*, 1985; Hudson *et al.*, 1986), which may be too late to provide adequate cover and opportunity for feeding for many young fish that hatched weeks previously. Thus, only two of six sites we sampled regularly from mid-April to late July, 1986, were in the river proper. Table 1 describes the habitat at each site, including two that were sampled occasionally.

126

*Table 1.* Habitat characteristics of study sites in the St. Clair River and adjacent waters, 1986.

| Sampling location | Habitat | Water clarity | Substrate | Cover |
|---|---|---|---|---|
| Bonny Doon | protected creek | turbid | sand, gravel | *Scirpus* sp. |
| Lake Huron | exposed lake-shore | clear | gravel, sand | none |
| Sarnia Beach | protected beach | clear | sand | none |
| Sarnia | exposed near-shore, river | turbid | rock, clay | *Myriophyllum* sp. |
| Stag Island | protected marsh, channels | clear | sand, clay | *Scirpus* sp. |
| Lambton Gen. Stn. | exposed shore, river | turbid | rock, concrete | none |
| Cundick Park | protected creek | turbid | ooze, gravel | *Carex* sp. |
| Whitebread Canal | protected canal | highly turbid | clay | marginal |

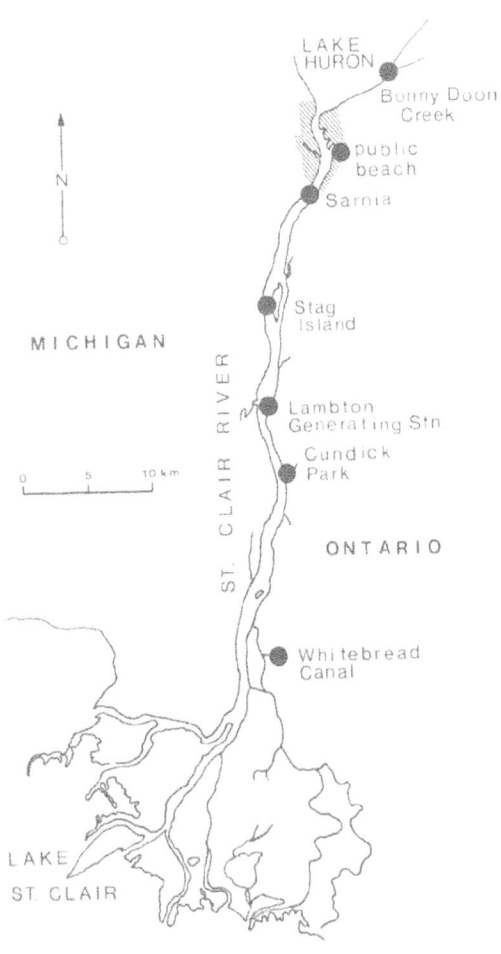

*Fig. 1.* Location of sampling sites for larval fish in the St. Clair River and environs, Ontario, 1986.

Bonny Doon is a small creek connected by municipal drains to the St. Clair River, but its main channel flows into Lake Huron (Fig. 1). A pool about 20 m from the outlet into the lake periodically slowly fills and discharges rapidly when accumulated debris from upstream breaks open. Lake Huron at the mouth of the creek is fully exposed to prevailing westerly winds. Open water at Sarnia was sampled about 30 m offshore.

Stag Island is 1 km long and 0.2 km wide, and is situated in the centre of the river. It has five parallel artificial channels originally constructed for a marina complex but long since (60 yr) aborted, marshy areas, and narrow sand beaches, all of which are subjected to drawdown and flooding when ships pass.

Lambton Generating Station, with a shoreline surface intake, draws 0.7 percent of the mean river flow for condenser cooling water, which is 'once-through'. It is located about mid-way between Lake Huron and the St. Clair Delta (Fig. 1). The shore at the power plant is stabilized with artificial fill (concrete and boulders).

Whitebread Canal is an agricultural drainage ditch that is connected to the river, several kilometers distant, and is about 15 m wide and 4.5 m deep; banks are steep and water depth generally maintained at about 1.5 m. Virtually no vegetative cover exists for fish, although the turbid, 'milky' water probably provides some cover.

Samples on an irregular basis were taken at a public beach in Sarnia during May and June. Similarly, a small ditch in Cundick Park (Fig. 1) was sampled twice in July, once each in September and October, 1986, and in January, 1987. These sites were added partly because of the possibility of detecting new or exotic species entering Canada from the United States or elsewhere. Such species are known to affect distribution and abundance of indigenous fish (Christie, 1974).

## 3. Methods

A beach seine (4 m long, 1 m wide, 0.3 mm mesh) designed to collect fish larvae was deployed in water 1.0 m deep, and filtered about 25 m³ in a standard 10 m sweep. Two such samples were taken at the edge of the pool at Bonny Doon Creek, along the adjacent shore on Lake Huron, and at Stag Island. At Sarnia, paired plankton nets (1.0 m × 0.5 m mouth opening, 2.5 m long, 0.3 mm mesh) were pushed at the surface. A single 0.3 m dia conical net (0.3 mm mesh) was also fished at a depth of 3 m. Sampling duration was 10 min against the current, after which nets were retrieved, rinsed, and their contents removed, and then pushed with the current for 10 min.

Similarly, at Stag Island, paired conical nets were pushed along one of the artificial channels for 5 or 10 min, then, after contents were removed, the nets were pushed over the same course in the reverse direction.

Power plants such as Lambton Generating Station that use 'once-through' cooling water for condensers are unbiased large-volume samplers of larval fish. Thus, we sampled a portion of intake cooling water, which is taken from the shore of the St. Clair River. Six nets in total were set simultaneously for 10 min, using two conical nets 2.5 m long, 1 m diameter with 0.5 mm mesh, two nets with rectangular mouths (1.0 m × 0.5 m) and 2.5 m in length with 0.5 mm mesh, and two conical nets with mouth openings of 0.3 m in diameter, 1 m long, and 0.3 mm mesh. After passive sampling for 10 min, all nets were retrieved, washed down and their contents removed and fixed with formalin. The six nets were then set again for 10 min and the process was repeated. Nets were placed at 1 and 3 m alternately and equidistant 3 m apart in the intake forebay. Mean abundance of larvae was determined from the mean volume filtered (1 112 m³) and mean catch in six nets in 10 min. Flow speed was determined with direct-reading current meters. Samples were taken from late April to mid-July, 1986. In addition, samples were taken at the discharge in mid-October with a beach seine 10 m long and 6 mm mesh opening, in order to catch fish present in waters artificially elevated in temperature.

Larval fish were collected in Whitebread Canal with a conical net fixed to the bow of a canoe pushed for 0.5 km. The contents of the net were removed, and the net was again pushed over the same course in the reverse direction. Sampler speed was monitored with a direct-reading current meter placed in front of the net. Four or five samples were also taken with the 0.3 mm mesh seine at the edge of the canal. On two occasions, a gill net (48 mm stretched mesh) was set across the canal. All other sampling was performed with a 3 m long seine of 3 mm mesh. The 0.3 mm mesh seine, a 3 mm mesh seine, and dip nets were used at Sarnia Beach and in Cundick Park. All fish were immediately fixed with 5 to 10 percent formalin, and later (within 2 mo) preserved with Davidson's B solution. Total length (TL) of fish 30 mm or less was measured to the nearest 0.2 mm; 30 larvae in large samples were measured, and in samples containing fewer than 30 larvae, all were measured. Fish larger than 30 mm were measured (TL) to the nearest 0.5 mm. Temperature, pH, and conductivity were measured at most sites (Table 1); climatological observations and general classification of aquatic and terrestrial vegetation at each site were recorded. Water velocity 1 m ahead of the net was determined with a direct-reading current meter to provide a basis for calculating the volume of water sampled. Nets were assumed to filter at 100 percent efficiency.

*Table 2.* Number of species in the main families of fish in the St. Clair River and adjacent waters, 1986.

| Family | St. Clair River | | | | Adjacent waters | | |
|---|---|---|---|---|---|---|---|
| | Sarnia | Lambton Gen. Stn. | Cundick Park | Stag Is. | Whitebread Canal | Bonny Doon Ck. | Lake Huron |
| Cyprinidae | 2 | 1 | 7 | 15 | 8 | 12 | 14 |
| Percidae | 2 | 3 | 0 | 5 | 2 | 4 | 2 |
| Catostomidae | 1 | 1 | 1 | 2 | 1 | 2 | 2 |
| Centrarchidae | 0 | 0 | 4 | 1 | 7 | 0 | 0 |
| Clupeidae | 1 | 1 | 0 | 1 | 1 | 2 | 2 |
| Others | 3 | 4 | 1 | 8 | 6 | 3 | 4 |
| Total species | 9 | 10 | 13 | 32 | 25 | 23 | 24 |
| Total families | 7 | 8 | 4 | 13 | 11 | 7 | 8 |

## 4. Results

### 4.1. Bonny Doon Creek

Of a total of 30 species collected in Bonny Doon Creek, 12 were cyprinids and 4 were percids (Table 2) including a single specimen of Canada's smallest fish, the blackside darter (*Percina maculata*). Cyprinids composed 84 percent of the total catch of 1050 fish at this site. Of this total catch, 37 percent were spottail shiner (*Notropis hudsonius*), 32 percent were bluntnose minnow (*Pimephales notatus*), and 16 percent were sand shiner (*Notropis stramineus*). It is unlikely that any of these cyprinids came from Lake Huron because the outflow of the creek is either a shallow trickle or a deluge when accumulated debris at the mouth of the stream suddenly opens. Several rainbow trout (*Salmo gairdneri*) parr were the first young fish caught (mid-April), succeeded by white sucker (*Catostomus commersoni*) early in May, and cyprinids in June and July.

Spottail shiner was the most abundant fish (mean 364 larvae 100 m$^{-3}$ on June 23), followed by bluntnose minnow (mean 220 larvae 100 m$^{-3}$ on May 21), and white sucker (mean 138 larvae 100 m$^{-3}$ in late June), which formed 14 percent of the total catch. Newly-hatched (10 mm long) white sucker larvae were first collected at 18.5 °C, and remained in the creek until mid-July, when they were 33 mm long. Several cyprinid species were not caught as larvae but were present as juveniles or adults.

### 4.2. Lake Huron

Yellow perch (*Perca flavescens*) larvae were first caught in early May, when the water temperature was 12 °C (Fig. 2) and were present in collections at the shore of Lake Huron until mid-July. Yellow perch was followed chronologically by white sucker and rainbow smelt (*Osmerus mordax*), both of which appeared in mid-May, and cyprinids within the following 6 weeks. Alewife

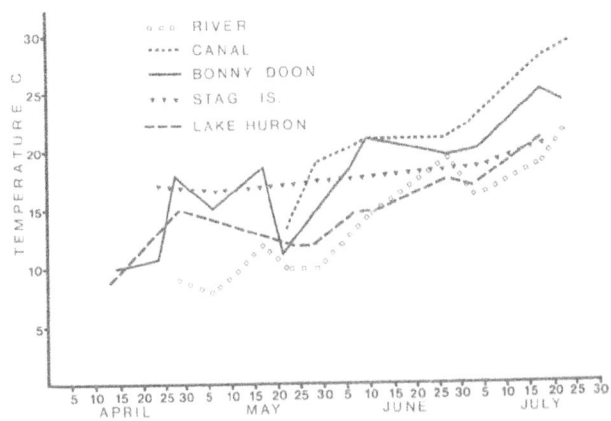

*Fig. 2.* Seasonal water temperature ( °C) at sampling sites in the St. Clair River and connecting waters, Ontario, 1986.

(*Alosa pseudoharengus*) was the most common species from early July to mid-September.

Variations in temperature at the shore of Lake Huron were less exaggerated than in Bonny Doon Creek, and as a result of wave action, temperature was usually lower than in the creek (Fig. 2). The different habitats at the shore of Lake Huron and in Bonny Doon Creek had few species in common. Fourteen cyprinid species were captured in Lake Huron, seven of which were also found in Bonny Doon Creek. Spottail shiner probably spawned at both sites because newly-hatched (5 mm) larvae were found in Bonny Doon Creek

3 or 4 weeks before they appeared in Lake Huron. They formed 25 percent of the total catch of 1 200 larvae in Lake Huron, and 50 percent of the catch of all cyprinids. Lake chub (*Couesius plumbeus*) and bigmouth shiner (*Notropis dorsalis*) were collected only in Lake Huron. Several cyprinids which we captured, e.g., carp (*Cyprinus carpio*) and bluntnose minnow, may have been flushed into the lake during periodic discharges of the stream because they are not usually found in relatively clear water along exposed shores of large lakes. Two other abundant species were alewife, which composed 31 percent of the total catch, and

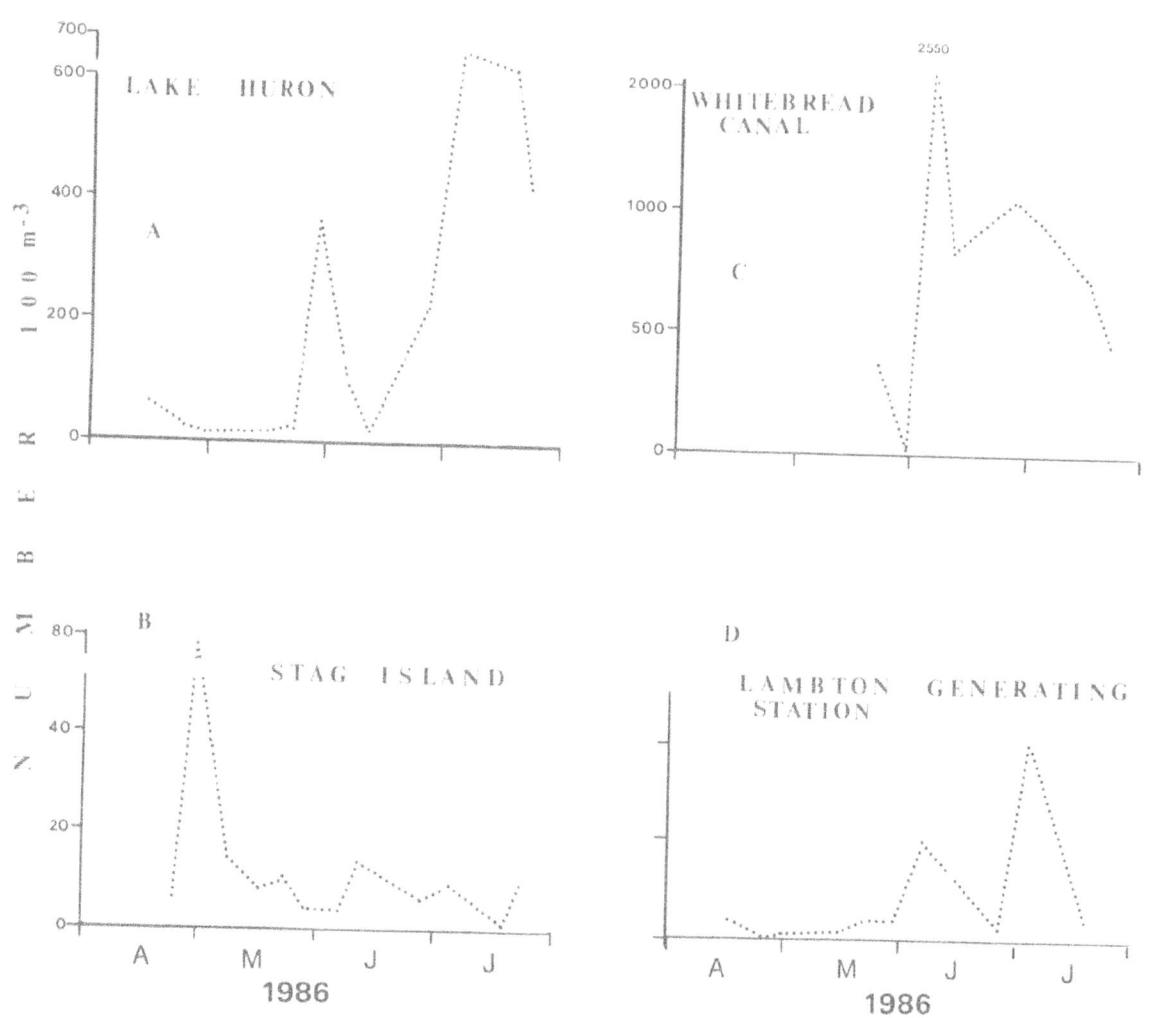

*Fig. 3.* Seasonal mean abundance (number 100 m$^{-3}$) of all larval fish at (A) Lake Huron, (B) Stag Island, (C) Whitebread Canal, and (D) Lambton Generating Station, 1986.

white sucker (14 percent); the former were present at the shore from early July until mid-September, and the latter evidently emigrated from Bonny Doon Creek or nearby streams in late May. Abundance of alewife in Lake Huron was higher than at other sites we sampled. On July 18 and 23, peak mean abundances were 130 and 150 larvae $100 \text{ m}^{-3}$, when the mean length of larvae was 15.8 and 19.4 mm, respectively. Similarly, the mean abundance of spottail shiner was highest in July (65 to 354 larvae $100 \text{ m}^{-3}$). The graph of mean abundance of all fish species (Fig. 3) mainly reflects dominance in late May of spottail shiner and in July of alewife.

## 4.3. Sarnia Beach

All species (6) collected were forage fish. Ninety percent of the catch of 200 young fish in May and June was emerald shiner (*Notropis atherinoides*), and most of the remainder were juvenile least darter (*Etheostoma microperca*). Fish were not likely to use this shallow area for spawning or as a nursery habitat due to continuous interference from the public.

## 4.4. Sarnia

Few species (9), low total catch (150 larvae) and low abundance (generally less than 1 larva $100 \text{ m}^{-3}$) characterized the fish community sampled in the St. Clair River at Sarnia (Fig. 1). Burbot (*Lota lota*) appeared first in late April, when water temperature was 6 °C; rainbow smelt and yellow perch next appeared in late May and early June, respectively, followed by white sucker and alewife in July. Uncommon species collected were a single adult specimen each of longnose gar (*Lepisosteus osseus*), lake sturgeon (*Acipenser fulvescens*), and slimy sculpin (*Cottus cognatus*). Relatively few species spawn in the river and recent surveys and the published lists of fish that occur as adults and larvae did not correspond closely (Goodyear *et al.*, 1982; Haas *et al.*, 1985; Hudson *et al.*, 1986; Edsall *et al.*, 1988). Larval gizzard shad (*Dorosoma cepedianum*), which were scarce in the river in 1974, 1977, and 1978 (Goodyear *et al.*, 1982) made up only 2 to 4 percent of the total catch in 1983 and 1984 (Muth *et al.*, 1986), and in the present study, were absent between Sarnia and Whitebread Canal (Fig. 1).

## 4.5. Stag Island

Water temperature between April and July at Stag Island remained almost constant at 17 °C. Larvae appeared mainly between April and mid-June at Stag Island, when temperature in sheltered waters reached 15 to 17 °C. Burbot and deepwater sculpin (*Myoxocephalus thompsoni*) were the first fish larvae sampled (in early May), and in May were succeeded by rainbow smelt, lake herring (*Coregonus artedii*), and pumpkinseed (*Lepomis gibbosus*). Most cyprinids occurred throughout the summer. No gizzard shad or alewife were found, which suggests either that no spawning habitat exists for them at Stag Island, or that a hydrologic barrier exists (UGLCC, 1989), which confines them to nearshore areas. Mean abundance of all fish collected on each sampling date was generally low (Fig. 3), which probably reflects the absence of larvae in areas with sparse submersed vegetation.

Cyprinids led all families at Stag Island in number of species (15) and abundance. Bluntnose minnow was the most abundant cyprinid (31 percent of a total of 836 fish). No other cyprinid contributed more than 4 percent of the total catch. Because of the swift currents that separate the island from both shores, it is probable that these cyprinids are residents for most of the year. Several uncommon species were present, including blackstripe topminnow (*Fundulus notatus*), a rare species in Canada, least darter, not reported elsewhere in the St. Clair River, and threespine stickleback (*Gasterosteus aculeatus*), only recently reported (Kelso & Leslie, 1979; Stedman & Bowen, 1985) to exist above Niagara Falls. The range of this species may be more extensive than previously thought (Scott & Crossman, 1973), because it was recently collected in Penetanguishene Harbour, Lake Huron (Leslie & Timmins, unpubl. data, 1988).

Most white sucker were caught on June 27 and on July 23 (mean peak abundances were 58 and 121 larvae 100 m$^{-3}$) at 15 to 22 °C and larvae were characteristically migrating downstream (Leslie & Moore, 1985). Virtually all white sucker larvae were collected at the shore of Stag Island; rainbow smelt apparently did not utilize the island as a nursery and catches were made almost exclusively in open water of the artificial channels at Stag Island. Few planktonic rainbow smelt were collected, suggesting that spawning occurred upstream, perhaps in tributaries or in southern Lake Huron. Rainbow smelt formed 16 percent of the total catch at Stag Island. Peak abundance in the catch (mean 8 larvae 100 m$^{-3}$) occurred on June 11 and larvae remained in the area until mid-July. Centrarchids doubtless occur in shallow areas north of Stag Island, where there is submersed vegetation, but neither young nor adult fish were caught. At Stag Island, newly-hatched (4 to 5 mm) pumpkinseed were collected at 15 to 21 °C, but were absent at higher temperatures. However, larger larvae and juveniles were observed in marshy areas of the island.

Stag Island, which had the most species (32) and families (12) of all sites we sampled, may be important as a spawning and nursery habitat for young fish, since large predators are unlikely to frequent the shallow channels and ponds on the island.

### 4.6. Lambton generating station

Few species were entrained in intake cooling water, and except for a brief appearance of rainbow smelt, were not abundant (Fig. 3). Ten taxa were collected as larvae, and ten as juvenile or adult species (Table 3). Larval deepwater sculpin (12.6 mm) were the first species to appear in late April, and alewife, mainly in mid-July, the last. Alewife larvae ranged in size from newly-hatched (3 mm) to 23 mm, and were in migration down the river. Rainbow smelt was the most common (88 percent of total catch) and sole abundant

species (mean 38 larvae 100 m$^{-3}$ in early July), whereas alewife arrived early in July and were the most abundant species in late July. Leslie *et al.* (1979) previously recorded similar results at Lambton Generating Station. Two of ten species caught at the discharge in mid-October were of particular interest, i.e., river redhorse (*Moxostoma carinatum*), classified as 'rare' in Canada (Campbell, 1987), and river shiner (*Notropis blennius*), which has not been reported previously in eastern Canada (Scott & Crossman, 1973).

### 4.7. Cundick Park

Seven cyprinids, of a total of 13 taxa, were concentrated in an area 1.0 m wide and 20 m long at the ditch in Cundick Park (Fig. 1). Dominant species were juvenile common shiner (*Notropis cornutus*) and blacknose shiner (*Notropis heterolepis*), both present during September and October. Although rock bass (*Ambloplites rupestris*) is numerically one of the most abundant juvenile and adult fish in the St. Clair River (Hudson *et al.*, 1986), we found it only in Cundick Park.

### 4.8. Whitebread Canal

First fish larvae hatched in the canal were carp, gizzard shad, and several centrarchids, all of which appeared in late May. The sole clupeid, gizzard shad, dominated all species; the assemblage of young fish is clearly one of clupeid-centrarchid.

Species diversity was higher in Whitebread Canal than in any other site sampled (Table 2), which suggests that the habitat was highly diverse or that many species were transient. Of 27 species recorded, 19 were probably spawned there and used it as a nursery; the size and time of capture of the other species indicated they were moving through the area. Total catch in Whitebread Canal was 4 200 larvae, most (72 percent) of which were gizzard shad; mean densities of 300 to 4 600 larvae 100 m$^{-3}$ gizzard shad occurred during peak appearance from early June to mid-

July. Mean abundance of all species collected in the canal was usually very high (Fig. 3); this was essentially a reflection of the abundance of gizzard shad. Larvae (3.0 mm) were first caught when temperature reached 14 °C, and juveniles (65 mm) were last sampled when the temperature was 17.5 °C. The turbid waters of the canal and other similar agricultural canals and ditches in the area are probably valuable spawning and nursery habitat for this important forage species, whose larvae were not found in 1986 at our study sites in Lake Huron or the St. Clair River proper. The high temperatures that prevailed in the canal (Fig. 2) were apparently suitable for the production, growth, and survival of gizzard shad larvae.

Cyprinids taken at Whitebread Canal formed 5 percent of the total catch. Three each of Cyprinidae and Centrarchidae hatched and remained in the canal throughout the spring and summer, whereas other species appeared briefly during September and October. Species of taxonomic interest include blackstripe topminnow, which was first noted in nearby Sydenham River in 1972 (McAllister, 1987), and river redhorse (*Moxostoma carinatum*), tadpole madtom (*Noturus gyrinus*), and green sunfish (*Lepomis cyanellus*), about which little is known in Canada (Scott & Crossman, 1973).

Although the highly turbid water, clayey bottom, and lack of vegetative cover in the canal suggest it is unsuitable spawning habitat for centrarchids, seven species were present (Table 2) and indeed, 20 percent of all larvae in the canal were pumpkinseed or bluegill (*Lepomis macrochirus*). Few species were permanent residents of the canal, but bluntnose minnow, white crappie (*Pomoxis annularis*), green sunfish, tadpole madtom, and blackstripe topminnow were collected in January, 1987, and thus were considered to overwinter there.

*Table 3.* Fish species at Lambton Generating Station sampled as larva (L) during April to July, 1986, and juvenile (J) or adult (A) on October 15, 1986.

| Species | Total catch | Occurrence | Range of TL (mm) | Habitat (preferred)[1] |
|---|---|---|---|---|
| *Osmerus mordax* Mitchill | 566(L) | Apr 26–Jul 4 | 4.3–30.2 | stream |
| *Alosa pseudoharengus* Wilson | 43(L) | Jul 4–Jul 18 | 3.3–22.7 | stream |
| *Myoxocephalus thompsoni* Girard | 10(L) | Apr 30–May 15 | 11.3–15.8 | offshore benthic |
| *Percina caprodes* Rafinesque | 9(L) | May 21–Jul 4 | 4.6–7.0 | shore |
| *Etheostoma nigrum* Rafinesque | 7(L) | Apr 15–Jul 18 | 5.6–36.5 | stream, shore |
| *Catostomus commersoni* Forster | 6(L) | Jun 6–Jul 18 | 11.2–13.3 | stream |
| *Notropis hudsonius* Clinton | 4(L) | Jun 27–Jul 18 | 4.6–9.3 | stream, lakeshore |
| *Lota lota* Linnaeus | 2(L) | Apr 26–May 21 | 4.3–4.6 | offshore, benthic |
| *Coregonus artedii* Lesueur | 1(L) | Apr 26 | 12.0 | offshore |
| *Etheostoma caeruleum* Storer | 1(L) | Jul 4 | 5.7 | stream, shore |
| *Dorosoma cepedianum* Lesueur (J) | | | | inshore |
| *Perca flavescens* Mitchill (J) | | | | inshore |
| *Hypentelium nigricans* Lesueur (J) | | | | stream |
| *Moxostoma carinatum* Cope (J)* | | | | stream |
| *Notropis atherinoides* Rafinesque (J,A) | | | | inshore |
| *Notemigonus crysoleucas* Mitchill (J,A) | | | | shore |
| *Morone americana* Gmelin (J,A) | | | | offshore |
| *Pimephales notatus* Rafinesque (J,A) | | | | shallows |
| *Pimephales promelas* Rafinesque (A) | | | | stream |
| *Notropis blennius* Girard (A)** | | | | stream |

[1] from Goodyear *et al.* (1982).
* Rare in Canada.
** New species in Canada.

## 5. Discussion

Compared with adjacent waters, the main body of the St. Clair River contained relatively few species and the abundance of fish larvae was generally low (Table 3; Fig. 3). Most of 21 species that were present in the river probably originated in Lake Huron. In contrast, there were more than 60 species in waters connected with and adjacent to the river. Burbot, deepwater sculpin, lake herring, lake whitefish, rainbow smelt and alewife are typical riverine species, and except for rainbow smelt and alewife, are ephemeral, numerically few, and usually appear only as yolk sac larvae. None seem to utilize the river as a nursery, nor to move into adjacent waters during early life. Rainbow smelt occur first in mid-spring and leave the river about the time alewife arrive in early July. Species that typify waters adjacent to the St. Clair River are gizzard shad, centrarchids, and cyprinids, which in general, remain in the same area in which they hatch. Thus, riverine species during early life do not appear to compete with those in adjacent waters.

The amount of suitable habitat available at some stage of early life may be chiefly responsible for determining the peak abundance of a given species (Rounsefell, 1975; Trautman, 1981). On the Michigan side of the St. Clair River, most of the shore is bulkheaded (Muth *et al.*, 1986), hence, fish spawning habitat is confined mainly to tributaries. In Ontario, much of the shore has remained undisturbed, in the physical sense, but impaired water quality persists as a result of industrial and municipal discharges, mainly from the Sarnia petrochemical complex (UGLCC, 1989). There is limited dilution of contaminant inputs, which tend to hug the Canadian shore and may impact directly on fish, especially during early life, when they are most sensitive and vulnerable to perturbation.

Macrophytes were present in stands of light to moderate density along most of the open shoreline of the river in Ontario (OME, 1979) and spawning and nursery habitat probably exist in these areas, as well as in the many small protected niches along the shoreline. The two riverine microhabitats that we sampled exemplified the considerable species diversity that may exist in confined areas.

In conclusion, in spite of chemical loading of the river, bulkheading of embayments, and back-filling of wetlands, there are presently many species of fish that utilize the various niches that remain for spawning and nursing in the St. Clair River. However, more studies are needed to determine the life history of fish and species interactions in the river and connecting waters, in order to wisely conserve, enhance, and indeed, create fish habitat.

## 6. Acknowledgements

We thank Jacqui Milne for assistance in the field, and John Fitzsimons and an anonymous person who reviewed and suggested changes to the manuscript.

## References

Campbell, R. R. (Ed.), 1987. Rare and endangered fishes and marine mammals of Canada: COSEWIC fish and marine mammal subcommittee status reports: 3. Can. Field-Nat. 101: 165–170.

Christie, W. J., 1974. Changes in the fish species composition of the Great Lakes. J. Fish. Res. Bd Can. 31: 827–854.

Derecki, J. A., 1984. St. Clair River physical and hydraulic characteristics. Great Lakes Env. Res. Lab. Contrib. No. 413, 10 pp.

Edsall, T. A., B. A. Manny & C. N. Raphael, 1988. The St. Clair River and Lake St. Clair, Michigan: an ecological profile. U.S. Fish Wildl. Serv. Biol. Rep. 85(7.3). 130 pp.

Goodyear, C. D., T. A. Edsall, D. M. O. Dempsey, G. D. Moss & P. E. Polanski, 1982. Atlas of the spawning and nursery areas of Great Lakes fishes, Vol. 6. U.S. Fish Wildl. Serv., Great Lakes Lab., Ann Arbor, Michigan.

Hass, R. D., W. C. Bryant, K. D. Smith & A. J. Nuhfer, 1985. Movement and harvest of fish in Lake St. Clair, St. Clair River, and Detroit River. U.S. Army Corps Engin., Detroit, Michigan. 141 pp.

Hatcher, C. O. & R. T. Nester, 1983. Distribution and abundance of fish larvae in the St. Clair and Detroit Rivers. U.S. Fish Wildl. Serv., Great Lakes Fish. Lab., Ann Arbor, Michigan. Admin. Rep 83–5. 41 pp.

Hudson, P. L., B. M. Davis, S. J. Nichols & C. M. Tomcko, 1986. Environmental studies of macrozoobenthos, aquatic macrophytes, and juvenile fishes in the St. Clair-Detroit

River system, 1983–1984. U.S. Fish Wildl. Serv., Great Lakes Fish. Lab., Ann Arbor, Michigan. Admin. Rep. 86–7. 303 pp.

Kelso, J. R. M. & J. K. Leslie, 1979. Entrainment of larval fish by the Douglas Point Generating Station, Lake Huron, in relation to seasonal succession and distribution. J. Fish. Res. Bd Can. 36: 37–41.

Leslie, J. K., 1986. Nearshore contagion and sampling of freshwater larval fish. J. Plankton Res. 8: 1137-1147.

Leslie, J. K & J. E. Moore, 1985. Ecology of young-of-the-year fish in Muscote Bay (Bay of Quinte), Ontario. Can. Tech. Rep. Fish. Aquat. Sci. 1377: 63 pp.

Leslie, J. K., R. J. Kozopas & W. H. Hyatt, 1979. Considerations of entrainment of larval fish by a St. Clair River, Ontario, power plant. Fish. and Marine Service Tech. Rept. 868. 25 pp.

McAllister, D. E., 1987. Status of the blackstripe topminnow, *Fundulus notatus* in Canada. Can. Field-Nat. 101: 219–225.

Muth, K. M., D. R. Wolfert & M. T. Bur, 1986. Environmental study of fish spawning and nursery areas in the St. Clair-Detroit River System. U.S. Fish. Wildl. Serv. Ntnl. Fish. Centre Admin. Rept. 86–6. 54 pp.

Ontario Min. Environment, 1979. St. Clair River organics study: Biological surveys 1968 and 1977. Wat. Resources Assess. Unit, Tech. Support Sect., Southwest Region. 90 pp.

Rounsefell, G. A., 1975. Ecology, Utilization, and Management of Marine Fisheries. C.V. Mosby Company, Saint Louis, Mo. 516 pp. ISBN 0-8016-4203-5.

Schloesser, D. W., T. A. Edsall & B. A. Manny, 1985. Growth of submersed macrophyte communities in the St. Clair-Detroit River system between Lake Huron and Lake Erie. Can. J. Bot. 63: 1061–1065.

Scott, W. B. & E. J. Crossman, 1973. Freshwater Fishes of Canada. Fish. Res. Bd Can. Bull. No. 104. 966 pp.

Stedman, R. M. & C. A. Bowen (2nd), 1985. Introduction and spread of the threespine stickleback (*Gasterosteus aculeatus*) in Lake Huron and Michigan. J. Great Lakes Res. 11: 508–511.

Texas Instruments Inc., 1975. Report of fish and macrozooplankton studies on the St. Clair River in the vicinity of the proposed Belle River Power Plant. Prep. for Detroit Edison. Texas Instruments Inc., Dallas, Texas. 147 pp.

Trautman, M. B., 1981. The Fishes of Ohio, 2nd ed. Ohio State Univ. Press, Columbus. 782 pp. ISBN 0-8142-0213-6.

UGLCC (Upper Great Lakes Connecting Channels), 1989. Final report of the Upper Great Lakes Connecting Channels Study. Toronto, Ontario. Vol. 2. 625 pp.

*Hydrobiologia* **219**: 135–142, 1991.
*M. Munawar & T. Edsall (eds),*
*Environmental Assessment and Habitat Evaluation of the Upper Great Lakes Connecting Channels.*
© 1991 *Kluwer Academic Publishers.*

# Distribution and abundance of young fish in Chenal Ecarte and Chematogen Channel in the St. Clair River delta, Ontario

J.K. Leslie & C.A. Timmins
*Great Lakes Laboratory for Fisheries and Aquatic Sciences, 867 Lakeshore Road, Burlington, Ontario, L7R 4A6, Canada*

*Key words:* larval fish, delta, succession, abundance, size

## Abstract

Samples of fish larvae collected in 1983, 1984, and 1986 in two distributary channels of the St. Clair River delta were characteristically rich in species (a total of 48) and low in abundance (generally less than a mean of 5 100 m$^{-3}$ of water filtered). Most species were residents of the delta; others apparently hatched in tributaries of the St. Clair River or in southern Lake Huron, and drifted into the delta. Highest species diversity was nearshore, although largest catches of larvae were of rainbow smelt, gizzard shad, and alewife, which were found mainly in mid-channel. Cyprinids (17 species) were better represented than other families.

## 1. Introduction

Fish produced in St. Clair River habitats have supported major fisheries in the river and connecting waters for more than a century (Smith & Snell, 1890; Koelz, 1926; Baldwin & Saalfeld, 1962; Edsall *et al.*, 1988), but the contribution of channel and wetland habitat in the St. Clair River Delta to this fishery is virtually unknown. Ecological studies are required of immature life stages of fish in the delta, mainly because the area may contain important spawning and nursery habitat for many species (Goodyear *et al.*, 1982; Herdendorf *et al.*, 1986; UGLCC, 1989). In addition, information is required to help determine effects on critical life stages of fish in the delta of industrial and municipal spills and discharges into the St. Clair River. Fish are themselves the best judge of habitat, hence their occurrence in any particular locale is an index of habitat suita-

bility. Therefore, we conducted surveys to determine the distribution and abundance of young fish in 1983, 1984, and 1986 in the Canadian waters of the delta at Walpole Island and to evaluate the significance of the spawning and nursery habitat in that area.

## 2. Study sites

The St. Clair River (Fig. 1) extends 43 km from its origin in Lake Huron to the St. Clair Delta. Four main creeks are tributary to the river on the Canadian side and three major rivers enter on the American side. Shallows at two small islands in the river provide spawning and nursery areas for fish, but more extensive habitats exist in the delta (Goodyear *et al.*, 1982). Walpole Island has the largest area (6 300 ha) of wetlands in the St. Clair System (Chapman & Putnam, 1966) and is dis-

136

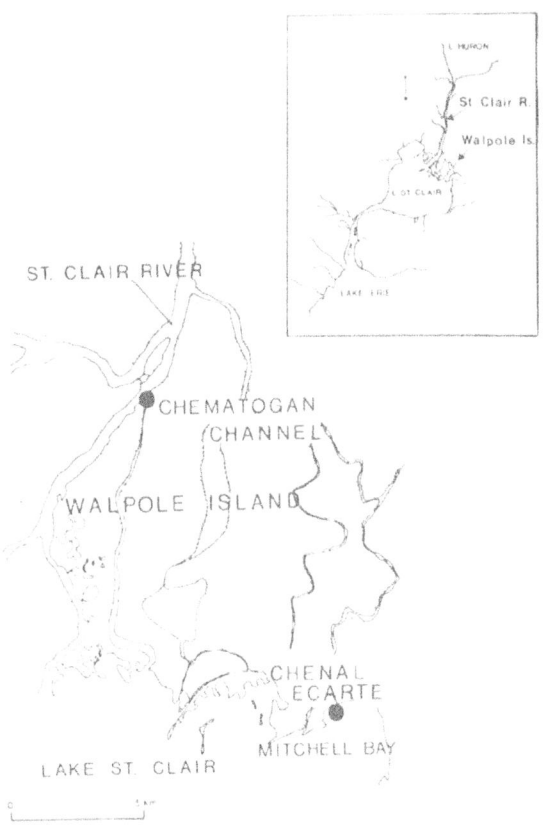

*Fig. 1.* Location of study sites at Walpole Island, Ontario.

sected by several distributaries of the St. Clair River, two of which, Chenal Ecarte and Chematogan Channel, we sampled for young fish. Walpole Island contains extensive stands of cattail (*Typha* spp.), waterweed (*Elodea canadensis*), and pondweeds (*Potamogeton* spp.). Substrate in marsh waters is mainly organic debris, sand, silt, and clay.

Chenal Ecarte (Fig. 1) enters the northeast sector of Lake St. Clair through a dredged channel into Mitchell Bay, and at its origin has an average discharge of $256 \, m^3 \, s^{-1}$, or 5 percent of the average flow of the St. Clair River (Herdendorf *et al.*, 1986). Width of the channel at the sampling site was 50 m, and depth was 1 m. Although channel waters are seldom subjected to effects of wind, small boat traffic creates considerable turbulence, turbidity, and wash. Cattail, water milfoil (*Myriophyllum spicatum*) and wild celery (*Vallisneria americana*) grow along the shore.

Chematogan Channel (Fig. 1) extends at least two-thirds the length of the western side of Walpole Island, and varies in width from about 40 to 300 m, with a depth of 1 m. Three percent of the average discharge of the St. Clair River enters Chematogan Channel (Herdendorf *et al.*, 1986). Cattail, bulrush (*Scirpus* sp.), and reed grass (*Phragmites australis*) dominate vegetation that borders the channel.

## 3. Methods

In 1983 we sampled young fish in Chenal Ecarte with two identical conical nets (1 m in diameter and 2.5 m long with 0.5 mm mesh) each at the side of the bow of a 6 m boat, pushed through the water at 1 to $2 \, m \, s^{-1}$. Speed was measured with a hand-held, direct reading current meter located 1 m in front of the net. On each sampling date, four samples were obtained from pushing the two nets upstream in the centre of the channel for 5 or 10 min, washing and removing the contents of the nets, then repeating the process in the downstream direction. The nets were assumed 100 percent efficient, and total volume of water filtered in 10 min was $1170 \, m^3$. Larvae were immediately fixed in 10 percent formalin, and within two months, they were preserved with Davidson's B solution. In the laboratory, these samples were first processed individually for purposes of comparison (larvae caught in the upstream vs downstream direction; total catch in the port vs starboard net) then combined and processed as a single sample. The sampling regime in 1984 was the same as in 1983, except that two identical nets of rectangular mouth ($1.0 \times 0.5$ m) and 2.0 m long (0.5 mm mesh) were used, and total volume filtered in 10 min was $720 \, m^3$. Sampling occurred from April to October in 1983 and from May to August in 1984. Samples were taken weekly in the spring, and thereafter, every two to four weeks. Abundance estimates were expressed as number of larvae $100 \, m^{-3}$ of water filtered.

Sampling in Chematogan Channel in 1986 commenced with replicated, paired drift samples taken at the surface on three occasions. However, this technique was abandoned because of low flow rate (0.2 to 0.3 m s$^{-1}$) in the channel. Regular sampling occurred weekly from mid-April to late July. Thereafter, samples were taken twice in September, and once each in October and in January, 1987. Techniques and equipment for sampling in open water were as in 1983 and 1984, but at least 50 additional samples were taken with a seine (3 m long, 1 m wide, 0.4 mm mesh) at wadable depths near bulrushes, over a substrate of sand and organic debris. Contents of the seine were removed and fixed with 10 percent formalin. Duplicate samples were taken on each occasion, with approximately 25 m$^3$ of water filtered per sample. Temperature, conductivity, and pH, all taken at 0.5 m, were routinely observed at sampling stations in all years.

## 4. Results

### 4.1. Species occurrence

Forty eight species were collected, including 35 in Chenal Ecarte and 36 in Chematogan Channel (Table 1). Most (27) were early hatching species that at some stage of development are associated with aquatic vegetation. Total number of young

*Table 1.* Occurrence (x) of fish larvae or juveniles at Walpole Island, 1983, 1984, and 1986. Percentage of total catch (in parentheses) of common species.

| Species | 1983 | 1984 | 1986 |
|---|---|---|---|
| Raibow smelt (*Osmerus mordax* Mitchill) | x (3) | x (2) | x (28) |
| Lake herring (*Coregonus artedii* Lesueur) | x | x | x |
| Lake whitefish (*C. clupeaformis* Mitchill) | | x | |
| Gizzard shad (*Dorosoma cepedianum* Lesueur) | x (12) | x (22) | |
| Alewife (*Alosa pseudoharengus* Wilson) | x (61) | x (67) | x |
| Yellow perch (*Perca flavescens* Mitchill) | x | x | x (2) |
| Logperch (*Percina caprodes* Rafinesque) | x | x | x |
| Johnny darter (*Etheostoma nigrum* Rafinesque) | x | | x |
| Rainbow darter (*E. caeruleum* Storer) | | | x |
| Bluegill (*Lepomis macrochirus* Rafinesque) | | x | x |
| Pumpkinseed (*L. gibbosus* Linnaeus) | x | x | |
| Largemouth bass (*Micropterus salmoides* Lacépède) | | x | |
| Smallmouth bass (*M. dolomieui* Lacépède) | x | | |
| Black crappie (*Pomoxis nigromaculatus* Lesueur) | x | | x |
| White crappie (*P. annularis* Rafinesque) | | x | |
| White perch (*Morone americana* Gmelin) | x (7) | x | x |
| White bass (*M. chrysops* Rafinesque) | | x | x |
| Burbot (*Lota lota* Linnaeus) | x | | x (2) |
| Deepwater sculpin (*Myoxocephalus thompsoni* Girard) | | x | x |
| Mottled sculpin (*Cottus bairdi* Girard) | | | x |
| Trout-Perch (*Percopsis omiscomaycus* Walbaum) | x | | |
| White sucker (*Catostomus commersoni* Forster) | x | x | x (11) |
| Lake chubsucker (*Erimyzon sucetta* Lacépède) | | | x |
| Quillback (*Carpiodes cyprinus* Lesueur) | x | x | |
| River redhorse (*Moxostoma carinatum* Cope) | x | x | |
| Brook silversides (*Labidesthes sicculus* Cope) | x (6) | x | x |
| Longnose gar (*Lepisosteus osseus* Linnaeus) | | x | |
| Central mudminnow (*Umbra limi* Kirtland) | | | x |
| Banded killifish (*Fundulus diaphanus* Lesueur) | x | x | |
| Black bullhead (*Ictalurus melas* Rafinesque) | | | x |
| Stonecat (*Noturus flavus* Rafinesque) | | | x |
| Rosyface shiner (*Notropis rubellus* Agassiz) | x | | x |
| Striped shiner (*N. chrysocephalus* Rafinesque) | | x | x |
| Emerald shiner (*N. atherinoides* Rafinesque) | x | x | x (12) |
| Blacknose shiner (*N. heterolepis* Eigenmann & Eigenmann) | | x | x |
| Sand shiner (*N. stramineus* Cope) | | | x |
| Spottail shiner (*N. hudsonius* Clinton) | x (4) | x | x |
| Blackchin shiner (*N. heterodon* Cope) | | x | x |
| Bridle shiner (*N. bifrenatus* Cope) | | | x |
| Spotfin shiner (*N. spilopterus* Cope) | | | x |
| Mimic shiner (*N. volucellus* Cope) | x | x | x |
| Golden shiner (*Notemigonus crysoleucas* Mitchill) | x | x | x (41) |
| Goldfish (*Carassius auratus* Linnaeus) | | | x |
| Carp (*Cyprinus carpio* Linnaeus) | x | x | |
| Bluntnose minnow (*Pimephales notatus* Rafinesque) | x | x | x |
| Stoneroller (*Campostoma anomulum* Rafinesque) | | | x |
| Creek chub (*Semotilus atromaculatus* Mitchill) | | | x |
| Blacknose dace (*Rhinichthys atratulus* Hermann) | | | x |

138

fish collected in 1983, 1984, and 1986 was 1 800, 6 000, and 4 200, respectively. Nineteen species were taken both in 1983 and 1984, five only in 1983 and nine only in 1984. Twenty nine species in 1986 were found at the shore, 14 in the channel, and 9 were common both to the shore and the open channel. Thirteen species of immature fish were found in addition to 48 known or presumed to use wetlands of the St. Clair River and Lake St. Clair (Herdendorf et al., 1986).

*4.2. Seasonality and abundance in Chematogan Channel, 1986*

We collected burbot (*Lota lota*), deepwater sculpin (*Myoxocephalus thompsoni*), and lake herring (*Coregonus artedii*) only in May. Little, if any, spawning habitat exists for any of these species in shallow Chematogan Channel, and thus, they were considered 'drifters' that probably originated in Lake Huron. Similarly, rainbow smelt (*Osmerus mordax*) likely spawned in streams and shores in southeast Lake Huron, and larvae were entrained by the St. Clair River. Because there are three distinct 'panels' of water (Canadian shoreline, mid-river, and U.S. shoreline) in the river (UGLCC, 1989), larvae may have followed a definite nearshore course on the Canadian side of the river. Rainbow smelt were first collected in mid-May at 10 °C, reached peak abundance in late May, and were absent from the channel two weeks later. White sucker (*Catostomus commersoni*) and several cyprinid taxa appeared between early June and late July, both in mid-channel and at the shore. Twelve cyprinids were present only at the shore, as were eight other species. Nearshore lake spawners such as burbot, deepwater sculpin, and white bass (*Morone chrysops*), and stream- or shore-spawning mottled sculpin (*Cottus bairdi*) and logperch (*Percina caprodes*) were found only in mid-channel.

Few species were found simultaneously in mid-channel and at the shore; on average, peak abundance of all species found both at the shore and in the channel was five times higher at the shore. Mean peak abundance (7 larvae 100 m$^{-3}$) of

rainbow smelt occurred on May 20; other species in mid-channel appeared briefly and in low abundance (mean less than 4 larvae 100 m$^{-3}$). Highest mean density of young fish (each at 100 larvae 100 m$^{-3}$) were schools of juvenile blackchin shiner (*Notropis heterodon*) and sand shiner (*Notropis stramineus*) which occurred only in October. Sand shiners probably were transients from nearshore sandy areas in the St. Clair River, whereas blackchin shiners likely were residents of the delta (Scott & Crossman, 1973).

Sampling with plankton nets in the middle of Chematogen Channel yielded few larvae of any species. More species were found at the shore of the channel but overall abundances were low, and many reflect the relatively low abundance and species diversity of the St. Clair River.

*4.3. Seasonality and abundance in Chenal Ecarte, 1983–1984*

In 1983, a single burbot embryo was captured at 8 °C on April 22 and one lake herring was caught in late April, and a second one in early June. Burbot and lake herring probably hatched in Lake Huron and then drifted down the St. Clair River. They were succeeded in May by ten species, the most abundant of which were yellow perch (*Perca flavescens*) and white perch (*Morone americana*); 13 other species appeared in June and early July. Muth et al. (1986) found that rainbow smelt, with 16 percent of the total catch in the St. Clair River in 1983, ranked second in abundance. Simultaneously in the delta, our catches of rainbow smelt represented only three percent of larvae collected, a difference that may be related to characteristics of flow regime, and thus the dispersion of larvae, in the St. Clair River and distributaries. Generally, cyprinids and brook silversides (*Labidesthes sicculus*) occurred throughout the summer, whereas seasonality of most other species was restricted to less than two months. Largest catches of gizzard shad (*Dorosoma cepedianum*) were in early June and mid-August, but mean abundance was low (5 larvae 100 m$^{-3}$). Alewife (*Alosa pseudoharengus*) formed an in-

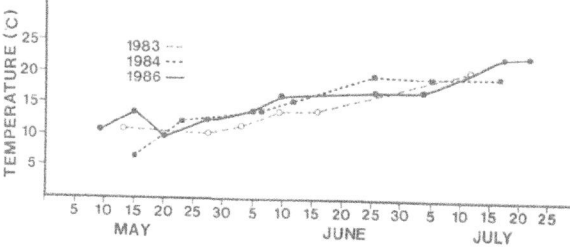

*Fig. 2.* Water temperatures in Chenal Ecarte, May–July, 1983 and 1984, and in Chematogan Channel, 1986.

creasing percentage of the total catch after late June, and was most abundant (32 larvae 100 m$^{-3}$) in mid-August (Fig. 3).

In 1984, the first fish larvae were caught in mid-May, marking the initial and final appearance of lake whitefish (*Coregonus clupeaformis*) and deepwater sculpin. Yellow perch, gizzard shad, and rainbow smelt also occurred in May, but most (16) species first appeared in June, when water temperatures were 15 to 20 °C (Fig. 2). Gizzard shad, which formed 22 percent of the total catch, was dominant throughout June, as was alewife (67 percent) during July and August. Abundance of gizzard shad fluctuated during early June to August, but peaked in late June (Fig. 3) just as alewife first appeared. Alewife were highly concentrated (205 larvae 100 m$^{-3}$)

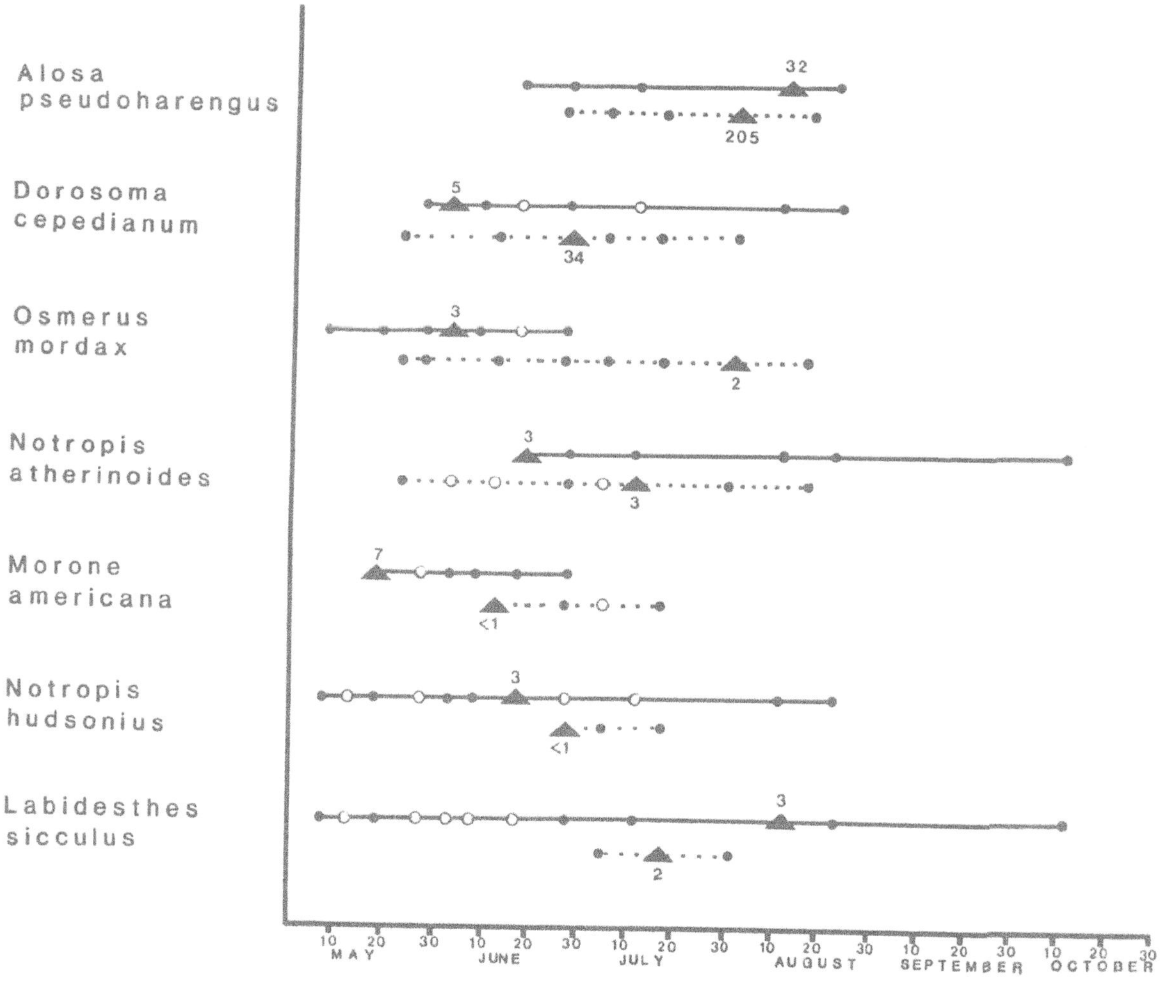

*Fig. 3.* Occurrence and peak abundance (larvae 100 m$^{-3}$) of selected species of immature fish in Chenal Ecarte, 1983 (——) and 1984 (.....). Symbols denote presence (●), absence (○), and peak (▲).

140

on August 1, but were otherwise not dominant. In 1984, brook silversides were present only for several weeks, mainly in mid-July as a pulse of larvae at 20 °C. In general, seasonal succession in Chenal Ecarte during 1984 proceeded from early hatching species, such as coregonids, cottids, and percids, to gizzard shad, cyprinids and centrarchids, and finally to alewife.

There were differences both between succession and abundance in 1983 and 1984 of dominant larval fish (Fig. 3), due in part to an ice jam in the St. Clair River in 1984 that caused a 0.4 m drop in water level, followed by flooding in upstream areas (Hudson *et al.*, 1986; Muth *et al.*, 1986). Seasonality and abundance of rainbow smelt were similar in 1983 and 1984, in spite of lower temperatures in early spring, 1984, that resulted in a lag of three weeks for spawning and incubation of eggs. Nevertheless, similarities in chronology suggest southern Lake Huron was the origin of many rainbow smelt larvae. Average abundances in 1984 of gizzard shad and alewife were, respectively, six and four times that in 1983.

Differences in fish species composition exist between Chematogan Channel and Chenal Ecarte, but were unlikely due to slight differences in water quality as indicated by mean conductivity (242 $\mu$S) and mean pH (7.8). However, the rate of water flow in Chematogan Channel is higher and the main species of macrophyte are emergents, such as cattail and bulrush, whereas beds of submersed vegetation prevail at the mouth of Chenal Ecarte. Hence, a few taxa in Chenal Ecarte, e.g., carp (*Cyprinus carpio*), black crappie (*Pomoxis nigromaculatus*), banded killifish (*Fundulus diaphanus*), pumpkinseed (*Lepomis gibbosus*), and smallmouth bass (*Micropterus dolomieui*), are usually associated more with the littoral zone of lakes than with channels. On the other hand, several cyprinids, mottled sculpin, and rainbow darter (*Etheostoma caeruleum*) were present only in unvegetated water in Chematogan Channel, whereas central mudminnow (*Umbra limi*), lake chubsucker (*Erimyzon sucetta*), and black bullhead (*Ictalurus melas*) were specific to vegetated, shallow areas on the fringe of the delta at Lake St. Clair. Cyprinids were much more common in

Chematogan Channel (55 percent of the total catch) than in Chenal Ecarte (5 percent of total catch).

### 4.4. Seasonal growth of selected species

Gizzard shad were 4.0 mm on June 3, 1983, when they were first caught. They probably hatched in turbid waters of agricultural ditches connected to Chenal Ecarte, from which they began emigrating in August at 35 mm. In contrast, in 1984 they were present in the channel throughout July and growth appeared comparatively faster, perhaps due to a faster rate of warming of water than in 1983. Length data for gizzard shad (Fig. 4) suggest that protracted spawning did not occur in Chenal Ecarte in 1983 or 1984, and that tempera-

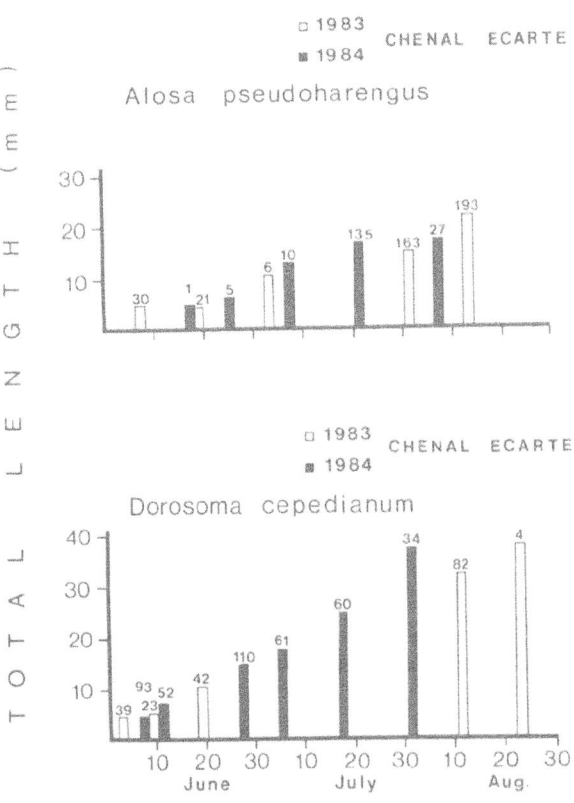

*Fig. 4.* Mean total length (mm) of larval *Alosa pseudoharengus* and *Dorosoma cepedianum* in Chenal Ecarte, 1983–84.

ture and other habitat conditions in calm waters of the delta were similar in both years.

Brook silversides mainly frequent shores of lakes (Scott & Crossman, 1973), and we found no larvae in Chematogan Channel, although several juveniles were collected. In Chenal Ecarte, brook silversides averaged 40 mm in 1983; growth was typically rapid, and increased from 9.1 mm in mid-July, to 43.0 mm in mid-August. In mid-July, 1984, larvae were 7.1 to 22.6 mm long (mean = 16.0 mm). These data are similar to those recorded in eutrophic Bay of Quinte, Lake Ontario, in 1981 (Leslie & Moore, 1985), when fish were 5 mm in early June and 36 mm in mid-August.

Shortly after hatching in mid-June (Fig. 4), alewife 7 to 10 mm long commence emigration to Lake St. Clair. However, some larvae remain in the delta until late August, when they are about 40 mm. White sucker hatch in the delta and emigrate to Lake St. Clair. They were caught in mid-channel at 17.5 °C during late June, when they were 15.9 mm long, and at 14.4 mm a month later at the shore, when water temperature was 23.5 °C, suggesting irregular emigration from natal streams into Chematogan Channel. The presence in the delta of larval cyprinids 5 to 20 mm long in June and July indicated that suitable spawning habitat existed for this group. The most common cyprinids in the delta were golden shiner (*Notemigonus crysoleucas*), emerald shiner (*Notropis atherinoides*), spottail shiner (*Notropis hudsonius*), and mimic shiner (*Notropis volucellus*).

## 5. Discussion and conclusions

Collections of immature fish in distributaries of the St. Clair River produced 48 species, indicating high species richness. However, abundance was for most species generally lower than elsewhere in the Great Lakes, and except for clupeids, ranged typically from a mean of 0.1 to 5 larvae 100 m$^{-3}$. These values probably are an underestimation of average population densities throughout the delta, as many larval fish may frequent areas such as vegetated nearshore habitats and sheltered backwaters that we did not sample.

On the basis of presence of many juvenile fish and the absence of their larval counterpart, and the movement through the delta to Lake St. Clair of rainbow smelt, lake herring, burbot, and others, we consider the delta more a nursery than a spawning area. In any case, the delta provides valuable habitat for a wide variety of species, many of which form a route for transfer of energy through the food chain to the St. Clair River and Lake St. Clair (Herdendorf et al., 1986).

Because there is only one published report of larval fish in deltas and wetlands of the Great Lakes (see Chubb & Liston, 1986), the opportunity for comparison is limited. There were distinct differences between larval fish assemblages in the St. Clair River Delta and Pentwater Marsh, a coastal wetland on Lake Michigan. Fish larvae in Pentwater Marsh were dominated by carp, which formed 80 percent of the total catch in 1982–84 (Chubb & Liston, 1986), whereas in the present study, carp formed less than 1 percent of the total catch in 1983 and 1984, and were absent in 1986. Alewife and gizzard shad together comprised 73 percent of the catch in 1983 and 89 percent in 1984 in channel waters of Walpole Island (Table 1), yet formed only 0.6 percent of the total catch in all waters of Pentwater Marsh (Chubb & Liston, 1986). Apparently, clupeids do not utilize the delta extensively; more suitable habitat is available elsewhere in the St. Clair system (Goodyear et al., 1982; Hudson et al., 1986; Poe et al., 1986).

Although human activities have a major influence on water quality and habitat structure in the delta (Herdendorf, 1987; Edsall et al., 1988; UGLCC, 1989), conversion of wetland for agriculture is undoubtedly the main threat to remaining fish habitat (Bardecki, 1984; McCullough, 1985). Species such as central mudminnow, lake chubsucker, muskellunge (*Esox masquinongy*), and longnose gar (*Lepisosteus osseus*), whose niches are already restricted, are among those that would be most impacted by further wetland loss.

Improved water quality in Chematogan Channel and Chenal Ecarte will reflect remediation of pollution in the St. Clair River, from which most industrial and municipal pollutants enter the delta

142

(IJC, 1984), but in the final analysis, preservation of fish spawning and nursery habitat in Ontario rests mainly with the people of Walpole Island Indian Reserve.

## 6. Acknowledgements

Thanks are due J.E. Moore, S. Bray, and S. Avery for technical help, and to L. Montour and Aaron Sony of the Walpole Island Indian Reservation for allowing access to sampling areas. Reviews by Dr. U. Borgmann, and especially an anonymous colleague, greatly improved the manuscript.

## References

Baldwin, N. R. & R. W. Saalfeld, 1962. Commercial fish production in the Great Lakes, 1867–1960. Great Lakes Fish. Comm. Tech. Rep. No. 3, 166 pp.

Bardecki, M. J., 1984. What value wetlands? J. Soil Wat. Conserv. 39: 166–169.

Chapman, L. J. & D. F. Putnam, 1966. The Physiography of Southern Ontario, 2nd edition, Univ. Toronto Press, Toronto, 386 pp.

Chubb, S. L. & C. R. Liston, 1986. Density and distribution of larval fishes in Pentwater Marsh, a coastal wetland on Lake Michigan. J. Great Lakes Res. 12: 332–343.

Edsall, T. A., B. A. Manny & C. N. Raphael, 1988. The St. Clair River and Lake St. Clair, Michigan: an ecological profile. U.S. Fish Wildl. Serv. Biol. Rep. 85(7.3), 130 pp.

Goodyear, C. D., T. A. Edsall, D. M. O. Dempsey, G. D. Moss & P. E. Polanski, 1982. Atlas of the spawning and nursery areas of Great Lakes fishes. Volume 13: Reproduction characteristics of Great Lakes fishes. U.S. Fish Wildl. Serv. Washington, D.C. FWS/OBS-82/52.

Herdendorf, C. E., 1987. The ecology of the coastal marshes of western Lake Erie: a community profile. U.S. Fish Wildl. Serv. Biol. Rep. 85(7.9), 171 pp.

Herdendorf, C. E., C. N. Raphael & E. Jaworski, 1986. The ecology of Lake St. Clair wetlands: a community profile. U.S. Fish Wildl. Serv. Biol. Rep. 85, 187 pp.

Hudson, P. L., B. M. Davis, S. J. Nichols & C. M. Tomcko, 1986. Environmental studies of macrozoobenthos, aquatic macrophytes, and juvenile fish in the St. Clair-Detroit River system. U.S. Fish. Wildl. Serv., Great Lakes Fish. Lab. Admin. Rep. 86–7, 303 pp.

IJC, 1984. Report on Great Lakes Water Quality. International Joint Commission. Windsor, Ontario.

Koelz, W., 1926. Fishing industry of the Great Lakes. Dept. Commerce Bur. Fish. Doc. 1001. Washington, D.C.: 53–617.

Leslie, J. K. & J. E. Moore, 1985. Ecology of young-of-the-year fish in Muscote Bay (Bay of Quinte), Ontario. Can. Tech. Rep. Fish. Aquat. Sci. 1377: 63 pp.

McCullough, G. B., 1985. Wetland threats and losses in Lake St. Clair. In; H. H. Prince & F. M. D'Itri (eds), Coastal wetlands. Lewis Publishers, Chelsea, Mich.: 202–208.

Muth, K. M., D. R. Wolfert & M. T. Bur, 1986. Environmental study of fish spawning and nursery areas in the St. Clair-Detroit River System. U.S. Fish. Wildl. Serv. Natnl. Fish. Centre Admin. Rept. 86–6, 54 pp.

Poe, T. P., C. O. Hatcher, C. L. Brown & D. W. Schloesser, 1986. Comparison of species composition and richness of fish assemblages in altered and unaltered littoral habitats. J. Freshwat. Ecol. 3: 525–536.

Scott, W. B. & E. J. Crossman, 1973. Freshwater Fishes of Canada. Fish. Res. Bd. Can., Bull. 184, Ottawa, 966 pp.

Smith, H. M. & M. M. Snell, 1890. Review of the fisheries of the Great Lakes in 1885, with introduction and description of fishing vessels and boats, by J. W. Collins. Government Printing Office, Washington, D.C., 333 pp.

UGLCC (Upper Great Lakes Connecting Channels), 1989. Final report of the Upper Great Lakes Connecting Channels study. Vol. 2. Toronto, Ontario, 625 pp.

*Hydrobiologia* **219**: 143–164, 1991.
*M. Munawar & T. Edsall (eds),*
*Environmental Assessment and Habitat Evaluation of the Upper Great Lakes Connecting Channels.*
© 1991 *Kluwer Academic Publishers.*

# Environmental quality assessment of the St. Clair River as reflected by the distribution of benthic macroinvertebrates in 1985

Ronald W. Griffiths
*Aquatic Ecostudies Limited, 1221 Weber St. East, Kitchener, Ontario, Canada N2A 1C2; Present address: Ministry of the Environment, Water Resources Assessment Unit, 985 Adelaide St. South, London, Ontario, Canada N6E 1V3*

*Key words:* macroinvertebrates, environmental impact, sediment quality, water quality, pollution, Great Lakes system

## Abstract

A benthic macroinvertebrate and sediment chemistry study of the St. Clair River from Lake Huron to Lake St. Clair was conducted in the spring of 1985. The purpose of the study was to evaluate the environmental quality of the nearshore areas and assess the effectiveness of industrial and municipal abatement programs that have been implemented since 1977.

A total of 112 macroinvertebrate taxa was collected from the river. Classification analysis indicated that 7 macroinvertebrate communities were evident in the river. Discriminant analysis suggested that physical habitat characteristics explained the distribution of 4 benthic communities, while sediment contaminants explained the distribution of 3 benthic communities. These analyses showed that the environmental quality of a 12 km stretch of the river along the Canadian shoreline had been degraded, probably by industrial waste discharges and spills. Toxic conditions were evident along the waterfront of Dow Chemical Canada Inc., probably a result of the combined effects of chlorinated organics, oils and greases, and mercury (historical contaminant) in the sediments. In contrast, the invertebrate fauna throughout the remainder of the St. Clair River reflected meso-eutrophic conditions, typical of a large, unstressed river.

A comparison of the environmental quality as reflected by the benthic invertebrate fauna in 1985 with that in 1977 suggests that the abatement programs implemented over the past decade have improved the environmental quality along the Canadian side of the river. The total length of river adversely affected by waste discharges from Canadian industries and municipalities decreased from 21 km in 1977 to 12 km in 1985.

## Introduction

The St. Clair River is a major navigable waterway approximately 63 km long with an average width of about 600 m that separates Canada and the United States. It flows southward from Lake Huron to Lake St. Clair (Fig. 1) at an average speed of 3.5 km h$^{-1}$ (Derecki, 1984). The average

discharge of the river is in the order of 5100 m$^3$ s$^{-1}$ with little seasonal variation (Quinn & Kelly, 1983), although during the mid-1980's the flow averaged 6000 m$^3$ s$^{-1}$ (Hudson *et al.*, 1986). Water entering the river from Lake Huron is of high quality; total phosphorus concentrations are typically below 10 $\mu$g l$^{-1}$, nitrate below 400 $\mu$g l$^{-1}$, hardness around 100 mg l$^{-1}$, tur-

144

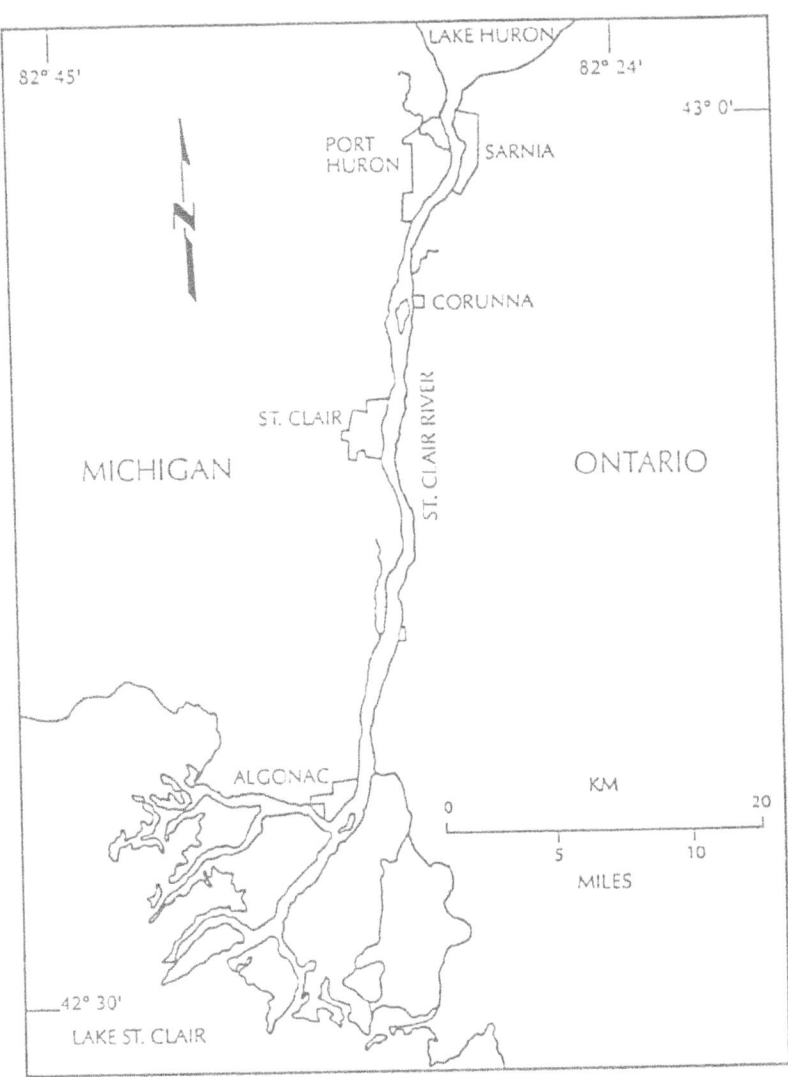

*Fig. 1.* Map of the St. Clair River.

bidity below 6 FTU and colour less than 5 TCU (OME, 1986a).

The St. Clair River is both a source for potable water and industrial cooling and process water and a receiver for wastes from municipalities and industries. About 2.6 million m³ of industrial effluent, containing a variety of metals, inorganic and organic compounds, is discharged daily into the river (EC & US EPA, 1989), 1.7 million m³ originate from the large petrochemical complex (the 'Chemical Valley') south of Sarnia, Ontario (EC & OME, 1986). In addition, about 60 000 m³ of treated sewage is discharged daily to the river from Canadian municipalities (OME, 1986b) and a slightly greater quantity from American municipalities (Edsall *et al.*, 1988). Contaminants discharged to the river, however, do not readily mix with water in the main channel but remain near the shoreline as they move downstream; Chan *et al.* (1986) found that the plume of contaminants discharged by industries in the Chemical Valley was still mainly confined to within 300 m of the Canadian shore about 34 km downstream, and Oliver & Pugsley (1986) showed that sediments contaminated with chlorinated hydrocarbons discharged by industries in the Chemical Valley were similarly confined along the Canadian shoreline. The nearshore benthic macroinvertebrate fauna downstream of Sarnia has been impaired along the Canadian shoreline as a consequence of these discharges (Griffiths, 1978; OME, 1979, Thornley, 1985). The International Joint Commission has designated the St. Clair River as an 'Area of Concern' because of the impairment to water uses and aquatic life (IJC, 1982).

The purpose of this study was to evaluate the nearshore environmental quality of the St. Clair River in 1985 and identify any temporal changes from 1977 (Griffiths, 1978; OME, 1979) in response to the industrial and municipal abatement programs that have been implemented since the late 1970's (EC & OME, 1986; EC & US EPA, 1989). Thus, a benthic macroinvertebrate and sediment chemistry study of the river was conducted in the spring of 1985. Benthic macroinvertebrates were used as indicators of environmental quality because:

a) they are abundant, living on or in the substrate;
b) they are readily collected and identified;
c) they show a wide range of tolerances to various degrees and types of pollutants;
d) they usually remain in a localized area because of their restricted mobility and habitat preference;
e) they are continuously subjected to the full rigor of the local environment throughout their aquatic life-cycle, which may vary from weeks to years;
f) they reflect past (historical) as well as present environmental conditions of a site; and
g) they occupy an intermediate trophic level in aquatic food-webs and are an important source of food for animals in higher trophic levels, most notably fish and waterfowl.

Thus, unlike surface sediment information, which only indicates the total concentration of contaminants in the environment at a point-in-time, benthic macroinvertebrate information represents the integrated effect of all environmental variables (biotic and abiotic) during the period of time that they have lived in the habitat. Furthermore, the use of *in situ* organisms circumvents the need for any assumptions about the toxicity of contaminants in order to surmise the effects on aquatic life. The benthic macroinvertebrate fauna, therefore, can be used to directly measure the effects of environmental stresses on aquatic systems, regardless of the frequency or intensity, while sediment chemistry can aid with elucidating possible causes for observed biological impacts.

## Methods

### Sampling design

The benthic macroinvertebrate fauna at 78 sampling stations in the St. Clair River was sampled from May 22 to June 12, 1985 (Fig. 2). The sampling stations were located primarily along the nearshore area of the river (depth < 8 m) upstream and downstream from specific discharge sources.

A ponar grab, which enclosed an area of 0.052 m², was used to collect the benthic fauna. Three samples were collected at each station. Each sample was washed in a sieve pail containing a No. 30 (U.S. Standard Sieve Series) mesh screen (aperture 0.60 mm); the remaining sediment, debris, and organisms were then placed into labelled 1 l jars and stored on ice. These samples were transported to a field laboratory where the organisms were sorted live from the sediment and debris in white enamel trays using forceps. All invertebrates, except oligochaetes, which were preserved with 10 percent formalin (4% formaldehyde), were then placed in 30-ml bottles and preserved with 80 percent ethanol.

All benthic invertebrates were identified by the author. Generally, insects and clams were identified to the generic level, mature annelids, crustaceans, snails, and mussels to the specific level, and other invertebrates to class. All organisms were identified except when large numbers of oligochaetes or chironomids were present. In samples with a large number of worms, all

146

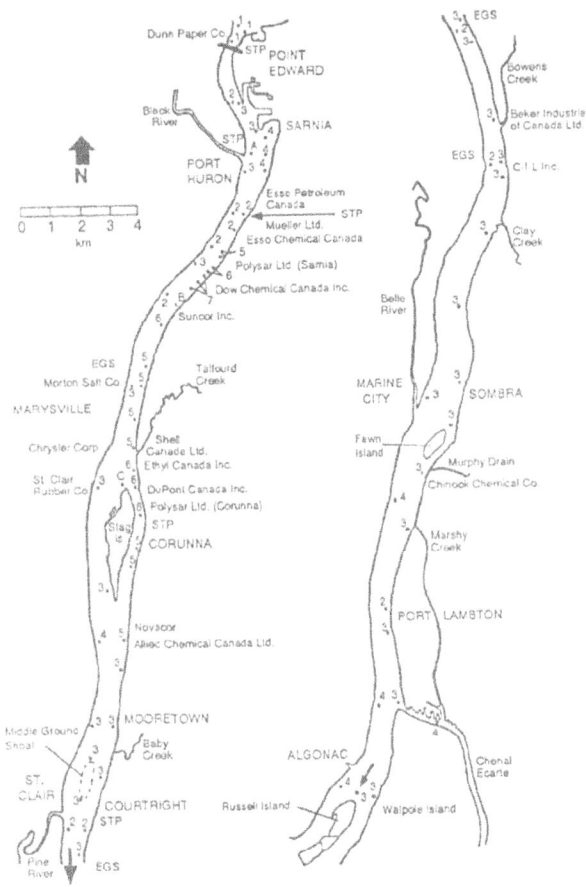

*Fig. 2.* Distribution of benthic invertebrate Communities in the St. Clair River, May 1985. See Table 2 for taxonomic composition of Communities 1 to 7. STP denotes sewage treatment plants; EGS denotes electrical generating stations; Dots represent sampling stations.

*Chironomus, Procladius, Crytochironomus,* were enumerated and removed from the sample. The remaining individuals were sorted into three groups: Chironominae, Tanypodinae and Orthocladiinae (including Diamesinae). A random sample of not less than 10 percent of the individuals, up to a maximum of 50 individuals, was removed from each group for identification. The subsamples of worms and midges were mounted on glass slides in a clearing medium, left for 24 h at 60 °C, then identified using a compound microscope.

A Shipek grab was used, in conjunction with the invertebrate sampling, to collect surface sediment samples for grain size analysis and chemical analyses. At each sampling station, the top 3 cm of sediment from three Shipek samples were composited, subsamples of this composite were placed in 500 ml wide-mouth glass jars with pulp-lined screw caps for heavy metals and grain size analyses, or in solvent-rinsed 500-ml jars with foil-lined caps for chlorinated organic compounds (OME, 1985).

The sediment samples were analyzed for total metals (iron, manganese, aluminum, arsenic, cobalt, copper, chromium, mercury, nickel, lead, and zinc), nutrients (total Kjeldahl nitrogen, total phosphorus), organics (loss-on-ignition, total organic carbon), oils and greases (solvent extractables), chlorinated organics (aldrin, lindane, mirex, chlordane, oxychlordane, dieldrin, endrin, thiodane, DDT and metabolites, heptachlor, heptachlor epoxide, polychlorinated biphenyls, hexachlorobenzene, octachlorostyrene) and grain size. Grain size analysis was made by dry-sieving for the larger particle fractions and by measurement of density changes using a hydrometer for the silt and clay fractions. All chemical analyses were conducted according to OME (1983).

*Data analysis*

Non-hierarchical classification analysis (Gauch, 1982) was used to identify species assemblages (benthic communities) among the sampling stations. The analysis was conducted using the

individuals of species that could be identified using a dissecting microscope, e.g. *Stylodrilus heringianus* Claparede, *Spirosperma ferox* Eisen, *Potamothrix moldaviensis* Vejdovsky & Mrazek, were enumerated and removed from the sample. The remaining individuals were sorted into two groups: those with hair chaetae and those without hair chaetae. A random sample of not less than 20 percent of the individuals from each group, up to a maximum of 100 individuals, was removed from each group for identification.

Similarly, in samples with a large number of midges, all individuals of species that could be identified using a dissecting microscope, e.g.

Fortran program Compclus (Gauch, 1979). The ln($x + 1$) transformed mean abundance of each taxon was used as descriptors of the sampling stations. Each sampling station therefore was represented by a single sample.

Discriminant analysis (Legendre & Legendre, 1983) was used to elucidate the relationship between the defined benthic communities and the measured physical and chemical sediment variables. Nineteen environmental variables were used to discriminate between the benthic communities: Gravel, sand, and mud (silts and clays) represented physical habitat characteristics; iron, manganese, aluminum, arsenic, cobalt, copper, chromium, mercury, nickel, lead, and zinc represented metal contamination; phosphorus and nitrogen represented nutrients; loss-on-ignition (LOI) and total organic carbon (TOC) represented organic content; and solvent extractables represented oil and grease contamination of the sediments. Specific chlorinated organics could not be used because the majority of values were below the detection limits of the analytical procedures. To satisfy statistical requirements of the analysis (Green, 1979), all variables were log-transformed.

Since invertebrate species have specific habitat requirements with respect to sediment grain size, current speed, etc. (Hynes, 1970), different benthic communities can be expected in areas of the river with different physical characteristics. Therefore, applying Occam's razor, physical and other 'non-pollution' variables were given priority for explaining the distribution of communities in the river. Zones of environmental quality were delineated based on the distribution of the benthic communities in the river.

## Results and discussion

### Benthic macroinvertebrate fauna

A total of 112 macroinvertebrate taxa was collected from the river. Twenty-four of these taxa, which included species of insects, crustaceans, snails, clams, worms, and flatworms,

*Table 1.* Occurrence of common benthic macroinvertebrate taxa at 78 sampling sites in the St. Clair River, May 1985.

| | % occurrence |
|---|---|
| CADDISFLIES: | |
| *Hydropsyche* | 50 |
| *Cheumatopsyche* | 40 |
| MAYFLIES: | |
| *Hexagenia limbata* | 45 |
| *Caenis* | 40 |
| TRUE FLIES: | |
| *Cryptochironomus* | 67 |
| *Polypedilum* | 65 |
| *Procladius* | 65 |
| *Tribelos* | 45 |
| AMPHIPODS: | |
| *Gammarus fasciatus* | 68 |
| SNAILS: | |
| *Amnicola* | 77 |
| *Elimia livenscens* | 68 |
| *Physella gyrina* | 67 |
| *Valvata tricarinata* | 49 |
| *Valvata piscinalis* | 46 |
| *Gyraulus* | 41 |
| CLAMS: | |
| *Pisidium* | 77 |
| WORMS: | |
| *Limnodrilus hoffmeisteri* | 83 |
| *Spirosperma ferox* | 74 |
| *Quistadrilus multisetosus* | 59 |
| *Limnodrilus udekemianus* | 42 |
| *Potamothrix moldaviensis* | 42 |
| *Limnodrilus claparedianus* | 40 |
| *Stylodrilus heringianus* | 40 |
| FLATWORMS: | |
| Turbellaria | 63 |

occurred at more than 40 percent of the nearshore sites (Table 1).

The classification analysis identified seven macroinvertebrate assemblages (benthic communities) that occurred at two or more sampling sites (Table 2). Community 1 occurred only at the head of the river (Fig. 2) and had the lowest invertebrate density and species richness (Table 2) of any Community. The midges *Chernovskiia*, *Cryptochironomus*, and *Saetheria*, dominated the fauna at these sites.

Community 2 occurred throughout the upper river (i.e. upstream of the delta), but was chiefly found in the headwater section from Sarnia to

*Table 2.* Taxonomic composition (mean number 0.052 m$^{-2}$) of benthic Communities 1 to 7 in the St. Clair River, May 1985. P denotes a mean density of less than 1 individual per sample. Figure 2 shows the distribution of the Communities in the river.

| | Benthic Communities | | | | | | |
|---|---|---|---|---|---|---|---|
| | 1 | 2 | 3 | 4 | 5 | 6 | 7 |
| INSECTA: | | | | | | | |
| AQUATIC MOTHS: | | | | | | | |
| Pyralidae | | P | P | P | P | P | |
| BEETLES: | | | | | | | |
| Elmidae: | | | | | | | |
|   *Dubiraphia* | | P | P | P | | | |
| BUGS: | | | | | | | |
| Corixidae: | | | | | | | |
|   *Sigara lineata* | | | P | | | | |
| CADDISFLIES: | | | | | | | |
| Hydropsychidae: | | | | | | | |
|   *Cheumatopsyche* | | 7.7 | P | P | P | P | P |
|   *Hydropsyche* | | 1.8 | P | P | P | P | |
| Glossosomatidae: | | | | | | | |
|   *Protoptila maculata* | | 1.0 | P | | | | |
| Lepidostomatidae: | | | | | | | |
|   *Lepidostoma* | | | P | P | | | |
| Leptoceridae: | | | | | | | |
|   *Ceraclea* | | | P | | | | |
|   *Mystacides* | | | P | | | | |
|   *Oecetis* | | P | P | P | | | |
|   *Setodes* | | | P | | | | |
|   *Triaenodes* | | | P | | | | |
| Polycentropodidae: | | | | | | | |
|   *Neureclipsis* | | P | P | P | | | |
|   *Phylocentropus* | | | P | | | | |
| DAMSELFLIES: | | | | | | | |
| Coenagrionidae: | | | | | | | |
|   *Argia* | | | | | | P | |
|   *Coenagrion* | | | | | | P | |
| DRAGONFLIES: | | | | | | | |
| Gomphidae: | | | | | | | |
|   *Gomphurus* | | | P | P | | | |
|   *Stylurus* | | | P | P | | | |
| MAYFLIES: | | | | | | | |
| Baetiscidae: | | | | | | | |
|   *Baetisca* | | P | P | P | P | P | |
| Caenidae: | | | | | | | |
|   *Caenis* | | 1.2 | 4.5 | 1.2 | P | | |
| Ephemerellidae: | | | | | | | |
|   *Ephemerella* | | P | P | P | P | | |
|   *Serratella* | | P | | | P | | |
| Ephemeridae: | | | | | | | |
|   *Ephemera simulans* | | P | P | | | | |
|   *Hexagenia limbata* | | P | 2.8 | 12.3 | P | P | P |
| Heptageniidae: | | | | | | | |
|   *Stenonema* | | 1.2 | P | P | P | | P |
| Leptophlebiidae: | | | | | | | |
|   *Paraleptophlebia* | | | P | | | | |

*Table 2.* (Continued).

| | Benthic Communities | | | | | | |
|---|---|---|---|---|---|---|---|
| | 1 | 2 | 3 | 4 | 5 | 6 | 7 |
| **STONEFLIES:** | | | | | | | |
| Perlidae: | | | | | | | |
| *Perlesta placida* | | P | P | | | | |
| Perlodidae: | | | | | | | |
| *Isoperla bilineata* | | P | | | | | |
| **TRUE FLIES:** | | | | | | | |
| Ceratopogonidae | | P | P | 2.9 | | | |
| Chironomidae: | | | | | | | |
| *Chernovskiia* | 3.4 | P | P | | | | |
| *Chironomus* | P | | 1.8 | 13.3 | | | |
| *Cryptochironomus* | 2.1 | P | 8.8 | 33.0 | P | P | P |
| *Demicryptochironomus* | | 1.1 | P | 2.4 | P | | |
| *Dicrotendipes* | | P | P | P | P | | |
| *Harnischia* | | | 1.0 | 4.6 | P | P | |
| *Microtendipes* | | P | | | | | |
| *Parachironomus* | | P | P | P | P | | |
| *Paracladopelma* | | P | P | 2.6 | P | | |
| *Paratanytarsus* | | P | P | | | | |
| *Paratendipes* | | P | | P | | P | |
| *Phaenopsectra* | | P | 2.6 | 1.0 | P | P | |
| *Polypedium* | | P | 16.2 | 30.7 | P | P | |
| *Pseudochironomus* | | P | P | | | | |
| *Rheotanytarsus* | | | P | 2.0 | | | |
| *Saetheria* | 3.0 | | | | | | |
| *Stenochironomus* | | | | | P | | |
| *Stictochironomus* | | | 2.7 | 7.9 | | | |
| *Tanytarsus* | | P | 3.5 | 15.3 | P | | |
| *Tribelos* | | P | 18.0 | 13.7 | P | P | |
| *Xenochironomus ( = Axarus)* | | P | P | | | | |
| *Potthastia* | | | P | P | | P | |
| *Cricotopus* | | P | P | P | P | 1.2 | P |
| *Epoicocladius* | | | P | P | | | |
| *Heterotrissocladius* | | | | P | | | |
| *Nanocladius* | | P | P | P | | P | |
| *Parakiefferiella* | P | P | P | P | | | |
| *Monodiamesa* | P | P | 1.0 | 4.0 | P | | |
| *Ablabesmyia* | | | P | 3.1 | P | | |
| *Djalmabatista* | | P | P | | | | |
| *Procladius* | | P | 11.3 | 67.1 | 6.7 | 3.6 | P |
| *Conchapelopia* | | P | P | | P | P | |
| Empididae | P | 1.2 | P | P | P | P | |
| Psychodidae | | | | | | P | |
| | | | | | | P | |
| **CRUSTACEA:** | | | | | | | |
| **AMPHIPODS:** | | | | | | | |
| Gammaridae: | | | | | | | |
| *Gammarus fasciatus* | | 1.4 | 6.2 | 4.4 | 1.2 | P | |
| Haustoriidae: | | | | | | | |
| *Pontoporeia hoyi* | | P | P | P | | | |
| Taltridae: | | | | | | | 3.6 |
| *Hyalella azteca* | | P | P | P | P | P | |

*Table 2.* (Continued).

| | Benthic Communities | | | | | | |
|---|---|---|---|---|---|---|---|
| | 1 | 2 | 3 | 4 | 5 | 6 | 7 |
| **ISOPODS:** | | | | | | | |
| Asellidae: | | | | | | | |
| *Asellus* (caecidotea) | | P | P | P | P | P | |
| *Lirceus* | | | P | | | | |
| **MOLLUSCA:** | | | | | | | |
| **CLAMS:** | | | | | | | |
| Sphaeridae: | | | | | | | |
| *Pisidium* | P | P | 6.2 | 5.5 | 11.3 | 2.3 | |
| *Sphaerium* | | P | P | P | P | | |
| Unionidae: | | | | | | | |
| *Villosa iris* | | | P | | | | |
| **SNAILS:** | | | | | | | |
| Ancylidae: | | | | | | | |
| *Ferrissia* | | | P | P | | | |
| Hydrobiidae: | | | | | | | |
| *Amnicola* | | P | 9.4 | 2.2 | 21.9 | 7.1 | P |
| *Probythinella lacustris* | | | P | P | | | |
| Lymnaeidae: | | | | | | | |
| *Fossaria* | | | P | | P | | |
| *Pseudosuccinea columella* | | | P | | P | P | |
| Physidae: | | | | | | | |
| *Physella gyrina* | | 1.0 | 2.7 | P | 14.4 | 4.0 | P |
| Planorbidae: | | | | | | | |
| *Gyraulus* | | P | 1.8 | P | P | P | |
| *Promenetus* | | P | | | | | |
| Pleuroceridae: | | | | | | | |
| *Elimia livescens* | | 6.4 | 2.6 | P | 4.2 | P | 2.4 |
| *Pleurocera acuta* | | | P | | | | |
| Valvatidae: | | | | | | | |
| *Valvata piscinalis* | | P | 3.2 | | 13.3 | 5.1 | |
| *Valvata tricarinata* | | P | 3.6 | P | 12.7 | 1.6 | |
| **ANNELIDA:** | | | | | | | |
| **LEECHES:** | | | | | | | |
| Erpobdellidae: | | | | | | | |
| *Mooreobdella microstoma* | | | | | 2.0 | P | |
| Glossiphonidae: | | | | | | | |
| *Actinobdella inequiannulata* | | | P | P | | | |
| Pisicolidae | | P | P | | | | |
| **WORMS:** | | | | | | | |
| Lumbricidae | | P | P | P | | | |
| Lumbriculidae: | | | | | | | |
| *Stylodrilus heringianus* | | P | 1.2 | P | 1.5 | | |

*Table 2.* (Continued).

| | Benthic Communities | | | | | | |
|---|---|---|---|---|---|---|---|
| | 1 | 2 | 3 | 4 | 5 | 6 | 7 |
| **Naididae:** | | | | | | | |
| *Nais* | | P | P | | P | 2.0 | |
| *Ophidonais serpentina* | | P | 1.2 | P | 2.4 | 3.9 | |
| *Specaria josinae* | | P | P | | P | | |
| *Stylaria lacustris* | | | | | | P | |
| *Uncinais uncinata* | | P | 1.8 | P | 1.6 | P | |
| *Wapsa mobilis* | | P | 2.3 | 3.3 | | | |
| **Tubificidae:** | | | | | | | |
| *Aulodrilus americanus* | | | P | P | | | |
| *Aulodrilus pluriseta* | | | P | | P | | |
| *Isochaetes freyi* | | | P | P | | | |
| *Ilyodrilus templetoni* | | | 1.8 | 1.2 | 4.8 | 9.2 | |
| *Limnodrilus angustipenis* | | | P | P | | | |
| *Limnodrilus cervix* | | | 1.5 | | P | 45.7 | |
| *Limnodrilus claparedianus* | | P | 3.7 | 3.9 | 1.8 | 1.2 | |
| *Limnodrilus hoffmeisteri* | | 3.8 | 31.2 | 12.3 | 229.4 | 327.7 | 7.7 |
| *Limnodrilus maumeensis* | | | P | 4.9 | | | |
| *Limnodrilus udekemianus* | | P | 1.4 | 3.1 | 2.2 | 15.3 | P |
| *Potamothrix moldaviensis* | | | 1.1 | 3.4 | P | P | |
| *Potamothrix vejdovskyi* | | | P | P | | | |
| *Quistadrilus multisetosus* | | P | 2.6 | 8.9 | 20.7 | 15.5 | P |
| *Spirosperma ferox* | | P | 17.3 | 9.6 | 4.4 | P | |
| *Tubifex ignotus* | | | | | | P | |
| *Tubifex tubifex* | | | | | | 10.5 | |
| immature | 0.2 | 4.6 | 22.2 | 38.4 | 113.6 | 139.9 | 0.2 |
| **NEMERTEA:** | | | | | | | |
| **PROBOSCIS WORMS:** | | | | | | | |
| *Prostoma* | | P | P | P | P | P | P |
| **PLATYHELMINTHES:** | | | | | | | |
| **FLATWORMS:** | | | | | | | |
| Turbellaria | | 4.5 | 4.3 | 2.5 | 3.1 | P | |
| Mean richness (No. taxa 0.052 m$^{-2}$) | 2.6 | 8.9 | 15.3 | 17.5 | 12.9 | 9.4 | 3.0 |
| Mean density | 9.9 | 47.3 | 217 | 332 | 490 | 602 | 17.1 |

Marysville (Fig. 2). *Cheumatopsyche, Elimia* ( = *Goniobasis*) *livescens* Menke, *Limnodrilus hoffmeisteri* Claparede and flatworms numerically dominated the fauna at these sites (Table 2). *Cheumatopysche, Hydropsyche, Protoptila maculata* Hagen, *Stenonema, E. livescens,* Empididae, and flatworms were more abundant at these sites than elsewhere in the river.

Community 3, the most common macroinvertebrate assemblage, occurred throughout the river (Fig. 2). Species richness averaged 15.3 taxa per sample, with insects, notably *Polypedilum, Tribelos,* and *Procladius,* and the worms *L. hoffmeisteri* and *Spirosperma ferox,* numerically dominating the fauna. *Caenis, Tribelos, Phaenopsectra, Gammarus fasciatus* Say, *Gyraulus, Uncinais uncinata* Orsted and *S. ferox* were more abundant at these sites than elsewhere in the river.

Community 4 also occurred throughout the river, but was chiefly confined to Sarnia Bay and

152

*Fig. 2.* (Continued).

the lower section of the river from Fawn Island to Lake St. Clair (Fig. 2). Species richness was highest among all Communities, averaging 17.5 taxa per sample, with *Hexagenia limbata* Serville, *Chironomus*, *Cryptochironomus*, *Polypedilum*, *Tanytarsus*, *Tribelos*, *Procladius*, and *L. hoffmeisteri* numerically dominating the fauna (Table 2). Eighteen different taxa, including *H. limbata*, *Chironomus*, *Cryptochironomus*, *Harnischia*, *Stictochironomus*, *Polypedilum*, and *P. moldaviensis*, were more abundant at these sites than elsewhere (Table 2).

Community 5 occurred along the Canadian shoreline between Esso Chemical Canada and Allied Chemical Canada Ltd (Fig. 2). Species richness and invertebrate density were high, averaging 12.9 and 490 per sample, respectively. *Pisidium*, *Probythinella lacustris* Baker, *Physella gyrina* Say, *Valvata piscinalis* Muller, *Valvata tricarinata* Say. and *L. hoffmeisteri* dominated the fauna (Table 2). *Pisidium*, *P. gyrina*, *P. lacustris*, *V. piscinalis*, *V. tricarinata*, *Mooreobdella microstoma* Moore, *S. heringianus*, and *Quistadrilus multisetosus* Smith were more abundant at these sites than elsewhere in the river.

Community 6 occurred along the Canadian shoreline between Esso Chemical Canada and Corunna (Fig. 2). Invertebrate density, averaging

602 per sample, was highest among all communities, with worms, principally *Limnodrilus cervix* Brinkhurst, *L. hoffmeisteri*, *L. udekemianus*, *Q. multisetosus*, and *Tubifex tubifex* Muller, the dominant component of the fauna (Table 2). *Cricotopus*, *Nais*, *Ophidonais serpentina* Muller, *Ilyodrilus templetoni* Southern, *L. cervix*, *L. hoffmeisteri*, *L. udekemianus*, *Tubifex ignotus* Stolc. and *T. tubifex* were the most abundant at these sites.

Community 7 was confined to the waterfront of Dow Chemical Canada Inc. (Fig. 2). Species richness and invertebrate density were low; *L. hoffmeisteri*, *E. livescens*, and *Pontoporeia hoyi* Smith were the most abundant taxa.

The macroinvertebrate fauna at three sites: Community A, just downstream of the Port Huron sewage treatment plant, Community B, downstream of Dow Chemical Canada, and Community C, at the head of the Stag Island (Fig. 2), were unique and not associated with the benthic fauna at the other sites. Community A averaged only 17 organisms per sample and had a species richness of 6.7 taxa per sample. Immature tubificids, *P. moldaviensis* and *Limnodrilus angustipenis* Brinkhurst & Cook accounted for half the organisms, while *Cyptochironomus*, *Demicryptochironomus*, and *Monodiamesa* were the most abundant insects. Single individuals of *Cheumatopsyche*, *Hydropsyche*, *Ceraclea*, *Ephemerella*, *Amnicola*, *Gyraulus*, and *E. livescens* were also found.

Community B averaged 20 organisms per sample and had a species richness of 7.3 taxa per sample. Snails, particularly *E. livescens*, accounted for more than half the organisms, while the naidids, *Nais*, *O. serpentina*, and *Specaria josinae* Vejdovsky, composed about 25 percent of the Community. A few individuals of *Ephemerella*, *Polypedilum*, *Cricotopus*, *Pisidium*, and immature tubificids made up the remainder of the fauna.

Community C averaged just 8 organisms per sample and had a species richness of 4.0 taxa per sample. The true flies, *Polypedilum*, *Cricotopus*, *Paracladopelma*, and empidids accounted for more than 80 percent of the fauna. Single individuals of *Hydropsyche*, Pisicolidae, *S. herin-*

*gianus*, and immature tubificids composed the remainder of the community.

## Environmental quality evaluation

The discriminant analysis suggests that benthic Communities 1 to 7 were each associated with different environmental conditions (Fig. 3). The first discriminant axis (DA 1), which separated Communities 5, 6, and 7 from the others, accounted for 36.0 percent of the total variation and was correlated with Hg, Zn, oils and greases, TOC, and LOI (Table 3). This axis suggests, therefore, that the concentration of sediment contaminants, e.g. metals, oils and greases, organic matter, may have accounted for the distribution of Communities 5 to 7 in the river.

The second discriminant axis (DA 2), which separated Communities 1, 2, 3, and 4 (Fig. 3), accounted for 30.8 percent of the total variation and was positively correlated with Fe, Mn, As, Co, Cr, Ni, and Cu and negatively correlated with the sand content of the sediments (Table 3). Sediment texture and metal concentrations are frequently correlated (e.g. Griffiths, 1987; 1989); coarser sediments, characteristic of erosional

*Table 3.* Correlations (*r*) between the physicochemical sediment variables and the first two discriminant functions.

| | Discriminant Functions | |
|---|---|---|
| | 1 | 2 |
| Iron | −0.019 | 0.495 |
| Manganese | 0.114 | 0.331 |
| Aluminum | −0.037 | 0.275 |
| Arsenic | 0.085 | 0.402 |
| Cobalt | 0.127 | 0.311 |
| Chromium | 0.023 | 0.319 |
| Copper | 0.195 | 0.391 |
| Mercury | 0.653 | −0.064 |
| Nickel | 0.151 | 0.380 |
| Lead | 0.223 | 0.279 |
| Zinc | 0.421 | 0.282 |
| Oils and greases | 0.334 | 0.148 |
| Loss-on-ignition | 0.328 | 0.265 |
| Total organic carbon | 0.323 | 0.107 |
| Total phosphorus | 0.154 | 0.139 |
| Total Kjeldahl nitrogen | 0.174 | 0.104 |
| Particle Size: gravels | 0.037 | 0.163 |
| sands | 0.066 | −0.323 |
| silts and clays | 0.208 | −0.064 |

areas, generally have lower metal concentrations than finer sediments, characteristic of depositional areas. However, because of the greater surface areas of the finer sediments, a greater quantity of the metals can be adsorbed and bound to particles, thus reducing their bio-availability and affect on aquatic life. DA 2 suggests, therefore, that sediment texture (i.e. habitat characteristics) probably accounted for the distribution of Communities 1 to 4 in the river.

An assessment of the community structure and composition of the seven Communities supports the suggestions of the discriminant analysis. The low species richness and taxonomic composition of Community 1 is consistent with a habitat composed almost solely of sand and low in organic content (Hynes, 1970). *Saetheria*, *Chernovskiia*, and *Cryptochironomus*, which dominated this Community, and *Parakiefferiella* and *Monodiamesa* are typically found in oligotrophic, unstable sand deposits (i.e. the psammon habitat) of large rivers and lakes (Pinder & Reiss, 1983; Saether, 1977; Winnell & Jude, 1984; Soluk,

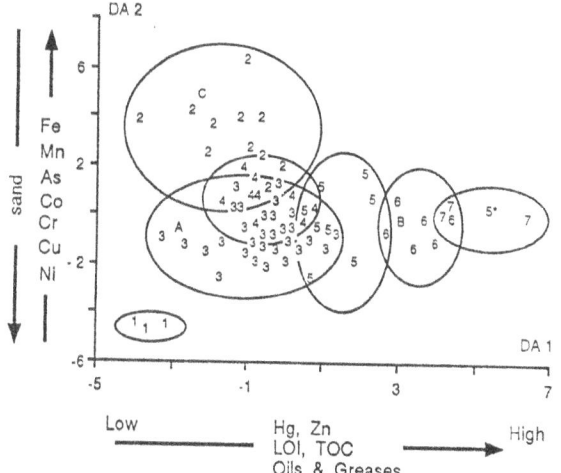

*Fig. 3.* Plot of benthic invertebrate Communities in discriminant space as defined by the first two discriminant functions. Each number or letter represents a single sampling station. * considered a misclassified station.

1985). The small size, body width $<< 0.6$ mm, of these specialized taxa account for the low abundance of animals collected from this habitat. Contaminants probably had little effect on this fauna as sediment concentrations of metals, oils and greases, and nutrients at these sites were the lowest found anywhere in the river (Table 4).

The macroinvertebrate fauna of Communities 2 to 4 is indicative of meso-eutrophic conditions, described by Merritt et al. (1984) as typical for large, unstressed rivers, and show little evidence of stress from pollution. Taxa characteristic of mesotrophic conditions, e.g. Stictochironomus, Demicryptochironomus, Tribelos, Physella, E. livescens, G. fasciatus, S. ferox, P. moldaviensis (Cook & Johnson, 1974; Lauritsen et al., 1985), and tolerant of eutrophic conditions, e.g. Cheumatopsyche, Hydropsyche, L. hoffmeisteri, Chironomus, Cryptochironomus, Procladius (Roback, 1962; Carr & Hiltunen, 1965; Brinkhurst et al., 1968; Lauritsen et al., 1985), numerically dominated these Communities (Table 1).

The high species richness of insects (Table 2) and the high relative abundance of insects (38–66%) and annelids (23–42%) in these communities suggest that the heavy metals, oils and greases, and pesticides in the sediments (Tables 4 & 5) had little adverse effect on the fauna (Wiederholm, 1984). Furthermore, the abundance of species intolerant of low dissolved oxygen concentrations, e.g. stoneflies, snails, Stenonema, Hexagenia, Caenis, P. maculata, Cheumatopysche, Tanytarsus, S. heringianus, and flatworms (Britt, 1955; Harman, 1974; Cook & Johnson, 1974; Pennak, 1978; Hilsenhoff, 1982; Rabeni et al., 1985), and the low relative abundance of worms (Thornley, 1985) indicate little stress from nutrient or organic enrichment.

Differences in the community structure and composition among Communities 2 to 4 thus reflect differences in habitat characteristics in the river. The high proportion of scrapers, notably P. maculata, Stenonema, and E. livescens, and filter-feeders, such as Cheumatopsyche and Hy-

Table 4. Mean values of physicochemical sediment variables associated with benthic Communities 1 to 7 from the St. Clair River, May 1985. All values expressed in mg kg$^{-1}$ unless otherwise stated. See Table 2 for the taxonomic composition of the Communities. See Fig. 2 for the distribution of the Communities in the St. Clair River.

| | Benthic Communities | | | | | | |
|---|---|---|---|---|---|---|---|
| | 1 | 2 | 3 | 4 | 5 | 6 | 7 |
| Iron (g kg$^{-1}$) | 3.87 | 16.34 | 8.18 | 12.63 | 8.31 | 8.33 | 10.11 |
| Manganese (g kg$^{-1}$) | 0.09 | 0.21 | 0.16 | 0.25 | 0.19 | 0.17 | 0.20 |
| Aluminum (g kg$^{-1}$) | 1.36 | 5.90 | 4.05 | 6.28 | 4.39 | 3.79 | 2.48 |
| Arsenic | 1.83 | 7.46 | 4.15 | 6.56 | 5.70 | 5.03 | 5.17 |
| Cobalt | 2.19 | 6.49 | 4.80 | 6.19 | 5.35 | 5.43 | 6.63 |
| Chromium | 6.37 | 18.88 | 13.59 | 17.53 | 16.24 | 8.82 | 20.90 |
| Copper | 4.54 | 43.08 | 16.92 | 29.80 | 19.40 | 29.90 | 74.44 |
| Mercury | 0.02 | 0.19 | 0.43 | 0.11 | 2.11 | 4.78 | 15.03 |
| Nickel | 2.53 | 13.67 | 8.25 | 12.14 | 7.94 | 8.81 | 21.79 |
| Lead | 2.22 | 34.92 | 16.52 | 17.47 | 22.28 | 69.66 | 21.89 |
| Zinc | 11.32 | 45.15 | 34.94 | 54.75 | 65.65 | 67.64 | 96.73 |
| Oils and greases (g kg$^{-1}$) | 0.07 | 0.52 | 0.48 | 0.88 | 0.63 | 2.54 | 2.03 |
| Loss-on-ignition (g kg$^{-1}$) | 2.91 | 12.78 | 8.92 | 17.62 | 13.30 | 25.49 | 18.12 |
| Total organic carbon (g kg$^{-1}$) | <5.00 | 8.28 | 7.39 | 13.62 | 11.06 | 19.52 | 13.41 |
| Total phosphorus (g kg$^{-1}$) | 0.10 | 0.22 | 0.21 | 0.29 | 0.23 | 0.26 | 0.27 |
| Total Kjeldahl nitrogen (g kg$^{-1}$) | 0.13 | 0.36 | 0.35 | 0.71 | 0.45 | 0.69 | 0.37 |
| Particle size: % gravels | 0.60 | 67.00 | 0.70 | 0.00 | 25.30 | 7.40 | 45.20 |
| % sands | 98.40 | 32.00 | 79.20 | 51.80 | 35.80 | 52.00 | 46.90 |
| % silts and clays | 1.00 | 1.00 | 20.10 | 48.20 | 38.90 | 40.60 | 7.90 |

*Table 5.* Occurrence (%) and maximum measured concentration ($\mu$g kg dry mass) of chlorinated organics in sediments associated with benthic Communities 3 to 7 in the St. Clair River, May 1985. Number of sampling sites noted in brackets below each Community. See Fig. 2 for distribution of Communities in the river.

| | Benthic Communities | | | | | | | | | |
| | 3 (24) | | 4 (9) | | 5 (9) | | 6 (6) | | 7 (3) | |
| | % | max | % | max | % | max | % | max | % | max |
|---|---|---|---|---|---|---|---|---|---|---|
| Hexachlorobenzene | 88 | 756 | 89 | 148 | 88 | 1131 | 100 | 1871 | 100 | *3280 |
| Octachlorostyrene | 83 | 116 | 22 | 34 | 100 | 438 | 83 | 588 | 100 | *1128 |
| DMDT Methoxychlor | 8 | 15 | 0 | N.D. | 11 | 10 | 33 | 100 | 33 | *170 |
| Heptachlorepoxide | 0 | N.D. | 0 | N.D. | 0 | N.D. | 67 | 10 | 33 | *26 |
| pp'-DDT | 0 | N.D. | 0 | N.D. | 11 | 5 | 0 | N.D. | 33 | *25 |
| pp'-DDD | 4 | 5 | 11 | 5 | 0 | N.D. | 0 | N.D. | 33 | *50 |
| pp'-DDE | 33 | 19 | 33 | 2 | 50 | *49 | 83 | 22 | 100 | 3 |
| Endosulfan II | 0 | N.D. | 0 | N.D. | 11 | *98 | 17 | 50 | 33 | 8 |
| Endosulfan sulfate | 4 | 14 | 11 | 5 | 11 | 5 | 50 | *40 | 33 | 10 |
| Dieldrin | 63 | 12 | 56 | 6 | 78 | 5 | 67 | 12 | 33 | 13 |
| Aldrin | 13 | 3 | 0 | N.D. | 38 | 2 | 33 | 3 | 0 | N.D. |

N.D. means not detected

* denotes the greatest measured concentration

*dropsyche*, in Community 2 (Fig. 4) is indicative of an erosional habitat with coarse sediments and swift current velocities. The low density of invertebrates in this Community (Fig. 4) probably was related to the shallow substrate depth, poor penetration of the grab sampler, and low concentration of organic matter (Table 4). The increase in the abundance of invertebrates and the proportion of gatherers, and reduction in the proportion of scrapers, filterers, and shredders from Communities 2 through 4 (Fig. 4) suggest a shift in site characteristics from an erosional to intermediate (run) to depositional habitat, respectively. The presence of species characteristic of oligotrophic conditions, e.g. *Tanytarsus*, *Monodiamesa*, *Paracladopelma*, *Heterorissocladius*, and *S. heringianus* (Brinkhurst *et al.*, 1968; Saether, 1979), in Communities 3 and 4 suggest that the current speed was slow enough for mineral sediments (sands) to settle from the river (Warwick, 1980). The presence of ceratopogonids and *Harnischia* in Community 4 are also indicative of slow flows and sand accumulation. The abundance of bur-

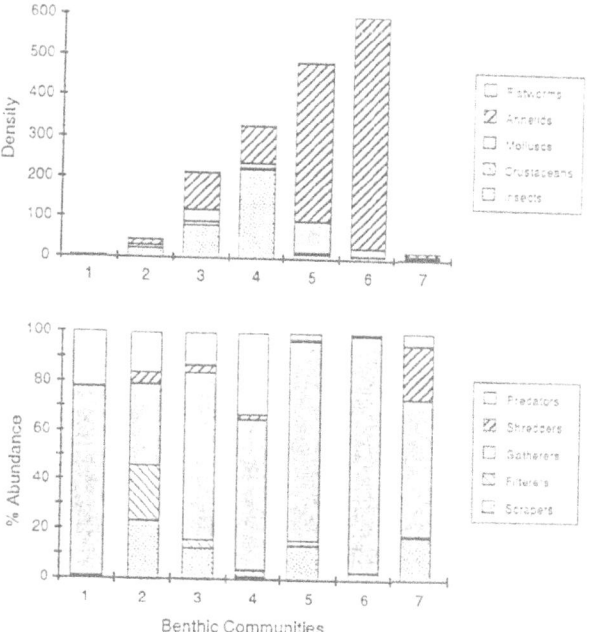

*Fig. 4.* Density (No. 0.052 m$^{-2}$) of invertebrates by taxonomic category (upper) and relative abundance of invertebrates by functional-feeding group (lower) in benthic Communities 1 to 7 in the St. Clair River, May 1985.

rowing organisms, such as worms, *Hexagenia*, *Chironomus*, *Polypedilum*, *Stictochironomus*, *Tanytarsus*, *Tribelos*, and *Pisidium*, and the predators, *Procladius*, *Cryptochironomus*, and *Ablabesmyia*, in Community 4 indicate a more lentic environment with slow current velocities, fine sediments, and organic accumulations, typical of the central basin of Lake St. Clair (Griffiths, 1987). Measurements of sediment texture, nutrients, and organic carbon (Table 4) and current speed observations (J.D. Westwood, Ont. Min. Env., pers. observ.) at the sampling sites confirm the habitat characteristics implied from the faunal composition: Community 2 was found at sites with strong current speeds and coarse sediments, low in nutrients and organic matter; Community 4 at sites with slow current speeds and fine sediments, higher in nutrients and organic matter, and Community 3 at sites with intermediate values.

The community structure and composition of Communities 5 to 7, in contrast, show the effects of pollution. The higher density of worms and relative abundance of gatherers, and the lower abundance of insects and species richness of Communities 5 and 6 relative to Communities 3 and 4 (Fig. 4; Table 2), which have similar habitat characteristics, are indicative of degraded environmental conditions (Goodnight & Whitley, 1960; Carr & Hiltunen, 1965). Since oligochaetes and snails are considered more sensitive than insects to heavy metals (Henderson, 1949; Brinkhurst & Cook, 1974; Harman, 1974), the concentrations of heavy metals in the sediment (Table 4) probably had little effect on the aquatic biota of Communities 5 and 6. Oils and greases, and pesticides, however, may have accounted for the reduced diversity and abundance of insects. Oils and chlorinated organics, such as BHC, DDT, toxaphene, and methoxychlor, have been shown to reduce the diversity and abundance of insects in streams and increase the relative abundance of oligochaetes and snails (Hynes, 1961; Grzenda *et al.*, 1964; Keenleyside, 1967; Cuffney *et al.*, 1984; Harrel, 1985). A greater diversity, concentration, and occurrence of chlorinated organics were observed at sites with Communities 5 and 6 than with Communities 3 or 4

(Table 5), and the concentration of oils and greases was markedly higher at sites with Community 6 than with Community 3 (Table 4). Furthermore, the high abundance of *L. hoffmeisteri* and the presence of *L. cervix*, *Q. multisetosus*, and *T. tubifex* in Community 6 indicate that the sediments at these sites were highly enriched with organic matter (Lauritsen *et al.*, 1985).

The extremely low species richness and invertebrate density of Community 7, relative to Communities 3 and 4, and the dominance of *L. hoffmeisteri* (Fig. 4; Table 2) indicate that Community 7 occurred at sites with severely degraded environmental conditions. Sediment concentrations of Hg, Zn, Ni, Cu, and Cr were higher at these sites than elsewhere in the river (Table 4). Furthermore, the highest concentrations of several chlorinated organics: DDT, DDD, hexachlorobenzene, octachlorostyrene, methoxychlor, and heptachlorepoxide were measured at these sites (Table 5). The sum of these variables appears to have produced an environment toxic to macroinvertebrate life.

Communities A, B, and C were not used in the discriminant analysis, but were plotted afterwards in discriminant space, using the defined discriminant functions, to evaluate their relationships with the other Communities. Figure 3 shows that Communities A and C were associated with the unimpaired communities, while Community B was associated with Community 6, a degraded community. The sediment concentrations of metals, oils and greases and nutrients at the sites of Communities A and C were less than the averages found for Community 2 (Table 4), while the sediment texture was similar, with gravel composing 85 percent of the sediments at the site of Community A and 67 percent at the site of Community C, the remainder of the sediments being largely sand at both sites. The low species richness and invertebrate abundance, and occurrence of hydropsychids in these Communities thus were probably related to the habitat characteristics at these sites – swift currents (J.D. Westwood, Ont. Min. Env., pers. observ.) and coarse sediments low in organic matter (TOC $<5$ mg l$^{-1}$). While the sediments at the site of

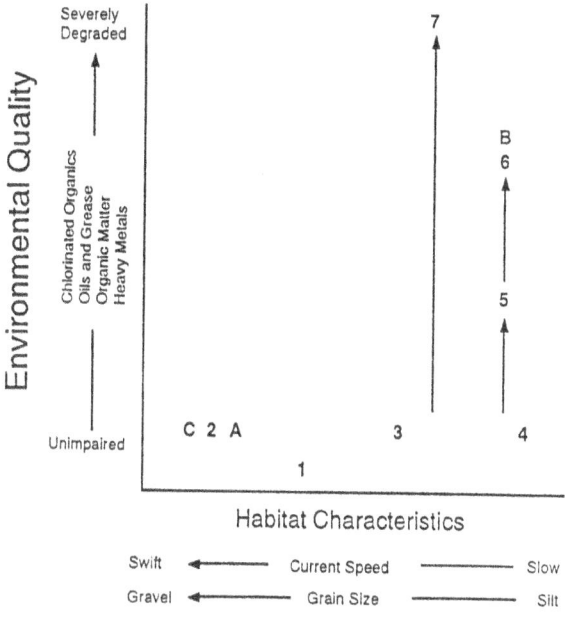

*Fig. 5.* Environmental quality and habitat characteristics reflected by benthic invertebrate Communities 1 to 7 in the St. Clair River, May 1985.

matter (depositional habitat). In contrast, these analyses suggest that the distribution of Communities 5, 6, 7 and B was related to pollutional stresses (Fig. 5). Community 5, which reflected slightly degraded (impaired) environmental conditions, occurred at sites with elevated sediment concentrations of Hg, Pb, Zn, and several chlorinated organic compounds, while Community 6, which reflected degraded environmental quality conditions, was found at sites with higher sediment concentrations of Hg, Pb, Zn, several chlorinated organic compounds, oils and greases, and organic matter relative to Communities 3 and 4. Community 7, which reflected seriously degraded and probably toxic environmental conditions, occurred at sites that had the highest measured sediment concentrations of several chlorinated organics, oils and greases, and several heavy metals, particularly Hg, anywhere in the river.

Community B were also mainly composed of gravel (58%) with sand (40%), sediment concentrations of mercury ($5.3$ mg kg$^{-1}$) and oils and greases ($1.0$ g kg$^{-1}$) were higher than at the sites of Community 2. The erosional conditions at this site may account for the low richness and abundance of invertebrates of Community B, however, the almost complete lack of insects, especially caddisflies, and the virtual absence of infaunal organisms suggest that sediment contaminants also affected the faunal composition.

In summary, these analyses suggest that the distribution of Communities 1, 2, 3, 4, A, and C was related to habitat characteristics (Fig. 5). Community 1 occurred at sites with unstable sandy sediments, poor in nutrients and organic matter (psammon habitat); Communities 2, A and C were found at sites with swift currents and coarse sediments, low in nutrients and organic matter (erosional habitat); Community 3 was found at sites with intermediate current speeds and sediment texture (run habitat); and Community 4 occurred at sites with slow currents and fine sediments, relatively rich in nutrients and organic

*Environmental quality zones*

Environmental quality zones were delineated in the river (Fig. 6) based on the distribution of the benthic communities. A severely degraded (toxic) environmental quality zone (sites with Community 7) occurred along the waterfront of Dow Chemical Canada. The low abundance and poor diversity of macroinvertebrates collected from this area indicated that toxic conditions prevailed in the sediments, probably as a result of wastes discharged from Dow Chemical's First Street sewer which include phenols, chlorinated organics, benzene, ethylbenzene, Hg, Cu, Ni, and chlorides (EC & OME, 1986; EC & US EPA, 1989). The highest concentration of several chlorinated organics, including hexachlorobenzene, methoxychlor, octachlorostyrene, DDT, DDD, and heptachlorepoxide, and mercury ($51$ mg kg$^{-1}$ dry mass) was measured in the sediments of this area. Periodic spills of acids, petrochemicals, oils, caustics, etc. from Dow Chemical and upriver industries (EC & OME, 1986; Edsall *et al.*, 1988) have probably contributed to the toxic conditions evident in this area. Modifications to

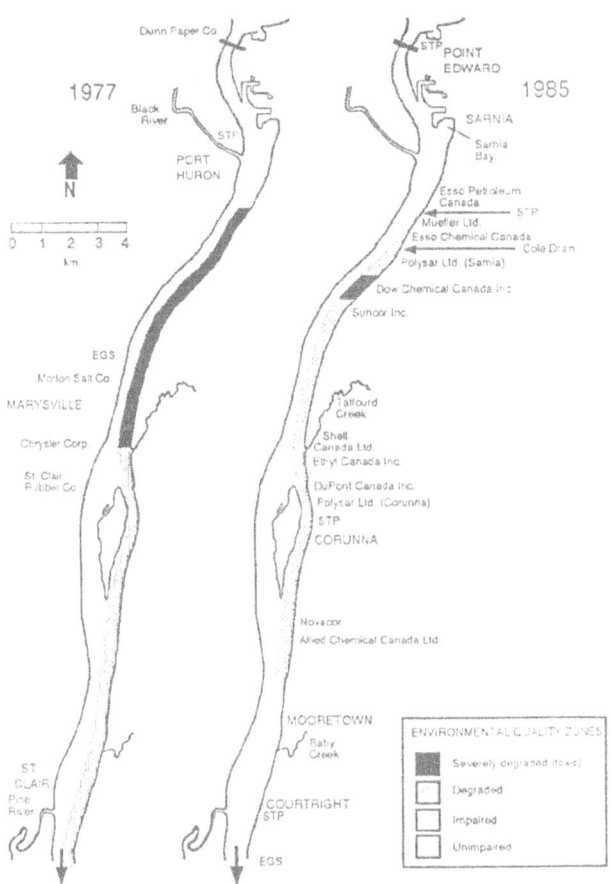

*Fig. 6.* Distribution of environmental quality zones in the spring of 1977 (from Griffiths, 1978) and 1985 along the St. Clair River. The lower part of the river was unimpaired each year.

the shoreline (bulkheading, filling), as evidenced by the high proportion of gravels in the sediment of sites (Community 7 in Table 4), probably have also contributed to the disturbance of the aquatic biota.

Zones of degraded environmental quality (sites with Community 6 and B) occurred along the waterfronts of Polysar Ltd. (Sarnia), and Suncor Inc., and from the mouth of Talfourd Creek to the town of Corunna (Fig. 6). Tubificids dominated the benthic fauna at these sites suggesting organic enrichment and possibly an impact of chlorinated organics. Waste discharges and spills from adjacent industries (EC & OME, 1986; EC & US EPA, 1989) were the most likely source of the contaminants degrading the environmental quality of these areas. Wastes discharged by Polysar

Ltd. (Sarnia), for example, included phenols, ammonia, cobalt, benzene, isoprene, styrene, and chlorinated organics. In addition, Polysar Ltd. (Sarnia) accounted for 34 percent of the total organic carbon discharged to the river, while the Cole Drain (township ditch), which empties to the river along the waterfront of Polysar Ltd. (Sarnia), accounted for 41 percent of the total oils and greases discharged to the river. Wastes from Ethyl Canada accounted for 66 percent of the lead and 17 percent of the volatile hydrocarbons discharged to the river. The highest concentration of lead was measured in sediments (330 mg kg$^{-1}$ dry mass) just downstream of Ethyl Canada.

Zones of impaired environmental quality (sites with Community 5) occurred along the upstream half of the waterfront of Polysar Ltd. (Sarnia), from Suncor Inc. to the mouth of Talfourd Creek, and from Corunna to downstream end of Allied Chemical Canada Ltd. (Fig. 6). Tubificids were abundant at these sites but snails were a noted component of the fauna. The impaired environmental quality along Polysar's upstream waterfront was probably the result of discharges and spills from upriver industries. Similarly, the nearshore environmental quality from Suncor to Talfourd Creek was probably impaired as a result of discharges from Polysar Ltd. (Sarnia) and Dow Chemical, since this section of the river only received diluted wastes from upriver industries. Similarly, the nearshore environmental quality from Corunna to Allied Chemical was probably impaired as a result of discharges from Shell Canada Ltd. and Ethyl Canada Inc.; discharges from Novacor and Allied Chemical appeared to have little additional effect on the invertebrate fauna of the river. However, the high proportion of gravels in the sediments (Community 5 in Table 4) along these sections indicates that shoreline modifications have physically altered the benthic habitat, which may partially account for the high snail abundance.

The environmental quality conditions of the remainder of the St. Clair River were considered to be typical of that of a large, unstressed river. Thus, throughout most of the river, the macroinvertebrate fauna simply reflected the physical attributes of the habitat.

*Changes in environmental quality from 1977*

Environmental quality zones were delineated in the St. Clair River in 1977 (Griffiths, 1978; OME, 1979) using similar methods to those in this study. Association analysis (Williams & Lambert, 1959) was used to define the benthic communities among the 60 sampled sites and ANOVA was used to relate the communities to various morphological, sediment, and water chemical variables. Seven benthic communities were identified in the river; three of which occurred at most of the sites (Table 6). The community structure and composition of Community 2 in 1977 was similar

to that of Communities 3 and 5 in 1985 and reflected little or slight environmental impairment. Community 5 was similar to Community 6 in 1985 and indicated degraded environmental conditions. Community 7 in 1977 was characterized by low species richness and invertebrate density. In nearshore areas (depth < 8 m), this community probably reflected severely degraded environmental conditions similar to Community 7 in 1985, whereas in deeper, offshore areas (depth > 8 m), which were not sampled in the 1985 study, this community probably reflected erosional characteristics of the habitat and the physical stress (high flows, propeller wash of commercial ships) associated with the main channel of the river.

The occurrence of Communities 5 and 7 at nearshore sites from the discharge point of the Sarnia sewage treatment plant to the mouth of Talfourd Creek indicated that a zone of severely degraded environmental quality existed along this part of the Canadian shoreline in 1977 (Fig. 6). Overall, the species richness at these sites was low, averaging 2.7 taxa per sample. Tubificids, primarily *L. hoffmeisteri*, *Q. multisetosus*, and *L. udekemianus*, accounted for 94 percent of the average 64 invertebrates $0.052 m^{-2}$. The occurrence of communities 2 and 5 at nearshore sites from the mouth of Talfourd Cr. to the electrical generating station of Ontario Hydro indicated that a zone of degraded environmental quality existed along this stretch of river. Overall, species richness averaged just 5.8 taxa per sample. Invertebrate density averaged 178 individuals $0.052 m^{-2}$, with the tubificids, *L. hoffmeisteri*, *L. cervix*, *Q. multisetosus*, *L. udekemianus*, and *T. tubifex* accounting for 91 percent and the snails, *P. gyrina*, *E. livescens*, *V. sincera*, and *V. tricarinata*, composing 4 percent of the fauna.

A comparison of the distribution of environmental quality zones in the spring of 1985 with those of 1977 suggests that environmental quality conditions along the Canadian side of the river significantly improved over the eight-year period (Fig. 6). The lower zone of unimpaired environmental quality advanced upstream approximately 8 km, while the headwater zone of unimpaired environmental quality extended downstream

*Table 6.* Taxonomic composition (mean number $0.052 \text{ m}^{-2}$) of benthic Communities 2, 5 and 7 along the Canadian shoreline of the St. Clair River, May 1977. P denotes a density of less than 1 individual per sample.

| | Benthic Communities | | | |
|---|---|---|---|---|
| | 2 | 5 | 7a (>8 m) | 7b (<8 m) |
| INSECTA: | | | | |
| AQUATIC MOTHS: | | | | |
| Pyralidae | P | | | |
| BEETLES: | | | | |
| Elmidae | P | | | |
| BUGS: | | | | |
| Corixidae | P | | P | |
| CADDISFLIES: | | | | |
| Hydroptilidae: | | | | |
| Agraylea | | | P | P |
| Hydropsychidae: | | | | |
| Cheumatopsyche | | P | P | |
| Hydropsyche | | | P | |
| Leptoceridae: | | | | |
| Ceraclea | P | | | P |
| Oecetis | P | | P | P |
| Polycentropodidae: | | | | |
| Polycentropus | | | P | |
| MAYFLIES: | | | | |
| Caenidae: | | | | |
| Caenis | P | P | | |
| Ephemerellidae: | | | | |
| Ephemerella (s.l.) | | P | P | P |
| Ephemeridae: | | | | |
| Hexagenia limbata | P | | | |
| Heptageniidae: | | | | |
| Stenonema | | P | P | |
| TRUE FLIES: | | | | |
| Athericidae: | | | | |
| Atherix | | | P | |
| Ceratopogonidae | P | | | |
| Chironomidae: | | | | |
| Cardiocladius | | | P | |
| Chironomini | P | P | P | |
| Chironomus | 9.1 | P | P | |
| Cryptochironomus | 3.2 | P | P | |
| Dicrotendipes | | P | | |
| Parachironomus | P | | P | |
| Paracladopelma | P | | | |
| Paralauterborniella | P | | P | |
| Paratendipes | P | | | |
| Phaenopsectra | P | | | |
| Polypedilum | 9.5 | P | P | |
| Stenochironomus | P | | | |
| Stictochironomus | P | | | |

*Table 6.* (Continued)

| | Benthic Communities | | | |
|---|---|---|---|---|
| | 2 | 5 | 7a (>8 m) | 7b (<8 m) |
| Tanytarsini | P | P | P | |
| *Tribelos* | 1.8 | | | |
| *Potthastia* | P | | | |
| *Corynoneura* | | P | | |
| *Cricotopus* | P | P | 1.1 | |
| *Heterotrissocladius* | | | P | |
| *Psectrocladius* | P | P | P | |
| Pentaneurini | 1.1 | P | | P |
| *Procladius* | 10.6 | | | P |
| **CRUSTACEA:** | | | | |
| **AMPHIPODS:** | | | | |
| Gammaridae: | | | | |
| *Gammarus* | 15.8 | 1.2 | P | P |
| Haustoriidae: | | | | |
| *Pontoporeia hoyi* | | | P | |
| Taltridae: | | | | |
| *Hyalella azteca* | P | | P | |
| **ISOPODS:** | | | | |
| Asellidae: | | | | |
| *Asellus* (Caecidotea) | P | P | | |
| **DECAPODS:** | | | | |
| *Orconectes propinquus* | P | | | |
| **MOLLUSCA:** | | | | |
| **CLAMS:** | | | | |
| Sphaeridae: | | | | |
| *Pisidium* | 10.2 | P | P | |
| **SNAILS:** | | | | |
| Hydrobiidae: | | | | |
| *Amnicola* | 3.3 | P | P | P |
| Lymnaeidae | | | P | |
| Physidae: | | | | |
| *Physella gyrina* | 1.7 | 2.5 | P | P |
| Planorbidae: | | | | |
| *Gyraulus* | P | P | | |
| Pleuroceridae: | | | | |
| *Elimia livescens* | 1.7 | P | P | 2.1 |
| Valvatidae: | | | | |
| *Valvata sincera* | 2.7 | 1.0 | P | |
| *Valvata tricarinata* | P | P | P | |
| **ANNELIDA:** | | | | |
| **LEECHES:** | | | | |
| Hirudinea | P | P | | P |
| **WORMS:** | | | | |
| Lumbriculidae: | | | | |
| *Stylodrilus heringianus* | 1.0 | 1.0 | P | 1.5 |

*Table 6.* (Continued)

| | Benthic Communities | | | |
|---|---|---|---|---|
| | 2 | 5 | 7a (>8 m) | 7b (<8 m) |
| Naididae: | | | | |
| *Nais* | | P | P | |
| *Ophidonais serpentina* | 2.7 | | | |
| Tubificidae: | | | | |
| *Aulodrilus pluriseta* | P | | | |
| *Ilyodrilus templetoni* | P | | | P |
| *Limnodrilus angustipenis* | | | P | |
| *Limnodrilus cervix* | 27.9 | 16.5 | | |
| *Limnodrilus claparedianus* | 4.7 | 1.3 | P | |
| *Limnodrilus profundicola* | P | | | |
| *Limnodrilus hoffmeisteri* | 97.9 | 87.6 | | |
| *Limnodrilus udekemianus* | 7.3 | 2.3 | | |
| *Potamothrix moldaviensis* | 2.4 | P | P | |
| *Quistadrilus multisetosus* | 23.9 | 10.5 | P | |
| *Spirosperma ferox* | 28.3 | 1.4 | P | |
| *Tasserkidrilus superiorensis* | | P | | |
| *Tubifex tubifex* | 6.8 | 1.3 | | |
| immature | 93.5 | 45.5 | P | P |
| PLATYHELMINTHES: | | | | |
| FLATWORMS | | | | |
| Turbellaria | 4.7 | P | P | P |
| Mean richness (No. taxa 0.052 m$^{-2}$) | 14.1 | 6.1 | 3.1 | 2.2 |
| Mean density | 387 | 178 | 9.6 | 6.2 |

about 1 km. Thus the total length of river affected by waste discharges from Canadian industries and municipalities decreased from 21 km in 1977 to 12 km in 1985. Furthermore, the occurrence of Community 5 along this 12 km stretch of the river in 1985 suggests that the overall environmental quality of this neashore area has improved since 1977.

The improvement in the structure and organization of the nearshore macroinvertebrate fauna corresponds to the reduction of waste loadings to the river through improvements in wastewater treatment, changes in industrial processes and better housekeeping practices (EC & US EPA, 1989). Abatement programs that have been implemented over the past decade, thus appear to have had a beneficial effect on the environmental quality along the Canadian side of the St. Clair River. Although a 12 km stretch of the river along the Canadian side was still affected by industrial and municipal wastes in 1985, about 9 km of river has recovered since 1977.

## Acknowledgements

Water Resources Branch, the Southwestern Region and the Detroit/St. Clair/St. Mary's River Project of the Ontario Ministry of the Environment provided funding and assistance during this study. S. Thornley and T. Edsall provided helpful comments for the final manuscript. The Great Lakes Section of the Ontario Ministry of the Environment provided the survey vessel and B. Hawkins and J.D. Westwood of the Southwestern Region collected the benthic and sediment samples. The Lambton Area Water Treatment Plant kindly provided space for a field laboratory. The field crew consisted of Lucy Sardella (supervisor), Jim Carbone, Cindy Faulkner, and Caroline Van Roestel.

# References

Brinkhurst, R. O. & D. G. Cook, 1974. Aquatic earthworms. In: C. W. Hart Jr. & S. L. H. Fuller (eds). Pollution Ecology of Freshwater Invertebrates. pp. 143–155. Academic Press, New York.

Brinkhurst, R. O., A. L. Hamilton & H. B. Herrington, 1968. Components of the bottom fauna of the St. Lawrence Great Lakes. Great Lakes Inst., Univ. Toronto. PR 33: 1–49.

Britt, N. W., 1955. Stratification in western Lake Erie in the summer of 1953: effects on the *Hexagenia* population. Ecology 36: 239–244.

Carr, J. F. & J. K. Hiltunen, 1965. Changes in the bottom fauna of western Lake Erie from 1930 to 1961. Limnol. Oceanogr. 10: 551–569.

Chan, C. H., Y. L. Lau & B. G. Oliver, 1986. Measured and modelled chlorinated contaminant distributions in the St. Clair River water. Wat. Pollut. Res. J. Can. 21: 332–343.

Cook, D. G. & M. G. Johnson, 1974. Benthic macroinvertebrates of the St. Lawrence Great Lakes. J. Fish. Res. Bd Can. 31: 763–782.

Cuffney, T. F., J. B. Wallace & J. R. Webster, 1984. Pesticide manipulation of a headwater stream: Invertebrate response and their significance for ecosystem processes. Freshwat. Invert. Biol. 3: 153.

Derecki, J. A., 1984. St. Clair River physical and hydraulic characteristics. Ntnl. Oceanic Atmos. Admin., Great Lakes Env. Res. Lab. Open File Report. Ann Arbor, Michigan. 8 pp.

EC & OME (Environment Canada & Ontario Ministry of the Environment), 1986. St. Clair River pollution investigation (Sarnia Area). Report of the Review Board of the Canada-Ontario Agreement Respecting Great Lakes Water Quality. Ottawa, Canada 135 pp.

EC & US EPA (Environment Canada & U.S. Environmental Protection Agency), 1989. Upper Great Lakes Connecting Channels Study – Volume 2, 626 pp.

Edsall, T. A., B. A. Manny & C. N. Raphael, 1988. The St. Clair River and Lake St. Clair, Michigan: an ecological profile. U.S. Fish Wildl. Serv., Biological Report 85 (7.3). Ann Arbor, Michigan. 130 pp.

Gauch, H. G. Jr., 1979. Compclus. Ecology and systematics, Cornell Univ., Ithaca, New York. 59 pp.

Gauch, H. G. Jr., 1982. Multivariate Analysis in Community Ecology. Cambridge Univ. Press, Cambridge. 298 pp.

Goodnight, C. J. & L. S. Whitley, 1960. Oligochaetes as indicators of pollution. In: Proc. 15th Indust. Waste Conf., Purdue Univ. pp. 139–142.

Green, R. H., 1979. Sampling Design and Statistical Methods for Environmental Biologists. John Wiley and Sons Inc., New York. 257 pp.

Griffiths, R. W., 1978. Benthic communities as indicators of water quality in the St. Clair River. B.Sc. Thesis. Univ. Western Ont., London, Ontario, Canada. 40 pp.

Griffiths, R. W., 1987. Environmental quality assessment of Lake St. Clair in 1983 as reflected by the distribution of benthic invertebrate communities. Ont. Min. Env., London, Ontario, Canada. 35 pp.

Griffiths, R. W., 1989. The effect of in-place pollutants on the benthic invertebrate fauna of the Spanish River harbour. Ont. Min. Env., Sudbury, Ontario, Canada. 27 pp.

Grzenda, A. R., G. J. Lauer & H. P. Nicholson, 1964. Water pollution by insecticides in an agricultural river basin. II. The zooplankton, bottom fauna and fish. Limnol. Oceanogr. 9: 318–323.

Harman, W. N., 1974. Snails. In: C. W. Hart Jr. & S. L. H. Fuller (eds), Pollution Ecology of Freshwater Invertebrates. pp. 143–155. Academic Press, New York.

Harrel, R. C., 1985. Effects of a crude oil spill on water quality and macrobenthos of a southeast Texas stream. Hydrobiologia 124: 223–228.

Henderson, C., 1949. Value of the bottom sampler in demonstrating the effects of pollution on fish-food organisms and fish in the Shenandoah River. Progress. Fish-Cult. 11: 217–230.

Hilsenhoff, W. L., 1982. Using a biotic index to evaluate water quality in streams. Dept. Ntrl. Res., Tech. Bull. No. 132. Madison, Wisconsin. 22 pp.

Hudson, P. L., B. M. Davis, S. J. Nichols & C. M. Tomcko, 1986. Environmental studies of macrozoobenthos, aquatic macrophytes and juvenile fishes in the St. Clair-Detroit River system, 1983–1984. U.S. Fish Wildl. Serv., Great Lakes Fish. Lab. Admin. Rep. 86-7. Ann Arbor, Michigan. 116 pp.

Hynes, H. B. N., 1961. The effect of sheep dip containing the insecticide BHC on the fauna of a small stream including *Simulium* and its predators. Ann. Trop. Med. Parasit. 55: 192–196.

Hynes, H. B. N., 1970. The Ecology of Running Waters. Liverpool Univ. Press. Liverpool, England. 555 pp.

IJC (International Joint Commission), 1982. 1982 Report on Great Lakes Water Quality. Prepared by the Great Lakes Water Quality Board. Windsor, Ontario, Canada. 153 pp.

Keenleyside, M. H. A., 1967. Effects of forest spraying with DDT in New Brunswick on food of young Atlantic salmon. J. Fish. Res. Bd Can. 24: 107.

Lauritsen, D. D., S. C. Mozley & D. S. White, 1985. Distribution of oligochaetes in Lake Michigan and comments on their use as indices of pollution. J. Great Lakes Res. 11: 67–76.

Legendre, L. & P. Legendre, 1983. Numerical Ecology. Elsevier Sci. Publ. Co., Amsterdam. 419 pp.

Merritt, R. W., K. W. Cummins & T. M. Burton, 1984. The role of aquatic insects in the processing and cycling of nutrients. In: V. H. Resh & D. M. Rosenberg (eds), The Ecology of Aquatic Insects. pp. 134–163. Praeger Publishers, New York.

Oliver, B. G. & C. W. Pugsley, 1986. Chlorinated contaminants in the St. Clair River sediments. Wat. Pollut. Res. J. Can. 21: 368–379.

OME (Ontario Ministry of Environment), 1979. St. Clair River organics study: Biological surveys 1968 and 1977. Southwestern Region, Water Resources Assessment Unit. London, Ontario, Canada. 90 pp.

OME, 1983. Handbook of analytical methods for environmental samples. Lab. Serv. Br., Rexdale, Ontario, Canada.

OME, 1985. A guide to the collection and submission of samples for laboratory analysis. 5th ed. Lab. Serv. Br., Rexdale, Ontario, Canada.

OME, 1986a. Drinking water survey, St. Clair-Detroit River area. Update August 1986. Wat. Resources Br., Toronto, Ontario, Canada. 20 pp.

OME, 1986b. Report of the 1985 discharges from municipal wastewater treatment facilities in Ontario. Wat. Resources Br., Toronto, Ontario, Canada. 241 pp.

Pennak, R. W., 1978. Freshwater Invertebrates of the United States. 2nd edn. J. Wiley and Sons, New York. 803 pp.

Pinder, L. C. V. & F. Reiss, 1983. The larvae of Chironomidae of the Holarctic Region. Keys and diagnoses. Entomol. Scand. Suppl. 19: 293–435. Lund, Sweden.

Quinn, F. H. & R. N. Kelly, 1983. Great Lakes monthly hydrologic data. Ntnl. Oceanic Atmos. Admin. ERL GLERL-26. Ann Arbor, Michigan. 79 pp.

Rabeni, C. F., S. P. Davies & K. E. Gibbs, 1985. Benthic invertebrate response to pollution abatement: Structural changes and functional implications. Wat. Res. Bull. 21: 489–497.

Roback, S. S., 1962. Environmental requirements of tricoptera. In: Biological problems in water pollution. Third Seminar. U.S. Dept. health, Educ. Welf. Public Health Serv. Publ. No. 999-WP-25. 1965. 424 pp.

Saether, O. A., 1977. Taxonomic studies on Chironomidae: *Nanocladius*, *Pseudochironomus* and the *Harnischia* complex. Bull. Fish. Res. Bd Can. 196. Ottawa, Ontario, Canada. 143 pp.

Saether, O. A., 1979. Chironomid communities as water quality indicators. Holarctic Ecol. 2: 65–74.

Soluk, D. A., 1985. Macroinvertebrate abundance and production of psammophilous chironomidae in shifting sand areas of a lowland river. Can. J. Fish. aquat. Sci. 42: 1296–1302.

Thornley, S., 1985. Macrozoobenthos of the Detroit and St. Clair rivers with comparisons to neighbouring waters. J. Great Lakes Res. 11: 290–296.

Warwick, W. F., 1980. Palaeolimnology of the Bay of Quinte, Lake Ontario: 2800 years of cultural influence. Can. Bull. Fish. aquat. Sci. 206. 117 pp.

Wiederholm, T., 1984. Responses of aquatic insects to environmental pollution. In: V. H. Resh & D. M. Rosenberg (eds), The Ecology of Aquatic Insects. pp. 508–557. Praeger Publishers, New York.

Williams, W. T. & J. M. Lambert, 1959. Multivariate methods in plant ecology. J. Ecol. 47: 83–101.

Winnel, M. H. & D. J. Jude, 1984. Associations among chironomidae and sandy substrates in nearshore Lake Michigan. Can. J. Fish. aquat. Sci. 41: 174–179.

*Hydrobiologia* **219**: 165–170, 1991.
*M. Munawar & T. Edsall (eds),*
*Environmental Assessment and Habitat Evaluation of the Upper Great Lakes Connecting Channels.*
© 1991 *Kluwer Academic Publishers.*

# Growth and overwinter survival of the Asiatic clam, *Corbicula fluminea,* in the St. Clair River, Michigan

John R. P. French III & Donald W. Schloesser
*U.S. Fish and Wildlife Service, National Fisheries Research Center-Great Lakes, 1451 Green Road, Ann Arbor, MI 48105, USA*

*Key words:* Corbiculidae, temperatures, population dynamics, downstream distribution

## Abstract

We report the discovery in April 1986 of the first population of the Asiatic clam, *Corbicula fluminea,* known to occupy a lotic environment in the Laurentian Great Lakes system. This population occupied a 3.8 km long sandy shoal in the discharge plume of a steam-electric power plant on the St. Clair River (Michigan), the outflow of Lake Huron. Samples collected April 1986 to April 1987 revealed the growth of one-year-old *Corbicula* (1985 cohort) began after mid-May and ended by mid-November, while water temperatures were higher than 9 °C. Maximum growth (0.78 mm wk$^{-1}$) occurred between mid-August and mid-September, while water temperatures were about 16–23 °C. We recorded a substantial overwinter mortality of the 1986 cohort, but not the 1985 cohort; this was particularly evident at sampling locations more remote from the heated discharge of the power plant, suggesting low water temperature was the major mortality agent. The available information suggests low water temperature in the St. Clair River may limit the success of *Corbicula* in the river, including portions of populations inhabiting thermal plumes, by reducing growth, delaying the onset of sexual maturity and reproduction, and by causing heavy overwinter mortality in the first year of life.

## 1. Introduction

The Asiatic freshwater clam, *Corbicula fluminea* (Müller) is established in most of the major rivers in the United States (McMahon, 1982). Many of these populations reach high densities and clog water intakes, steam-electric power plant cooling systems, and sewage treatment plants, causing millions of dollars of damage annually (McMahon, 1983). In the northern midwest, which includes the Great Lakes, *Corbicula* is con-

fined to the heated discharge plumes of power plants (McMahon, 1982; Scott-Wasilk *et al.,* 1983a; White *et al.,* 1984) because it cannot survive water temperatures lower than 2.0 °C (Mattice & Dye, 1976; Rodgers *et al.,* 1977).

The first *Corbicula* reported from the Great Lakes was found by Clarke (1981) in Lake Erie. Subsequently *Corbicula* has been collected from power plant plumes in western Lake Erie (Scott-Wasilk *et al.,* 1983a) and in southeastern Lake Michigan (White *et al.,* 1984). In this paper we document the discovery of the first population of *Corbicula* in a lotic Great Lakes habitat and describe the growth and overwinter survival of this population.

---

This paper is contribution 730 of the National Fisheries Research Center-Great Lakes, U.S. Fish and Wildlife Service, 1451 Green Road, Ann Arbor, Michigan 48105.

166

## 2. Materials and methods

Samples of *Corbicula* on which this study is based were collected on a sandy shoal on the western side of the St. Clair River, 29.3 to 33.0 km downstream from the head of the river in Lake Huron April 1986 to April 1987 (Fig. 1). This 3.8-km long shoal, adjacent to the bulkheaded shoreline between the Detroit Edison St. Clair Power Plant and Marine City, is between 50 and 100 m wide with water depths of 2 to 4 m. We used a 484 cm$^2$ Ponar grab (Powers & Robertson, 1967) to collect ten samples monthly, April to November (except July) in 1986 and in April 1987 on the shoal at a location 30.1 km downstream from Lake Huron;

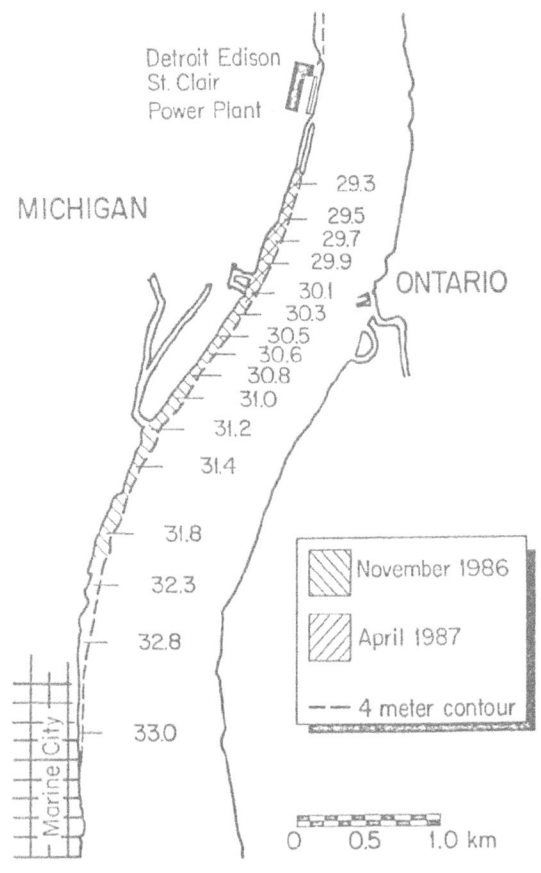

*Fig. 1.* Sampling locations and distribution (shaded areas) of *Corbicula fluminea* in the St. Clair River near the Detroit Edison St. Clair Power Plant, April–June and August–November 1986 and April 1987. Station numbers are distances (km) from Lake Huron.

this location is hereinafter referred to as station 30.1 (Fig. 1). The samples were washed through a U.S. Standard No. 30 sieve (0.65 mm mesh openings) and the residue containing *Corbicula* was preserved in 10 percent formalin. The preserved residue was examined under a magnification of 1.8 ×, the clams were removed, and shell lengths were measured to the nearest 0.1 mm. Growth rate was calculated from changes in mean shell length and is expressed as mm wk$^{-1}$ (Joy, 1985). Clams of the 1985 cohort collected from September to November were dissected to check for larvae in their demibranchs according to Britton & Morton (1982).

On November 15, 1986 and April 21–22, 1987 we collected between one and ten additional samples with the Ponar grab at each of 15 other stations located at 100 to 300 m intervals downstream from the power plant (Fig. 1). We used these samples to study the overwinter survival of *Corbicula*. The samples were washed through a sieve with 3 mm mesh and examined in the field without magnification to determine if clams were present. We also used mean densities (no. m$^{-2}$) of clams in samples collected at station 30.1 in November 1986 and April 1987 to estimate overwinter survival.

To determine if *Corbicula* was present at the Ontario Hydro Lambton Generating Station at river km 25.0 (i.e. 25 km downstream from Lake Huron), or at the Detroit Edison Marysville Power Plant at river km 10.0, we collected 10 Ponar grab samples at each location in October 1986. We washed these samples through a sieve with 3 mm mesh and examined them in the field without magnification.

We recorded water temperatures monthly at station 30.1 when we collected the Ponar grab samples; daily water temperature records were obtained from the Port Huron (Michigan) Water Filtration Plant intake located at river km 1.9.

## 3. Results

The *Corbicula* found at station 30.1 on April 17, 1986 are the first reported from the connecting

Table 1. Density (no. m$^{-2}$) of live *Corbicula* at station 30.1 in the St. Clair River, April 1986–April 1987.

| Sampling Dates | 1985 Cohort | | 1986 Cohort | |
|---|---|---|---|---|
| | Mean | SD | Mean | SD |
| 1986 | | | | |
| Apr. 17 | 19 | 18 | 0 | 0 |
| May 15 | 39 | 27 | 0 | 0 |
| Jun 17 | 101 | 33 | 0 | 0 |
| Aug. 12 | 81 | 49 | 0 | 0 |
| Sep. 15 | 35 | 41 | 4 | 13 |
| Oct. 15 | 43 | 46 | 109 | 115 |
| Nov. 17 | 31 | 33 | 1,454 | 1,171 |
| 1987 | | | | |
| Apr. 21[a] | 37 | 27 | 801 | 845 |

[a] On April 21, 1987, the density of dead *Corbicula* (not included in this table) was 4($\pm$9) m$^{-2}$ for the 1985 cohort

channels of the Great Lakes. Sampling in the St. Clair River, at the Marysville Power Plant and at the Lambton Station in October 1986 yielded no *Corbicula*.

Mean densities (no. m$^{-2}$) of the 1985 cohort at station 30.1 near the St. Clair Power Plant in-creased from 19 in April to 101 in June (Table 1) and then decreased to between 31 and 43 in the fall. Mean densities of the 1986 cohort increased rapidly from 4 on September 15 to 1,454 on November 15. Maximum density of the 1986 cohort (4 545 m$^{-2}$) was recorded for one Ponar grab sample collected on November 15.

In November 1986, we found live *Corbicula* at stations 29.3 to 31.8 on the shoal but in April 1987, they were present only at stations 29.3 to 30.1 (Fig. 1). Mean density at station 30.1 did not change from November 1986 to April 1987, for the 1985 cohort, but the 1986 cohort declined about 45 percent (Table 1). The average density of dead clams (empty shells) at stations 30.3 and 30.5 in April 1987 was 386 m$^{-2}$.

On April 17, and May 15, 1986, the mean shell length of clams of the 1985 cohort at station 30.1 was 1.1 mm (Table 2). Shell length of the cohort increased to 12.4 mm by November 17 and was unchanged from November 17, 1986 to April 17, 1987. Mean growth rate was low (0.10 mm wk$^{-1}$) during May 15 to June 15, when water temperatures were increasing from about 11.5 to about 15.0 °C and increased rapidly to 0.78 mm wk$^{-1}$ during August 12 to September 15, when water temperatures were falling from 22.5 to 16.0 °C. The mean shell length of the 1986 cohort did not

Table 2. Growth of *Corbicula* at station 30.1 in the St. Clair River, April 1986–April 1987.

| Sampling Dates | 1985 Cohort Shell length (mm) | | | Growth (mm wk$^{-1}$) | 1986 Cohort Shell length (mm) | | |
|---|---|---|---|---|---|---|---|
| | Range | Mean | SD | | Range | Mean | SD |
| 1986 | | | | | | | |
| Apr. 17 | 0.9–1.5 | 1.1 | 0.18 | | | | |
| May 15 | 0.8–1.6 | 1.1 | 0.24 | 0.0 | | | |
| Jun 17 | 1.1–2.3 | 1.6 | 0.33 | 0.11 | | | |
| Aug. 12 | 3.4–7.7 | 5.3 | 1.03 | 0.46 | | | |
| Sep. 15 | 7.8–10.9 | 9.1 | 0.93 | 0.78 | 1.1–1.7 | 1.4 | 0.42 |
| Oct. 15 | 9.5–12.8 | 11.3 | 0.84 | 0.51 | 0.8–3.1 | 1.3 | 0.50 |
| Nov. 17 | 11.3–13.3 | 12.4 | 0.58 | 0.23 | 0.7–4.9 | 1.3 | 0.55 |
| 1987 | | | | | | | |
| Apr. 21 | 11.3–13.8 | 12.4 | 0.74 | 0.00 | 0.8–6.4 | 1.4 | 0.63 |

168

change from September to November because recruitment into the free-living bottom population became increasingly heavy throughout the period, masking the growth of the earlier recruits (Table 1). The increase in length of a portion of the cohort to a maximum length of about 4.9 mm by November 17 suggests that some members of the cohort, probably the earliest recruits, did grow substantially between September 15 and November 17. The similarity of mean (and maximum) shell lengths in November 1986 and April 1987 indicated no overwinter growth of the 1986 cohort. The cessation of growth in mid-November, when water temperatures fall below 9.0 °C is in agreement with the results of other studies (Eng, 1977; Welch & Joy, 1984; Joy, 1985).

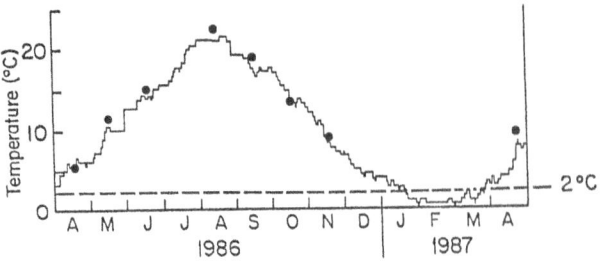

*Fig. 2.* Water temperatures of the St. Clair River, April 1986–April 1987. Solid line describes temperatures recorded at the Port Huron (Michigan) Water Filtration Plant intake located 1.9 km downstream from Lake Huron; dots are surface water temperatures recorded at station 30.1. Dashed line shows the incipient lower lethal temperature for *Corbicula.*

## 4. Discussion

Our study shows that a reproducing population of *Corbicula* is established in the St. Clair River immediately downstream from the Detroit Edison St. Clair Power Plant. W. Kovalak (Detroit Edison, *pers. communication,* 1987) did not find *Corbicula* in annual benthic surveys in the thermal plume of the St. Clair Power Plant in the early 1980's, and none were found in extensive benthic surveys throughout the St. Clair River from 1968 to 1985 (Hiltunen, 1980; Ontario Ministry of the Environment, 1979; Schloesser *et al.,* 1988). Furthermore, we did not find *Carbicula* in Ponar grab samples taken in October 1986 in the heated discharge plumes of two other power plants on the St. Clair River – the Ontario Hydro Lambton Generation Station and the Detroit Edison Marysville Power Plant. Thus, it appears that the invasion of St. Clair River by *Corbicula* occurred relatively recently, and that in 1986–87 the population may have been confined to the vicinity of the St. Clair Power Plant.

Low water temperatures from January to March (Fig. 2) may effectively restrict the distribution of *Corbicula* in the St. Clair River to thermal refugia provided by power plant discharges. The overwinter loss of *Corbicula* at stations 30.3 to 33.0, but not at stations 29.3 to 30.1 that were

closer to the heated discharge of the St. Clair Power Plant, suggests low winter water temperature was the major mortality agent. No data are available to describe water temperatures in the St. Clair Power Plant discharge plume between November 17, 1986 and April 21, 1987, but temperatures upriver at the Port Huron Water Filtration Plant intake were lower than 2 °C, the incipient lower lethal temperature for *Corbicula* (Cherry *et al.,* 1980; Rodgers *et al.,* 1977; Mattice & Dye, 1976), from late January through mid-March 1987 (Fig. 2).

The sharp reduction in density of the 1986 cohort and the apparent lack of change in the density of the 1985 cohort between November 17, 1986 and April 21, 1987 at station 30.1 suggests that the effect of low overwinter temperature on *Corbicula* may be substantially more severe in the first winter of life than in subsequent winters, even in thermal refugia like that provided by the St. Clair Power Plant discharge.

The relatively low water temperatures in the St. Clair River during the warmer months of the year (Fig. 2) may also limit the species in the river by restricting its growth rate. In 1986, water temperatures at the Port Huron Water Treatment Plant intake and at stations 29.3 to 30.3 did not reach 24–25 °C (Fig. 2), the optimum for growth of *Corbicula* (O'Kane, 1976; Mattice & Dye, 1979; Joy, 1985). The maximum growth rate of *Corbicula* (0.76 mm wk$^{-1}$) occurred between mid-

August and mid-September when water temperature was falling from 22.5 to 16.0 °C. This growth rate was lower than the maximum (0.95 mm wk$^{-1}$) reported in the Kanawha River, West Virginia, where water temperature peaked at about 30 °C in late August (Welch & Joy, 1984; Joy, 1985); however, it was almost identical to the growth rate of 0.8 mm wk$^{-1}$ reported by Scott-Wasilk et al. (1983b) from thermal plumes of two power plants near the mouth of the Maumee River at western Lake Erie from June to September when water temperatures were between 29 and 30 °C.

Low water temperatures in the St. Clair River may also delay the onset of sexual maturity and thus further limit the success of Corbicula. In the southern United States, Corbicula may grow to a shell length of 16 to 30 mm at the end of their first year, and they begin to reproduce at 15 mm (McMahon, 1983); the 1985 cohort grew more slowly at station 30.1 and only averaged 12.4 mm on November 17, 1986 (Table 2). Larvae of the 1986 cohort were absent from the inner demibranches of the 1985 cohort collected at stations 29.3 to 33.0 from September to November 1986, when the density of the 1986 cohort in Ponar grab samples was increasing rapidly. The source of the 1986 cohort may therefore have been larger individuals of the 1985 cohort residing in warmer water closer to the St. Clair Plant, or an unsampled 1984 cohort residing in that area.

## 5. Acknowledgements

We are indebted to S. J. Nichols, G. W. Kennedy, B. M. Davis, M. A. Ford, and others of the National Fisheries Research Center-Great Lakes for assisting in the field work and we thank the staff of the Port Huron Water Filtration Plant for providing water temperature data.

## References

Britton, J. C. & B. Morton, 1982. A dissection guide, field, and laboratory manual to the introduced bivalve Corbicula fluminea. Malacol. Rev. 15(Suppl. 3). 82 pp.

Cherry, D. S., J. H. Rodgers, Jr., R. L. Graney & J. Cairns, Jr., 1980. Dynamics and control of the Asiatic clam in the New River, Virginia. Virginia Wat. Resour. Res. Center, Virginia Polytech. Inst. State Univ. Blacksburg, Virginia. Bull. 123. 72 pp.

Clarke, A. H., 1981. Corbicula fluminea, in Lake Erie. Nautilus 95: 83–84.

Eng, I. I., 1977. Population dynamics of the Asiatic clam, Corbicula fluminea (Müller), in the concrete-lined Delat-Mendota Canal of central California. In; J. C. Britton (ed.), Proc. First Internat. Corbicula Symp. Texas Christian Univ. Res. Found., Fort Worth, Texas: 39–68.

Hiltunen, J. K., 1980. Composition, distribution, and density of benthos in the lower St. Clair River, 1976–77. Ntnl. Fish. Res. Center-Great Lakes, Ann Arbor, Michigan. Admin. Rep. 80–4. 28 pp.

Joy, J. E., 1985. A 40-week study on growth of Asian clam, Corbicula fluminea (Müller), in the Kanawha River, West Virginia. Nautilus 99: 110–116.

Mattice, J. S. & L. L. Dye, 1976. Thermal tolerance of adult Asiatic clam. In; E. W. Esch & R. W. McFarlene (eds), Thermal Ecology 2. U.S. Energy and Development Admin. Symp. Ser., Washington, D.C.: 130–135.

Mattice, J. S. & L. L. Dye, 1979. Growth of the Asiatic clam. Pap. Pres. 27th Ann. Meet. North Amer. Bentholog. Soc., Erie, Pennsylvania, April 18–20.

McMahon, R. F., 1982. The occurrence and spread of the introduced Asiatic freshwater clam, Corbicula fluminea (Müller), in North America: 1924–1982. Nautilus 96: 134–141.

McMahon, R. F., 1983. Ecology of an invasive pest bivalve, Corbicula. In; W. D. Russel-Hunter (ed.), Biology of Mollusca, Vol. 5, Ecology. Academic Press, New York, San Francisco London: 505–561.

O'Kane, K. D., 1976. A population study of the exotic bivalve Corbicula manilensis (Philipi 1841) in selected Texas reservoirs. M.S. Thesis, Texas Christian Univ. 197 pp.

Ontario Ministry of the Environment, 1979. St. Clair River organics study. Biological Surveys 1968 and 1977. London, Ontario. 90 pp.

Powers, C. F. & A. Robertson, 1967. Design and evaluation of an all-purpose benthic sampler. In; A. C. Ayers & D. C. Chandler (eds), Studies on the Environment and Eutrophication of Lake Michigan. Great Lakes Res. Div. Univ. Michigan, Ann Arbor, Spec. Rep. 30. pp. 126–131.

Rodgers, J. H., Jr., D. S. Cherry, K. L. Dickson & J. Cairns, Jr., 1977. Invasion, population dynamic and elemental accumulation of Corbicula fluminea in the New River at Glen Lyn, Virginia. In; J. C. Britton (ed.), Proc. First Internat. Corbicula Symp. pp. 100–110. Texas Christian Univ. Res. Found., Fort Worth, Texas.

Schloesser, D. W., B. A. Manny, S. J. Nichols & T. A. Edsall, 1988. Distribution and abundance of Hexagenia nymphs and oil laden sediments in the Upper Great Lakes Connecting Channels. Pap. Pres. 31st Conf. Internat. Assoc. Great Lakes Res. Hamilton, Ontario, Canada, May 17–20.

Scott-Wasilk, J., G. G. Downing & J. S. Lietzow, 1983a. Occurrence of the Asiatic clam, *Corbicula fluminea*, in the Maumee River and western Lake Erie. J. Great Lakes Res. 9: 9–13.

Scott-Wasilk, J., J. S. Lietzow, G. G. Downing & K. L. Clayton, 1983b. Growth of *Corbicula fluminea* in Lake Erie. Pap. Pres. 31st Ann. Meet. North Amer. Bentholog. Soc. LaCrosse, Wisconsin, April 27–29.

Welch, K. J. & J. E. Joy, 1984. Growth rates of the Asiatic clam, *Corbicula fluminea* (Müller), in the Kanawha River, West Virginia. Freshwat. Invert. Biol. 3: 139–142.

White, D. S., M. H. Winnell & D. J. Jude, 1984. Discovery of the Asiatic clam, *Corbicula fluminea*, in Lake Michigan. J. Great Lakes Res. 10: 329–331.

*Hydrobiologia* **219**: 171–185, 1991.
*M. Munawar & T. Edsall (eds),*
*Environmental Assessment and Habitat Evaluation of the Upper Great Lakes Connecting Channels.*
© 1991 *Kluwer Academic Publishers.*

# Deformities in larval *Procladius* spp. and dominant Chironomini from the St. Clair River

R. M. Dermott

*Department of Fisheries and Oceans, Canada Centre for Inland Waters, Box 5050 Burlington, Ontario, L7R 4A6 Canada*

*Key words:* chironomids, deformed mouthparts, petrochemicals, Great Lakes

## Abstract

The benthic community of the St. Clair River is impacted by the petrochemical complex near Sarnia, Ontario. Larvae of the common chironomid *Procladius* spp. and dominant Chironomini from various sections of the river were examined to determine if the incidence of morphological deformities in their mouth parts reflected the degree of chemical pollution. *Procladius* had a much greater (14%) incidence of deformed ligula downstream of the industrial section near Sarnia, than occurred in Lake St. Clair (3%), or at the mouth of Bear Creek, which drains agricultural land east of the St. Clair delta (7%). The incidence of deformed ligula at a control site in Lake Superior was 4 percent. The incidence of deformities in *Procladius* larvae was lower than that in *Chironomus* larvae from the same site, but greater than that in other chironomid genera.

## 1. Introduction

Chemical contamination of aquatic environments can change the structural of the normal community as less tolerant species disappear. However, there are few measures of the effect of chronic long term exposure to contaminant mixtures on organisms surviving in less contaminated environments. The occurrence of neoplasia in fish (Malins *et al.*, 1984), the development of malformed structures and eventual reproductive failures of aquatic species (Anon, 1986; Béland, 1988) have been related to chemical contamination. A number of studies have linked the occurrence of morphometrical abnormalities in benthic fauna to contaminants (Milbrink, 1983; Warwick, 1985). Some of these deformities can be induced during bioassays, appearing as delayed effects of sublethal exposure (Shealy & Sandifer, 1975; Warwick, 1985).

In the Great Lakes, deformed chironomids have been reported in the benthic faunas of Lake Erie (Brinkhurst *et al.*, 1968; Krieger, 1984), Georgian Bay (Hare & Carter, 1978) and Lake Ontario (Warwick, 1980; Warwick *et al.*, 1987). Warwick (1985) proposed that the incidence and severity of deformities in larvae of the common midge *Chironomus* be used to assess the impact of contaminants in freshwater ecosystems. Incidence of deformed larvae in other chironomid genera have also been noted, with deformed larvae of the predatory Tanypodinae genus *Procladius* being reported at a number of locations (Hamilton & Saether, 1971; Hare & Carter, 1976; Tennessen & Gottfried, 1983; Pettigrove, 1989).

The purpose of this study was to determine if

the dominant chironomids inhabiting the polluted sections of the St. Clair River system had increased structural deformities. The benthic fauna in this interconnecting river of the Great Lakes are known to be exposed to a large variety of contaminants (COARGLWQ, 1986). As a control, the incidence of deformities was assessed in *Procladius* and the dominant Chironomini larvae inhabiting a bay of eastern Lake Superior, an area receiving only atmospheric pollutants.

## 2. Study area

The St. Clair River flows from Lake Huron to a delta at Lake St. Clair. Average width is 0.7 km with depth in the channel of 8.2 m to 18 m. Average flow is 5100 $m^3$ $s^{-1}$ (Hendendorf *et al.*, 1986) with currents in many areas of greater than 1.0 m $s^{-1}$. As a result, the main river channel is scoured and composed of gravel, sand, and hard clay. In bays and along the 3 to 5 m contour, aquatic weeds grow in a narrow band of soft sediments within 30 m of the shore.

Several large petrochemical industries are situated along the eastern shore, between the city of Sarnia and Corunna. Water quality problems have occurred downstream from Sarnia, where significant environmental degradation of water, sediment, and biota has been documented (Kauss & Hamdy, 1985; Oliver & Bourbonniere, 1985), and high levels of several contaminants being detected in the sediment (Table 1) and the biota. The strong current confines the industrial effluent along the shores, while relatively clean water from Lake Huron flows down the centre of the river (COARGLWQ, 1986). As a result, limited transchannel mixing occurs upstream of the delta. The benthic fauna of the nearshore area has been seriously impaired for several kilometers downstream of Sarnia (Thornley, 1985).

Periodic chemical spills have occurred. The most noteworthy was on August 1985, when 11 000 l of perchloroethylene escaped from Dow Chemical, resulting in chemical puddles forming downstream on the river bottom. The composition of this material was: perchloroethylene

*Table 1.* Highest levels of the contaminants of concern found in sediments from the upper St. Clair River near Corunna and 24 km downstream at the Walpole water treatment plant. Data as ppm except where indicated, a dash indicates below detection limits (COARGLWQ, 1986).

| Chemical | Corunna | Walpole |
|---|---|---|
| Benzene | 23.0 | – |
| Dichlorobenzene | 31.0 | – |
| Trichlorobenzene | 2.4 | – |
| Tetrachlorobenzene | 33.0 | 0.1 |
| Pentachlorobenzene | 15.0 | 0.1 |
| Hexachlorobenzene | 1000. | 0.08 |
| Carbon tetrachloride | 6300. | 0.1 |
| Chloroform | 4.0 | – |
| Dichloroethane | 11.0 | – |
| Tetrachloroethane | 24.0 | 0.1 |
| Trichloroethane | 13.5 | 0.1 |
| Hexachloroethane | 290.0 | 0.06 |
| Hexachlorobutadiene | 430.0 | – |
| Perchloroethylene | 14000. | 0.5 |
| PAH's | 138.7 | + |
| PCB's | 76.0 | 0.2 |
| Octachlorostyrene | 52.0 | 0.03 |
| Octachlorodibenzodioxin | 0.45 (ppb) | – |
| Tetrachlorodibenzodioxin | 0.55 (ppb) | – |
| 2,3,7,8,-TCDD | ND | – |
| Hg | 43.0 | 0.3 |
| Pb | 330.0 | 12 |

56 percent, carbon tetrachloride 13 percent, with hexachloroethane, chloroform and benzene being other components. Concentrations of perchloroethylene up to 14 000 ppm were measured in sediment samples below the spill area. The sediments from the vicinity were acutely toxic to mayflies and amphipods (COARGLWQ, 1986). The acute lethal concentration of the puddle material to young carp at 15 °C (96 h. LC50) was 7.5 ± 6.9 ppm (Fisheries and Oceans, Canada, 1986).

## 3. Samples and methods

An initial survey of the bottom fauna was made during April 1986 along the Canadian shoreline from Lake Huron to the Walpole Island delta (Fig. 1). Ekman grabs (232 $cm^2$) were collected

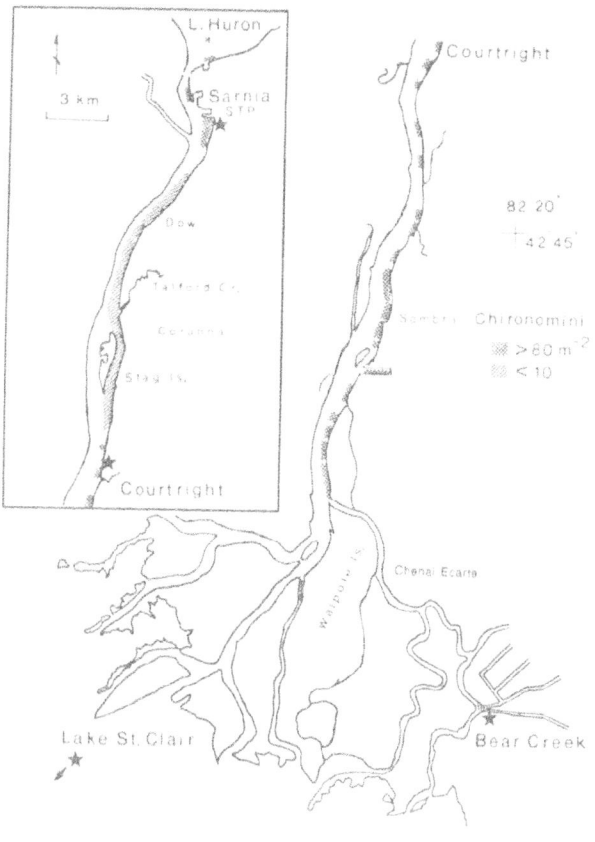

*Fig. 1.* Sampling sites and distribution of red Chironomini along the St. Clair River, May 1986. Area outlined by dots contains most of the chemical refineries. Striped area represents impacted community composed of annelids and a few *Procladius.*

abundance of 'red' chironomids. Areas having populations greater than 100 animals m$^{-2}$ were noted.

During May, 1986, four areas, having abundant chironomids, were re-sampled. One site (Sarnia Bay) was above the petrochemical industries (Fig. 1). This site is a former industrial harbour and receives treated sewage effluent. Two sites were downstream in the main channel: Courtright at 12 km below the main petrochemical industries, and a site in Lake St. Clair (42° 25.9′ N; 82° 43.5′ W). The fourth site, located at the junction of Bear Creek and the St. Clair River delta, was sampled to determine the incidence of deformities in water receiving mostly agricultural drainage from the east side of Lake St. Clair. In addition, during September, 1986, a total of 119 *Procladius* larvae in the III and IV instars were collected where Talford Creek enters the St. Clair River (Fig. 1). This creek drains a very industrialized area. Samples from Batchawana Bay, eastern lake Superior (45° 56′ N; 84° 27′ W), collected during May 1978 and 1986 were used as controls. Contamination into Batchawana Bay is mostly from atmospheric deposition and possible metal loadings during low pH runoff. (Johnson *et al.*, 1988).

About 3 m$^2$ of the bottom was collected at each site, using a 0.05 m$^2$ Ekman and sieved through a 500 $\mu$m mesh net. The retained material was preserved in 5 percent formalin, later the chironomids were sorted and preserved in 70 percent ethanol for later mounting. Each site had a different genus of dominant Chironomini (Table 2). The ubi-

along the 3 to 5 m contour of the river. The samples were sieved on site and the residue visually examined for the occurrence and relative

*Table 2.* Percent composition of chironomids at the 5 sampling sites. A dash indicates less than 1% of the Chironomidae.

|  | Lake Superior | Sarnia | Courtright | Bear Creek | Lake St. Clair |
|---|---|---|---|---|---|
| Latitude | 46° 56′ N | 42° 59′ | 42° 50′ | 42° 34′ | 42° 26′ N |
| Depth (m) | 5–18 | 3–6 | 3–4 | 2–3 | 5 |
| *Procladius* | 13 | 46 | 24 | 55 | 40 |
| *Chironomus* | 7 | 43 | 2 | 3 | 1 |
| *Phaenopsectra* | 2 | 9 | 72 | – | – |
| *Cryptochironomus* | 8 | – | – | 40 | – |
| *Harnischia* | – | – | – | – | 5 |
| *Dicrotendipes* | 16 | – | – | – | 53 |
|  |  |  |  |  | – |

quitous *Procladius* was the only genus common to all sites, and was examined in detail for morphological deformities.

Where available, 300 specimens were mounted from each site. Only 4th and where necessary, large 3rd instar larvae were mounted. Any larvae with an enlarged head capsule with respect to its body length, or weakly sclerotized heads indicating recent molting into the 4th instar were not mounted as their mouth parts are easily distorted. Larvae were mounted on slides in Canada balsam following the preparation used by Oliver and Roussel (1983) and Warwick (1985), except that cedarwood oil was replaced by clove oil. Prior to mounting, the first and last three segments were dissected from the body. The anterior and posterior segments were placed together into the same drop of balsam with the ventral side of the head capsule and the procerci facing up. A round coverslip was put on each specimen, with six specimens mounted per slide, and the slides labelled. To randomize the order of examination, after the slides had dried all slides from the same site were shuffled, then numbered sequentially. After numbering all the slides, they were re-arranged on slide trays so no more than two slides from the same site were examined in sequence.

The *Procladius* larvae were examined using a phase compound microscope with objectives of 10 to 40 power, beginning at the posterior proleg claws and ending at the antennal segments, with 21 items examined as in Table 3. The incidence and nature of the deformed specimens were recorded. When present, any broken ligula, antenna or posterior procerci were not enumerated as deformations. The degree of abnormality was rated as a point system listed in Table 3, with grossly deformed structures given 4 points. This classification was simpler than the Index of Morphological Response (IMR) proposed by Warwick (1985) to classify antennal deformities in *Chironomus*.

After examining all specimens, the slides from each site were subdivided into 3 groups with about 100 specimens per group. Total number ($n$) of *Procladius* in each group was enumerated and the percentage incidence of larvae with deformities in each group was calculated as:

$$\text{Percent} = \frac{\text{number of deformed specimens}}{n} \times 100.$$

*Table 3.* Rating of structural deformities in *Procladius* sp. larvae.

| | | | | |
|---|---|---|---|---|
| **Proleg** | | | **Paraligula combs** | |
| bifid claws | – 1 | | unsymmetrical number of teeth | – 1 |
| | | | fused or bifid teeth | – 1 |
| **Anal gills** | | | major deformity, maximum | – 3 |
| <2 pair | – 1 | | | |
| non conical | – 1 | | **Mandible** | |
| | | | wrinkled outer edge | – 1 |
| **Procerci** | | | bifid apical tooth | – 1 |
| unequal lengths, | | | reduced auxiliary tooth | – 1 |
| not 2–4 x width | – 1 | | major deformity, maximum | – 4 |
| < >8–10 long seta | – 1 | | | |
| | | | **Maxillary** | |
| **Ligula** | | | missing ring organ | – 1 |
| unsymmetrical | – 1 | | unequal length | – 1 |
| 4 or 6 teeth | – 1 | | major deformity | – 2 |
| bifid tips | – 1 | | | |
| major deformity, maximum | – 4 | | **Antenna** | |
| | | | misplaced ring organ | – 1 |
| **Paraglossa** | | | missing or reduced segments | – 1 |
| not flame shaped | – 1 | | antennal blade shortened | – 1 |
| unsymmetrical pair | – 1 | | major deformity, maximum | – 4 |
| bifid apex | – 1 | | | |

Each specimen was given 1 point for being *Procladius*, so that a specimen with a single minor deformity would have a simplified index (IMR) of 2. Therefore the sum of the IMR in each group was the sum of the deformed points plus the number of normal specimens in the group. The group's Index of Severity of All Deformities, using the same term as Warwick's (1985) Index of Severity of Antennal Deformities (ISAD) was calculated for each group as:

$$ISAD = \frac{\Sigma \, IMR}{n}.$$

This ISAD would range from 1.0 for groups with no deformities to greater than 2 for groups in which all specimens were deformed. For each site, the mean percent incidence of deformities, and ISAD with their respective 95 percent confidence intervals were calculated from the averages of the three groups at that site.

The frequencies of deformed ligula were compared for heterogeneity among sites using a G-test (Sokal & Rohlf, 1969) with the average incidence at the Lake Superior site as the expected frequency. A posteriori test for homogeneity in sets of replicate samples was also made using a G-test. The proportion of deformities in the III and IV instar larvae from the single sample of *Procladius* from Talford Creek were compared to the total observed incidence using the chi square distribution (Elliott, 1977).

A limited number of the dominant Chironomini genus from each site were also examined for deformities. The percent incidence at ISAD for these five genera of Chironomini were calculated. Obvious abnormalities in the hardened mentum and mandibles were used to compare the different genera.

# 4. Results

## 4.1. Procladius abnormalities

The most common differences in the *Procladius* larvae examined were the presence of serrated hair claws on the posterior prolegs, antennal blades being longer than the apical segments 3 to 5, and an enlarged inner tooth on the mandibles. These characters often occurred together, and were more common at certain sites, particularly Bear Creek and Batchawana Bay, Lake Superior. As these structures may represent differences in the separate species or races present at the sites, they were not tallied as deformities. The complex of species examined were *P. bellus, P. sublettei*, and a related species (B. Bilyj, Freshwater Institute, Winnipeg, personal communications).

*Procladius* head capsule structures that were deformed included the teeth on the ligula, paraligula, paralabial plates, mandibles and structures on the maxillaries and antenna. The most common deformity was an unequal number of teeth in the comb-like paralabial (hypopharyngeal) plates (Fig. 2k). Total number of paralabial teeth was instar or species specific, however pairs with 4 and 5, or 6 and 7 teeth were very common. Other abnormal paralabials had irregular teeth or teeth fused together at their base. The incidence of abnormal paralabials ranged from 6.5 percent in Lake Superior to 12.6 percent at Courtright.

The most obvious deformity in the *Procladius* larvae were abnormal ligula. These fork-like structures have five dark teeth, with the medial being the shortest (Fig. 2a). The medial tooth was often greatly reduced or missing in some specimens. The incidence of ligula with four normal looking teeth (Fig. 2c) ranged from 1.4 percent of the specimens from Batchawana Bay, Lake Superior to 3.3 percent at Bear Creek (Table 4). Several specimens had ligula with 6 normal or slightly deformed teeth (Fig. 2f), more common were unsymmetrical ligula or teeth with bifid tips, particularly in larvae from the St. Clair River (Fig. 2g-j).

The highest incidence and severity of deformed ligula occurred at the Courtright site (Fig. 2g, h, j, and Fig. 3). A G-test on these values (Table 5) indicated that deformed ligula were significantly more common ($p = 0.001$) at Courtright and Sarnia than at the Lake St. Clair and Lake Superior sites. The incidence of deformed ligula at Beak Creek was also greater ($p = 0.05$) than at

176

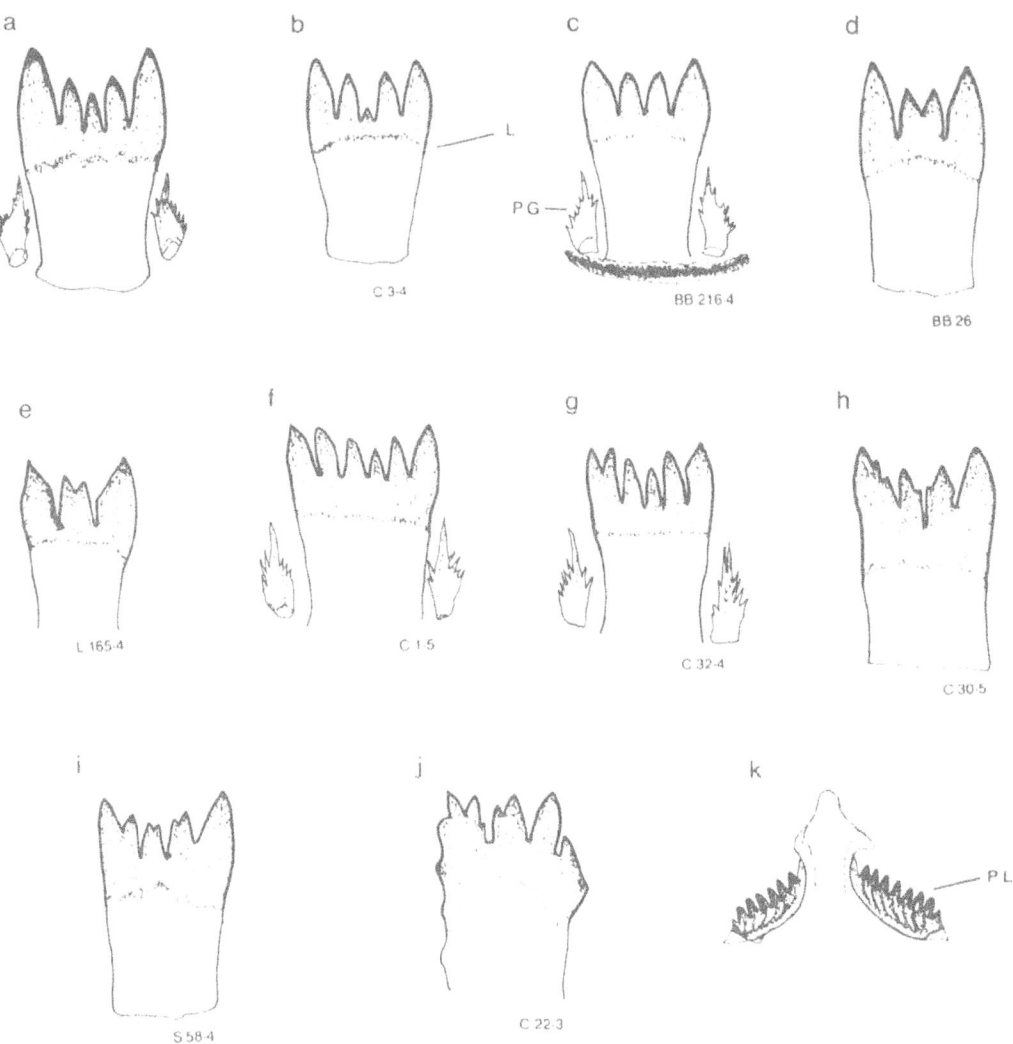

*Fig. 2.* Deformities in *Procladius* ligula (L), paraglossa (PG) and paralabial plates (PL): (a) normal; (b-k) deformed in numbered specimens.

*Table 4.* Percent incidence of various deformities in the *Procladius* larvae examined.

|  | Lake Superior | Sarnia | Courtright | Bear Creek | Lake St. Clair |
|---|---|---|---|---|---|
| Number examined | 293 | 291 | 190 | 300 | 300 |
| Paralabial combs | 6.5 | 7.5 | 12.6 | 10.6 | 7.0 |
| Total mandible | 2.0 | 2.7 | 6.3 | 10.6 | 4.0 |
| (wrinkled margin) | (1.4) | (1.0) | (4.2) | (6.7) | (2.3) |
| Total ligula | 3.9 | 11.6 | 14.5 | 6.6 | 3.1 |
| (4 toothed) | (1.4) | (3.1) | (3.1) | (3.3) | (1.4) |
| Antenna | 2.4 | 3.8 | 6.3 | 4.7 | 4.6 |

177

*Table 5.* G-tests for goodness of fit on the incidence of deformed ligula in *Procladius* spp.

| Site | Average % deformed | Number normal | Number deformed | Total | df | G | |
|---|---|---|---|---|---|---|---|
| Superior (LS) | 3.89 | 282 | 11 | 293 | 1 | 0.02 | |
| L. St. Clair (SC) | 3.14 | 291 | 9 | 300 | 1 | 0.69 | |
| Bear Creek (BC) | 6.59 | 280 | 20 | 300 | 1 | 5.13 | $p < 0.05$ |
| Sarnia (S) | 11.62 | 257 | 34 | 291 | 1 | 31.32 | $p < 0.001$ |
| Courtright (CR) | 14.47 | 162 | 28 | 190 | 1 | 35.79 | $p < 0.001$ |
| | | | G Total | | 5 | 72.94 | |
| | | | G heterogeneity | | 4 | 36.40 | $p < 0.001$ |

Homogeneous sets of replicates tested for goodness of fit (G-test) in *Procladius* ligula. Replicates overlapping both sets are underlined.

SET 1

| Replicate | SC1 | LS3 | BC3 | LS1 | SC3 | SC2 | LS2 | BC2 | | |
|---|---|---|---|---|---|---|---|---|---|---|
| % deformed | 0 | 1.9 | 3.1 | 3.8 | 3.8 | 5.5 | 5.9 | 8.0 | | |

SET 2

| Replicate | SC2 | LS2 | BC2 | BC1 | CR2 | S1 | S2 | S3 | CR3 | CR1 |
|---|---|---|---|---|---|---|---|---|---|---|
| % deformed | 5.5 | 5.9 | 8.0 | 8.6 | 8.9 | 9.9 | 12.2 | 12.7 | 13.4 | 21.2 |

the Lake St. Clair and Superior sites. The a posteriori G-test for homogeneity of deformed ligula in the triplicate replicates from each site indicated that all replicates from the two lakes, having less than 5 percent incidence were significantly different from the St. Clair River samples, which all had incidences of deformed ligula greater than 8 percent (Table 5). The Bear Creek samples were not different from either the lake or river samples.

The sickle shaped mandibles of *Procladius* are simple, with a dark apical tooth and one blunt inner auxiliary tooth (Fig. 4a, c). Deformed mandibles were more numerous at Bear Creek and Courtright than at the other sites (Table 4). The most prevalent mandible abnormalities were notches and wrinkles along the outer margin (Fig. 4b, c). These wrinkles were most abundant at Bear Creek occurring in 6.7 percent of the specimens. Other mandible deformations were rare, being mainly bifid tips or malformed apical teeth, with a few severely deformed mandibles found (Fig. 4e, f). Unlike the teeth on the internal ligula, the mandibles are exposed and used in seizing prey. Mandibles with worn or broken tips were common, however these were not included in calculating the incidence of deformities.

The incidence of antennal deformities ranged between 2.4 percent at Lake Superior to 6.3 percent of the specimens from Courtright (Table 4). Reduced length of apical antennal segments,

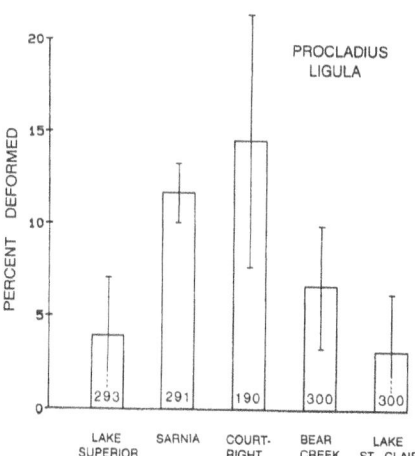

*Fig. 3.* Percent incidence of deformed ligula in the *Procladius* populations, interval bars represent 95% confidence intervals. Values in bars are the number of specimens examined.

178

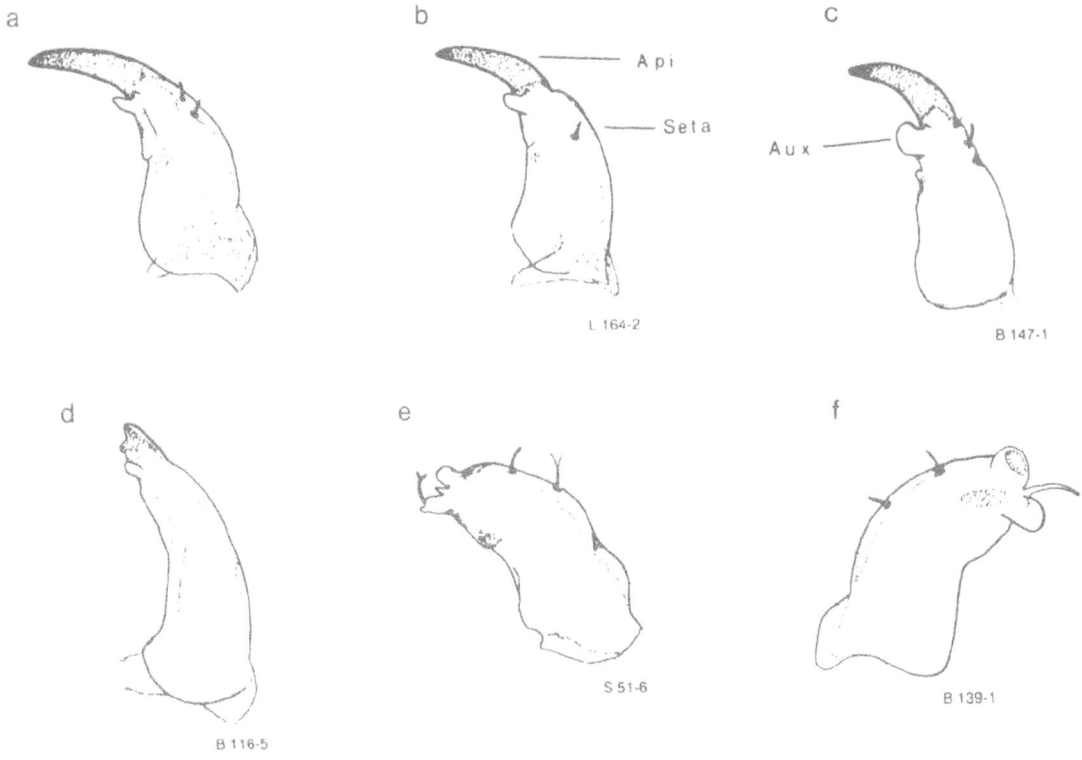

*Fig. 4.* Normal (a) and deformed (b-f) mandibles in *Procladius* spp. larvae: Api – apical tooth, Aux – auxiliary tooth, Seta – lateral seta.

short and deformed antennal blades and missing or misplaced ring organs on the first segment were the more common abnormalities. A few specimens had incompletely sclerotized antenna or badly malformed first antennal segments (Fig. 5b, c), with one specimen from Sarnia having badly malformed antennal segments (Fig. 5e). Other deformed structures were rare. A few *Procladius* had a pair of the flame shaped paraglossa (paralabial) which varied greatly in size, or had distinct bifid tips on the main apical branch (Fig. 2a, c, g). Other rare deformities included maxillary palpi of unequal length, anal tubules (gills) that were duplicated, and posterior proleg claws with bifid tips. A few shortened procerci were found, with only one specimen from Bear Creek having an extra tuft of seta on its procerci.

The incidence of total abnormalities in *Procladius* was greatest at the Courtright and Bear Creek sites (Fig. 6), with total incidence slightly greater in Lake St. Clair than in Lake Superior. The total index of severity (ISAD) was also greatest in larvae from the Courtright and Bear Creek sites (Fig. 7). However, the nature of the deformities was different at these two sites. The majority of deformities at Courtright, immediately below the major petrochemical industries, were abnormal ligula (14.5%). At Bear Creek, and Talford Creek, unlike the other sites examined, the majority of the deformities were abnormal mandibles (Tables 4 and 6).

Few chironomids other than *Procladius* (98% of chironomids) were present in the heavily polluted sediments at Talford Creek. The majority of these *Procladius* were small III instar larvae. Because of the differences in larval age and season of collection, the Talford Creek specimens are not directly comparable with the other samples from the St. Clair River. The incidence of total deformities in the limited samples from Talford

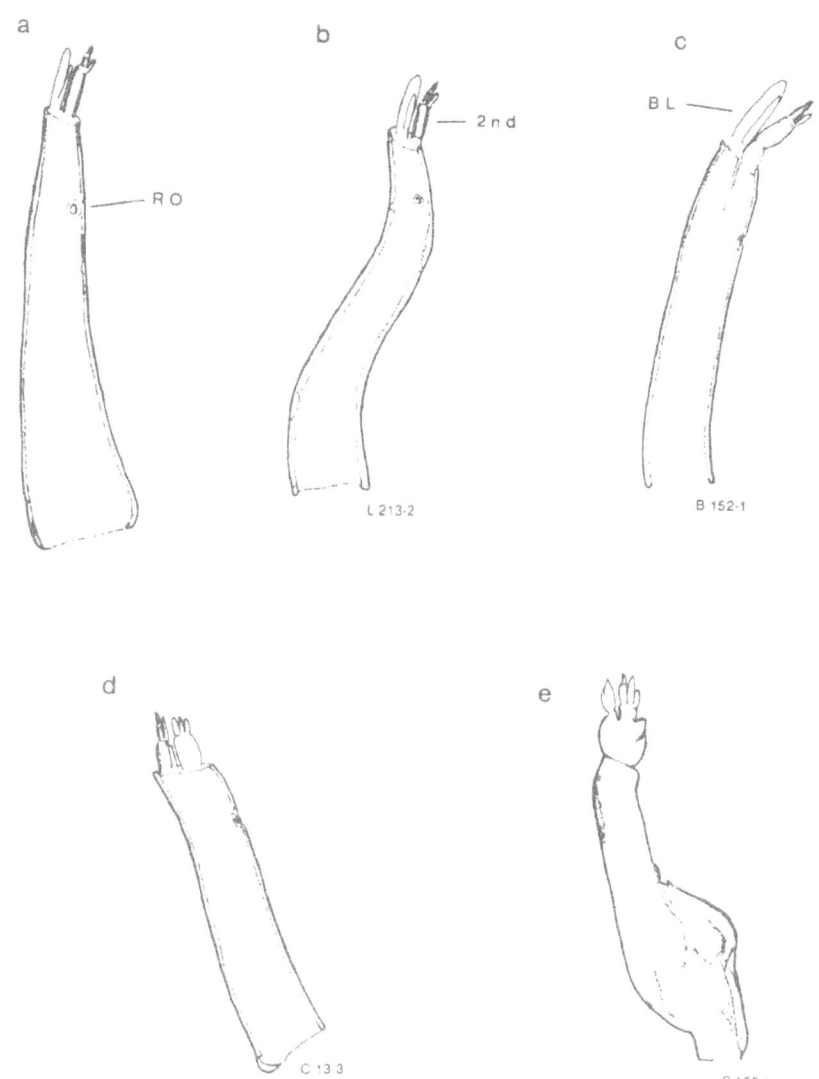

*Fig. 5.* Normal (a) and deformed (b-e) antenna in *Procladius* larvae: BL – antenna blade, RO – ring organ, 2nd – second antenna segment.

Creek was similar to that occurring farther downstream at Courtright (Table 6, Fig. 6). However, the incidence of deformed ligula and the ISAD of the Talford specimens was lower than that either upstream at Sarnia, or below at Courtright. Chi square analysis indicated that IV instar larvae had greater incidences of total deformities than the younger III instar larvae examined (Table 6).

### 4.2. Deformities in other genera

A variety of abnormalities were present in the most common Chironomini genus present at each site. The nature of deformities in *Chironomus* and *Cryptochironomus* have been outlined in Warwick (1988). Head capsule structures of *Phaenopsectra* and *Dicrotendipes* are similar to *Chironomus*, while those in *Harnischia* are similar to *Cryptochironomous*.

180

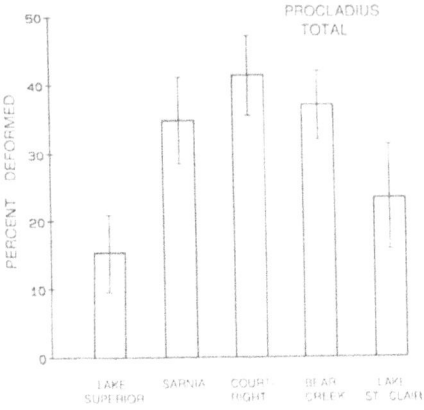

Fig. 6. Percent incidence of the total deformities in the *Procladius* populations.

*Table 6.* Percent incidence of deformities in *Procladius* larvae from the mouth of Talford Creek, St. Clair River, collected September, 1986. * indicates instar differences ($p < 0.05$) using Chi-square.

| Instar | III | IV | Total |
|---|---|---|---|
| Number examined | 55 | 64 | 119 |
| Percentage deformed | 23.6 | 44.0 | 40.3* |
| Paralabial combs | 10.9 | 18.7 | 15.4 |
| Total mandible | 7.2 | 15.6 | 12.6 |
| (wrinkled margin) | (5.4) | (12.5) | (8.4) |
| Total ligula | 7.7 | 3.1 | 5.0 |
| (4 toothed) | (1.8) | (0) | (0.8) |
| Antenna | 1.8 | 9.4 | 5.9* |
| ISAD | 1.27 | 1.68 | 1.54 |

The incidence of total deformities and the ISAD for these five genera followed the trend of that for *Procladius*. The severity of deformities in the *Chironomus anthracinus* group at Sarnia (Fig. 7) was much greater than that in the other Chironomini genera examined, with 40 percent of the larvae having abnormalities. The incidence of abnormalities in the *Harnischia* sp. from Lake St. Clair was very low, the only abnormality found being malformed 2nd antenna segments. The resulting ISAD for this population was 1.04. The only abnormality seen in the 62 *Dicrotendipes modestus* larvae examined from Batchawana Bay,

Lake Superior was a single specimen with a bifid medial tooth on its pecten epiharyngis. As a result the ISAD for this population was only 1.02.

As in *Procladius*, the most common and obvious deformities were the shape and number of teeth on the mentum and mandibles. Comparing only the abnormalities of the teeth on the heavily sclerotized menta or ligula and the mandibles, the *Chironomus* population at Sarnia clearly had the highest incidence of deformities (Fig. 8). That for the *Phaenopsectra* (*Tribelos*) *jucundus* at Courtright and the *Cryptochironomus digitatus* at Bear Creek was 9 and 7 percent respectively. No

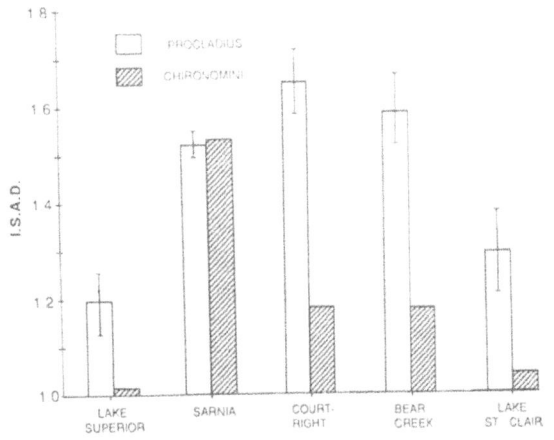

Fig. 7. Index of Severity of All Deformities (ISAD) in the larvae of *Procladius* spp. and dominant Chironomini genera at the 5 sites. The interval bars for the *Procladius* larvae are 95% confidence intervals.

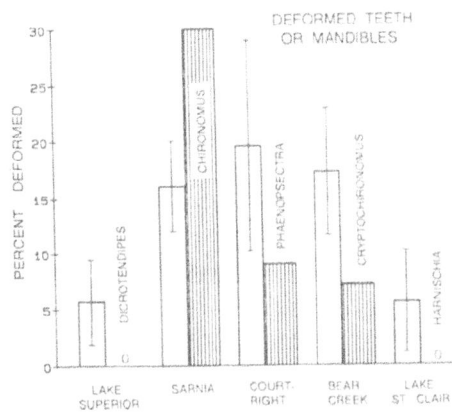

Fig. 8. Percent incidence of deformed teeth on the ligula, mentum or mandibles in the Chironomini genera (hatched) and *Procladius* (open bars) at the 5 sites.

specimens at *Harnishia* sp. from Lake St. Clair or *D. modestus* from Lake Superior were observed with abnormal teeth or mandibles. At all sites except Sarnia, the incidence of deformed mandibles and menta in the dominant Chironomini genus was lower than in comparable structures of *Procladius* spp. from the same site.

### 4.3. Impacted community response

The initial 3-day survey indicated that a rich macrobenthic community normally inhabited the narrow band of soft mud and *Elodea* along the 3 to 4 m contour of the St. Clair River. The occurrence of Chironomini at greater than 80 m$^{-2}$ is indicated in Fig. 1. Chironomini were virtually absent in the weedbeds along a 10 km section of the river between the industrial area downstream of Sarnia to south of Stag Island. The only organisms collected in this zone were oligochaetes (*Limnodrilus* sp. and *Quistadrilus*), leeches and a few *Procladius*. The Courtright site was chosen as it was the first area downstream of the impacted zone that had populations of chironomids greater than 120 m$^{-2}$. Downstream from Courtright, large populations of *Phaenopsectra*, together with *Chironomus*, Orthocladinae and *Gammarus* were again associated with the macrophyte beds.

## 5. Discussion

The incidence of deformed ligula in the control specimens from Lake Superior (3.9%) was not different from the 4.5 percent of atypical ligula found in *Procladius bellus* by Tennessen & Gottfried (1983). The apparent background incidence of deformed *Procladius* ligula is therefore about 4 percent. The incidence of abnormal ligula in the related Pentaneurini (Tennessen & Gottfried, 1983) was somewhat less (0.9%). Their study found no increase in the incidence of deformed Tanypodinae in coal strip mine ponds compared with man-made lakes. It is curious that the incidence of total or ligula deformities in Lake

St. Clair was not significantly greater than that in Lake Superior. Lake St. Clair is downstream from the major contaminant sources on the St. Clair River, and had its fisheries closed due to elevated levels of mercury. (Hamdy & Post, 1985). Apparently, mixing of the river in the delta and Lake St. Clair (COARGLWQ, 1986) dilutes the chemicals to levels below the dose response of *Procladius*.

The Lake Superior site suggests that the incidence of all possible abnormalities in *Procladius* populations from unstressed environments could be as high as 15 percent. Incidences of obvious deformed *Procladius* from other studies on the Great Lakes (Hamilton & Saether, 1971; Hare & Carter, 1976; Warwick, 1980) were near 1 percent, occurring mostly in harbours and near the mouths of major tributaries. Although these studies refer mainly to deformed ligula, their findings were still much lower than the 4 percent found in the Lake Superior population. The high incidence at the Lake Superior site is difficult to explain, given the excellent water quality of Batchawana Bay. Other than atmospheric deposition (Johnson *et al.*, 1988), the only known contaminants of the bay are the periodic treatments of the tributaries with the lampricides TFM and Bayer 73 (Fisheries and Oceans, Canada, 1980), which place a nonlethal stress on the invertebrate community.

Two inherent problems in making accurate identification of deformities are wear or damage to the structure and natural variability in morphology. Care must be made to correctly identify worn mouthparts which occur due to feeding activities in coarse sediment. Wear does not alter the symmetry of the mouth parts, which is normally evident in deformed specimens. Damaged antenna or mandibles resulting from the screening process usually have abrupt breaks that are readily visible and easily separated from deformed structures.

Variation in the normal structures of individuals, separate populations or individual species can only be suspected after examining a number of specimens from several sites. In this study, these suspected morphometric differences, which vary from site to site (ie. enlarged mandible auxiliary teeth, or antennal blade lengths), were not classified as deformities. However, the total abnormalities tallied may include less common variants. The high incidence of ligula with four symmetrical teeth at all of the sites examined, particularly Lake Superior (35% of the deformed ligula), may represent a natural variant rather than a morphological deformity. This abnormality was responsible for between 27 and 50 percent of the deformed ligula ranked at the sites. Exclusion of these specimens would have reduced the incidence of deformed ligula in Lake Superior to 2.5 percent. The studies by Tennessen & Gottfried (1983) and Warwick & Tisdale (1988) included these four toothed ligula as deformed specimens.

Before any differences are classified as deformities, care also has to be taken that the larvae are identified correctly by verifying other head capsule structures. The genera *Psectrotanypus* and *Djalmabatista* also have four teeth on their ligula but can be separated from *Procladius* by other structures, especially the mandibles and paraglossa.

The frequencies of total deformities in the *Procladius* from the St. Clair River (34 to 41%) and deformed ligula (12 to 15%) are very high compared to incidences reported elsewhere, reflecting that the St. Clair River is heavily impacted by a variety of pollutants (Oliver & Bourbonniere, 1986). Other authors have suggested a relation between the incidence of deformed chironomids and the degree of sediment contamination (Wiederholm, 1984; Warwick *et al.*, 1987). Similarly, Milbrink (1983) found a strong covariation between the frequency of severe deformities within populations of the oligochaete *Potomothrix hammoniensis* and industrial outflows in Lake Varern, Sweden.

Increased exposure to contaminants does not necessarily result in a higher incidence of deformities. Warwick (1985) found both the incidence of deformities in *Chironomus* and the severity of the response was inversely related to the exposure to insecticide. The lower incidence of deformed ligula and the index of severity in the contaminated Talford Creek samples, as compared to

immediately up or downstream, fits with Warwick's suggestion of a graded response at lower concentrations, but increased resistance or mortality occurring at higher exposure. Severe deformities such as those present at the sites on the St. Clair River (Fig. 2j, 4e) could be expected to reduce the survival of those larvae inflicted, especially considering the reorganization of tissues and sequence of events necessary for the larvae to molt through the four instars.

The high severity of deformation (ISAD) and grossly deformed ligula at the Courtright site is a definite response to the numerous contaminants stressing that area of the St. Clair River. The equally high ISAD but dissimilar deformities in the Bear Creek *Procladius* is a response to different chemical stresses. This population is exposed to few industrial chemicals but is exposed to a great variety of agricultural chemicals (Table 7). Agriculture pesticide use in the area surrounding Bear Creek averages 3 kg ha$^{-2}$, about double the provincial average (Ontario Ministry Agriculture and Food, 1984). The majority of these are herbicides (2.7 kg ha$^{-2}$) with lesser amounts of fungicides and insecticides (0.2 kg ha$^{-1}$). Many of the insecticides and herbicides reach their designed goal by interfering with the growth pattern of the target pests.

The present study cannot determine which of the numerous industrial and agricultural chemicals are inducing deformities in the larvae. In the

Table 7. Commonly used agriculture chemicals used near the Bear Creek watershed (Kent County) during 1983, in approximate order of use. Data from Ont. Agr. and Food 1984.

| Herbicides (71 Tonnes) | Fungicides (3.2 T) | Insecticides (2.7 T) |
|---|---|---|
| Atrazine | Chlorothalonil | Carbofuran |
| Alachlor | Mancozeb | Demeton |
| Allidochlor | Maneb | Carbaryl |
| Metolachlor | Metiram | Chlorpyrifos |
| Butylate | Captafol | Fonofos |
| Trifluralin | | Malathion |
| Metribuzin | | Terbufos |
| Linuron | | |
| 2,4-D | | |

St. Clair River, the effects of the toxicants are likely being manifested through somatic rather than genetic changes in the next generation, as the river currents displace the emerging pupa and egg masses downstream. Therefore, it is unlikely that the *Procladius* hatching at the Talford Creek site are genetically more resistant to the contaminants present. The lower incidence at that site compared to farther up or downstream of the heavily impacted river section, is a result of increased mortality of larvae sensitive to the contaminants. In the highly contaminated sediments at Talford Creek, fewer individuals survived to the IV instar than farther downstream at Courtright, resulting in a very low population of Tanypods in a community devoid of other chironomids.

The reported incidence of deformed larvae in other chironomid genera has been much higher than in the predatory genus *Procladius* (Hare & Carter, 1976; Wiederholm, 1984). The genus *Chironomus* appears to be the most sensitive to the causative agents, with reported deformities in excess of 30 percent. Both this genus and *Procladius* tolerate poorer environmental conditions than other chironomids. This greater tolerance by *Chironomus* and its ingestion of fine contaminated detritus, exposes the species to much greater levels of contaminants, particularly volatile organics (Hare & Carter, 1976) and insecticides (Hamilton & Saether, 1971; Warwick, 1986). The very high incidence in the *Chironomus* population at the Sarnia site (38%), upstream of the major petrochemical industries is a result of past and present uses of this bay on the river. The Sarnia site is a former industrial harbour and contains a large recreational marina, a source of toxic tributyltin, which has been shown to be a major cause of deformations in mollusks and crustaceans (Byran *et al.*, 1986; Weis *et al.*, 1987). This site also receives effluent from the combined storm-sewage drains from the Sarnia and some of the chemical industries situated farther downstream (COARGLWQ, 1986).

As a tool in assessing the morphometric response by benthic fauna to pollutants, the incidence of deformed mentum and mandibles in *Chironomus* larvae appears the most useful, due to the much greater incidence in this genus and its larger size. However, *Chironomus* may not always be present. *Procladius* larvae can be found in most areas, but their lower abundance requires extensive collecting to ensure adequate sample size. Only certain genera can be used as their morphometric response to contaminants varies greatly. Wiederholm (1984) had also noted that in areas having a high incidence of deformity in some genera, cohabiting genera were not deformed. The head capsules of the genera are often quite different, (i.e. the *Cryptochironomus* group and *Chironomus*), whereas many structures are not common to all chironomid families, thus making direct comparison of different genera unsound.

The incidence of deformed ligula in *Procladius*, is the simplest and most accurate of the morphometric measurements. An index of severity, such as Warwick's ISAD allows a measure of response by different genera or the same genera at different sites, where the type and degree of severity may differ. In spite of the merits of using antenna deformities in chironomid larvae as an index to pollution (Warwick, 1985), the delicate antenna are more easily damaged during sampling, or distorted during mounting than are the hardened mouth parts. Furthermore, the incidence of antennal deformities in *Procladius* was low and requires accurate examination at much greater magnifications.

Although this technique has value in evaluating nonlethal responses of the benthic fauna to pollution, the demanding nature of the work cannot be overlooked. A week is required to collect, sort and mount the required 300 larvae for each site, while examination of each specimen requires about 3 minutes.

Community responses to pollution usually occur at sites where contamination is present. The initial 3-day survey of the St. Clair River produced a simple description of the community response to the contamination. This sampling and that of Thornley (1985), showed that severe depopulation of the chironomid fauna occurs for 10 km downstream of the chemical industries.

184

Changes in the benthic community as well as the nature and odours of sediment samples provide a much more rapid and as accurate indicator of environmental degradation than does the occurrence of deformities, a conclusion also reached by Cushman (1984). The incidence of deformities can be useful to augment, but not replace the monitoring of community response to environmental quality. Unfortunately, even extensive laboratory bioassays cannot positively link the deformities or community response to the causative agents.

## 6. Acknowledgements

I thank J. K. Leslie and C. A. Timmins for their assistance in collecting the samples, and J. E. Milne for help in preparing the slide mounts. Drs. M. G. Johnson and U. Borgmann provided valuable comments on the manuscript. Drs. D. Oliver and B. Bilyj provided verification of the chironomid identifications.

## References

*Anon*, 1986. TBT linked to dogwelk decline. Mar. Pollut. Bull. 17(9): 390–391.

Béland, P., 1988. Witness for the prosecution. Nature Canada 17(4): 28–36.

Brinkhurst, R. O., A. L. Hamilton & H. B. Herrington, 1968. Components of the bottom fauna of the St. Lawrence Great Lakes. Great Lakes Inst., Univ. Toronto. PR. No. 33, 50 pp.

Bryan, G. W., P. E. Gibbs, L. G. Hummerstone & G. R. Burt, 1986. The decline of the gastropod *Nucella lapillus* around south-west England: evidence for the effect of tributyltin from antifouling paints. J. Mar. Biol. Assoc. U.K. 66(3): 611–640.

COARGLWQ (Canada Ontario Agreement Respecting Great Lakes Water Quality), 1986. St. Clair River pollution investigation (Sarnia area). Environment Canada – Ontario Ministry of the Environment, January 1986. 135 pp.

Cushman, R. M., 1984. Chironomid deformities as indicators of pollution from a synthetic, coal-derived oil. Freshwat. Biol. 14: 179–182.

Elliott, J. M., 1977. Some methods for the statistical analysis of samples of benthic invertebrates. Freshwat. Biol. Assoc., Sci. Publ. 25. 160 pp.

Fisheries and Oceans, Canada, 1980. Sea Lamprey Control Centre, Sault Ste. Marie. Annual Report 1980. 99 pp.

Fisheries and Oceans, Canada, 1986. Great Lakes Fisheries Research Branch, Burlington, Ontario. Quarterly Report. September, 1986: 6–7.

Hamdy, Y. & L. Post, 1985. Distribution of mercury, trace organics, and other heavy metals in Detroit River sediments. J. Great Lakes Res. 11(3): 353–365.

Hamilton, A. L. & O. A. Saether, 1971. The occurrence of characteristic deformities in the chironomid larvae of several Canadian lakes. Can. Entomol. 103(3): 363–368.

Hare, L. & J. C. H. Carter, 1976. The distribution of *Chironomus* (s.s.)? *cucini* (*salinarius* group) larvae (Diptera: Chironomidae) in Parry Sound, Georgian Bay, with particular reference to structural deformities. Can. J. Zool. 54(12): 2129–2134.

Henderdorf, C. E., C. N. Raphael & E. Jaworski, 1986. The ecology of Lake St. Clair wetlands, a community profile. US Fish. Wildl. Serv. Biol. Rep. 85(7.7) 187 pp.

Johnson, M. G., J. R. M. Kelso & S. E. George, 1988. Loadings of organochlorine contaminants and trace elements to two Ontario lake systems and their concentrations in fish. Can. J. Fish. Aquat. Sci. 42 (Suppl. 1): 170–178.

Kauss, P. B. & Y. S. Hamdy, 1985. Biological monitoring of organochlorine contaminants in the St. Clair and Detroit rivers using introduced clams, *Elliptio complanatus*. J. Great Lakes Res. 11(3): 247–263.

Krieger, K. A., 1984. Benthic macroinvertebrates as indicators of environmental degradation in the southern near-shore zone of the central basin of Lake Erie. J. Great Lakes Res. 10: 197–209.

Malins, D. C., B. B. McCain, D. W. Brown, S. L. Chan, M. S. Myers, J. T. Landahl, P. G. Prohaska, A. J. Friedman, L. D. Rhodes, D. G. Burrows, W. D. Gronlund & H. O. Hodgins, 1984. Chemical pollutants in sediments and diseases of bottom-dwelling fish in Puget Sound, Washington. Env. Sci. Technol. 18(9): 705–713.

Milbrink, G., 1983. Characteristic deformities in tubificid oligochaetes inhabiting polluted bays of Lake Vanern, Southern Sweden. Hydrobiology 106: 169–184.

Oliver, B. G. & R. A. Bourbonniere, 1985. Chlorinated contaminants in surficial sediments of Lakes Huron, St. Clair, and Erie: implications regarding sources along the St. Clair and Detroit rivers. J. Great Lakes Res. 11(3): 366–372.

Oliver, D. R. & M. E. Roussel, 1983. The genera of larval midges of Canada. The insects and arachnids of Canada Part 11. Agriculture Canada Publ. 1746 263 pp. ISSN 0706–7313.

Ontario Ministry of Agriculture and Food, 1984. Survey of pesticide use in Ontario, 1983. Economics and Policy Coordination Branch. Report No. 84–05.

Pettigrove, V., 1989. Larval mouthpart deformities in *Procladius paludicola* Skuse (Diptera: Chironomidae) from the Murray and Darling Rivers, Australia. Hydrobiologia 179(2): 111–117.

Shealy, M. H. & P. A. Sandifer, 1975. Effects of mercury on survival and development of the larval grass shrimp *Palaemonetes vulgaris*. Mar. Biol.33(1): 7–16.

Sokal, R. R. & F. J. Rohlf, 1969. Biometrics. W. H. Freeman and Comp. San Francisco. 776 pp.

Tennessen, K. J. & P. K. Gottfried, 1983. Variation in structure of ligula of Tanypodinae larvae (Diptera: Chironomidae). Entomol. News. 94(4): 109–116.

Thornley, S., 1985. Macrozoobenthos of the Detroit and St. Clair Rivers with comparisons to neighbouring waters. J. Great Lakes Res. 11(3): 290–296.

Warwick, W. F., 1980. Palaeolimnology of the Bay of Quinte, Lake Ontario: 2800 years of cultural influence. Can. Bull. Fish. Aquat. Sci. 206. 117 pp.

Warwick, W. F., 1985. Morphological abnormalities in Chironomidae (Diptera) larvae as measures of toxic stress in freshwater ecosystems: indexing antennal deformities in *Chironomus* Meigen. Can. J. Fish. Aquat. Sci. 42(12): 1881–1914.

Warwick, W. F. & N. A. Tisdale, 1988. Morphological deformities in *Chironomus, Cryptochironomus* and *Procladius* larvae (Diptera: Chironomidae) from two differentially stressed sites in Tobin Lake, Saskatchewan. Can. J. Fish. Aquat. Sci. 45(7): 1123–1144.

Warwick, W. F., J. Fitchko, P. M. McKee, D. R. Hart & A. J. Burt, 1987. The incidence of deformities in *Chironomus* spp. from Port Hope Harbour, Lake Ontario. J. Great Lakes Res. 13(1): 88–92.

Weis, J. S. & K. Kim, 1988. Tributyltin is a teratogen in producing deformities in limbs of the fiddler crab *Uca pugilator*. Arch. Environ. Contam. Toxicol. 17(5): 583–587.

Wiederholm, T., 1984. Incidence of deformed chironomid larvae (Diptera: Chironomidae) in Swedish lakes. Hydrobiology 109: 243–249.

*Hydrobiologia* **219**: 187–202, 1991.
*M. Munawar & T. Edsall (eds),*
*Environmental Assessment and Habitat Evaluation of the Upper Great Lakes Connecting Channels.*
© 1991 *Kluwer Academic Publishers.*

# Biota of Lake St. Clair: habitat evaluation and environmental assessment[1]

J.H. Leach

*Ontario Ministry of Natural Resources, Lake Erie Fisheries Station, R.R. No. 2, Wheatley, Ontario N0P 2P0, Canada*

*Key words:* phytoplankton, macrophytes, invertebrates, fishes, contaminants, wetlands

## Abstract

As a shallow, productive lake in the drainage system between Lake Huron and Lake Erie, Lake St. Clair provides habitat for a diverse biota including significant populations of fish and wildlife that are of use to man. Of the more than 70 species of native and migrant fishes, 43 use the lake for spawning. Peak numbers of waterfowl utilizing the lake and adjoining wetlands have been estimated at 60 000 in spring and 150 000 in autumn. In addition to recreational fishing and hunting, the lake is also used for swimming, boating and as a source of municipal water. It is located downstream from an industrial centre and adjacent to a population estimated at about four million. Mercury contamination closed the fisheries in 1970 and concentrations of the metal persist above consumption guidelines in some species. Almost all of the Michigan shoreline is urbanized and much of it altered through dyking and bulkheading. Coastal wetlands of the lake have declined 41 percent in the past century and only about one-half of the remaining area is open to the lake.

Despite impacts from the large urban population and users, ecosystem quality remains reasonably good. The flushing action of relatively clean water from Lake Huron has slowed the eutrophication process. Major habitat problems are toxic substances from industries located on the St. Clair River and the continued loss of shoreline and wetlands to urbanization and agriculture. Further research and monitoring of sources, fates and impacts of toxic substances in sediments and biota are required. In addition, there is a need for environmental and economic evaluations of shorelines and wetlands to prevent further losses of these important habitats.

## 1. Introduction

Lake St. Clair is a shallow productive lake in the drainage system between Lake Huron and Lake Erie. Although not designated as an Area of Concern by the International Joint Commission, it is situated between two major Areas of Concern, the St. Clair River upstream and the Detroit River downstream. Because the lake is vulnerable to potential impact from the St. Clair River, it was included in the Upper Great Lakes Connecting Channels Study. A third Area of Concern, the Clinton River, is a tributary to Lake St. Clair. However, impacts or impairments on the lake from this source are considered minimal (GLWQB, 1987).

---

[1]Contribution No. 90–16 of the Ontario Ministry of Natural Resources, Research Section, Fisheries Branch, Box 5000, Maple, Ontario.

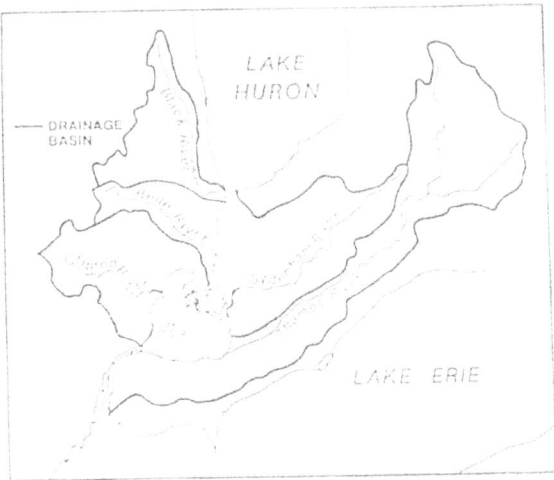

Fig. 1. Drainage basins of Lake St. Clair and the St. Clair River (Herdendorf *et al.*, 1986).

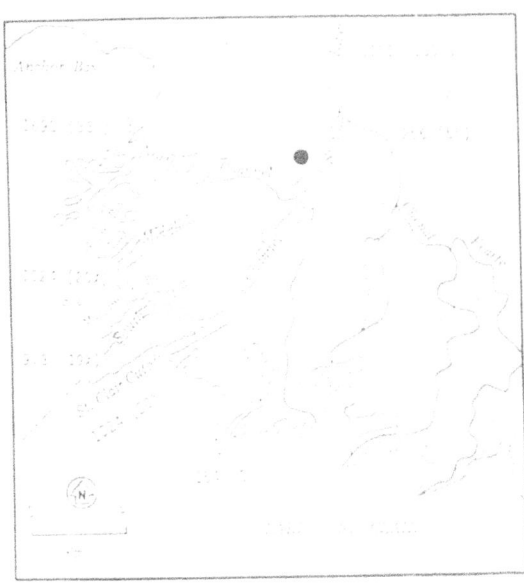

Fig. 2. Distribution channels of the St. Clair River through the St. Clair delta (Edsall *et al.*, 1988). Average flows are shown as m³ s⁻¹ and (%) of total flow.

Although Lake St. Clair is a relatively small water body in the Great Lakes system, it is productive and provides habitats for a diverse biota including invertebrates, fish, mammals, and waterfowl. As the lake is located adjacent to a population estimated at about four million, it is used extensively for recreational purposes and as a source of industrial and municipal water.

The objective of this paper is to provide an overview of the habitats of the lake and their respective biota and to discuss the effects of anthropogenic impacts on the ecosystem.

## 2. Description of the area

Lake St. Clair (42° 28' N, 82° 40' W) has a maximum length of 43 km, a width of 40 km, a surface area of about 1115 km², and a drainage basin area of 13 500 km², 77 percent of which is in Ontario (Fig. 1). It has a maximum natural depth of 6.4 m, a mean depth of only 3.0 m and a volume of 3.4 km³. A navigation channel 8.3 m deep is maintained by dredging between south channel in the delta and the head of the Detroit River. Mean St. Clair River inflow (5300 m³ sec⁻¹) represents about 98 percent of outflow (5400 m³ sec⁻¹) from the lake through the Detroit River. The other main tributaries are the Thames River and the

Sydenham River in Ontario and the Clinton River in Michigan. The St. Clair River enters the lake through the largest delta system in the Great Lakes with 53 percent flowing through the north and middle channels, 42 percent through the south and Bassett channels and 5 percent through Chenal Ecarte (Fig. 2). The mean hydraulic retention time of the lake has been estimated by Schwab & Clites (1986) at 9 days with a range of 2 to 30 days depending on wind conditions.

Water circulation in the lake is influenced by wind direction and inflow of water through the delta. Ayers (1964) modelled current patterns in Lake St. Clair under different wind directions and found that wind strongly influenced the lake's current structure. Leach (1980) used cluster analysis of physical and chemical data to delineate two fairly discrete water masses: a northwestern mass of mainly Lake Huron water flowing from the delta and a southeastern mass of more stable water enriched by nutrient loadings from Ontario tributaries and shoreline development. He found that wind direction and speed changed the margins of the masses but that overall discreteness of distribution was maintained. This distribution

*Table 1.* Percentage use of Lake St. Clair shorelines in Michigan and Ontario.

| Type of use | Michigan | Ontario |
|---|---|---|
| Residential/Commercial | 73.6 | 33.1 |
| Wetland/Recreational | 9.8 | 36.0 |
| Agricultural | 7.0 | 17.6 |
| Forest and Other | 9.6 | 13.3 |

was confirmed recently by numerical modelling of lake circulation (Ibrahim & McCorquodale, 1985). There is a significant difference in productivity between the two water masses. Since the southeastern mass is enriched from land drainage and has a longer retention time, its biological productivity is greater than that of the northwestern mass (Leach, 1972, 1973).

The Lake St. Clair shoreline (including islands) is 413 km long, divided between Michigan (230) and Ontario (183) (Robinson, 1977). Shoreline use is different between jurisdictions with most of the Michigan shoreline altered for urban uses (Table 1). In Ontario, wetlands and agricultural lands predominate.

The size distribution of surficial sediments have been mapped recently by Rukavina (1987). Muddy sand is the major bottom type particularly in the central part of the lake (Fig. 3). Well sorted

*Fig. 3.* Distribution of sediment types in Lake St. Clair (Rukavina, 1987).

sand is deposited as an outwash of the delta and a gravel-sand mixture occurs along the south shore. Clay-rich sediments occur in the west-central part of the basin. Average thickness of modern sediments is 13 cm with thicker layers of sand (30–50 cm) along the delta front and mud (20–30 cm) in the west-central part of the main basin.

## 3. Water quality

Physical and chemical characteristics of the lake have been discussed by Leach (1972), MWRC (1975) and reviewed by Edsall *et al.* (1988). Despite the lake's position downriver from the chemical valley and adjacent to a population estimated at four million, water quality is generally good due primarily to the large inflow of Lake Huron water. Gradients in concentrations of nutrients, major ions, plant pigments, and temperature increase from the northwest to the southeast. Nutrient inputs from agricultural drainage and urban development on the south shore together with greater stability in water mass contribute to more enriched conditions and hence more biological productivity in the southeastern part of the lake. Because the lake is shallow, turbidity caused by resuspension of particles is relatively high. Submersed vegetation in sheltered areas such as Anchor Bay and Mitchell Bay improve water clarity in those areas in summer.

Thermal stratification does not occur and dissolved oxygen concentrations throughout the water column are usually close to saturation. Oxygen depletion does not appear to be a problem even in shallow embayments under ice. In Anchor Bay, dissolved oxygen concentrations were high (9.8–16.4 mg l$^{-1}$) during the period of ice cover from January to March, 1979 (Werner & Manny, 1979). In Mitchell Bay, which supports heavy submersed macrophyte growth, mean dissolved oxygen remained above 75 percent (range: 58 to 89%) of saturation under ice cover during January and February, 1986 (Fig. 4) (Leach, unpublished data).

Relatively little is known about the impact of

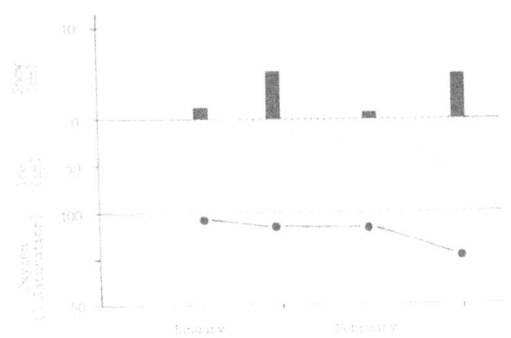

*Fig. 4.* Mean dissolved oxygen (*n* = 9) under ice during January and February, 1986. The ice broke up in early March following heavy rains.

chemical discharges from industries located on the St. Clair River and their potential threat to the health of the Lake St. Clair ecosystem. However, the efforts of the Great Lakes Institute and environmental agencies on both borders are attempting to quantify the degree and impact of past and current contamination (see below).

## 4. Biota and their habitats

### 4.1. Phytoplankton

Very little was known about the phytoplankton of the lake until the 1984 study by Munawar *et al.*, (1991). Highest seasonal biomass was measured in May and June when the algae were dominated by diatoms (67–90%) (Fig. 5). In the July to September sampling period, species of Diatomeae (34%) and Chrysophyceae (34%) were co-

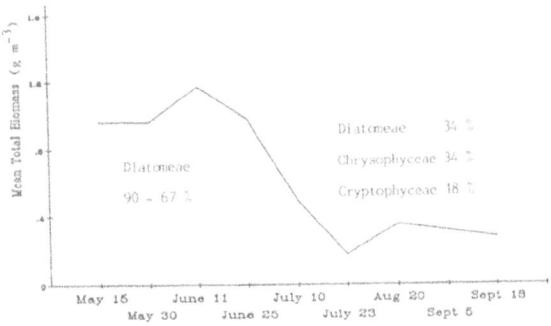

*Fig. 5.* Mean biomass of phytoplankton in Lake St. Clair, May to September, 1984 (Munawar *et al.*, 1991).

dominant followed by Cryptophyceae (18%). Only once, in late July, did Chlorophyta exceed 20 percent of the biomass. Biomass of Cyanophyta was relatively low in 1984 (Munawar *et al.*, 1991) but earlier (Winner *et al.*, 1970; MWRC, 1975) this group dominated the phytoplankton community in July and August. The predominant diatoms in spring samples (*Fragilaria*, *Melosira*, *Tabellaria*) are also common in Lake Huron (Vollenweider *et al.*, 1974). Johnston (1977) considered the phytoplankton community to be a eutrophic diatom association according to the classification of Hutchinson (1967).

Chlorophyll *a* was found by Leach (1972) to be distributed along a gradient increasing from northwest to southeast. The distribution of plant pigments was correlated with those of temperature, phosphorus, nitrate, and POC.

### 4.2. Macrophytes (submersed)

Schloesser & Manny (1982) found 12 taxa of submersed plants in Anchor Bay and the main

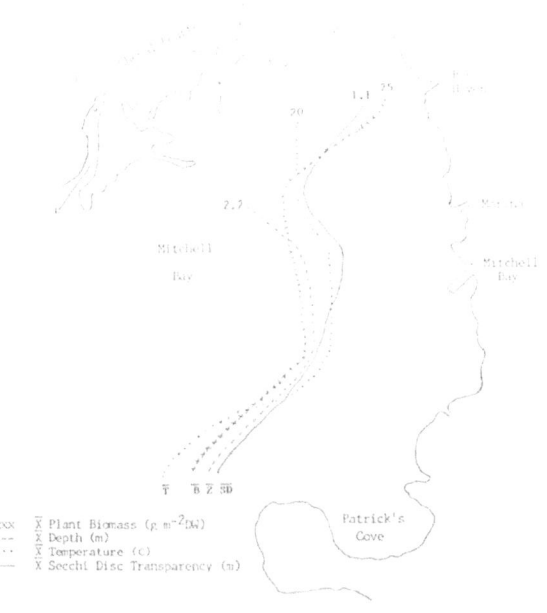

*Fig. 6.* Distribution of mean biomass of submersed macrophytes in relation to environmental factors in Mitchell Bay, Lake St. Clair in July, 1985.

basin of Lake St. Clair. Most common taxa were *Vallisneria americana, Chara* sp., *Myriophyllum spicatum, Najas flexilis, Potomogetan richardsonii,* and *Elodea canadensis.* In areas protected from strong wave action, such as Anchor Bay, the bottom is almost entirely populated with plants, whereas they are scarce in the main part of the lake where the bottom is scoured by waves. Most stands occur in the 0 to 3.7 m depth interval which in Lake St. Clair covers about 628 km². The frequency of occurrence of plants within this depth zone is about 35 percent.

*Vallisneria americana* is the overall predominant submersed plant in Lake St. Clair but in some areas, e.g. Sand Island in Anchor Bay (Schloesser *et al.*, 1985), *M. spicatum* contributes the most biomass. In Mitchell Bay, Leach (unpublished data) found 15 taxa present but only four species (*V. americana, Chara* sp., *Zanichellia palustris,* and *M. spicatum*) contributed over 90 percent to overall mean biomass (AFDW). *V. americana* was predominant but found in association with *Chara* sp. and *Z. palustris.* Biomass of all three species peaked in August. Biomass of *M. spicatum* increased in the fall as that of *V. americana* declined. Schloesser *et al.* (1985) found a similar seasonal distribution in Anchor Bay.

Distribution of mean biomass of macrophytes in Mitchell Bay was found (Leach, unpublished data) to be related to several environmental factors (Fig. 6). Isopleths of mean values of Secchi disc transparency, temperature, plant biomass, and number of taxa bisected the bay in approximately the same location as the mean depth isopleth. Differences in mean values of environmental factors, number of species, and biomass of plants between the two areas were significant ($P < 0.01$) (Table 2).

Production of submersed plants in Lake St. Clair was estimated by Edwards *et al.* (1989) at about 13 700 tonnes AFDW yr$^{-1}$. A substantial amount of this material drifts out of Lake St. Clair into the Detroit River and Lake Erie. Manny *et al.* (1988) calculated that the surface drift from April to October, 1986 was about 32 000 tonnes wet weight or about 1 600 tonnes AFDW. They con-

*Table 2.* Environmental characteristics and mean biomass of most abundant species of plants in inner and outer basins of Mitchell Bay, Lake St. Clair.

| | Inner Basin | Outer Basin | $P < 0.01$ |
|---|---|---|---|
| Mean Depth (m) | 2.0 | 2.7 | x |
| Mean Temperature (C) | 21.6 | 19.4 | x |
| $\bar{x}$ Secchi Disc Transparency (m) | 1.2 | 0.9 | x |
| Mean POC (mg C g$^{-1}$) | 14.3 | 9.6 | N.S. |
| Mean Plant Biomass g m$^{-2}$ (DW) | | | |
| All species | 39.1 | 2.3 | x |
| *Vallisneria americana* | 14.1 | 1.7 | x |
| *Chara* sp. | 12.6 | 0.07 | x |
| *Myriophyllum* sp. | 8.0 | 0.01 | x |
| *Najas flexilis* | 2.1 | 0.06 | x |
| *Heteranthera dubia* | 0.8 | 0.08 | x |
| *Zannichellia palustris* | 0.7 | 0.09 | x |
| Others | 0.8 | 0.29 | – |
| Mean number of Species | 5.5 | 1.4 | x |
| Total number of Species | 13.0 | 9.0 | – |

sidered that the drift of living plant material may be an important mechanism for redistribution of food resources but expressed concern that contaminants associated with the plants would also be dispersed into Lake Erie.

### 4.3. Macrophytes (emergent-wetlands)

The area of emergent plants in Lake St. Clair has been estimated at 9 170 ha (Lyon, 1979). *Typha latifolia, T. angustifolia, Scirpus validus, Phragmites communis* and *Eleocharis quadrangulata* are

*Fig. 7.* Distribution of Lake St. Clair coastal wetlands in 1873 and 1968 (Herdendorf *et al.*, 1986).

192

the predominant species. Principal stands of emergents occur in the delta area and along the eastern shoreline. Production of emergent aquatic plants was estimated by Edwards *et al.* (1989) to be about 61 000 tonnes AFDW yr$^{-1}$. The plants provide substrate for invertebrates and periphyton and cover and food for fish, birds and mammals.

The most significant historical aspect of Lake St. Clair wetlands has been their losses. Jaworski & Raphael (1976) estimated that 72 percent of

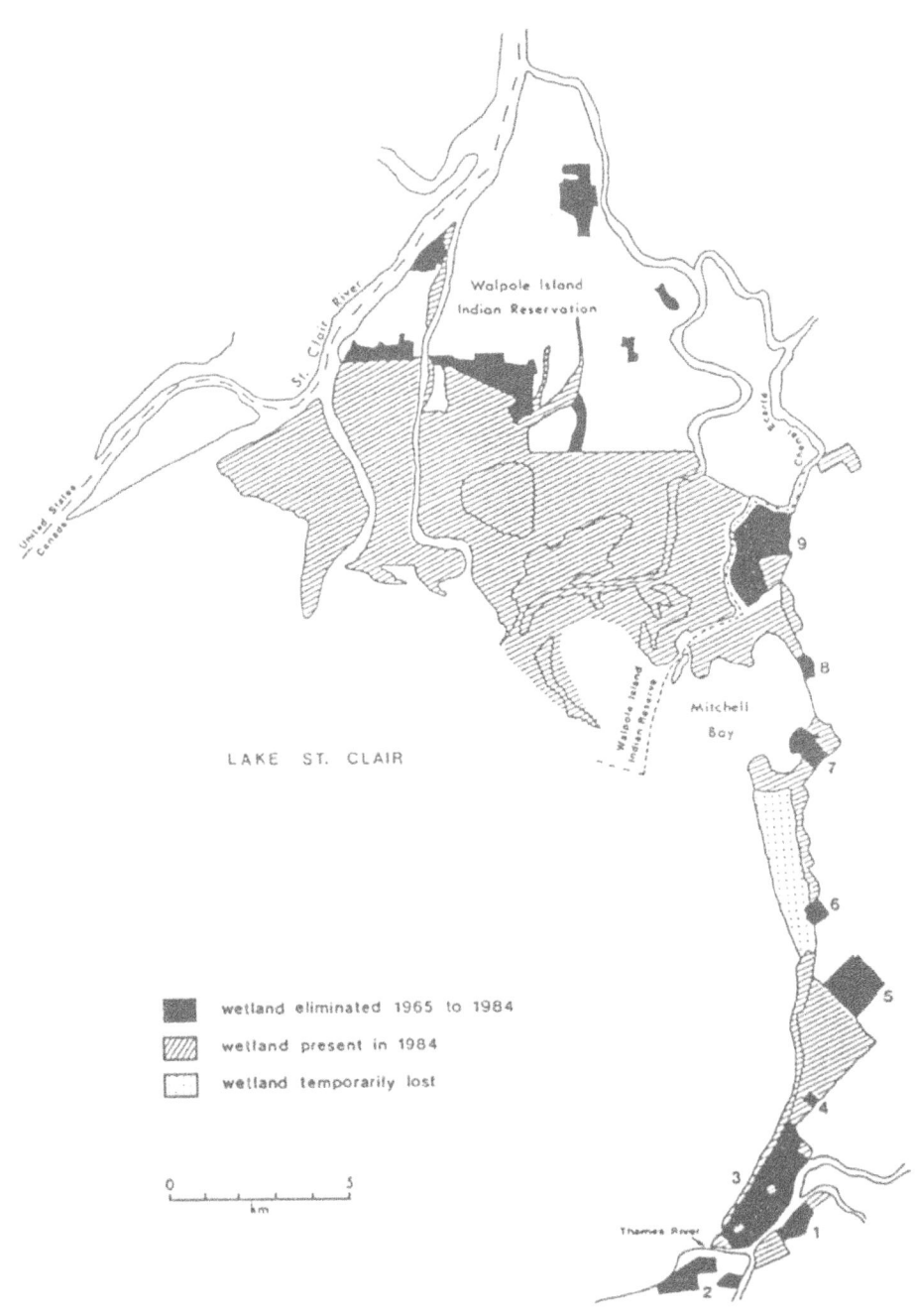

*Fig. 8.* Losses of Ontario's coastal wetlands in Lake St. Clair, 1965 to 1984 (McCullough, 1985).

wetlands (5 252 ha) bordering the Michigan side of Lake St. Clair were lost between 1873 and 1973 mostly due to development of shoreline for residential and commercial purposes (Fig. 7). In Ontario, wetland losses have been due almost entirely to drainage for agriculture (92%) with the remainder for cottage and marina developments. Between 1965 and 1984 over 30 percent of privately owned wetlands along the eastern shoreline were destroyed or converted to other uses (Fig. 8) (McCullough, 1985). Further wetland losses in this area have been slowed by a recent Provincial tax rebate system (effective January 1987) which brings the tax rate for marshland in line with that for agricultural land thereby eliminating the financial incentive to drain wetlands. This action, together with lower water levels, may result in an increase in emergent wetlands on the Ontario side of the lake.

## 4.4. Zooplankton

Zooplankton from different parts of Lake St. Clair has been surveyed sporadically (Birge, 1894; Jennings, 1894; Marsh, 1895; Winner et al., 1970; Leach, 1973; Bricker et al., 1976). In the most recent work, Sprules & Munawar (1991) found that abundance of zooplankton ranged from 35 to 93 organisms $l^{-1}$ and that biomass ranged from 500 to 1 500 $\mu g\ l^{-1}$ which is among the highest recorded for the Great Lakes. Cladocerans, particularly bosminids, were the predominant group in both abundance and biomass (Table 3). In the other Great Lakes, copepods are usually predominant with cladocerans contributing less than 20 percent of total zooplankton

*Table 3.* Mean zooplankton abundance and biomass in Lake St. Clair, June, 1984 (from Sprules & Munawar, 1991).

|  | $\overline{X}$ Abundance | $\overline{X}$ Biomass |
| --- | --- | --- |
| Total Zooplankton | 93 $l^{-1}$ | 1548 $\mu g\ l^{-1}$ |
| Cladocera | 50% | 43% |
| Calanoida | 18% | 30% |
| Cyclopoida | 14% | 21% |
| Rotifers | 17% | 4% |

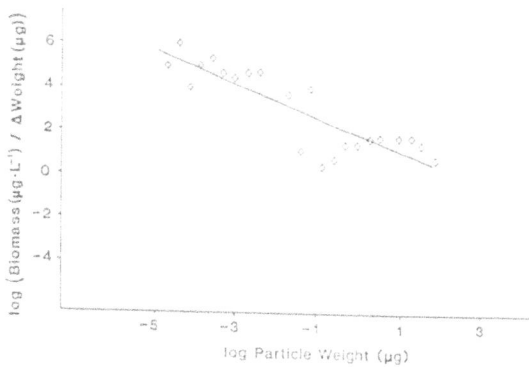

*Fig. 9.* Normalized plankton spectrum for Lake St. Clair, June 1984. Straight line is fitted by least squares regression. Coefficient of determination is 0.73 (Sprules & Munawar, 1991).

biomass (Sprules *et al.*, 1988). Lake St. Clair may be considered as more typical cladoceran habitat because it is shallow, productive, and does not harbour dense populations of planktivorous fish. Moreover, the high flushing rate of the lake may favour species with short generation times. For the same reasons rotifers are numerically important (Leach, 1973; Sprules & Munawar, 1991), but contribute little to zooplankton biomass.

Sprules & Munawar (1991) plotted plankton size against a normalized scale of biomass concentration (Fig. 9). In this type of plot, which describes community structure, variations around the line are indicative of ecosystem productivity and perturbations (Sprules *et al.*, 1983). The large variation in Fig. 9 suggests that the plankton communities in Lake St. Clair are not in steady state, perhaps due to the short flushing time or the influence of the different water masses or the effects of contaminants on the species. The low slope of the particle size spectrum is consistent with the high productivity of the lake.

## 4.5. Zoobenthos

The benthic macroinvertebrates of Lake St. Clair have been studied to various degrees by Hiltunen (1971), Hiltunen & Manny (1982), Thornley (1985), Hudson *et al.* (1986), Griffiths (1987),

*Table 4.* Mean abundance and biomass of major taxa of macroinvertebrates in Mitchell Bay, Lake St. Clair, May to October, 1985 (Leach, unpublished data).

| Taxa | No. m$^{-2}$ | Percent | AFDW (mg m$^{-2}$) | Percent |
|------|------|------|------|------|
| Diptera | 246 | 28.6 | 100 | 8.7 |
| Oligochaeta | 228 | 26.6 | 172 | 15.0 |
| Amphipoda | 158 | 18.4 | 82 | 7.2 |
| Gastropoda | 125 | 14.5 | 668 | 58.4 |
| Pelecypoda | 57 | 6.6 | 77 | 6.7 |
| Trichoptera | 16 | 1.9 | 20 | 1.7 |
| Ephemeroptera | 13 | 1.5 | 14 | 1.2 |
| Others | 17 | 1.9 | 12 | 1.1 |
| | 860 | | 1,145 | |

Nalepa & Gauvin (1988) and Leach (unpublished data). The benthic community is diverse. Griffiths (1987) found 101 taxa of invertebrates in a grid of 47 stations which covered the entire lake and lower St. Clair River. Hudson *et al.* (1986) found 65 taxa of invertebrates in Lake St. Clair which was lower than that of the St. Clair River (98 taxa) and the Detroit River (80 taxa). Leach (unpublished data) found 53 taxa in Mitchell Bay with chironomids and oligochaetes most abundant

(Table 4). In the main lake, Hudson *et al.* (1986) found oligochaetes, a polychaete and chironomids to be numerically predominant.

In general, the system supports organisms associated with relatively unpolluted waters. Thornley (1985) reported that 'nowhere in Lake St. Clair in 1983 did the percentage of tubificid worms exceed 60 percent, indicating good water quality.' Amphipods are common with *Hyalella* the most numerous taxa. *Hexagenia* is the most common mayfly with densities of 3 000 nymphs m$^{-2}$ measured by Hudson *et al.* (1986). Leach (unpublished data) found 15 species of Trichoptera in Mitchell Bay. Nalepa & Gauvin (1988) found 18 species of bivalves in the lake with a total biomass of 4.4 g DW m$^{-2}$. Both Hudson *et al.* (1986) and Griffiths (1987) found 10 taxa of gastropods in the lake; Leach (unpublished data) found 12 taxa in Mitchell Bay. The benthic fauna in Lake St. Clair is much more diverse than that of Lake Erie where tubificid worms dominated (75%) the benthos in the western basin in 1979 (Thornley, 1985).

Mean biomass of 1.1 g DW m$^{-2}$ in Lake St.

*Table 5.* Mean biomass (g AFDW m$^{-2}$) and net production (g AFDW m$^{-2}$y$^{-1}$) of food web components in Lake St. Clair, St. Clair River and Detroit River (Edwards *et al.*, 1989).

| Food Web Component | Lake St. Clair | | St. Clair River | | Detroit River | |
|------|------|------|------|------|------|------|
| | Biomass | Production | Biomass | Production | Biomass | Production |
| **Primary Producers** | | | | | | |
| Phytoplankton | 0.64 | 54 | 0.45 | 67 | 0.67 | 54 |
| Periphyton | 2.5 | 32 | 2.0 | 26 | 3.0 | 39 |
| Submersed macrophytes | 46* | 58 | 131* | 164 | 113* | 174 |
| Emergent macrophytes | 532* | 665 | 532* | 665 | 374* | 468 |
| Total | | 809 | | 922 | | 735 |
| **Secondary Producers** | | | | | | |
| Zooplankton | 0.44 | 7.9 | 0.56 | 10.1 | 0.46 | 8.3 |
| Macrobenthos | 1.1 | 6.8 | 1.0 | 7.4 | 0.75 | 5.4 |
| Total | | 14.7 | | 17.5 | | 13.7 |
| **Tertiary Producers** | | | | | | |
| Fish | 2.3 | 1.4 | 8.7 | 5.2 | 3.4 | 2.1 |

* Seasonal maximum standing crop.

Clair is similar to that $(1.0 \text{ g DW m}^{-2})$ found in the St. Clair River (Hudson *et al.*, 1986) but considerably less than the $2.5 \text{ g DW m}^{-2}$ estimated by Leach (unpublished data) for Mitchell Bay. Gastropoda was the largest component (58%) of zoobenthos biomass in Mitchell Bay where abundance of that group is probably associated with the density of macrophytes. Edwards *et al.* (1989) estimated annual production of benthic invertebrates in the lake at $6.8 \text{ g AFDW m}^{-2}$ which is similar to the estimate for the St. Clair River $(7.4 \text{ g AFDW m}^{-2})$ and higher than the $5.4 \text{ g AFDW m}^{-2}$ estimate for the Detroit River (Table 5). Benthic invertebrates are a predominant component in food webs leading to fish in riverine systems (Edsall *et al.*, 1988) and they probably fulfill this role in shallow, productive Lake St. Clair.

*Fig. 10.* Spawning and nursery areas for major species of sport fish in Lake St. Clair (Goodyear *et al.*, 1982).

196

## 4.6. Fish

The fish community of Lake St. Clair is diverse and abundant, consisting mainly of warm-water and mesothermic species. Cold-water species are also found in the lake but are not year-round residents. Some species move into Lake St. Clair from Lake Erie and historically were important to the fishery. Lake herring declined in the 1890's and lake whitefish in the 1900's. Lake trout ceased to be recorded in the Ontario commercial harvest in 1892.

Of the more than 70 species recorded as native or migrants, 34 use the lake for spawning (Goodyear et al., 1982). Most of the 28 native species spawn in shallow water along the delta (St. Clair Flats) or other shoreline areas or in tributaries to the lake (Fig. 10). Of the six exotic species, rainbow smelt and sea lamprey spawn in tributaries and alewives, carp, goldfish, and gizzard shad spawn in bays, marshes, and other shallow areas.

The lake supports an active year-round sport fishery and, historically, supported viable commercial fisheries. The Michigan commercial fishery exploited mainly lake whitefish, lake herring, walleye, and yellow perch from the early 1800's until it was closed in 1908 (Edsall et al., 1988). Catches peaked in the late 1800's and then declined substantially, possibly due to overfishing. In addition to the above species, the Ontario commercial fishery harvested lake sturgeon, northern pike, carp, suckers, and ictalurids (Table 6). The fishery was closed in 1970 following discovery of elevated levels of mercury in fish and re-opened in 1980 on a quota basis with no allotment for walleye. Recent catches consisted of catfish, bowfin, and carp (95%). Because of limited economic returns (and purchases of licenses by the Ontario government) the commercial fishery has almost ceased (OMNR, 1987).

Because of the proximity of Lake St. Clair to large urban populations, recreational fisheries have been active since the turn of the century. Haas et al. (1985) estimated annual boat angling effort in Michigan waters of Lake St. Clair in 1983

Table 6. Major fish species which contributed to the commercial fisheries of Lake St. Clair.

| Target species | Michigan (early 1800s–1908) | Ontario (early 1800s–1970) |
|---|---|---|
| Lake whitefish | x | x |
| Lake herring | x | x |
| Lake trout | x | x |
| Walleye | x | x |
| Yellow perch | x | x |
| Lake sturgeon | | x |
| Northern pike | | x |
| Carp | | x |
| Suckers | | x |
| Ictalurids | | x |

and 1984 at 1 486 000 hours and annual ice fishing effort at 467 000 hours. Boat angling effort was 2.5 times greater than that estimated by Krumholz & Carbine (1943, 1945) in the 1942–43 fishing seasons. Annual harvests in 1983 and 1984 from both fisheries totalled 1 197 000 fish of which 62 percent were taken by boat anglers. The boat fishery harvest was about 1.5 times greater than that estimated by Krumholz & Carbine (1943, 1945) in 1942 and 1943. Boat angling effort in Canadian waters during the 1978–81 period was only about one-quarter that in U.S. waters of the lake and ice fishing effort about one-third. Catch rate for the total fishery in Michigan waters averaged 0.61 fish per hour (boat 0.50; ice 0.98) which is greater than those in the St. Clair River and Detroit River fisheries. Yellow perch (59%) and walleye (18%) were the main species harvested by boat anglers; yellow perch contributed over 90 percent to the ice fishery harvest. Krumholz & Carbine (1943, 1945) also found that yellow perch was the main species harvested in 1942 and 1943.

Angling effort in Ontario waters of the lake in 1986 totalled 487 600 hours, 71 percent of which was from boats (OMNR, 1987). The boat fishery produced 170 000 fish which was 77 percent of total 1986 harvest. Catch rates were 0.49 fish per hour for the boat fishery and 0.36 for the ice fishery. Walleye (59%) was the main species taken by boat anglers followed by yellow perch (24%) and smallmouth bass (4.6%) (Table 7).

Table 7. Major fish species contributing to the Lake St. Clair sport fisheries in the 1980's.

| Target Species | Michigan | Ontario |
|---|---|---|
| Walleye | 20% | 60% |
| Yellow Perch | 60% | 25% |
| Smallmouth Bass | | 5% |
| | Total harvest exceeds 1.2 million fish (65% by boat anglers) | Total harvest exceeds 225 000 fish (75% by boat anglers) |

Table 8. Abundance of waterfowl observed by the Michigan Department of Natural Resources on Lake St. Clair during fall migration, October to December, 1974 (from Edsall et al., 1988).

| Type | Anchor Bay and adjoining marshes | Ontario | Total |
|---|---|---|---|
| Dabblers | | | |
| Mallard | 21 688 | 4 598 | 26 286 |
| Black duck | 5 415 | 2 937 | 8 352 |
| Wigeon | 17 | 1 830 | 1 847 |
| Pintail | 726 | | 726 |
| Teal | 369 | | 369 |
| Total | 28 215 | 9 365 | 37 580 |
| Divers | | | |
| Canvasback | 9 014 | 56 305 | 65 319 |
| Redhead | 6 519 | 3 896 | 10 415 |
| Scaup | 6 098 | 7 956 | 14 054 |
| Goldeneye | 4 361 | 382 | 4 743 |
| Bufflehead | 216 | 24 | 240 |
| Total | 26 208 | 68 563 | 94 771 |
| Coots | 720 | 612 | 1 332 |
| Swans | | 402 | 402 |
| Grand Total | 55 143 | 78 942 | 134 085 |

Yellow perch (43%) dominated the ice fishery followed by walleye (26%) and bluegills (16%).

Index fishing with pound nets in Ontario waters indicated above normal year classes of walleye in 1982 and 1984 which together with the relatively strong year classes of 1977 and 1980 have sustained elevated indices of abundance since 1979 (OMNR, 1987). Mean length at age has decreased in that period which may be a density-dependent response to increased abundance.

White perch, first observed at south index stations in 1977, has steadily increased in abundance and in 1986 was the predominant species in the index catch. White perch has expanded its range into Mitchell Bay where catches per unit of effort in 1986 were 10× those of 1981 and 1983.

### 4.7. Waterfowl

Lake St. Clair is one of the most important wetland areas for waterfowl in the Great Lakes system. In Ontario it ranks a close second to Lake Erie's Long Point Bay for use during migration (McCullough, 1985). Peak numbers of birds in the marshes have been estimated at 60 000 and 150 000 during spring and fall migrations respectively. Wetlands along the eastern shore of Lake St. Clair are currently Ontario's most important staging area for migrating mallards, black ducks, Canada geese, and tundra swans (McCullough, 1985). Marshes in Anchor Bay and Ontario are also used extensively by migrating diving ducks, particularly canvasback, redhead, and scaup

(Table 8). Although the primary importance of the St. Clair wetlands to waterfowl is in the provision of resting and feeding habitat, they are also used for nesting with recent production of ducklings estimated at 240 to 475 $km^{-2}$ in the delta area (Herdendorf et al., 1986). The submersed vegetation and associated macroinvertebrates provide important food for waterfowl.

### 4.8. Other Biota

In addition to 41 species of waterfowl, the Lake St. Clair area provides habitat for 15 species of shore birds, 11 species of gulls and terns, and 11 species of raptors (Herdendorf et al., 1986). The system supports 39 species of amphibians and reptiles, including frogs, toads, salamanders, snakes, lizards, and turtles (Edsall et al., 1988). Some of the more obvious mammals are muskrat, whitetailed deer, Virginia opossum, eastern cottontail, and striped skunk. Muskrats are

198

harvested as a furbearer but are also valuable in providing open-water habitat for nesting waterfowl by reducing stands of cattails and rushes.

## 5. Contaminants in Biota

### 5.1. Benthic Macroinvertebrates

During 1983 the Great Lakes Institute of the University of Windsor sampled populations of a native clam (*Lampsilis radiata siliquoidea*) and associated sediments in the Lake Huron to Lake Erie corridor to monitor distribution patterns of lead, cadmium, octachlorostyrene (OCS) and PCBs (GLI, 1986; Pugsley *et al.*, 1985). In Lake St. Clair, concentrations of lead, cadmium and OCS were generally highest in clams taken from an area adjacent to the south channel outlet (Fig. 11) suggesting that the St. Clair River may be a primary source of those contaminants. Oliver & Bourbonniere (1985) also implicated the St. Clair River as a source of OCS and several other chlorinated contaminants found in Lake St. Clair sediments.

In contrast, highest concentrations of PCBs were found in clams near the western shore of the lake, indicating that the Clinton River area may be an important source of PCB contamination (Fig. 11). Concentrations of cadmium, PCBs and OCS were higher in clams than surrounding sediments, but a positive correlation between concentrations in clams and sediments was observed only for OCS (Pugsley *et al.*, 1985).

### 5.2. Fish

Since mercury was discovered in Lake St. Clair fish in 1970, the Ontario Ministry of the Environment has been routinely monitoring contaminants

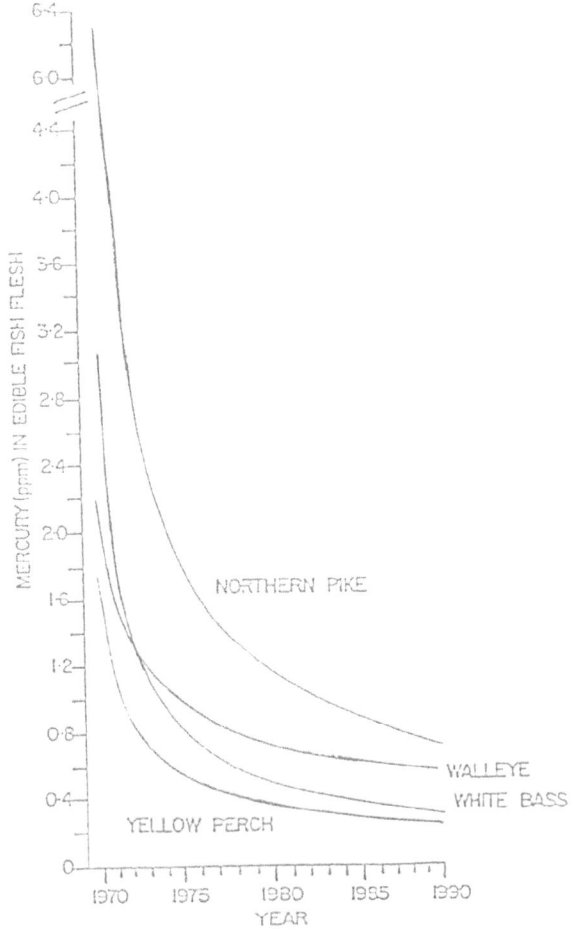

*Fig. 12.* Trends in mercury concentrations in edible portions of major species of sport fish from Lake St. Clair. Trend lines have been fitted by geometric regression (EC & OMOE, 1986).

*Fig. 11.* Zones of highest concentrations of contaminants in clams (*Lampsilis radiata siliquoidea*) (after Pugsley *et al.*, 1985; GLI, 1986).

in fish from the Lake St. Clair system. Discharges of mercury from the Dow chlor-alkali plant to the St. Clair River were controlled shortly after the discovery of the contaminant in fish flesh and since then, concentrations of mercury have generally been declining steadily (Fig. 12). Current mean concentrations of mercury in walleye, northern pike, and carp fillets are less than 25 percent and yellow perch and white bass less than 20 percent of 1970 levels (OMOE & OMNR, 1987). On the other hand, mercury concentrations in muskellunge did not decline between 1975 and 1985 (Lundgren, 1986).

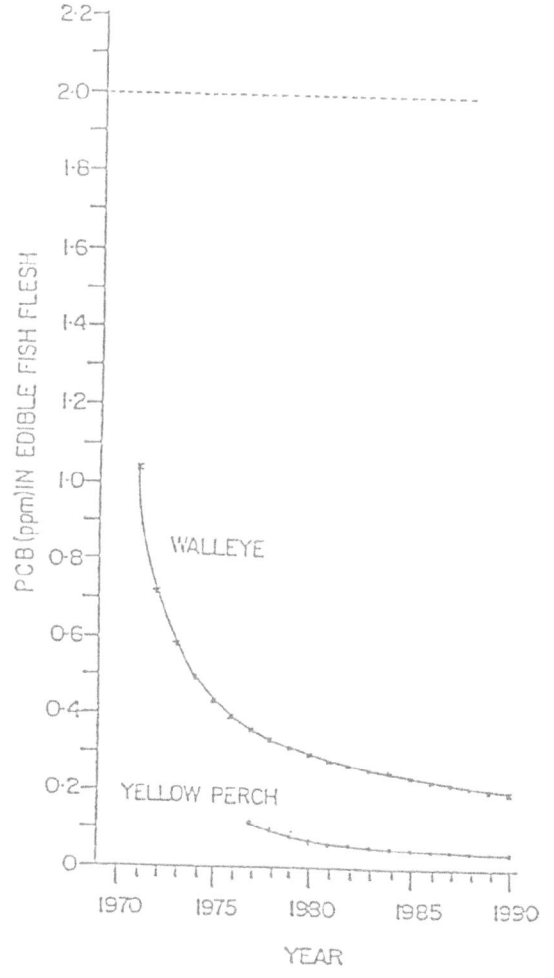

*Fig. 13.* Trends in PCB concentrations in edible portions of walleye and yellow perch from Lake St. Clair. Trend lines have been fitted by geometric regression (EC & OMOE, 1986).

PCBs in edible portions of Lake St. Clair fish have also declined generally and, with the exception of carp, channel catfish, and muskellunge, all species have not exceeded the Health and Welfare Canada guideline of 2.0 ppm (Fig. 13). Mean PCB concentrations in muskellunge have increased since 1980 (Lundgren, 1986).

Measurements of DDT in Lake St. Clair fish never exceeded the Health and Welfare guideline of 5 ppm in any of the 13 species tested. As with the PCB data, highest concentrations were detected in channel catfish and carp and lowest in yellow perch.

Concentrations of HCB and OCS in channel catfish from Lake St. Clair are greater than those from southern Lake Huron. In recent years, OCS concentrations in southern Lake Huron carp are greater than in Lake St. Clair carp. Chlorinated dioxins and dibenzofurans have been detected in Lake St. Clair channel catfish and carp but not in walleye (Lundgren, 1986).

The spottail shiner contaminant survey (Suns, undated) indicates that the heaviest contaminant loadings are associated with the south channel of the St. Clair River. Shiners from Mitchell Bay were less impacted and those from the southeastern part of the lake near the mouth of the Thames River were not measurably impacted by contaminant loadings from the St. Clair River.

## 6. Conclusions and the future

Lake St. Clair is productive and provides favourable habitat for a large and diverse biota. The flushing action of relatively clean water from Lake Huron has slowed the eutrophication process which normally would rapidly destroy such a shallow lake.

The fish community has been one of the most stable in the Great Lakes. For example, out of 21 major walleye stocks in the Great Lakes, the Lake St. Clair-Southern Lake Huron stock is one of only five which have remained relatively stable since 1800 (Schneider & Leach, 1979). Walleye and yellow perch were the main species exploited by now defunct commercial fisheries and remain as principal target species for active year-round sport fisheries.

In general, the health of the ecosystem appears to be good; but with a large chemical industry upstream and an area population of about four million the lake cannot escape some cultural stresses. The two major habitat problems are: (1) the impact of toxic substances, primarily from industries located on the St. Clair River and (2) the alteration of shorelines and loss of wetlands.

Most sport fishes have been contaminated with mercury and organochlorines. Although concentrations of mercury have declined substantially since loadings to the system ceased in 1970, consumption advisories issued by the Ontario government remain in effect for larger specimens of 15 species of sport fish. Some of the more common organochlorines such as PCBs and DDT are present in the food web and occur in fish but usually at levels below federal government guidelines. Less is known about HCB, OCS, chlorinated dioxins, and dibenzofurans and very little about causes of lesions and tumours. Additional information about sources, fates, and impacts of these and other chemical substances on sediments and biota is required.

Access between open lake and wetlands continues to be lost due to dyking and bulkheading for urbanization and drainage projects. This loss affects spawning and nursery habitat for sport and prey species of fish. Moreover, wetland losses affect not only fish species but also waterfowl. Between 1968 and 1982 use by dabbling ducks declined 79 percent in Lake St. Clair wetlands during spring migrations and 41 percent in fall migrations (Herdendorf et al., 1986). The wetland losses are primarily to urbanization in Michigan and to agriculture in Ontario. Hopefully, wetland loss in Ontario will be checked by the recently announced policy to equalize tax assessment on marshland and agricultural land. However, we need environmental and economic evaluation of shorelines and wetlands to prevent further losses of these important habitats for fish and wildlife.

## 7. Acknowledgements

I am grateful to the members of the Biota Work Group of the Upper Great Lakes Connecting Channels Study for input and valuable discussions. I thank Tom Edsall, Doug Haffner, Steve Nepszy, John Collins, and Mohiuddin Munawar for suggestions which improved the manuscript.

## References

Ayers, J. C., 1964. Currents and related problems at Metropolitan Beach, Lake St. Clair. Univ. Michigan, Great Lakes Res. Div. Rep. No. 20, Ann Arbor. 55 pp.

Birge, E. A., 1894. A report on a collection of Cladocera, mostly from Lake St. Clair, Michigan, appendix 2. In: J. E. Reighard, A Biological Examination of Lake St. Clair. Bull. Michigan Fish. Comm. No. 4.

Bricker, K. S., F. J. Bricker & J. E. Gannon, 1976. Distribution and abundance of zooplankton in the U.S. waters of Lake St. Clair, 1973. J. Great Lakes Res. 2: 256–271.

EC & OMOE (Environment Canada & Ontario Ministry of the Environment), 1986. St. Clair River pollution investigation (Sarnia area). 135 pp.

Edsall, T. A., B. A. Manny & C. N. Raphael, 1988. The St. Clair River and Lake St. Clair Michigan: an ecological profile. U.S. Fish Wildl. Serv., Biol. Rep. 85 (7.3). 130 pp.

Edwards, C. J., P. L. Hudson, W. G. Duffy, S. J. Nepszy, C. D. McNabb, R. C. Hass, C. R. Liston, B. Manny & W.-D. Busch, 1989. Hydrological, morphometrical, and biological characteristics of the connecting rivers of the International Great Lakes: a review. In D. P. Dodge (ed.), Proc. Internat. Large River Symp., Can. Spec. Publ. Fish. Aquat. Sci. 106.

GLI (Great Lakes Institute), 1986. A case study of selected toxic contaminants in the Essex Region. (Final Report DSS Contract No. UP-175). Univ. Windsor, Windsor, Ontario.

GLWQB (Great Lakes Water Quality Board), 1987. Report on Great Lakes Water Quality. Appendix A. Progress in developing remedial action plans for areas of concern in the Great Lakes basin. Internat. Joint Commiss., Windsor, Ontario.

Goodyear, C. D., T. A. Edsall, D. M. Demsey, G. D. Moss & P. E. Polanski, 1982. Atlas of spawning and nursery areas of Great Lakes fishes. U.S. Fish Wildl. Serv. Ann Arbor, MI. FWS/OBS – 82/52. 164 pp.

Griffiths, R. W., 1987. Environmental quality assessment of Lake St. Clair in 1983 as reflected by the distribution of benthic invertebrate communities. Aquatic Ecostudies, Ltd. Kitchener, Ontario. 35 pp.

Haas, R. C., W. C. Bryant, K. D. Smith & A. J. Nuhfer, 1985.

Movement and harvest of fish in Lake St. Clair, St. Clair River, and Detroit River. U.S. Army Corps Engin., Detroit, Mich. 141 pp.

Herdendorf, C. E., C. N. Raphael & E. Jaworski, 1986. The ecology of Lake St. Clair wetlands: a community profile. U.S. Fish Wildl. Serv. Biol. Rep. 85(7.7): 187 pp.

Hiltunen, J. K., 1971. Limnological data from Lake St. Clair, 1963 and 1965. U.S. Dep. Commer. NOAA Ntnl. Mar. Fish. Serv. Data Rep. 54: 45 pp.

Hiltunen, J. K. & B. A. Manny, 1982. Distribution and abundance of macrozoobenthos in the Detroit River and Lake St. Clair, 1977. U.S. Fish Wildl. Serv., Great Lakes Fish. Lab. Admin. Rep. 82-2, Ann Arbor, Mich. 87 pp.

Hudson, P. L., B. M. Davis, S. J. Nichols & C. M. Tomcko, 1986. Environmental studies of macrozoobenthos, aquatic macrophytes, and juvenile fish in the St. Clair-Detroit River system. U.S. Fish Wildl. Serv., Great Lakes Fish. Lab. Admin. Rep. 86-7. 303 pp.

Hutchinson, G. E., 1967. A Treatise on Limnology. Vol. 2. Introduction to Lake Biology and the Limnoplankton. John Wiley and Sons, Inc., New York, N.Y. 1115 pp.

Ibrahim, K. A. & J. A. McCorquodale, 1985. Finite element circulation model for Lake St. Clair. J. Great Lakes Res. 11(3): 208–222.

Jaworski, E. & C. N. Raphael, 1976. Modification of coastal wetlands in southeastern Michigan and management alternatives. Mich. Acad. 8: 303–317.

Jennings, H. S., 1894. A list of the Rotatoria of the Great Lakes and of some of the inland lakes of Michigan. Bull. Michigan Fish. Comm. No. 3.

Johnston, D. A., 1977. Population dynamics of walleye (*Stizostedion vitreum vitreum*) and yellow perch (*Perca flavescens*) in Lake St. Clair, especially during 1970–76. J. Fish. Res. Bd Can. 34: 1869–1877.

Krumholz, L. A. & W. F. Carbine, 1943. The results of the cooperative creel census in the connecting waters between Lake Huron and Lake Erie in 1942. Mich. Dep. Conserv., Fish. Res. Report No. 879, Ann Arbor, MI. U.S.A.

Krumholz, L. A. & W. F. Carbine, 1945. Results of the cooperative creel census on the connecting waters between Lake Huron and Lake Erie, 1943. Mich. Dep. Conserv., Fish. Res. Report No. 997, Ann Arbor, MI. USA.

Leach, J. H., 1972. Distribution of chlorophyll *a* and related variables in Ontario waters of Lake St. Clair. Proc. 15th Conf. Great Lakes Res. 1972: 80–86.

Leach, J. H., 1973. Seasonal distribution, composition, and abundance of zooplankton in Ontario waters of Lake St. Clair. Proc. 16th Conf. Great Lakes Res. 1973: 54–64.

Leach, J. H., 1980. Limnological sampling intensity in Lake St. Clair in relation to distribution of water masses. J. Great Lakes Res. 6: 141–145.

Lundgren, R. N. (ed.), 1986. Fish contaminant monitoring in Michigan. U.S. EPA 205j Grant. Mich. Dept. Ntrl. Resourc. Lansing, Michigan.

Lyon, J. G., 1979. Remote sensing analyses of coastal wetland characteristics: The St. Clair Flats, Michigan. Proc. 13th Symp. Remote Sensing of Environment. Mich. Sea Grant Rep. MICHU-56-80-313.

Manny, B. A., D. W. Schloesser, S. J. Nichols & T. A. Edsall, 1988. Drifting submersed macrophytes in the upper Great Lakes Channels. U.S. Fish Wildl. Serv., Ntnl. Fish. Center-Great Lakes. Unpubl. MS.

Marsh, C. D., 1895. On the *Cyclopidae* and *Calanidae* of Lake St. Clair, Lake Michigan, and certain of the inland lakes of Michigan. Bull. Michigan Fish. Comm. No. 5.

McCullough, G. B., 1985. Wetland threats and losses in Lake St. Clair. In: H. P. Prince & F. M. D'Itri (eds), Coastal wetlands. pp. 201–208 Lewis Publishing Co., Chelsea, Mich.

Munawar, M., I. F. Munawar & W. G. Sprules, 1991. The plankton ecology of Lake St. Clair, 1984. Hydrobiologia 219: 203–227.

MWRC (Michigan Water Resources Commission), 1975. Limnological survey of the Michigan portion of Lake St. Clair, 1973. Mich. Dep. Ntrl. Resourc. 59 pp.

Nalepa, T. F. & J. M. Gauvin, 1988. Distribution, abundance, and biomass of freshwater mussels (*Bivalvia*: *Unionidae*) in Lake St. Clair. J. Great Lakes Res. 14: 411–419.

Oliver, B. G. & R. A. Bourbonniere, 1985. Chlorinated contaminants in surficial sediments of Lakes Huron, St. Clair, and Erie: implications regarding sources along the St. Clair and Detroit rivers. J. Great Lakes Res. 11: 366–372.

OMNR (Ontario Ministry of Natural Resources), 1987. Lake St. Clair Fisheries Report 1986. Prep. for Lake Erie Committee Meeting, Great Lakes Fish. Commiss., March 24–25, 1987. Ont. Min. Ntnl. Resources, Tilbury, Ont. 35 pp.

OMOE & OMNR (Ontario Ministry of the Environment & Ontario Ministry of Natural Resources), 1987. Guide to eating Ontario sport fish. Toronto. 296 pp.

Pugsley, C. W., P. D. N. Hebert, G. W. Wood, G. Brotea & T. W. Obal, 1985. Distribution of contaminants in clams and sediments from the Huron-Erie corridor. I-PCBs and Octachlorostyrene. J. Great Lakes Res. 11: 275–289.

Robinson, J. R., 1977. Coordinated Great Lakes physical data. Can./U.S. Coord. Committ. Great Lakes Basic Hydraul. Hydrolog. Data, Dept. Mines Can. MS 32 pp.

Rukavina, N. A., 1987. Status Report on UGLCCS Lake St. Clair bottom sediment data. Level 1 report to the IJC. NWRI-CCIW, Burlington, Ont.

Schloesser, D. W. & B. A. Manny, 1982. Distribution and relative abundance of submersed aquatic macrophytes in the St. Clair-Detroit River ecosystem. U.S. Fish Wildl. Serv., Great Lakes Fish. Lab., USFWS-GLFL/AR-82-7. Ann Arbor, Mich. 49 pp.

Schloesser, D. W., T. A. Edsall & B. A. Manny, 1985. Growth of submersed macrophyte communities in the St. Clair-Detroit River system between Lake Huron and Lake Erie. Can. J. Bot. 63: 1061–1065.

Schneider, J. C. & J. H. Leach, 1979. Walleye stocks in the Great Lakes, 1800–1975: Fluctuations and possible causes. Great Lakes Fish. Comm. Tech. Rep. 31. 51 pp.

Schwab, D. J. & A. E. Clites, 1986. The effect of wind-induced circulation on retention time in Lake St. Clair.

Proc. 29th. Conf. Great Lakes Res., Internat. Assoc. Great Lakes Res. Abstract.

Sprules, W. G. & M. Munawar, 1991. Plankton community structure in Lake St. Clair, 1984. Hydrobiologia 219: 229–237.

Sprules, W. G., J. M. Casselman & B. J. Shuter, 1983. Size distribution of pelagic particles in lakes. Can. J. Fish. aquat. Sci. 40: 1761–1769.

Sprules, W. G., M. Munawar & E. Jin, 1988. Plankton community structure and size spectra in the Georgian Bay and North Channel ecosystems. In: M. Munawar (ed.) Limnology and Fisheries of Georgian Bay/North Channel Ecosystems. Hydrobiologia 163: 135–140.

Suns, K. undated. Organic contaminants in young-of-the-year spottail shiners from the St. Clair River, Lake St. Clair and the Detroit River. Ont. Min. Env. Toronto. (unpublished M.S.)

Thornley, S., 1985. Macrozoobenthos of the Detroit and St. Clair Rivers with comparisons to neighboring waters. J. Great Lakes Res. 11: 290–296.

Vollenweider, R. A., M. Munawar & P. Stadelmann, 1974. A comparative review of phytoplankton and primary production in the Laurentian Great Lakes. J. Fish. Res. Bd Can. 31: 739–762.

Werner, M. T. & B. A. Manny, 1979. Fish distributions and limnological conditions under ice cover in Anchor Bay, Lake St. Clair, 1979. U.S. Fish Wildl. Serv. Ann Arbor, MI. Admin. Rep. No. 80-1. 58 pp.

Winner, J. M., A. J. Oud & R. G. Ferguson, 1970. Plankton productivity studies in Lake St. Clair. Proc. 13th Conf. Great Lakes Res. pp. 640–650.

*Hydrobiologia* **219**: 203–227, 1991.
*M. Munawar & T. Edsall (eds),*
*Environmental Assessment and Habitat Evaluation of the Upper Great Lakes Connecting Channels.*
© 1991 *Kluwer Academic Publishers.*

# The plankton ecology of Lake St. Clair, 1984

M. Munawar[1], I.F. Munawar[2] & W.G. Sprules[3]
[1] *Fisheries & Oceans Canada, 867 Lakeshore Road, P.O. Box 5050, Canada Centre for Inland Waters, Burlington, Ontario, Canada L7R 4A6*; [2] *Plankton Canada, Burlington, Ontario*; [3] *Erindale College, University of Toronto, Ontario*

*Key words:* plankton, phytoplankton, zooplankton, species, biomass, primary productivity, Great Lakes

## Abstract

Lake St. Clair phytoplankton and zooplankton abundance and composition was analyzed during the period of May to September 1984. In addition, size-fractionated primary productivity and other limnological parameters were measured. Highest phytoplankton biomass was observed during spring (May) with high values for the southern and southeastern regions of the lake. Seasonally, the mean phytoplankton biomass ranged between 0.17 and 1.18 g m$^{-3}$ with high values recorded during spring (May, June) compared to summer. In the spring the phytoplankton was dominated by Diatomeae followed by Chrysophyceae and Cryptophyceae. During the summer the diatoms showed a decreasing trend due to the relative prevalence of Chrysophyceae, Cryptophyceae, and Chlorophyta. The species composition was oligotrophic-mesotrophic with mixed occurrence of some eutrophic species. The phytoplankton size composition indicated dominance of microplankton/netplankton ($> 20 \ \mu$m) and ultraplankton ($< 20 \ \mu$m) during spring and summer respectively. On an overall basis ultraplankton contributed overwhelmingly to primary productivity, as much as 75 percent in the summer.

The mean zooplankton biomass ranged from 173.0 to 1306.0 mg l$^{-1}$ dominated by Cladocerans (bosminids) in contrast to the other Great Lakes. Statistical evaluation of the phytoplankton – nutrient-contaminant interactions revealed positive correlations with heavy metals, suggestive of a physiological adaptation to contamination from the chemical valley. Based on low biomass, high Production/Biomass ratio, dominance of ultraplankton, characteristic species composition and plankton spectra, the lake appears to be an oligotrophic-mesotrophic perturbed ecosystem.

## Introduction

The plankton dynamics of Lake St. Clair, including both phytoplankton and zooplankton, are among the least known in the Great Lakes (Winner *et al.*, 1970; Leach, 1972, 1973; Bricker *et al.*, 1976). Lake St. Clair is the smallest lake in the Laurentian Great Lakes system with maximum length of 43 km, a width of 40 km, a mean depth of 3 m and a surface area of 1115 km$^2$

(Leach, 1991). About 98 percent of the water flowing into the lake is from the St. Clair River in the north originating from Lake Huron, while in the south it is connected to the Detroit River which drains into Lake Erie. The Thames River, Sydenham River (Ontario) and Clinton River (Michigan) are the main tributaries of the lake and are enriched with nutrients from land drainage and domestic sewage.

Lake St. Clair was studied as early as the

nineteenth century by Reighard (1894) but no further investigations were made until 1969 when the phytoplankton was studied by Winner *et al.* (1970) and zooplankton by Leach (1973). However, the phytoplankton analysis was problematic since it used Sedgewick-Rafter counting chambers which excluded the smallest organisms such as ultraplankton/picoplankton in the phytoplankton enumeration. In 1984 extensive lakewide cruises were conducted by Fisheries and Oceans Canada to conduct an in-depth study dealing with the plankton biomass (phytoplankton, zooplankton), their composition, and primary productivity, as well as other limnological variables about which very little was known.

## Methods and materials

### Phytoplankton

In 1984, an extensive study of the lake, consisting of nine bi-monthly cruises, was conducted from May to September. Ten stations distributed over the entire lake were chosen (Fig. 1). Five of the stations fanned the delta of the St. Clair River mouth and two were located around the outlet towards the Detroit River. The water samples for plankton analysis and primary productivity experiments were collected usually at two to three depths using a Van Dorn bottle aboard the research vessel C.S.S. Advent. These samples were mixed in a large carboy to ensure homogeneity. A portion of the sample was preserved in modified Lugol's solution. Taxonomic identification and enumeration was carried out by means of an inverted microscope (Wild Model M40) equipped with phase contrast illumination. The enumeration procedure followed is described in Munawar *et al.* (1974) and Munawar & Munawar (1978). The size analysis of phytoplankton into various size fractions was based on microscopic measurements with the help of a computer program which included picoplankton (Munawar *et al.*, 1978). Based on the maximum dimension the biomass was divided into three size fractions namely: a) $< 2\ \mu m$, b) $2$–$20\ \mu m$; and c) $> 20\ \mu m$.

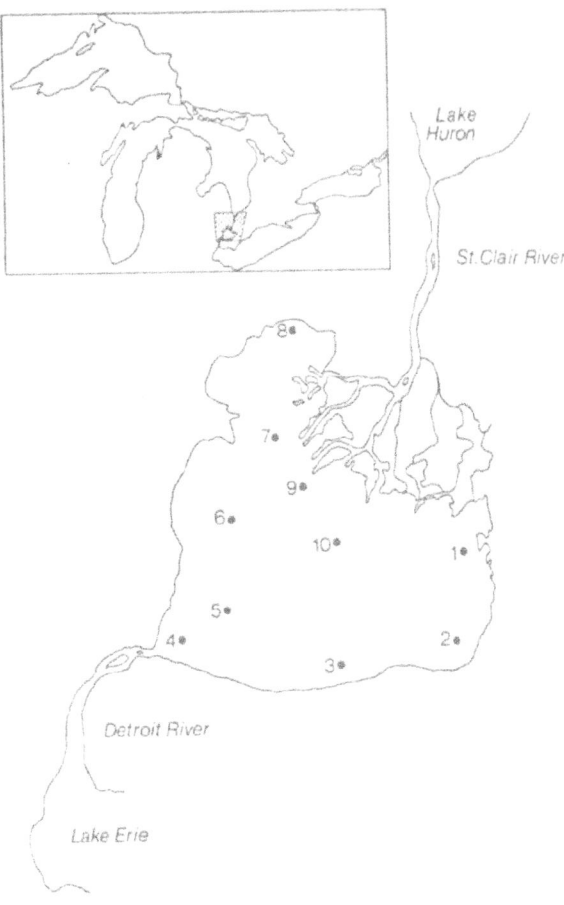

*Fig. 1.* Sampling locations in Lake St. Clair, 1984.

### Primary productivity

Primary productivity experiments were conducted as described by Vollenweider *et al.* (1974). An incubator was used with constant light intensity (30 000 lux) and temperatures approximating the surface temperature of the lake. Samples of 250 ml were poured into polycarbonate incubation bottles. Four replicates were used for each experiment and each bottle was inoculated with $10\ \mu Ci\ ml^{-1}$ of $NaH^{14}CO_3$. The samples were incubated for four hours in the incubator and at the end of the incubation period the samples were size fractionated (Munawar *et al.*, 1987) into three size fractions described below:

$< 2\ \mu m$ .... picoplankton
$2$–$20\ \mu m$ .... ultraplankton
$> 20\ \mu m$ .... microplankton/netplankton

Hourly Activity Coefficient or Production/Biomass quotient was calculated in the following way where phytoplankton biomass is based on microscopic enumeration (Vollenweider et al., 1974).

$$\frac{mgC\ m^{-3}\ h^{-1}}{mg\ m^3} = mgC\ mg^{-1}\ h^{-1}$$

### Chemical analysis
Water samples were also collected for chemical analysis of nutrients and contaminants according to the methodology described by the Water Quality Branch Inland Waters Directorate (1979).

### Zooplankton
Zooplankton samples were collected during daylight hours with a tow net of 30 cm mouth diameter and a 110 $\mu$m nylon mesh (filtering efficiency of 70%) which was hauled from 1 to 2 m above the bottom of the lake to the surface. Samples were preserved in a four percent formalin solution and subsamples were examined microscopically (Sprules & Munawar, 1990). The length of each organism was measured with microcomputer-based calipers (Sprules et al., 1981) and individual masses were computed using a program incorporating length-wet weight regressions pooled from the literature (Sprules, 1984).

The zooplankton were not uniformly identified to species, but were classified into 10 groups and included copepod nauplii, rotifers, daphniids, chydorids, Holopedium gibberum, bosminids, Diaptomus spp., Senecella calanoides, cyclopoid copepods, Limnocalanus macrurus/Epischura lacustris, Mysis relicta, Leptodora kindtii, and Chaoborus spp. These groups are considered to be more significant for ecological studies than just as taxonomic information (Sprules, 1984).

### Statistical analysis
The data were transformed using a log transformation. The data were analyzed using a principle components analysis on the correlations (Cooley & Lohnes, 1971; Nie et al., 1975; Munawar & Wilson, 1978). The number of principle components were limited to that number which covered

at least 75 percent of the total variance. The variables were chosen on a basis of including as many as possible without severely limiting the size of the data set. Variables listed at the bottom of the tables are the most important variables making up each factor; those not listed are of lesser importance. The absolute size of the loading for each variable is a measure of relative importance within each component (Munawar & Munawar, 1988).

## Results

### Phytoplantkon biomass

#### Taxonomic composition
The spatial distribution of phytoplankton biomass for each cruise is presented in Fig. 2. It is apparent that higher concentrations were observed during the spring period compared to the summer. Furthermore, relatively high values were recorded for the southern and southeastern regions of the lake during the month of May. During June most of the stations revealed high concentration of biomass although the western section of the lake harboured relatively more biomass. During the summer the biomass distribution, with low concentration, was less variable than during spring.

The spatial distribution of major taxonomic groups namely Chrysophyceae, Diatomeae and Cryptophyceae, is shown in Figs. 3 to 5. The Chrysophyceae and Cryptophyceae revealed homogeneous distribution on a lakewide basis with minor exceptions. On the other hand the Diatomeae exhibited pronounced variability during the spring period (Fig. 4). During May, in the southern and southeastern regions of the lake, a high concentration of diatoms was found which spread to other parts of the lake in early June. The midlake station (Station 10) had the highest abundance during late June. With the exception of Stations 1 and 4 in early July the diatoms were uniformly distributed throughout the summer with low concentrations across the lake. The northwestern region near the North Channel

206

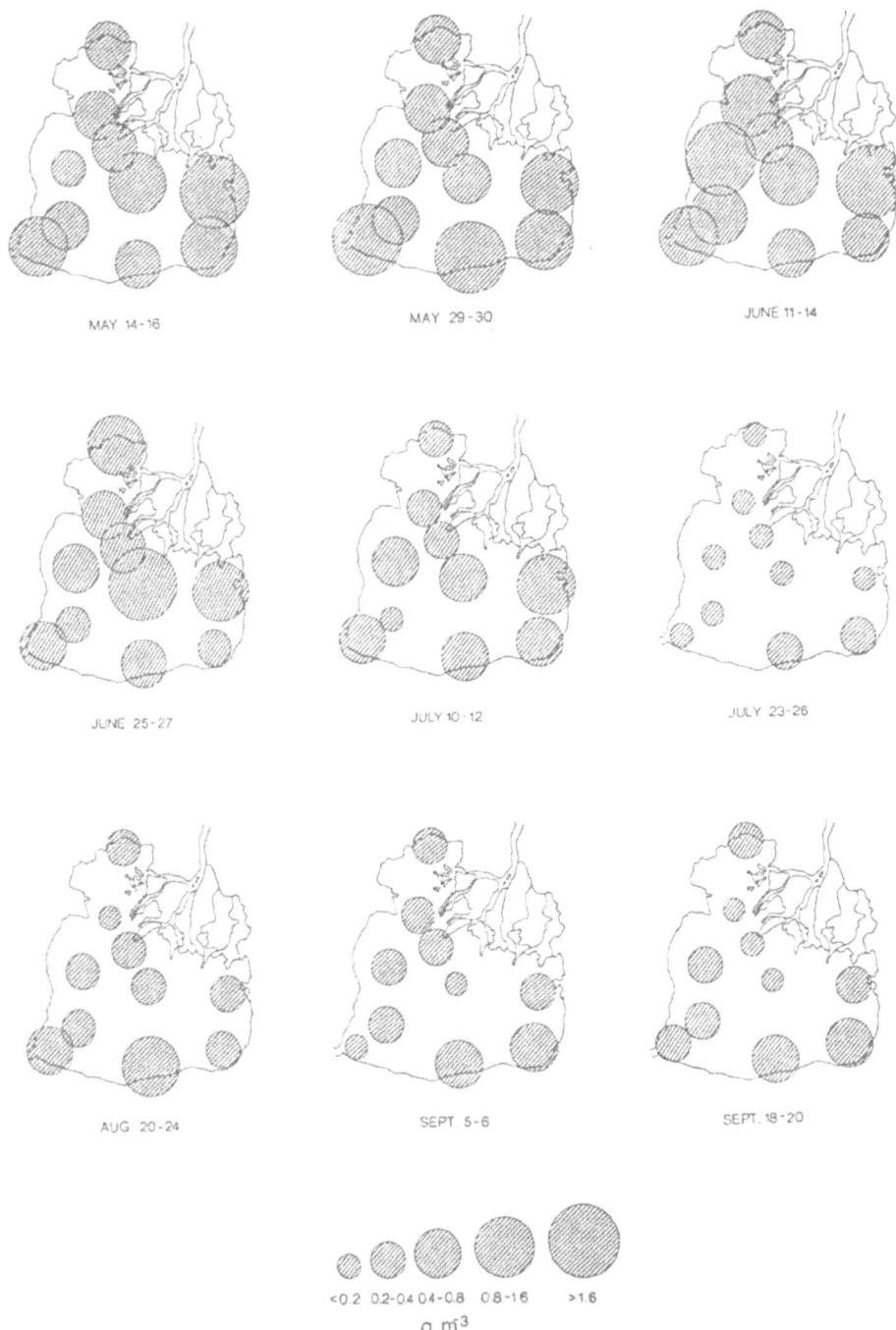

MAY 14-16

MAY 29-30

JUNE 11-14

JUNE 25-27

JULY 10-12

JULY 23-26

AUG. 20-24

SEPT. 5-6

SEPT. 18-20

<0.2 0.2-04 04-0.8 0.8-16 >1.6

g m³

*Fig. 2.* Spatial distribution of phytoplankton biomass in Lake St. Clair, 1984.

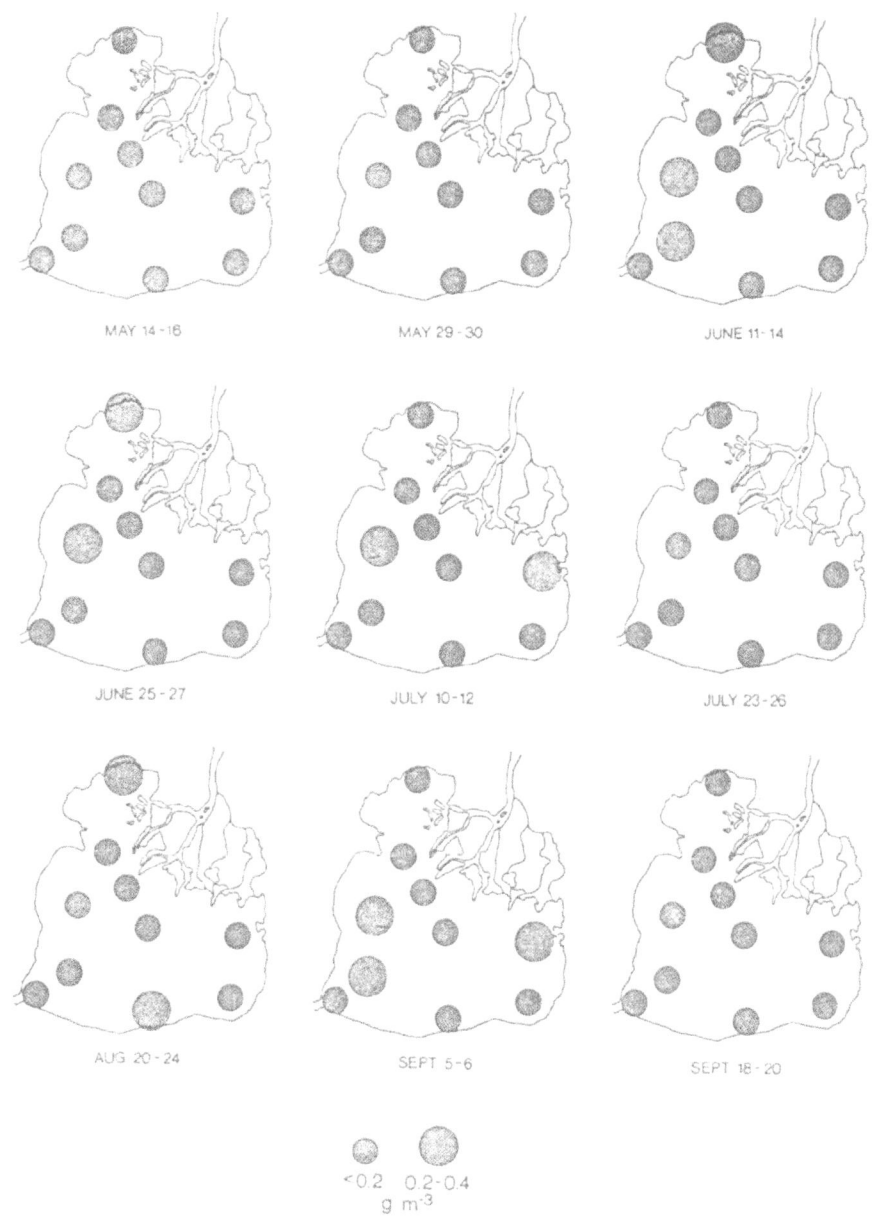

*Fig. 3.* Spatial distribution of Chrysophyceae in Lake St. Clair, 1984.

usually harboured low concentrations of diatoms.

The mean biomass per station is shown in Fig. 6 indicating homogeneous distribution across the lake with slightly higher concentrations at Stations 1 and 10 located at the eastern and midlake regions of the lake. The mean taxonomic composition of the biomass is also shown in Fig. 6. It is apparent that Diatomeae dominated

by contributing from 48 to 80 percent to the phytoplankton biomass followed by Chrysophyceae (7 to 33%) and Cryptophyceae (2 to 22%). The midlake station (Station 10) had the highest percentage of diatoms (80%). The contributions of Cyanophyta, Chlorophyta, and Dinophyceae were minor.

208

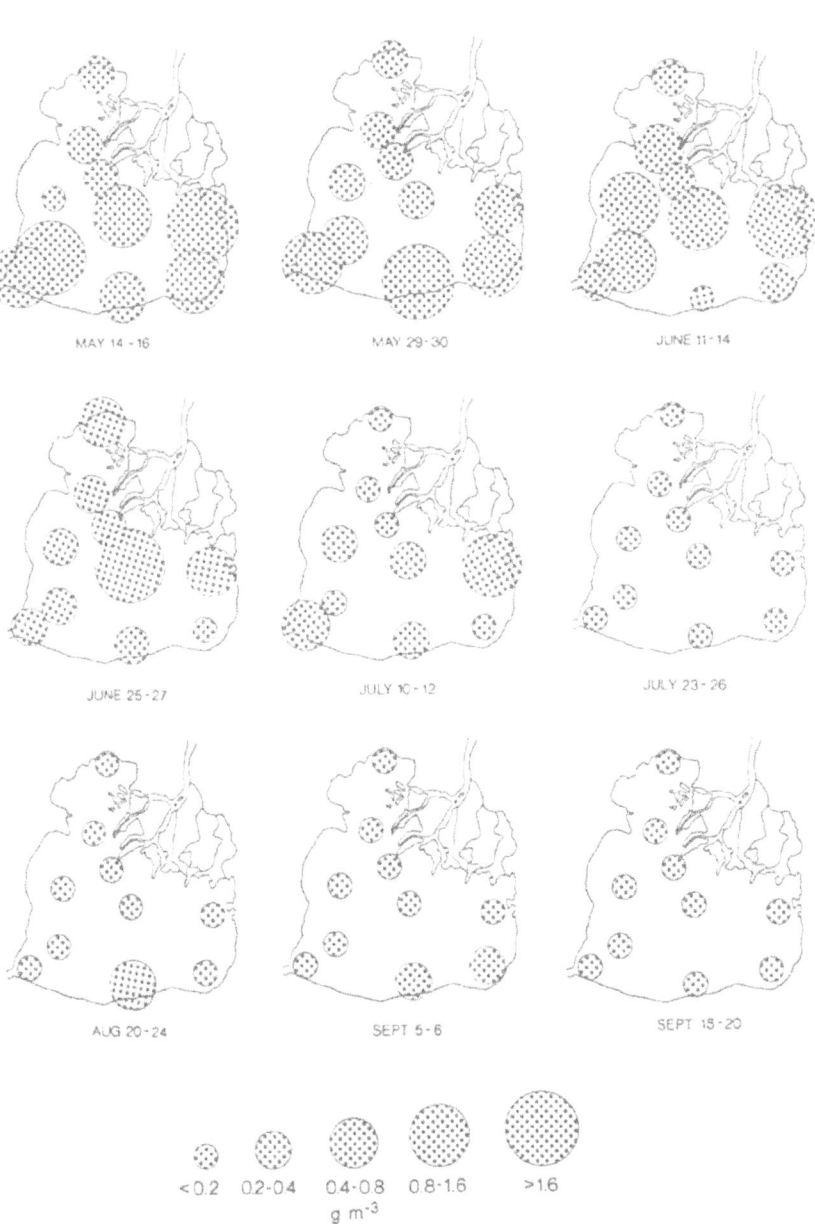

*Fig. 4.* Spatial distribution of Diatomeae in Lake St. Clair, 1984.

*Seasonality*

Seasonal fluctuations of phytoplankton biomass (the mean of 10 stations per cruise) is presented in Fig. 7 and ranged from 0.17 to 1.18 g m$^{-3}$. Relatively high biomass concentrations were recorded during spring (May, June) compared to the summer. Of particular interest is the low representation of Cyanophyta, Chlorophyta, and Dinophyceae. These groups did not play any significant role in the phytoplankton seasonality of

CRYPTOPHYCEAE

*Fig. 5.* Spatial distribution of Cryptophyceae in Lake St. Clair, 1984.

Lake St. Clair. Phytoplankton composition was dominated during the spring by Diatomeae (67 to 90%), while the co-dominant groups were Chrysophyceae (5 to 43%) and Cryptophyceae (3 to 24%) phytoflagellates. During summer (July to September), Diatomeae exhibited variable contributions ranging from 25 to 59 percent, they decreased steadily from July to September being

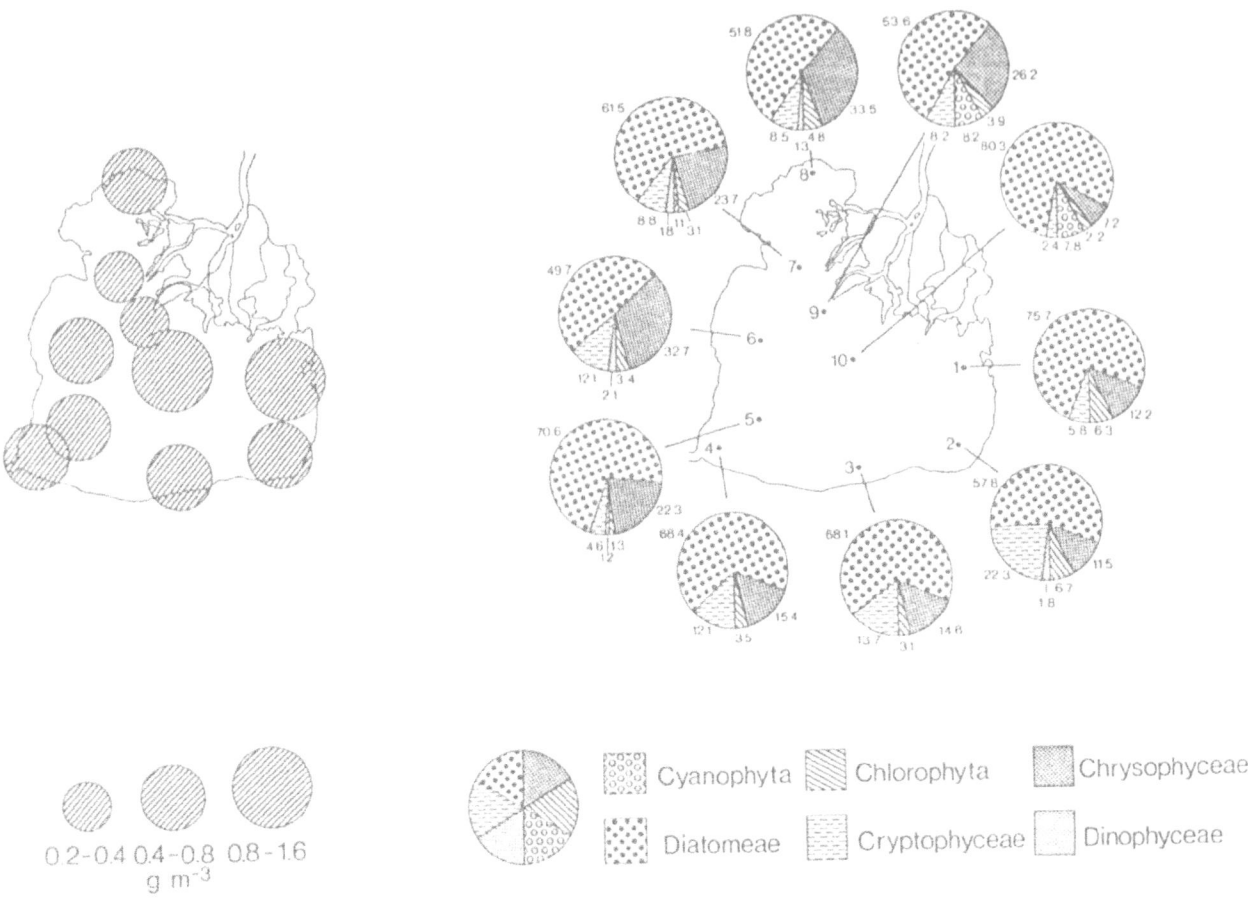

0.2-0.4  0.4-0.8  0.8-1.6
g m⁻³

Cyanophyta    Chlorophyta    Chrysophyceae

Diatomeae    Cryptophyceae    Dinophyceae

*Fig. 6.* Mean phytoplankton biomass and its composition in Lake St. Clair, 1984.

replaced partially by Chrysophyceae and Cryptophyceae. The only increased abundance of Chlorophyta was found during the end of July (24%).

*Seasonal succession of species*
Table 1 lists the species which contributed a least one percent to the total biomass. It is apparent that the largest number of species were represented by Diatomeae being followed by Chrysophyceae and Cryptophyceae. Some of the species were seasonal whereas others were perennial. The seasonality of most common species is shown in Figs. 8 and 9 indicating the dominance of species such as *Ochromonas* spp., *Chromulina minuta* Dolfein, *Cryptomonas erosa* Ehrenberg, *Katablepharis ovalis* Skuja, *Rhodomonas minuta* Skuja, *R. minuta* var. *nannoplanctica* Skuja. Chrysophy-

ceae species such as *Ochromonas* spp., occurred throughout the period of study with increasing contributions from July to September. The most important Cryptophyceae phytoflagellates were *Cryptomonas erosa*, *Rhodomonas minuta*, and *R. minuta* var. *nannoplanctica* (Fig. 8). Amongst the diatoms *Fragilaria crotonensis* Kitton contributed overwhelmingly to the biomass in all samples with a summer peak in July. Other codominant species included *Tabellaria flocculosa* (Roth) Kuetzing (May-July), *Melosira islandica* O. Muller (perennial), *Nitzschia acicularis* W. Smith (May-June) and others (Fig. 9).

*Seasonal composition by size*
The seasonal fluctuations of phytoplankton size assemblages is given in Fig. 10. The spring peak

*Table 1.* Phytoplankton species composition in Lake St. Clair, 1984 ( > 1.0% of biomass); number represents non-weighted mean percent occurence per cruise.

| Species | MY | MY | JN | JN | JY | JY | AG | ST | ST |
|---|---|---|---|---|---|---|---|---|---|
| | ← SPRING → | | | | ← SUMMER → | | | | |
| **Cyanophyta** | | | | | | | | | |
| *Anabaena circinalis* Rabenhorst | | | | 2 | | | | | |
| *Aphanothece gelatinosa* (Henn.) Lemmermann | | | | | | | 3 | | |
| *A. stagnina* (Speng.) A. Braun | | | | | | | | 5 | |
| **Chlorophyta** | | | | | | | | | |
| *Ankistrodesmus* spp. | | | | 1 | | | | | |
| *Chlorella* spp. | | | | | | 14 | | | |
| *Crucigenia tetrapedia* (Kirck.) West & West | | 2 | | | | 2 | | | |
| *Dictyosphaerium pulchellum* Wood | | 1 | | | | | | | |
| *Gloecystis planctonica* (West & West) Lemmermann | | | | | | | 3 | | |
| *Pediastrum simplex* (Meyen) Lemmermann | | | | | | 1 | | | |
| *Planctonema lauterbornii* Schmidle | | | | 3 | | | | | |
| *Westella botryoides* (W. West) de Wildemann | | | | | | | 2 | | |
| **Chrysophyceae** | | | | | | | | | |
| *Chromulina minuta* Doflein | 2 | 8 | 3 | 2 | 2 | 1 | | | |
| *Chrysophaerella rhodel* Skuja | | | | | | | 8 | 15 | 13 |
| *Dinobryon cyclindricum* Imhof | | | 3 | 1 | | | | | |
| *D. sociale* Ehrenberg | | 2 | | | | | | | |
| *Mallomonas pumillo var. canadensis* Holmgren | | | | | | 1 | | | |
| *Ochromonas* spp. | | 4 | 6 | 17 | 19 | 13 | 21 | 26 | 32 |
| *Rhizochrysis* spp. | | | | | | | 2 | 1 | |
| **Diatomeae** | | | | | | | | | |
| *Asterionella formosa* Hassall | | 2 | 1 | 3 | 3 | | | | 1 |
| *A. gracillima* (Hantz.) Heiberg | 2 | 2 | | 1 | | | | | |
| *Cyclotella compta* (Ehr.) Kuetzing | 2 | 2 | | 2 | | | | | |
| *Cymbella ventricosa* (Ehr.) Kuetzing | | 1 | | | | | | | |
| *Fragilaria capucina* Desmazieres | | | | | | 3 | 2 | | |
| *F. construens* (Ehr.) Grunow | 9 | 2 | 17 | 4 | | | | | |
| *F. crotonensis* Kitton | 21 | 18 | 18 | 22 | 38 | 22 | 8 | 6 | 8 |
| *F. intermedia* Grunow | | 2 | | | | | | | |
| *Fragilaria* spp. | | | 1 | | | | | | |
| *Gyrosigma* spp. | | | | | | | 1 | | |
| *Melosira binderana* Kuetzing | 2 | 4 | 9 | | | | | | |
| *M. islandica* O. Muller | 18 | 14 | 4 | 2 | 6 | 7 | 11 | 7 | 3 |
| *M. islandica subsp helvetica* O. Muller | 1 | | | | | | 1 | 5 | 1 |
| *M. granulata* (Ehr.) Ralfs | | | | | | | 1 | | |
| Navicula spp. | | | | | | | | | 2 |
| *Nitzschia acicularis* W. Smith | 8 | 7 | 1 | | | | | | |
| *N. dissipata* (Kuetz.) Grunow | | | | | | | | | 2 |
| *N. palea* (Kuetz.) W. Smith | | 3 | | | | | | | |
| *Rhizosolenia longiseta* Zacharias | 2 | | | | | | | | |
| *Stephanodiscus astraea* (Ehr.) Grunow | 1 | | | 2 | | | | | |
| *Synedra acus* Kuetzing | | | | | | | 1 | 3 | 2 |
| *S. acus var. radians* (Kuetz.) Hustedt | 2 | | | | | | | | 2 |
| *Tabellaria fenestrata* (Lyngb.) Kuetzing | | | | | | | 2 | 4 | 3 |
| *T. flocculosa* (Roth) Kuetzing | 14 | 16 | 12 | 21 | 8 | 4 | | | |

*Table 1.* (Continued)

| Species | SPRING | | | | SUMMER | | | | |
|---|---|---|---|---|---|---|---|---|---|
| | MY | MY | JN | JN | JY | JY | AG | ST | ST |
| **Cryptophyceae** | | | | | | | | | |
| *Cryptomonas erosa* Ehrenberg | | | | 2 | 6 | 5 | 2 | 2 | 3 |
| *C. erosa var. reflexa* Marsonii | | | | | 1 | | | | |
| *C. reflexa* (Marsson) Skuja | | | | | 1 | | | | |
| *C. pusilla* Bachmann | | | 3 | | | | | | |
| *Katablepharis ovalis* Skuja | | 1 | 3 | 2 | | 3 | 2 | 1 | |
| *Rhodomonas minuta* Skuja | | | | | 1 | 1 | 2 | 2 | 2 |
| *R. minuta var. nannoplanctica* Skula | | | | | | 2 | 10 | 5 | 11 |
| **Dinophyceae** | | | | | | | | | |
| *Gymnodinium ordinatum* Skuja | | | | | | 1 | | | |
| *G. varians* Maskell | | | | | | | | 1 | |
| *Glenodinium borgei* (Lemm.) Schiller | | | | | | | | 3 | 3 |
| *G. pascheri* Suchlandt | | | | | | | 2 | | 2 |

Note: MY - May, JN - June, JY - July, AG - August, ST - September

was mainly comprised of microplankton/netplankton ($> 20 \mu m$) ranging from 49 to 76 percent. On the other hand, the late summer community was dominated by ultraplankton ($< 20 \mu m$) with contributions of 43 to 81 percent to the total biomass. The contribution of the picoplankton ($< 2 \mu m$) was quite insignificant (0.1 to 1.4%) and has been incorporated for the sake of convenience in the ultraplankton biomass.

*Biomass-nutrient-contaminant interactions*
The seasonal fluctuations of temperature and nutrients is given in Fig. 11. The Soluble Reactive Phosphorus (SRP) and Total Phosphorus showed considerable variability in their concentration whereas $NO_3$, $NO_2$, and $NH_4$ were more or less constant. The $SiO_2$ was relatively low and constant during the spring and early summer but very high values were recorded during both the collections of September (Fig. 11). The seasonal variations of metals is given in Table 2 which indicated some fluctuations in the concentrations of Cu, Cr, Ni, Zn, Fe, and Mn.

The results of the factor analysis are shown for various taxonomic groups, nutrients, metals as well as temperature in Table 3. The factor analy-

*Table 2.* Mean concentration of heavy metals in water (mg $l^{-1}$) of Lake St. Clair, 1984.

| Metal | MY 29–30 | JN 11–14 | JN 25–27 | JY 10–12 | JY 23–26 | AG 20–24 | ST 5–6 | ST 18–20 |
|---|---|---|---|---|---|---|---|---|
| Cd | <0.001 | 0.002 | 0.002 | <0.001 | <0.001 | <0.001 | <0.001 | <0.001 |
| Cu | 0.005 | 0.002 | 0.002 | 0.003 | 0.003 | 0.026 | 0.003 | 0.005 |
| Co | 0.001 | 0.001 | <0.001 | <0.001 | <0.001 | <0.001 | <0.001 | <0.001 |
| Cr | 0.002 | <0.001 | 0.001 | 0.017 | 0.003 | <0.001 | <0.001 | 0.001 |
| Pb | 0.002 | 0.002 | 0.001 | 0.007 | 0.004 | <0.001 | <0.001 | 0.002 |
| Ni | 0.003 | 0.002 | 0.002 | 0.016 | 0.001 | <0.001 | <0.001 | 0.001 |
| Zn | 0.013 | 0.007 | 0.002 | 0.137 | 0.005 | 0.006 | 0.004 | 0.006 |
| Fe | 3.870 | 0.227 | 0.009 | 0.538 | 0.224 | 0.261 | 0.227 | 0.198 |
| Mn | 0.023 | 0.010 | 0.415 | 0.018 | <0.010 | 0.009 | <0.005 | 0.009 |

Note: MY - May, JN - June, JY - July, AG - August, ST - September

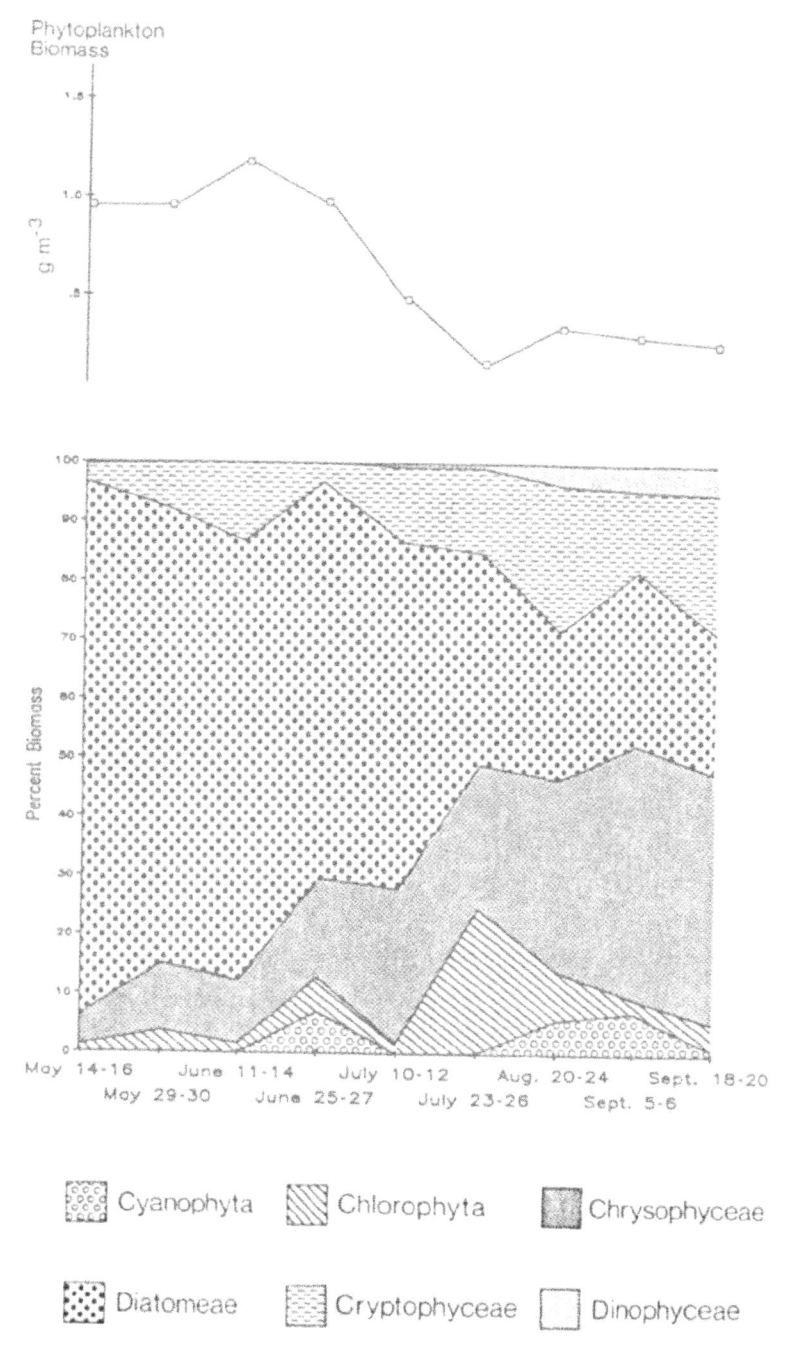

*Fig. 7.* Seasonality of phytoplankton biomass and its composition in Lake St. Clair, 1984.

sis was made in three categories, namely whole year data, spring, and summer separately. Considering all the data together, the first factor was heavily loaded with only abiotic factors and no biotic factor emerged in this group. The second factor contained Diatomeae, Dinophyceae, $SiO_2$,

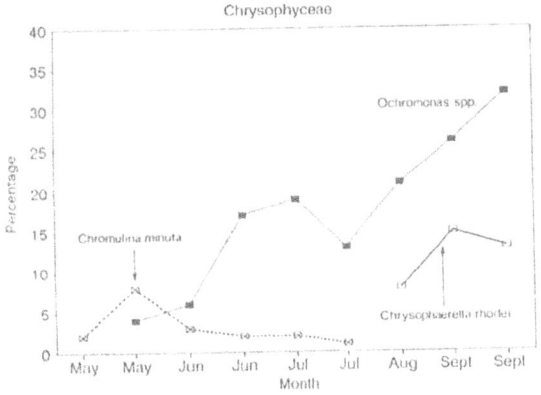
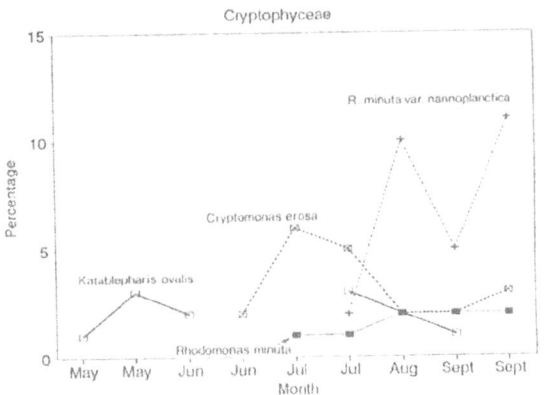

*Fig. 8.* Species succession of Chrysophyceae and Cryptophyceae in Lake St. Clair, 1984.

and Cu. The positive correlation of silica and diatoms is not surprising since silica never depleted to extremely low or limiting concentrations, remaining well above $1.0 \, \text{mg} \, l^{-1}$. The positive correlation with Cu is enigmatic since Cu is known to be toxic. Its concentration remained more or less unchanged ranging from 0.002 to $0.005 \, \text{mg} \, l^{-1}$ although it showed highest value of $0.026 \, \text{mg} \, l^{-1}$ once during late August. The third factor indicated negative correlation between

*Table 3.* Varimax rotated factor matrix of 18 variables in Lake St. Clair, 1984.

| Season | Whole Season ($n = 20$) | | | Spring ($n = 16$) | | | Summer ($n = 20$) | | | |
|---|---|---|---|---|---|---|---|---|---|---|
| Factor | 1 | 2 | 3 | 1 | 2 | 3 | 1 | 2 | 3 | 4 |
| Variance | 3.693 | 2.459 | 2.02 | 5.196 | 2.454 | 1.758 | 4.248 | 2.83 | 2.658 | 1.695 |
| % of Total Variance | 21.7 | 14.5 | 11.9 | 32.5 | 15.4 | 11 | 25 | 16.6 | 15.6 | 10 |
| Factor | COMPONENT LOADING | | | | | | | | | |
| Cyanophyta: Cyan | −0.446 | 0.355 | −0.199 | | | | −0.360 | 0.733 | 0.043 | −0.082 |
| Chlorophyta: Chlor | 0.460 | −0.059 | 0.255 | 0.201 | 0.363 | −0.061 | 0.578 | −0.048 | 0.012 | −0.531 |
| Chrysophyceae: Chrys | −0.024 | −0.424 | −0.531 | −0.441 | −0.540 | 0.185 | −0.283 | 0.532 | 0.300 | 0.324 |
| Diatomeae: Diat | 0.379 | −0.733 | −0.225 | −0.342 | 0.567 | −0.340 | 0.274 | 0.425 | 0.627 | −0.177 |
| Cryptophyceae: Cryp | −0.069 | 0.325 | 0.259 | 0.015 | 0.076 | 0.913 | 0.311 | 0.706 | −0.370 | −0.341 |
| Dinophyceae: Dino | −0.472 | 0.593 | 0.185 | | | | −0.299 | 0.066 | −0.483 | 0.509 |
| Temp: Temperature | −0.191 | 0.018 | 0.858 | −0.509 | 0.630 | 0.338 | 0.540 | −0.527 | −0.403 | −0.231 |
| DOC: Dissolved Organic Carbon | 0.451 | 0.397 | 0.044 | 0.565 | −0.016 | −0.481 | 0.417 | −0.075 | −0.002 | 0.540 |
| NO₂/NO₃ | 0.536 | −0.139 | 0.433 | 0.113 | 0.692 | 0.086 | 0.638 | 0.358 | −0.436 | 0.253 |
| NH₄ | 0.272 | −0.272 | 0.253 | −0.122 | 0.080 | −0.156 | 0.293 | 0.250 | −0.775 | 0.323 |
| SRP: Soluble Reactive Phosphorus | 0.666 | 0.384 | 0.257 | 0.766 | 0.266 | 0.086 | 0.811 | 0.103 | −0.251 | −0.108 |
| TP: Total Phosphorus | 0.577 | 0.036 | −0.135 | 0.730 | 0.250 | 0.180 | 0.294 | 0.555 | −0.346 | −0.149 |
| SiO₂ | −0.207 | 0.563 | −0.473 | 0.601 | −0.264 | 0.497 | −0.493 | 0.618 | 0.118 | 0.092 |
| Cu | 0.360 | 0.562 | −0.220 | 0.853 | −0.149 | −0.074 | 0.170 | 0.218 | 0.145 | −0.267 |
| Ni | 0.699 | 0.066 | −0.038 | 0.834 | 0.196 | 0.152 | 0.614 | −0.166 | 0.472 | 0.402 |
| Zn | 0.556 | 0.192 | 0.007 | 0.620 | −0.413 | −0.233 | 0.717 | 0.094 | 0.399 | 0.309 |
| Fe | 0.748 | 0.284 | −0.367 | 0.785 | −0.244 | 0.055 | 0.760 | 0.257 | 0.499 | 0.095 |
| Pb | | | | 0.564 | 0.573 | −0.144 | | | | |

| Factor Interpretation | | | | | | | | | | |
|---|---|---|---|---|---|---|---|---|---|---|
| Variables with Factor Loading > .500 | NO₂/NO₃ SRP TP Ni Zn Fe | -Diat Dino SiO₂ Cu | Temp -Chrys | -Temp DOC SRP TP SiO₂ Cu Ni Zn Fe Pb | Temp -Chrys Diat NO₂/NO₃ Pb | Cryp | Temp Chlor NO₂/NO₃ SRP Ni Zn Fe | -Temp Cyano Chrys Cryp TP SiO₂ | Diat -NH₄ | -Chlor Dino DOC |

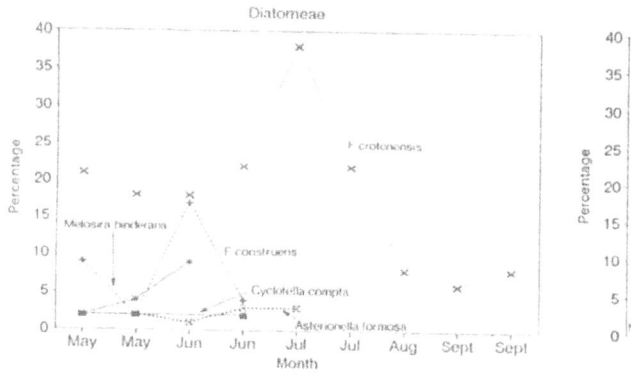
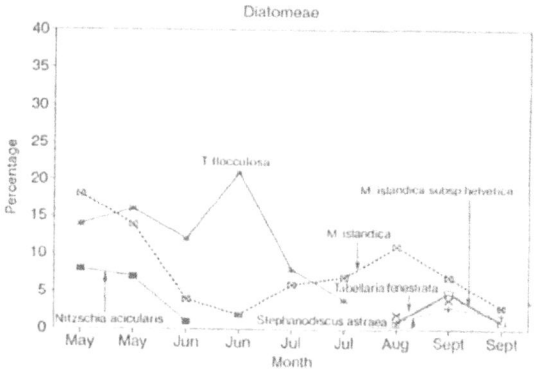

*Fig. 9.* Species succession of Diatomeae in Lake St. Clair, 1984.

Chrysophyceae and temperature. When the data was divided into seasons some interesting correlations were observed. For example, during the spring the second factor (explaining 11% of variance) was loaded with a combination of biotic and abiotic factors. Positive correlation between diatoms and $NO_2/NO_3$, Pb, and temperature but negative correlation with Chrysophyceae was observed. The Chrysophyceae revealed negative correlation with the temperature. Statistical analysis of the summer data was most interesting. The first factor (explaining 25% of variance) was

*Table 4.* Varimax rotated factor matrix of 18 variables in Lake St. Clair, 1984.

| Season | Whole Season (n = 35) | | | | Spring (n = 16) | | | Summer (n = 19) | | |
|---|---|---|---|---|---|---|---|---|---|---|
| Factor | 1 | 2 | 3 | 4 | 1 | 2 | 3 | 1 | 2 | 3 |
| Variance | 3.374 | 2.452 | 1.802 | 1.441 | 5.213 | 2.465 | 1.649 | 4.197 | 2.237 | 2.036 |
| % of Total Variance | 24.1 | 17.5 | 12.9 | 10.3 | 37.2 | 17.6 | 11.8 | 30 | 16 | 14.5 |
| Factor | COMPONENT LOADING | | | | | | | | | |
| Size <2 μm | 0.343 | 0.026 | 0.274 | 0.388 | | | | 0.571 | 0.385 | -0.119 |
| Size 2-20 μm | 0.100 | -0.681 | -0.454 | 0.013 | -0.149 | 0.823 | -0.280 | -0.123 | 0.719 | 0.201 |
| Size >20 μm | 0.058 | -0.820 | -0.118 | -0.359 | -0.606 | 0.416 | -0.305 | 0.574 | 0.368 | 0.383 |
| Temp | -0.219 | 0.003 | 0.843 | 0.268 | -0.507 | 0.317 | 0.743 | 0.391 | -0.525 | -0.567 |
| DOC | 0.541 | 0.270 | 0.129 | 0.028 | 0.583 | 0.136 | -0.177 | 0.407 | -0.151 | 0.030 |
| $NO_2/NO_3$ | 0.520 | -0.588 | 0.283 | 0.252 | 0.023 | 0.779 | 0.173 | 0.734 | 0.278 | -0.363 |
| $NH_4$ | 0.251 | -0.678 | 0.002 | 0.268 | -0.165 | 0.414 | -0.580 | 0.386 | 0.308 | -0.643 |
| SRP | 0.749 | 0.135 | 0.243 | 0.304 | 0.782 | 0.315 | 0.191 | 0.794 | -0.027 | -0.322 |
| TP | 0.582 | -0.167 | -0.289 | 0.419 | 0.717 | 0.112 | 0.298 | 0.352 | 0.550 | -0.330 |
| $SiO_2$ | 0.010 | 0.477 | -0.518 | 0.359 | 0.629 | -0.353 | 0.201 | -0.319 | 0.672 | 0.243 |
| Cu | 0.497 | 0.324 | -0.273 | -0.063 | 0.829 | -0.143 | -0.386 | 0.153 | 0.123 | 0.268 |
| Ni | 0.645 | 0.074 | 0.156 | -0.508 | 0.819 | 0.249 | 0.190 | 0.604 | -0.399 | 0.453 |
| Zn | 0.619 | 0.081 | 0.313 | -0.538 | 0.627 | -0.196 | -0.248 | 0.792 | -0.193 | 0.454 |
| Fe | 0.789 | 0.216 | -0.305 | -0.067 | 0.818 | -0.118 | -0.193 | 0.805 | 0.024 | 0.462 |
| Pb | | | | | 0.507 | 0.589 | 0.279 | | | |
| Factor Interpretation | | | | | | | | | | |
| Variables with Factor Loading > .500 | DOC $NO_2/NO_3$ SRP TP Ni Zn Fe | -2-20 ->20 -$NO_2/NO_3$ -$NH_4$ | Temp -$SiO_2$ | -Ni -Zn | -Temp ->20 DOC SRP TP $SiO_2$ Cu Ni Zn Fe Pb | 2-20 $NO_2/NO_3$ Pb | Temp -$NH_4$ | <2 >20 $NO_2/NO_3$ SRP Ni Zn Fe | -Temp 2-20 TP $SiO_2$ | -Temp -$NH_4$ |

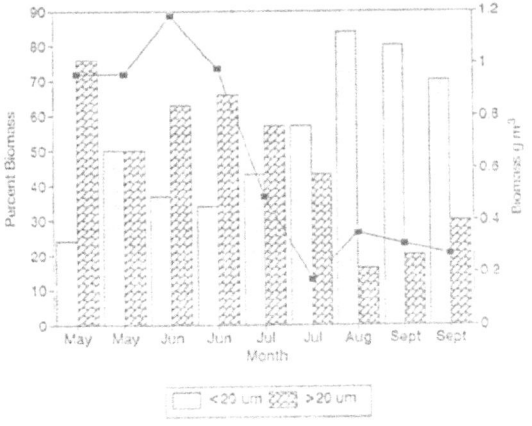

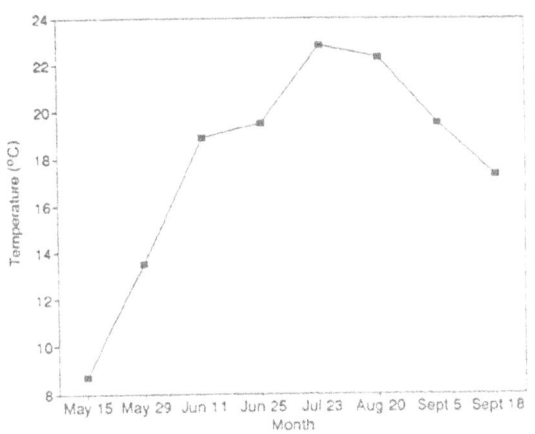

Fig. 10. Seasonal fluctuations of phytoplankton biomass and its size composition in Lake St. Clair, 1984.

heavily loaded with variables such as Chlorophyta, temperature, $NO_2/NO_3$, SRP, NI, Zn, and Fe. This suggests that the Chlorophyta population increased when the temperature and the concentrations of the above-mentioned nutrients and heavy metals were elevated. This correlation indicates an unusual adaptation of Chlorophyta to increasing contaminants like metals. Since the lake is located downstream from a major industrial centre on St. Clair River, such an adaptation is ecologically possible.

The results of factor analysis of phytoplankton biomass fractionated into size categories with temperature, nutrients, and metals is presented in Table 4. When the whole-year data was considered, the second factor showed negative correlation of ultraplankton (2 to 20 $\mu$m) and microplankton/netplankton with $NO_2/NO_3$ and $NH_4$. During the spring the first factor was heavily loaded, revealing positive correlation of microplankton/netplankton with several chemical variables but negative correlation with temperature. Since the spring phytoplankton was predominantly diatoms, it can be inferred that large-sized diatoms play a key role in the springtime ecology of Lake St. Clair phytoplankton. Similarly, the summer factor analysis indicated that the picoplankton ($<2$ $\mu$m) and microplankton/netplankton were positively correlated with nutrients like $NO_2/NO_3$, SRP and contaminants such as

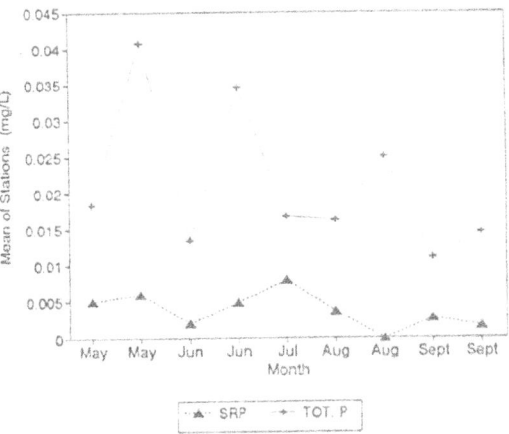

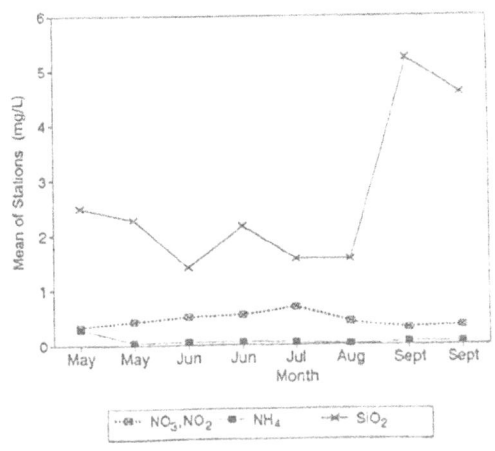

Fig. 11. Seasonal fluctuations of temperature and nutrients in Lake St. Clair, 1984.

217

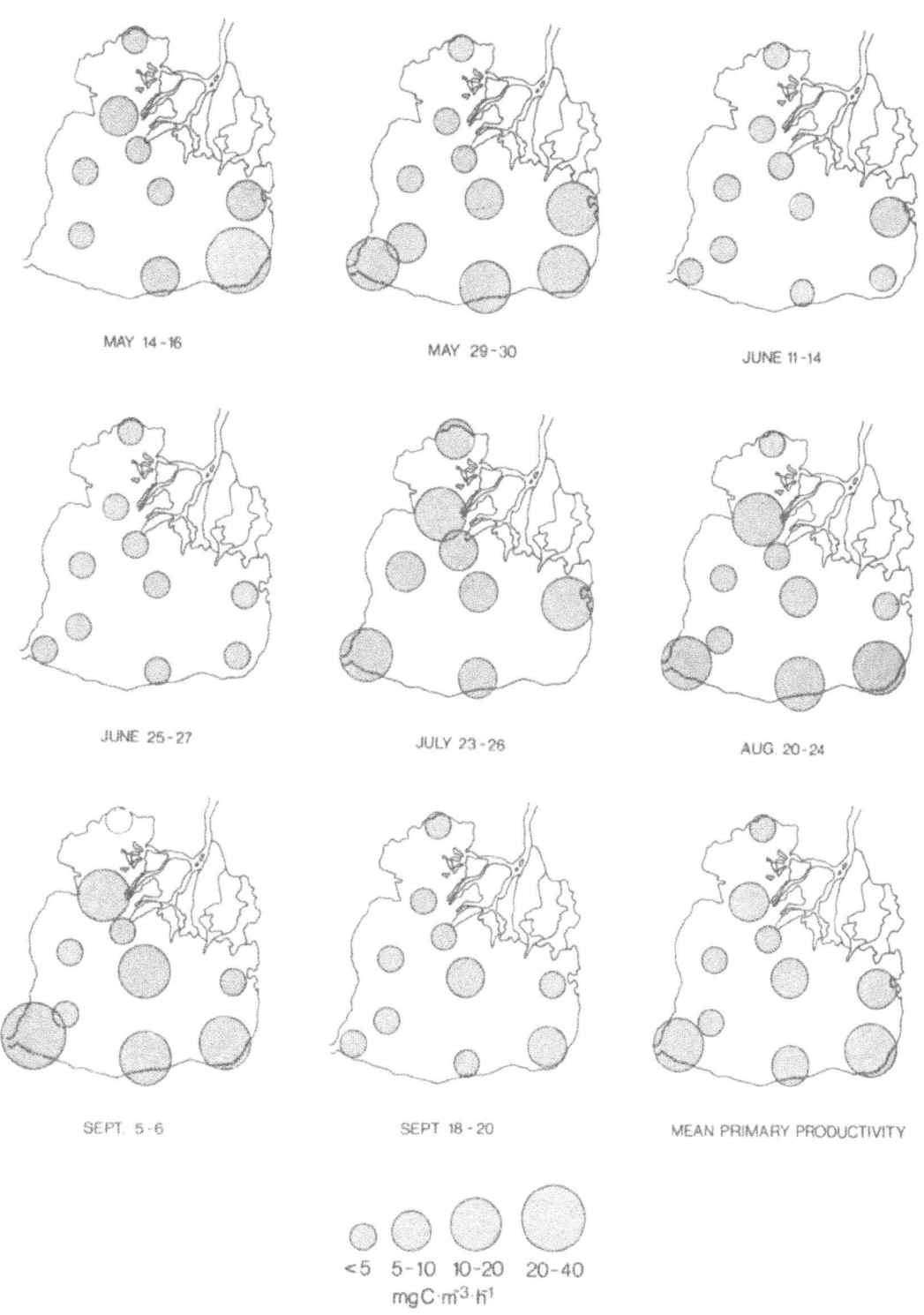

MAY 14-16     MAY 29-30     JUNE 11-14

JUNE 25-27     JULY 23-26     AUG. 20-24

SEPT. 5-6     SEPT. 18-20     MEAN PRIMARY PRODUCTIVITY

<5   5-10   10-20   20-40
$mgC \cdot m^{-3} \cdot h^{-1}$

*Fig. 12.* Spatial distribution of primary productivity in Lake St. Clair, 1984.

Ni, Zn, and Fe. Once again this frequently observed positive correlation of biotic factors with contaminants like heavy metals is suggestive of physiological adaptation to contamination originating from the chemical valley located upstream in the industrial complex.

*Primary productivity*

*Size fractionation*

The spatial distribution of primary productivity is shown in Fig. 12. The productivity rate showed little variability compared to the biomass, but with minor exceptions. For example, a higher rate was observed during May in the southern and southeastern part of the lake. Relatively high productivity rates were recorded during the summer (late July to early September). Usually the offshore or midlake stations exhibited lower rates than the nearshore areas.

The spatial composition of size-fractionated primary productivity is presented in Figs. 13 and 14. During the spring (May/June) both the ultraplankton (2 to 20 $\mu$m) as well as microplankton/netplankton ($>20$ $\mu$m) contributed more or less equally to the primary productivity with variable percent contribution from station to station. The summer productivity was overwhelmingly dominated by ultraplankton with the exception of Station 3 during July in which the microplankton/netplankton contributed about 71 percent to the primary productivity (Fig. 10). On the whole, the contribution by the picoplankton ($<2$ $\mu$m) to the primary productivity was minor in both the seasons.

The seasonal trends of size-fractionated primary productivity is shown in Fig. 15 which reveal an overall dominance of ultraplankton. Early spring (May) and summer (July and September) peaks of productivity were observed. The spring peak was mainly made up of ultraplankton (54 to 62%). The summer peaks were also dominated by ultraplankton by contributing over 75 percent to the total productivity. The picoplankton share of the productivity was very low (2 to 7%) and showed no seasonal trends.

*Activity Coefficient (P/B: Production/Biomass)*

Activity Coefficient (mg C mg$^{-1}$ h$^{-1}$) was low during the spring when diatoms and microplankton/netplankton dominated (Fig. 15). During this period, species like *Fragilaria crotonensis*, *F. construens*, *Tabellaria flocculosa*, *T. fenestrata*, *Stephanodiscus astraea* and *Nitzschia acicularis* dominated. Relatively high Activity Coefficients were observed during the summer. Two peaks were recorded during late July and early September. The phytoplankton biomass was the lowest during July and was mainly composed of a mixture of Diatomeae, Chrysophyceae, Cryptophyceae and Chlorophyta. Of the three size fractions, the ultraplankton/picoplankton ($<20$ $\mu$m) were most prevalent. The species composition was made up of *Ochromonas* spp., *Cryptomonas erosa* and *Fragilaria crotonensis*. The second Activity Coefficient pulse of September was dominated by Chrysophyceae, Cryptophyceae and Diatomeae with overwhelming prevalence of ultraplankton/picoplankton. The September pulse was mainly comprised of *Ochromonas* spp., *Chrysochromulina parva*, *Rhodomonas minuta*, *R. minuta* var. *nannoplanktica*, *Cryptomonas erosa*, *Fragilaria crotonensis*, *Stephanodiscus astraea*, *Melosira islandica* subsp. *helvetica* and *M. islandica*.

*Phytoplankton–Zooplankton interactions*

Figure 16 provides a comparative picture of the phytoplankton and zooplankton biomass (see Sprules & Munawar, 1990 for details). The zooplankton is briefly discussed here to examine possible phytoplankton and zooplankton relationships in Lake St. Clair. The comparison is made only for those stations where zooplankton were analyzed, since the zooplankton data were not as extensive as the phytoplankton data. It is apparent that the zooplankton biomass was at its height during late June, followed by a decreasing trend from early July through late August and mid-September. The phytoplankton biomass also showed a similar high to low decreasing trend for the same period. The June zooplankton peak was mainly composed of Cladocerans, Calanoids, and rotifers. During this period the phytoplankton

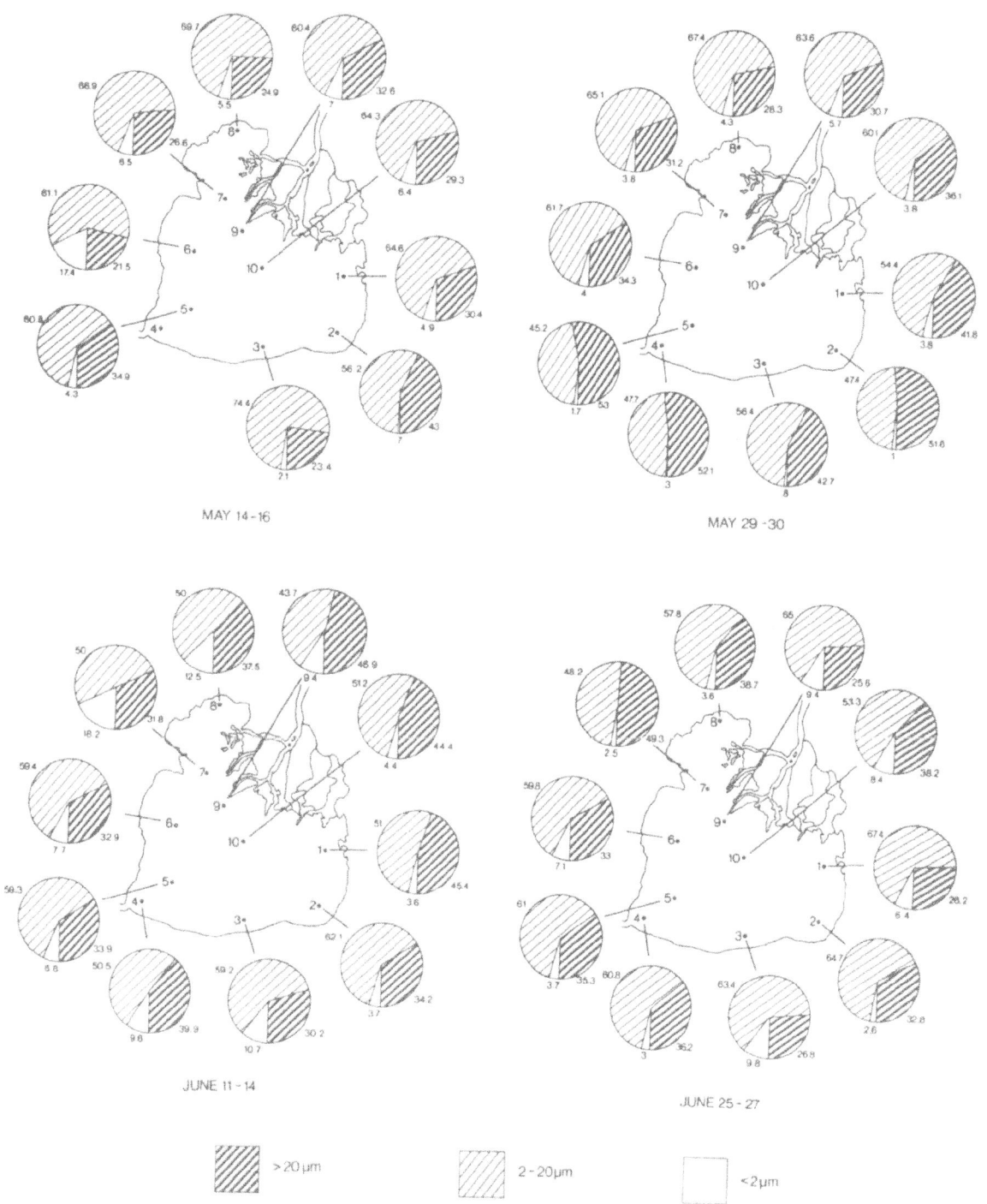

Fig. 13. Spatial distribution of size fractionated primary productivity during spring. Data given as percent combination by various size assemblages.

220

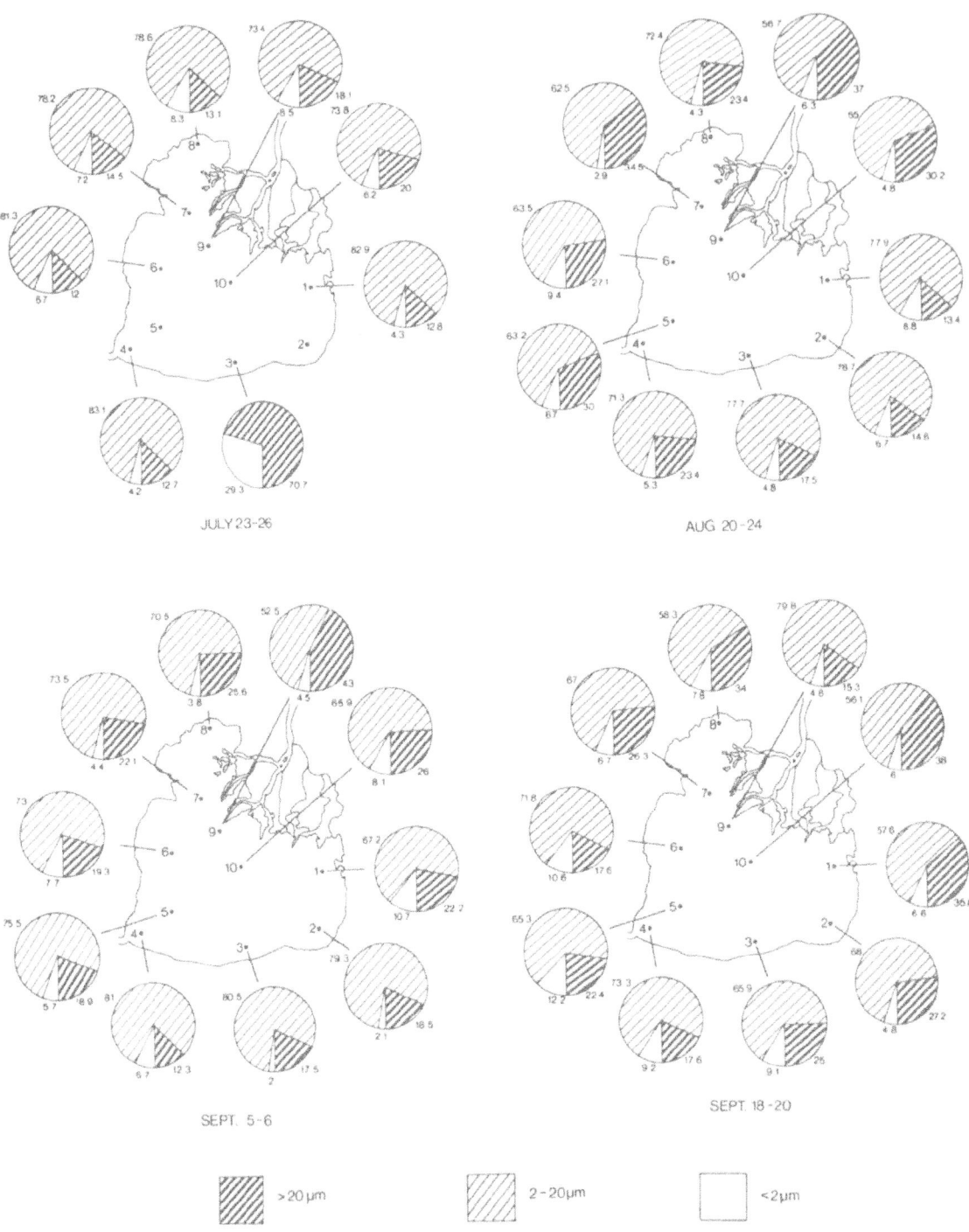

Fig. 14. Spatial distribution of size fractionated primary productivity during summer. Data given as percent combination by various size assemblages.

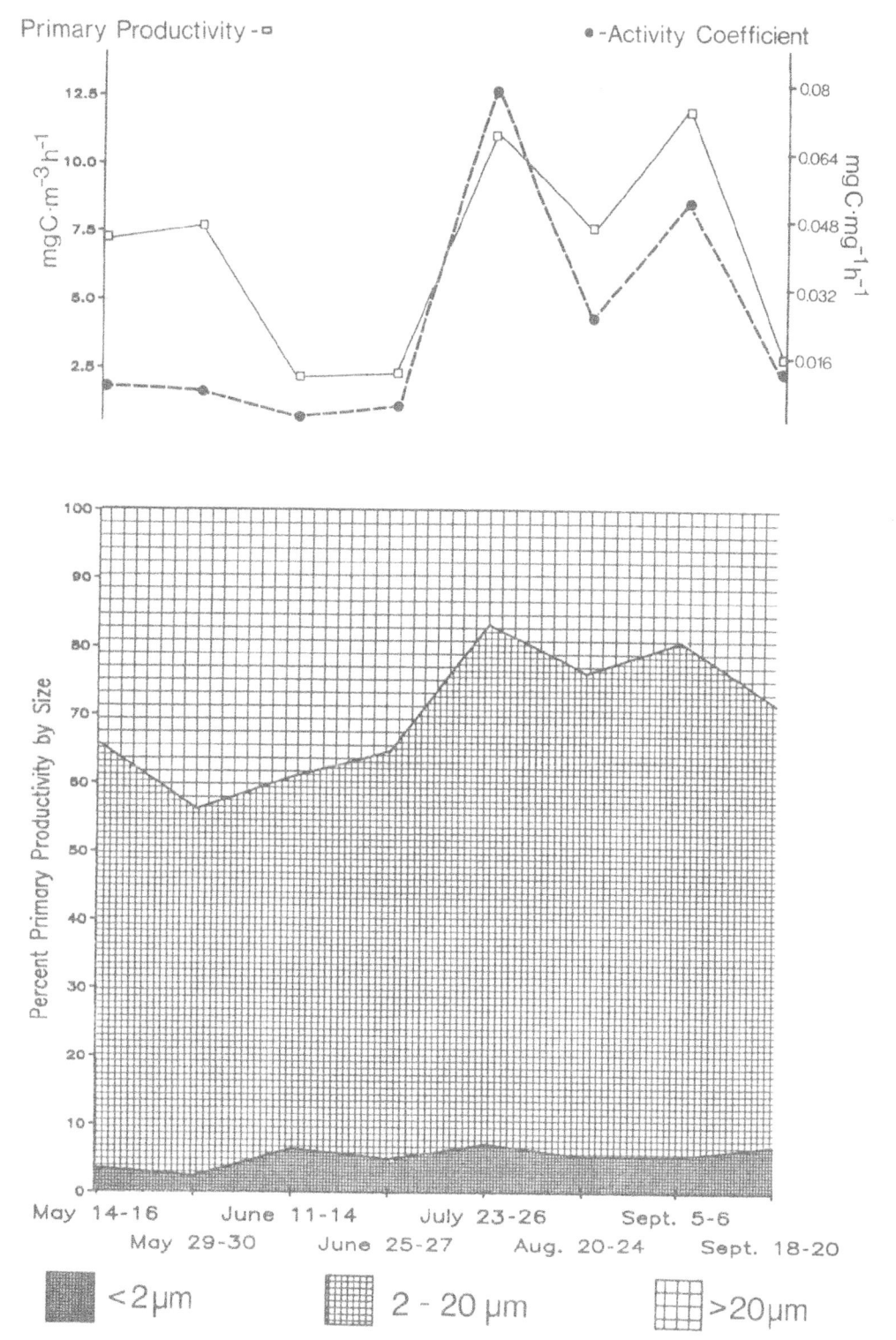

*Fig. 15.* Seasonal fluctuations of activity coefficient (Production/Biomass), primary productivity and contributions by various size assemblages.

222

was dominated by microplankton/netplankton composed mainly of diatoms.

The mean zooplankton biomass ($\mu g\, l^{-1}$) in Lake St. Clair ranged from 173.0 to 1306.0. Sprules & Munawar (1991) report that mean seasonal zooplankton biomass in Lakes Ontario, Erie, Huron, and Superior, were 601, 1070, 484, 33 respectively. Thus, Lake St. Clair mean zooplankton biomass (663) is among the highest in the Great Lakes.

Of the five major groups of zooplankton, cladocerans were dominant numerically. Bosminids constituted the majority of cladoceran numbers and biomass (Sprules & Munawar, 1991). *Holopedium* and *Leptodora* were never very abundant. On a biomass basis, rotifers are of minor importance although numerically they were important.

Calanoids are generally more or less equal to cyclopoids in numbers and biomass, with *Epischura lacustris* and *Diaptomus* spp. constituting the majority of these calanoid copepods. The cold stenothermic species of *Mysis*, and *Senecella*, were not recorded in Lake St. Clair (Sprules & Munawar, 1991).

The statistical analysis of phytoplankton and zooplankton interactions revealed some interest-ing correlations (Tables 5 and 6). On a yearly basis, the first factor was mainly loaded by the biotic factors alone. Diatoms were positively correlated with cladocerans, calanoids, and cyclopoids, whereas Dinophyceae indicated negative relationship with these variables. The third factor exhibited a negative correlation of Chlorophyta to rotifers.

When the seasonal interactions are evaluated, more interesting relationships were apparent. During the spring, the first factor was predominantly influenced by various phytoplankton groups. However, the second factor indicated that the fluctuations of Chlorophyta were negatively related to those of cladocerans, calanoids, and rotifers. The third factor continued to show negative correlation of Chlorophyta to rotifers but its fluctuations were similar to those of cyclopoids.

Most dynamic interactions between various components of the plankton were observed during the summer and they appear to influence the overall results of the statistical analyses. The relative abundance of cladocerans, calanoids, and cyclopoids (Fig. 16) decreased during the summer, thereby reducing the competition and predation pressure on rotifers. This seems to result in an increase of rotifers during September.

*Table 5.* Varimax rotated factor of 10 variables in Lake St. Clair, 1984.

| Season | Whole Season ($n = 22$) | | | | Spring ($n = 7$) | | | Summer ($n = 15$) | | | |
|---|---|---|---|---|---|---|---|---|---|---|---|
| Component | 1 | 2 | 3 | 4 | 1 | 2 | 3 | 1 | 2 | 3 | 4 |
| Variance | 3.846 | 1.902 | 1.34 | 1.017 | 3.294 | 2.923 | 1.615 | 3.733 | 1.884 | 1.345 | 1.173 |
| % of Total Variance | 38.5 | 19 | 13.4 | 10.2 | 36.6 | 32.5 | 17.9 | 37.3 | 18.8 | 13.4 | 11.7 |
| Factor | COMPONENT LOADING | | | | | | | | | | |
| Cyanophyta: Cyan | −0.370 | 0.727 | 0.403 | 0.133 | 0.791 | 0.185 | 0.348 | −0.749 | 0.450 | 0.343 | 0.193 |
| Chlorophyta: Chlor | −0.287 | 0.500 | −0.577 | −0.210 | 0.314 | −0.728 | −0.508 | −0.414 | 0.418 | −0.561 | −0.357 |
| Chrysophyceae: Chrys | 0.143 | −0.628 | 0.164 | −0.700 | −0.836 | −0.061 | 0.528 | 0.049 | −0.630 | −0.518 | 0.451 |
| Diatomeae: Diat | 0.765 | 0.225 | −0.326 | −0.331 | 0.777 | −0.464 | 0.356 | 0.765 | 0.002 | −0.500 | 0.004 |
| Cryptophyceae: Cryp | −0.406 | −0.676 | 0.230 | 0.219 | −0.908 | 0.317 | −0.211 | 0.035 | 0.357 | −0.051 | 0.868 |
| Dinophyceae: Dino | −0.650 | −0.248 | −0.132 | 0.258 | | | | −0.062 | −0.686 | 0.526 | 0.066 |
| Cladocerans: Clado | 0.860 | −0.101 | 0.075 | 0.245 | −0.013 | 0.896 | −0.291 | 0.770 | 0.352 | 0.055 | 0.137 |
| Calanoids: Cala | 0.930 | −0.036 | 0.080 | 0.197 | 0.469 | 0.805 | 0.061 | 0.875 | 0.233 | 0.225 | 0.000 |
| Cyclopoids: Cyclo | 0.911 | 0.013 | 0.080 | 0.228 | 0.443 | 0.491 | −0.606 | 0.761 | 0.393 | 0.235 | −0.076 |
| Rotifers: Roti | 0.006 | 0.385 | 0.789 | −0.300 | 0.162 | 0.590 | 0.575 | −0.686 | 0.422 | −0.071 | 0.150 |
| Factor Interpretation | | | | | | | | | | | |
| Variables with Factor Loading > 0.500 | Diat -Dino Clado Cala Cyclo | Cyan Chlor -Chrys -Cryp | -Chlor Roti | -Chrys | Cyan -Chrys Diat -Cryp | -Chlor Clado Cala Roti | -Chlor Chrys -Cyclo Roti | -Cyan Diat Clado Cala Cyclo -Roti | -Chrys -Dino | -Chlor -Chrys -Diat Dino | Cryp |

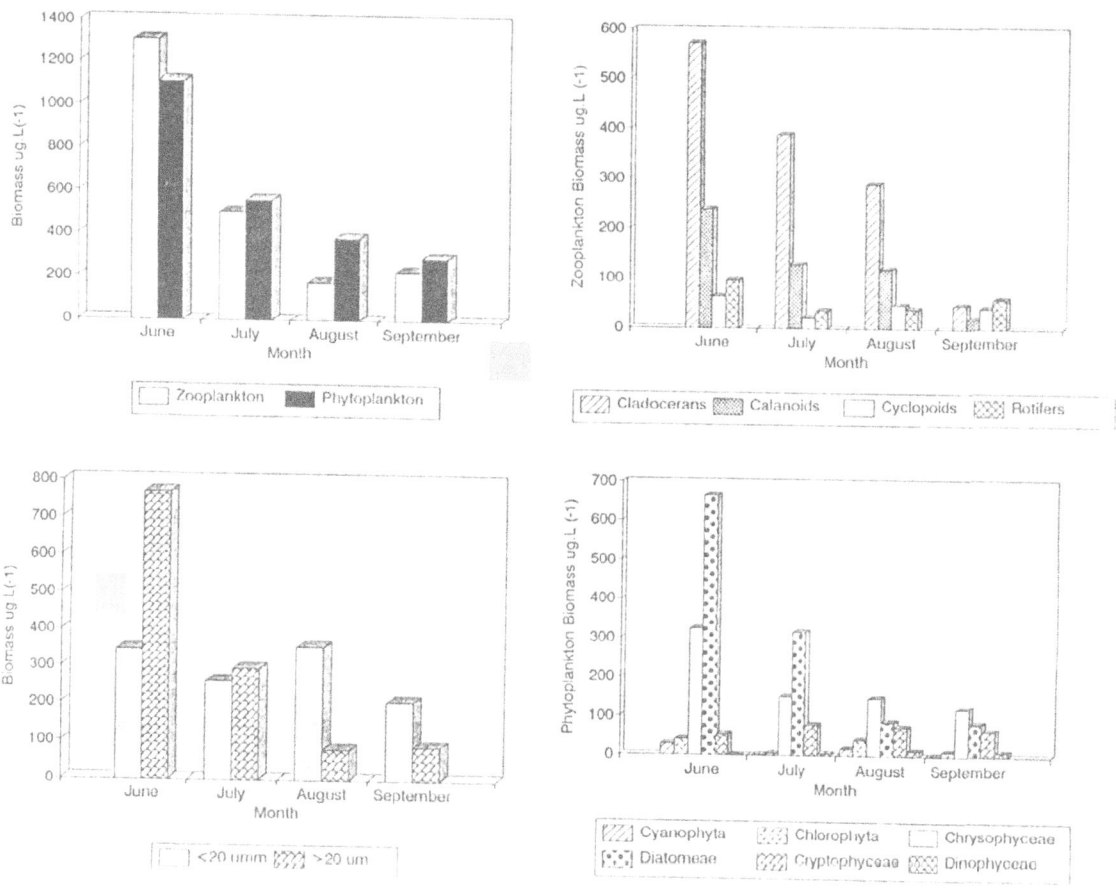

*Fig. 16.* Mean phytoplankton and zooplankton biomass and composition from June to September, 1984.

Since rotifers preferred food is small-sized Cyanophyta, a similar correlation was observed between them. The cladocerans, calanoids, and cyclopoids showed a positive correlation with diatoms and negative correlation with Cyanophyta.

The factor analysis of zooplankton with phytoplankton size categories for the whole year revealed negative correlation of picoplankton and positive correlation of microplankton/netplankton with cladocerans, calanoids, and cyclopoids (Table 6). Furthermore, the rotifers were found to be positively correlated to both ultraplankton (2–20 $\mu$m) and microplankton/netplankton size assemblages. This observed correlation of microplankton/netplankton appear to be indirect and

not of ecological significance. For example, it is well known that zooplankton feed on the bacterial growth associated with the abundance and decay of larger plankton which are not suitable for grazing. The second factor was loaded with the rotifers, ultraplankton and once again microplankton/netplankton.

Seasonally, similar positive correlations of rotifers, ultraplankton, and microplankton/netplankton was apparent which appear to have influenced yearly results. During the summer, the cladocerans, calanoids, and cyclopoids showed negative correlation with picoplankton but positive correlation with microplankton/netplankton. Rotifers and picoplankton revealed similar relationships during the summer.

*Table 6.* Varimax rotated factor matrix of 7 variables in Lake St. Clair, 1984.

| Season | Whole Season (n = 22) | | | Spring (n = 7) | | | Summer (n = 15) | | |
|---|---|---|---|---|---|---|---|---|---|
| Factor | 1 | 2 | 3 | 1 | 2 | 3 | 1 | 2 | 3 |
| Variance | 3.213 | 1.217 | 1.002 | 2.458 | 1.86 | 1.004 | 3.349 | 1.141 | 0.847 |
| % of Total Variance | 45.9 | 17.4 | 14.3 | 41 | 31 | 16.7 | 47.8 | 16.3 | 12.1 |
| Factor | COMPONENT LOADING | | | | | | | | |
| Size $<2$ $\mu$m | −0.667 | 0.121 | 0.190 | | | | −0.602 | 0.230 | 0.440 |
| Size 2-20 $\mu$m | 0.255 | 0.560 | −0.652 | −0.439 | 0.599 | 0.663 | −0.119 | −0.769 | 0.596 |
| Size $>20$ $\mu$m | 0.519 | 0.612 | −0.097 | 0.230 | 0.928 | −0.222 | 0.725 | −0.405 | −0.219 |
| Cladocerans: Clado | 0.870 | −0.137 | 0.155 | 0.949 | −0.096 | −0.058 | 0.803 | 0.111 | 0.249 |
| Calanoids: Cala | 0.942 | −0.120 | 0.042 | 0.744 | 0.326 | 0.480 | 0.907 | 0.156 | 0.154 |
| Cyclopoids: Cyclo | 0.888 | −0.172 | 0.184 | 0.719 | −0.042 | 0.238 | 0.764 | 0.425 | 0.359 |
| Rotifers: Roti | 0.008 | 0.671 | 0.686 | 0.492 | 0.722 | −0.474 | −0.630 | 0.340 | 0.189 |

| Factor Interpretation | | | | | | | | | |
|---|---|---|---|---|---|---|---|---|---|
| Variables with Factor Loading >.500 | -<2 >20 Clado Cala Cyclo | 2-20 >20 Roti | -(2-20) Roti | Clado Cala Cyclo | 2-20 >20 Roti | 2-20 | -<2 >20 Clado Cala Cyclo -Roti | -(2-20) | 2-20 |

## Discussion

Lake St. Clair is a shallow temperate lake with no thermal or chemical stratification. The inputs of oligotrophic Lake Huron through the St. Clair River via the delta is responsible for good water quality conditions. (Leach, 1972; EC & EPA, 1988a; Edsall *et al.*, 1988). This complex ecosystem receives nutrients and contaminants including pesticides from the St. Clair River, tributaries, agricultural drainage, and urban runoff (EC & EPA, 1988b; Leach, 1991). The physics of the lake, including the short flushing time of five to seven days, appear to be the key factors which determine the nutrient and contaminant dynamics in both water and sediments. The water-mass movement and circulation patterns are wind-driven and determine the mixing of different types of water masses such as northwestern and southeastern. Other variables which might affect the intricate plankton dynamics and food-web interactions include turbidity due to resuspension of particulates, lack of thermal stratification, and contaminated sediments.

Indeed, the phytoplankton/zooplankton assemblages of Lake St. Clair are quite unique since they are similar in some respects but different in others when compared to the adjacent Lake Huron, the sole source of its water. For example the phytoplankton biomass and its taxonomic group composition is more or less similar to that of Lake Huron study of 1971 (Munawar & Munawar, 1982). On the other hand the species composition of phytoplankton is somewhat different in both the lakes. For instance, the species like *Fragilaria crotonensis*, *Melosira islandica*, *M. granulata*, *M. binderana*, *Stephanodiscus astraea*, *Tabellaria fenestrata*, *T. flocculosa*, *Synedra acus*, *S. acus* var. *radians* and *Asterionella formosa*, and *A. gracillima* were commonly found in both Lakes Huron and St. Clair. On the other hand, the variety of *Cyclotella* species recorded in Lake Huron was distinctly missing in Lake St. Clair. Furthermore the well-developed seasonality characteristic of the Lower Great Lakes was observed in Lake St. Clair in contrast to Lakes Huron and Superior (Munawar & Munawar, 1978; 1982). Most of Lake St. Clair had a high concentration of phytoflagellates belonging to Chrysophyceae and Cryptophyceae as opposed to Lake Huron where these phytoflagellates did not thrive well. Another example of contrast between the two lakes is the relative prevalence of Cyanophyta in Lake Huron which was not the case in Lake St. Clair during 1984. However, earlier studies have reported Cyanophyta to be dominant in the lake (Winner *et al.*, 1970; MWRC, 1975).

The species composition of Lake St. Clair phytoplankton during 1984 appear to be oligotrophic-mesotrophic with mixed abundance of some eutrophic species such as *Melosira binderana*, *M. granulata* and *Fragilaria capucina* (Table 1). Conversely, an earlier report by Johnston (1977) considered the diatom community of the lake to be eutrophic.

The phytoplankton-nutrient-contaminant interaction provides an interesting scenario of an oligotrophic Lake Huron phytoplankton passing through the St. Clair River, a contaminated flow-through system, to a shallow, turbid but nutrient-rich ecosystem. Due to the large number of data, it was essential to subject the data statistically to factor analysis to achieve a better insight of the plankton ecology. The positive correlation between diatoms and silica was obvious in contrast to Wallen (1979) who reported silica depletion. Furthermore, positive correlations were observed with heavy metals which are known to be toxic in the Great Lakes (Wong *et al.*, 1978; Munawar *et al.*, 1988). These results suggest ecological adaptation of the phytoplankton community to contaminants originating from the chemical valley upstream. This observation is also supported by the fact that Lake St. Clair sediments are known to be contaminated with Ni, Cr, Cu, and Zn (EC & EPA, 1988a) which are usually resuspended due to the shallow depth and wind-driven circulation patterns of the lake. In spite of these observed complexities, the good water quality of the lake is attributable to the large influx of oligotrophic Lake Huron water and the fast flushing rate. This is more or less similar to the observations made for the trophic status of North Channel (Munawar *et al.*, 1988b) which acts like

a fast flow-through system for the Lake Superior phytoplankton.

The mean zooplankton biomass ($663 \, \text{mg} \, l^{-1}$) is the second highest concentration in the Great Lakes, only next to Lake Erie. The zooplankton of Lake St. Clair was different in composition when compared to Lake Huron from whence it receives most of its water. For example, cladocerans, mainly bosminids, are prevalent in Lake St. Clair as opposed to calanoid copepods in Lake Huron. Lake St. Clair is a more typical cladoceran habitat than the other Great Lakes because it is shallow, more productive, and may not contain a dense population of planktivorous fish to which cladocerans are vulnerable.

The statistical analyses of the data showed some interesting interactions between phytoplankton and zooplankton groups. On a yearly basis, diatoms were positively correlated with cladocerans, calanoids and cyclopoids. During the spring the fluctuations of Chlorophyta were negatively correlated with those of cladocerans, calanoids, and rotifers. Most dynamic interactions between various components of the plankton were observed during the summer. The cladocerans, calanoids, and cyclopoids were positively correlated to diatoms but negatively related to Cyanophyta. The fluctuations of rotifers and small-sized Cyanophyta were similar. The factor analysis of zooplankton with phytoplankton size assemblages for the whole year, revealed a negative correlation of picoplankton ($<2 \, \mu m$) and a positive correlation of microplankton/netplankton ($>20 \, \mu m$) with cladocerans, calanoids, and cyclopoids. Furthermore, rotifers were found to be positively correlated to both ultraplankton ($2-20 \, \mu m$) and microplankton/netplankton ($>20 \, \mu m$) size assemblages. The summer results indicated interesting and complex interactions between various size assemblages of phytoplankton and zooplankton groups. For example, similar correlations were observed between cladocerans-calanoids-cyclopoids and microplankton/netplankton on the one hand, and between rotifers and picoplankton on the other.

High biomass and a different community structure of zooplankton was attributed to a productive and isolated southeastern water mass on the Ontario side of the lake. In addition, the rapid flushing rate together with the impact of contaminants was suggested to be responsible for the growth of small-bodied organisms with reduced taxonomic diversity (Sprules & Munawar, 1991).

Normalized biomass size spectra have been used recently as an indicator of ecosystem stress (Sprules & Munawar, 1986; Sprules *et al.*, 1988; Sprules & Munawar, 1991). Lake St. Clair size spectra are characteristic of mesotrophic lakes but low explained variance in the annual normalized spectrum is indicative of a perturbed system (Sprules & Munawar, 1991).

A comparison of Lake St. Clair with other Great Lakes will be interesting and informative. Vollenweider *et al.* (1974) have classified the Great Lakes based on yearly primary production, chlorophyll *a*, and phosphorus loadings. Munawar & Munawar (1982) proposed a classification of lake trophic status based on mean biomass concentration. Based on the mean biomass basis, Lake St. Clair could be tentatively classified as an oligotrophic (0.63 g m$^{-3}$) ecosystem.

Finally, it will be useful to compare the Activity Coefficient or Production/Biomass ratio (mgC mg$^{-1}$ h$^{-1}$) as an indicator of photosynthetic efficiency. This quotient has been extensively used in the Great Lakes and is believed to possess considerable potential in flagging community efficiency or stress (Munawar & Munawar, 1988). For example, comparison of the mean hourly Production/Biomass quotient indicated that Lake St. Clair had relatively a high quotient (0.022), second only to Lake Superior (0.025), and considerably higher than Lakes Huron (0.006), Erie (0.004), and Ontario (0.003). Consequently, based on our results Lake St. Clair could be defined as an oligotrophic-mesotrophic and perturbed ecosystem with small-sized plankton. Furthermore, the Lake St. Clair plankton community contains photosynthetically efficient organisms adapted to contaminants/nutrients. Due to its unique location and proximity to an industrial complex, agricultural activities, high flushing rates, and an oligotrophic nature of its origin, it offers a tremendous opportunity for research and experimentation for a better understanding of the physiological ecology of stressed ecosystems.

## Acknowledgements

We are grateful to Dr. H. Duthie for his assistance and advice in phytoplankton analyses and Dr. M. Legner for the constructive review of the manuscript. Thanks are also die to the Biota Work Group of the Upper Great Lakes Connecting Channels for their comments on the draft manuscript. We appreciate the cheerful assistance and hard work of Lynda McCarthy in data processing and preparation of this paper. We thank the following personnel who assisted in the project: Captain and Crew of C.S.S. Advent, D. Myles, W. Page, W. Norwood, L. Keeler, J. Jones, E. Jin, A. Aujla,,M. Burley, S. Nielson, P. Desoroche, V. Hall and J. Wotherspoon. The assistance of Mr. H.F. Nicholson for the technical editing of the manuscript is gratefully acknowledged.

## References

Bricker, K. S., F. J. Bricker & J. E. Gannon, 1976. Distribution and abundance of zooplankton in the U.S. waters of Lake St. Clair, 1973. J. Great Lakes Res. 2(2): 256–271.

Cooley, W. W. & P. R. Lohnes, 1971. Multivariate data analysis. J. Wiley & Sons, Inc. N.Y. 504 pp.

EC & EPA, (Environment Canada & U.S. Environmental Protection Agency) 1988a. Upper Great Lakes Connecting Channels Study. Final Report, Vol. 2. Chicago, Illinois & Toronto, Ontario, 626 pp.

EC & EPA, 1988b. Upper Great lakes Connecting Channels Study. Executive Summary, Vol. 1, Chicago, Illinois & Toronto, Ontario. 50 pp.

Edsall, T. A., P. B. Kauss, D. Kanega, J. Leach, M. Munawar, T. Nalepa & S. Thornley, 1988. Lake St. Clair biota and their habitats: A geographic Area Report of the Biota Work Group, Upper Great Lakes Connecting Channels Study. Report submitted to Upper Great Lakes Connecting Channels Activity Integration Committee. 66 pp.

Johnston, D. A., 1977. Population dynamics of walleye (*Stizostedion vitreum vitreum*) and yellow perch (*Perca flavescens*) in Lake St. Clair, especially during 1970–76. J. Fish. Res. Bd Can. 34: 1869–1877.

Leach, J. H., 1972. Distribution of chlorophyll *a* and related variables in Ontario waters of Lake St. Clair. Proc. 15th Conf. Great Lakes Res. :80–86.

Leach, J. H., 1973. Seasonal distribution, composition and abundance of zooplankton in Ontario waters of Lake St. Clair. Proc. 16th Conf. Great Lakes Res.: 54–64.

Leach, J. H., 1991. Biota of Lake St. Clair: habitat evaluation and environmental assessment. In: M. Munawar & T. Edsall (Eds), Environmental Assessment and Habitat Evaluation of the Upper Great Lakes Connecting Channels. Hydrobiologia 219: 187–202.

Munawar, M., 1989. Ecosystem health evaluation of Ashbridges Bay environment with a battery of tests. Report to M.O.E. (Ministry of the Environment) under MISA (Municipal Industrial Strategy for Abatement). Great Lakes Lab. Fish. aquat. Sci. Burlington, Ontario, Canada. 43 pp.

Munawar, M. & I. F. Munawar, 1978. Phytoplankton of Lake Superior J. Great Lakes Res. 4(3–4): 415–422.

Munawar, M. & I. F. Munawar, 1986. The seasonality of phytoplankton in the North American Great Lakes, a comparative synthesis. In: M. Munawar & J. F. Talling (eds), Seasonality of Freshwater Phytoplankton: A Global Perspective. Hydrobiologia. 138: 85–115.

Munawar, M. & J. B. Wilson, 1978. Phytoplankton-zooplankton associations in Lake Superior: A statistical approach. J. Great Lakes Res. 4(3–4): 497–504.

Munawar, M., P. Stadelmann & I. F. Munawar, 1974. Phytoplankton biomass, species composition, and primary production at a nearshore and a midlake station of Lake Ontario during IFYGL. Proc. 17th Conf. Great Lakes Res: 629–652.

Munawar, M., I. F. Munawar, L. R. Culp & G. Dupuis, 1978. Relative importance of nannoplankton in Lake Superior phytoplankton biomass and community metabolism. J. Great Lakes Res. 4(3–4): 462–480.

Munawar, M., I. F. Munawar, P. E. Ross & C. Mayfield, 1987. Differential sensitivity of natural phytoplankton size assemblages to metal mixture toxicity. In: M. Munawar (ed.), Proc. Internat. Symp. on Phycology of Large Lakes of the World. Arch. Hydrobiol. Beih. Ergebn. Limnol. Advances in Limnology, 25: 123–139.

Munawar, M., I. F. Munawar & L. H. McCarthy, 1988a. Seasonal succession of phytoplankton size assemblages and its ecological implications in the North American Great Lakes. Verh. int. Ver. Limnol. 23: 659–671.

Munawar, M., I. F. Munawar, L. H. McCarthy & H. C. Duthie, 1988b. Phycological studies in the North Channel, Lake Huron. In: M. Munawar (ed.), Limnology & Fisheries of Georgian Bay and the North Channel Ecosystems. Hydrobiologia 163: 119–134.

Munawar, M., P. T. S. Wong & G.-Y. Rhee, 1988c. The effects of contaminants on algae: an overview. In: N. W. Schmidtke (Ed.), Toxic Contamination in Large Lakes. pp. 113–160. Lewis Publishers Inc. Chelsea, Michigan.

MWRC (Michigan Water Resources Commission), 1975. Limnological survey of the Michigan portion of Lake St. Clair, 1973. Mich. Dep. Ntrl. Resourc. 59 pp.

Nie, N. H., C. H. Hull, J. H. Jenkins, K. Streinbrenner & D. H. Bent, 1975. SPSS. Statistical Package for the Social Sciences. 2nd Edition, McGraw-Hill Co. New York. 675 pp.

Reighard, J. E., 1894. A biological examination of Lake St. Clair. Bull. Michigan Fish Comm. 4: 60.

Sprules, W. G., 1984. Towards an optimal classification of zooplankton for lake ecosystem studies. Verh. int. Ver. Limnol. 22: 320–325.

Sprules, W. G. & M. Munawar, 1986. Plankton size spectra in relation to ecosystem productivity, size, and perturbation. Can. J. Fish. aquat. Sci. 43: 1789–1794.

Sprules, W. G. & M. Munawar, 1991. Plankton community structure in Lake St. Clair, 1984. In: M. Munawar & T. Edsall (eds), Environmental Assessment and Habitat Evaluation of the Upper Great Lakes Connecting Channels. Hydrobiologia 219: 229–237.

Sprules, W. G., L. B. Holtby & G. Griggs, 1981. A microcomputer-based measuring device for biological research. Can. J. Zool. 59: 1611–1614.

Sprules, W. G., M. Munawar & E. H. Jinn, 1988. Plankton community structure and size spectra in the Georgian Bay and the North Channel ecosystems. In: M. Munawar (ed.), Limnology and Fisheries of Georgian Bay/North Channel ecosystems. Hydrobiologia 163: 135–140.

Vollenweider, R. A., M. Munawar & P. Stadelmann, 1974. A comparative review of phytoplankton and primary production in the Laurentian Great Lakes. J. Fish. Res. Bd Can. 31: 739–762.

Wallen, D. G., 1989. Nutrients limiting phytoplankton production in ice-covered Lake St. Clair. J. Great Lakes Res. 5(2): 91–98.

Water Quality Branch, Inland Wayters Directorate. 1979. Analytical methods manual. Environment Canada, Ottawa, Ontario, Canada.

Winner, J. M., A. J. Oud & R. G. Ferguson, 1970. Plankton productivity studies in Lake St. Clair. Proc. 13th Conf. Great Lakes Res.: 640–650.

Wong, P. T. S., Y. K. Chau & P. L. Luxon, 1978. Toxicity of a mixture of metals on freshwater algae. J. Fish. Res. Bd Can. 35: 479–481.

*Hydrobiologia* **219**: 229–237, 1991.
*M. Munawar & T. Edsall (eds),*
*Environmental Assessment and Habitat Evaluation of the Upper Great Lakes Connecting Channels.*
© *1991 Kluwer Academic Publishers.*

# Plankton community structure in Lake St. Clair, 1984

W. Gary Sprules & M. Munawar[1]
*Department of Zoology, University of Toronto, Erindale College, Mississauga, Ontario, L5L 1C6,*
*Canada;* [1]*Great Lakes Fisheries Research Branch, Fisheries and Oceans Canada, Canada Centre for*
*Inland Waters, Burlington, Ontario, L7R 4A6, Canada*

*Key words:* phytoplankton, zooplankton, community structure, size distribution, Great Lakes

## Abstract

Spatial and seasonal patterns in phytoplankton and zooplankton communities of Lake St. Clair from June through September, 1984 are described. Phytoplankton biomass averages 586 $\mu$g l$^{-1}$ with the Diatomae and Chrysophyceae predominating. Zooplankton biomass averages 663 $\mu$g l$^{-1}$ with small bosminid Cladocera being the most abundant organisms. Lake St. Clair zooplankton biomass is second only to that of Lake Erie amongst the St. Lawrence Great Lakes. Biomass size spectra are typical in structure for mesotrophic lakes but low explained variance in the annual normalized spectrum is indicative of a perturbed system. Since 1972/1973 there appears to have been a slight decrease in zooplankton abundance in the lake accompanied by a shift from dominance of rotifers to dominance of cladocerans. We hypothesize that high flushing rate and seasonal variability coupled with contaminant loadings have resulted in a plankton community reduced in taxonomic diversity and dominated by small-bodied species.

## 1. Introduction

Lake St. Clair is a mesotrophic lake located in the St. Lawrence Great Lakes system between Lake Huron to the north and Lake Erie to the south. Drainage is north-south entering via the St. Clair River and exiting via the Detroit River. Extensive wetlands support large waterfowl communities and the lake sustains a high level of recreational activity (Edsall *et al.*, 1988). Comparatively few plankton data are available for the lake, with the studies by Bricker *et al.* (1976), Leach (1973) and Winner *et al.* (1970) being the most recent. As a basis for comparison with other Great Lakes and with earlier data from Lake St. Clair, we present information on composition and size distribution of pelagic phytoplankton and zooplankton com-

munities at a variety of sites in Lake St. Clair that were sampled through the mid-summer period of 1984.

## 2. Methods

Zooplankton samples were collected during daylight hours with a tow net of 30 cm mouth diameter and 110 $\mu$m nylon mesh (filtering efficiency 70%) that was hauled from 1 to 2 m off bottom to the surface. Samples were preserved in 4 percent formalin solution and subsamples examined microscopically. The length of each zooplankton encountered was measured with a microcomputer-based caliper (Sprules *et al.*, 1981) and individual masses simultaneously computed from

230

*Fig. 1.* Map of Lake St. Clair showing sampling stations.

niids, chydorids, *Holopedium gibberum*, bosminids, *Diaptomus* spp., cyclopoid copepods, *Limnocalanus macrurus/Epischura lacustris*, and *Leptodora kindtii*. These groups are considered to be of more significance than purely taxonomic ones for ecological studies (Sprules, 1984).

Phytoplankton samples were collected with an integrating sampler (Schroeder, 1969) hauled from near bottom to the surface during the day. These samples were preserved in Lugol's solution, settled and examined under an inverted light microscope according to the Utermöhl technique (Munawar & Munawar, 1981).

A total of eight sampling sites was established for Lake St. Clair (Fig. 1). In this paper we present data on seven stations each for cruises during June 25–27 and September 18–20, and eight stations each for cruises during July 10–11, and August 21–22, 1984.

length-wet weight regressions pooled from the literature (Sprules, 1984). Animals were not uniformly identified to species, but were classified into 10 groups – copepod nauplii, rotifers, daph-

## 3. Results and discussion

Except in July, cladocerans predominated over copepods in terms of relative numerical density (Table 1). However, copepod biomass was usu-

*Table 1.* Mean Abundances (nos. $1^{-1}$) and proportionate composition of zooplankton in functional groups. Means are weighted by the volume of water collected for each sample.

|  | June | July | August | September | Mean |
|---|---|---|---|---|---|
| Copepod nauplii | 5.9 | 9.2 | 10.7 | 4.9 | 7.8 |
| Rotifers | 13.2 | 11.5 | 15.2 | 25.3 | 15.3 |
| Daphniids | 4.9 | 4.5 | 0.2 | 0.1 | 3.1 |
| Chydorids | <0.1 | <0.1 | 0. | 0. | <0.1 |
| *Holopedium* | 0.1 | 0.1 | <0.1 | 0. | 0.1 |
| Bosminids | 40.5 | 9.1 | 20.3 | 15.5 | 25.5 |
| *Diaptomus* | 4.8 | 3.2 | 1.2 | 1.4 | 2.9 |
| Cyclopoids | 12.4 | 5.2 | 3.8 | 2.8 | 7.1 |
| *Limnocalanus/Epischura* | 4.6 | 0.6 | <0.1 | 0.1 | 1.9 |
| *Leptodora* | 0.1 | 0.1 | <0.1 | <0.1 | 0.1 |
| Total | 86.5 | 43.5 | 51.5 | 50.3 | 63.8 |
| Number of samples | 7 | 8 | 8 | 7 | 30 |
| Standard deviation | 32.0 | 33.1 | 31.0 | 58.1 | 19.2 |
| Cladocerans (%) | 52.5 | 31.4 | 39.9 | 31.1 | 45.1 |
| Calanoids (%) | 13.8 | 17.6 | 7.5 | 6.6 | 12.5 |
| Cyclopoids (%) | 18.2 | 24.2 | 23.1 | 11.9 | 18.4 |
| Rotifers (%) | 15.2 | 26.4 | 29.5 | 50.4 | 24.0 |

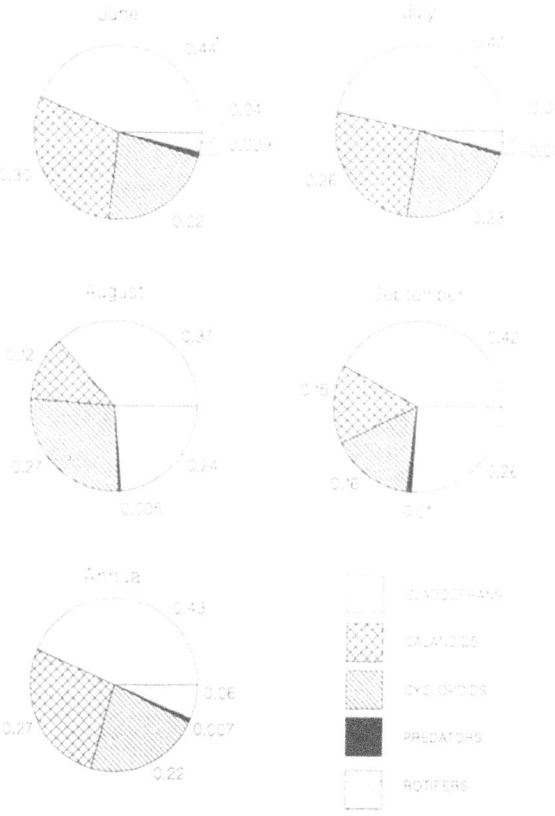

Fig. 2. Biomass composition of zooplankton communities in different months and for all stations combined. Predators comprise *Leptodora*.

Fig. 3. Seasonal trends in mean biomass of phytoplankton and zooplankton. Standard errors of the mean are shown except when they were smaller than the symbol size.

to the zooplankton of Lake St. Clair (Table 1), but on a biomass basis they were important only later in the season (Fig. 2), reaching a peak in September at station 3 of 183 $\mu$g l$^{-1}$ (Table 5).

The Diatomae and Chrysophyceae were the two most important components of phytoplankton biomass (Tables 2 to 5) although the Cryptophyceae can be co-dominant, particularly in August and September (Tables 4 and 5). Diatom dominance is typical of Great Lakes phytoplankton communities, but chrysophyte co-dominance is not usual (Munawar & Munawar, 1981). Cyanophyta, Chlorophyta, and Dinophyceae

ally greater than cladoceran biomass, although on an annual basis the difference was slight (Fig. 2, Tables 2 to 5). Bosminids comprised the majority of cladoceran numbers and biomass. *Holopedium gibberum*, *Leptodora kindtii* and the Chydoridae were only minor components of the Cladocera in Lake St. Clair. Cyclopoid and calanoid copepods are roughly equal components of copepod numbers and biomass in the lake (Fig. 2, Tables 1 to 5). *Epischura lacustris* and *Diaptomus* spp. constitute the majority of calanoid copepods, with *Diaptomus* normally predominating over *Epischura*. *Limnocalanus macrurus* was found occasionally, presumably washed in from Lake Huron, but other cold stenothermic species such as *Mysis relicta* and *Senecella calanoides* were not recorded.

Numerically, rotifers contributed substantially

Fig. 4. Biomass size spectra for complete plankton communities in different months and for all data combined. Phytoplankton and zooplankton comprise the major biomass peaks on the left and right respectively of each panel.

*Table 2.* Biomass concentration ($\mu$g l$^{-1}$) of zooplankton and proportionate composition of zooplankton and phytoplankton. Zooplankton means are weighted by the volume of water taken for each sample.

| Station | June 25–27, 1984 | | | | | | | |
|---|---|---|---|---|---|---|---|---|
| | 2 | 3 | 4 | 6 | 8 | 9 | 10 | Mean |
| **ZOOPLANKTON** | | | | | | | | |
| Copepod nauplii | 2.0 | 15.0 | 7.8 | 3.6 | 2.1 | 13.4 | 12.1 | 7.4 |
| Rotifers | 1.3 | 169.2 | 90.1 | 32.6 | 14.8 | 59.2 | 37.7 | 46.4 |
| Daphniids | 223.7 | 540.5 | 51.6 | 4.3 | 6.6 | 134.0 | 146.1 | 174.9 |
| Chydorids | 0. | 0. | 0. | 0. | 0. | 0. | 0.5 | 0.1 |
| *Holopedium* | 0. | 0. | 0. | 0. | 1.3 | 0. | 8.7 | 1.6 |
| Bosminids | 340.2 | 261.7 | 835.7 | 404.9 | 204.2 | 794.5 | 232.4 | 394.0 |
| *Diaptomus* | 349.4 | 389.0 | 235.7 | 45.2 | 59.8 | 77.5 | 61.7 | 188.8 |
| Cyclopoids | 314.2 | 531.8 | 260.4 | 103.6 | 185.6 | 264.3 | 301.1 | 285.6 |
| *Limnocalanus/Epischura* | 0. | 129.4 | 479.2 | 283.6 | 59.9 | 205.1 | 464.3 | 195.2 |
| *Leptodora* | 0. | 86.2 | 0. | 0. | 14.8 | 0. | 0. | 12.1 |
| Biomass ($\mu$g l$^{-1}$) | 1230.8 | 2122.8 | 1960.6 | 877.8 | 549.1 | 1548.1 | 1264.6 | 1306.1 |
| Cladocerans (%) | 45.8 | 37.8 | 45.3 | 46.6 | 38.6 | 60.0 | 30.6 | 43.7 |
| Calanoids (%) | 28.5 | 24.7 | 36.7 | 37.7 | 21.9 | 18.6 | 42.0 | 29.7 |
| Cyclopoids (%) | 25.6 | 25.5 | 13.4 | 11.9 | 34.1 | 17.6 | 24.4 | 22.1 |
| Rotifers (%) | 0.1 | 8.0 | 4.6 | 3.7 | 2.7 | 3.8 | 3.0 | 3.6 |
| **PHYTOPLANKTON (%)** | | | | | | | | |
| Cyanophyta | 0. | 0. | 2.3 | 0. | 0. | 0. | 15.7 | 2.6 |
| Chlorophyta | 11.9 | 0. | 0. | 0. | 7.8 | 4.4 | 0.9 | 3.6 |
| Chrysophyceae | 28.5 | 31.9 | 39.9 | 50.2 | 24.3 | 28.2 | 1.5 | 29.2 |
| Diatomeae | 43.3 | 62.9 | 48.9 | 48.7 | 67.7 | 65.6 | 81.9 | 59.8 |
| Cryptophyceae | 16.2 | 5.2 | 8.9 | 1.7 | 0.2 | 1.6 | 0.0 | 4.8 |
| Dinophyceae | 0. | 0. | 0. | 0. | 0. | 0. | 0. | 0. |
| Biomass ($\mu$g l$^{-1}$) | 348.3 | 417.7 | 472.0 | 638.3 | 1029.6 | 535.8 | 4347.4 | 1112.7 |

made up smaller fractions of the algal biomass. Munawar *et al.* (1991) discuss seasonal and taxonomic trends in phytoplankton in greater detail.

Zooplankton biomass was generally highest in the southern part of Lake St. Clair, especially at Stations 3 and 4 and to a lesser extent Stations 10 and 2 (Tables 2 to 5, Fig. 1). Patterns were less consistent for phytoplankton, although in July through September biomass tended to be higher at southerly Stations 3 or 4 and lower at central and northern Stations 8, 9 and 10 (Tables 3 to 5). These patterns are consistent with other data for the lake, identifying a southeastern water mass less subject to the flushing action from the St. Clair river and enriched by nutrient input from the Thames River and adjacent land drainage (Leach,

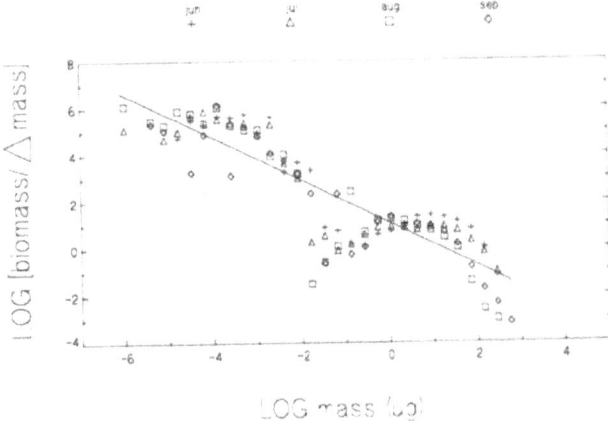

*Fig. 5.* Normalized biomass spectrum for plankton data from all months. The least squares linear regression line through the data is shown. $Y = 1.24 - 0.92X$, mean square deviation from regression = 6.92.

*Table 3.* Biomass concentration ($\mu$g l$^{-1}$) of zooplankton and proportionate composition of zooplankton and phytoplankton. Zooplankton means are weighted by the volume of water taken for each sample.

| Station | July 10–12, 1984 | | | | | | | | |
|---|---|---|---|---|---|---|---|---|---|
| | 1 | 2 | 3 | 4 | 6 | 8 | 9 | 10 | Mean |
| **ZOOPLANKTON** | | | | | | | | | |
| Copepod nauplii | 2.8 | 7.8 | 13.3 | 10.8 | 1.6 | 0.4 | 3.2 | 43.5 | 9.5 |
| Rotifers | 13.8 | 10.0 | 4.8 | 20.8 | 7.2 | 3.3 | 12.8 | 101.4 | 19.6 |
| Daphniids | 36.9 | 172.4 | 425.0 | 248.4 | 53.0 | 43.0 | 166.4 | 39.4 | 152.9 |
| Chydorids | 0.8 | 0.3 | 0. | 2.9 | 0. | 0. | 0. | 0. | 0.5 |
| *Holopedium* | 0. | 0. | 0. | 2.0 | 0. | 4.2 | 13.4 | 7.0 | 3.4 |
| Bosminids | 76.9 | 24.1 | 55.3 | 89.0 | 75.8 | 172.0 | 44.9 | 144.1 | 80.0 |
| *Diaptomus* | 6.3 | 4.7 | 34.6 | 184.1 | 33.2 | 76.7 | 23.4 | 365.0 | 80.8 |
| Cyclopoids | 17.2 | 16.5 | 144.4 | 391.8 | 35.3 | 116.4 | 75.6 | 138.6 | 111.3 |
| *Limnocalanus/Epischura* | 10.2 | 14.6 | 179.5 | 0. | 28.9 | 72.0 | 0. | 44.4 | 42.8 |
| *Leptodora* | 13.3 | 2.6 | 0. | 0. | 0. | 0. | 0.3 | 8.9 | 2.7 |
| Biomass ($\mu$g l$^{-1}$) | 178.1 | 253.0 | 885.3 | 949.8 | 235.0 | 487.9 | 340.0 | 892.2 | 503.5 |
| Cladocerans (%) | 63.9 | 77.7 | 54.2 | 35.7 | 54.8 | 44.9 | 66.1 | 11.0 | 47.1 |
| Calanoids (%) | 9.5 | 8.2 | 27.9 | 19.6 | 26.8 | 30.5 | 7.1 | 37.8 | 25.5 |
| Cyclopoids (%) | 11.0 | 9.0 | 17.3 | 42.2 | 15.3 | 23.9 | 23.0 | 18.0 | 23.4 |
| Rotifers (%) | 7.8 | 4.0 | 0.5 | 2.2 | 3.1 | 0.7 | 3.8 | 32.6 | 3.9 |
| **PHYTOPLANKTON (%)** | | | | | | | | | |
| Cyanophyta | 0. | 0.2 | 0. | 0. | 0. | 0. | 0. | 0. | 0.03 |
| Chlorophyta | 3.5 | 0.3 | 0. | 3.5 | 0.1 | 0.0 | 0.4 | 0. | 1.0 |
| Chrysophyceae | 20.6 | 19.3 | 21.7 | 12.5 | 52.0 | 32.4 | 32.8 | 24.5 | 27.0 |
| Diatomeae | 73.4 | 22.6 | 47.2 | 75.2 | 40.6 | 61.6 | 63.3 | 70.9 | 56.9 |
| Cryptophyceae | 2.4 | 57.5 | 31.0 | 8.7 | 4.3 | 1.8 | 2.7 | 4.6 | 14.1 |
| Dinophyceae | 0. | 0.2 | 0.1 | 0. | 2.9 | 4.1 | 0.9 | 0. | 1.0 |
| Biomass ($\mu$g l$^{-1}$) | 1100.4 | 448.0 | 641.2 | 539.8 | 597.0 | 306.4 | 309.7 | 518.9 | 557.7 |

1972, 1973). By contrast, phytoplankton biomass in June tended to be highest at central station 10 and at Stations 6, 8, and 9 in the northwestern part of the lake (Table 2). For the whole lake both phytoplankton and zooplankton biomass decreased from maxima in June of 1113 and 1306 $\mu$g l$^{-1}$ respectively to level out at values under 400 $\mu$g l$^{-1}$ in August and September (Fig. 3).

The size distribution of planktonic organisms in Lake St. Clair, expressed in the form of biomass size spectra (Fig. 4), were broadly similar to those found in other lakes (Sprules *et al.*, 1983, Sprules & Munawar, 1986). There were two major biomass peaks corresponding to phytoplankton on the left of each panel and zooplankton on the right with some minor peaks,

particularly within phytoplankton, reflecting component size groups. The seasonal decrease in plankton biomass is evident in these spectra, as is a shift in peak algal abundance to smaller sizes in August/September (ca 0.0001 $\mu$g or 6 $\mu$m equivalent spherical diameter) compared to June/July (ca 0.002 $\mu$g or 15 $\mu$m) (Fig. 4). Of particular note is the small size of zooplankton in Lake St. Clair, with the largest organism being about 600 $\mu$g in mass or just over 1.5 mm in length (Fig. 4). The small size of zooplankton is a reflection of the relatively high biomass of rotifers and bosminid cladocerans in the lake.

In unperturbed ecosystems with high size diversity of component organisms, the biomass spectrum on a normalized scale should be close to linear (Sprules & Munawar, 1986). For Lake St.

*Table 4.* Biomass concentration ($\mu g \, l^{-1}$) of zooplankton and proportionate composition of zooplankton and phytoplankton. Zooplankton means are weighted by the volume of water taken for each sample.

| Station | August 20–24, 1984 | | | | | | | | |
|---|---|---|---|---|---|---|---|---|---|
| | 1 | 2 | 3 | 4 | 6 | 8 | 9 | 10 | Mean |
| **ZOOPLANKTON** | | | | | | | | | |
| Copepod nauplii | 15.8 | 37.5 | 4.4 | 22.4 | 4.4 | 5.3 | 13.5 | 7.8 | 12.7 |
| Rotifers | 93.7 | 63.3 | 54.5 | 25.9 | 11.9 | 31.0 | 23.7 | 19.7 | 40.7 |
| Daphniids | 0. | 4.2 | 0.4 | 10.5 | 0.5 | 0.1 | 1.5 | 0. | 1.9 |
| Chydorids | 0. | 0. | 0. | 0. | 0. | 0. | 0. | 0. | 0. |
| *Holopedium* | 0. | 0. | 0. | 0. | 0. | 0. | 1.9 | 0. | 0.2 |
| Bosminids | 134.5 | 114.4 | 33.6 | 215.2 | 9.1 | 16.8 | 26.7 | 4.9 | 61.3 |
| *Diaptomus* | 7.3 | 2.5 | 1.7 | 61.3 | 9.8 | 25.3 | 42.6 | 7.1 | 16.6 |
| Cyclopoids | 45.8 | 33.0 | 20.2 | 99.4 | 16.6 | 19.1 | 76.5 | 20.7 | 38.1 |
| *Limnocalanus/Epischura* | 0. | 0. | 0. | 0. | 0. | 0. | 0. | 0. | 0.8 |
| *Leptodora* | 1.3 | 2.2 | 1.0 | 1.9 | 0.1 | 0. | 0.1 | 0. | 0.8 |
| Biomass ($\mu g \, l^{-1}$) | 298.3 | 257.1 | 115.9 | 436.6 | 52.4 | 97.6 | 193.2 | 60.3 | 173.1 |
| Cladocerans (%) | 45.1 | 46.2 | 29.4 | 51.7 | 18.2 | 17.4 | 15.6 | 8.2 | 36.6 |
| Calanoids (%) | 3.0 | 2.1 | 1.5 | 15.7 | 21.7 | 28.4 | 27.7 | 14.4 | 12.3 |
| Cyclopoids (%) | 20.1 | 26.3 | 21.2 | 26.2 | 37.2 | 22.5 | 44.4 | 44.7 | 27.4 |
| Rotifers (%) | 31.4 | 24.6 | 47.1 | 5.9 | 22.8 | 31.8 | 12.3 | 32.7 | 23.5 |
| **PHYTOPLANKTON (%)** | | | | | | | | | |
| Cyanophyta | 2.7 | 0.5 | 10.0 | 1.8 | 0.2 | 0. | 20.7 | 0.9 | 4.6 |
| Chlorophyta | 22.0 | 13.6 | 2.6 | 28.5 | 0.1 | 4.5 | 0. | 13.6 | 10.6 |
| Chrysophyceae | 41.3 | 28.0 | 7.1 | 22.2 | 60.2 | 66.1 | 47.0 | 38.0 | 38.7 |
| Diatomeae | 27.5 | 10.6 | 26.3 | 27.8 | 28.0 | 21.8 | 16.6 | 24.1 | 22.8 |
| Cryptophyceae | 6.6 | 39.0 | 51.5 | 19.7 | 6.2 | 5.7 | 15.6 | 11.4 | 19.5 |
| Dinophyceae | 0. | 8.3 | 2.5 | 0. | 5.2 | 1.9 | 0. | 12.0 | 3.7 |
| Biomass ($\mu g \, l^{-1}$) | 228.5 | 379.8 | 885.6 | 395.5 | 243.0 | 315.9 | 352.7 | 259.3 | 382.5 |

Clair, this distribution for all data combined has a linear regression slope of $-0.92$ which is within the characteristic range for more productive lakes (Sprules & Munawar, 1986) (Fig. 5). However, the relatively high mean square deviation from regression (6.92) suggests that the Lake St. Clair plankton community on an annualized basis has low size diversity which could be indicative of a perturbed system. The major departure from linearity is due to the rather narrow unimodal size distribution of zooplankton (Fig. 4) leading to lower and higher than expected abundance at the tails and centre of the distribution respectively (Fig. 5, right-hand cluster of points). The very seasonal nature of Lake St. Clair plankton communities, an unusually high flushing rate of nine days (Bricker *et al.*, 1976), and contaminant loading from the St. Clair River have created highly variable conditions to which, we hypothesize, a reduced assemblage of zooplankton species has adapted.

Our zooplankton data differ somewhat from those collected during 1972 and 1973 (Leach, 1973; Bricker *et al.*, 1976), although comparisons must be interpreted cautiously because our 110 $\mu$m plankton nets would underestimate small rotifers relative to the 80 $\mu$m mesh used by Leach (1973) and Bricker *et al.* (1976). Leach (1973) collected zooplankton in the Canadian waters of Lake St. Clair (our Stations 1 to 4, and 10, Fig. 1) from April to November, 1972 and determined a seasonal mean abundance for zooplankton, excluding Protozoa, of 286 $l^{-1}$. Rotifers, copepods and cladocerans made up 73, 21, and 6.6

*Table 5.* Biomass concentration ($\mu$g l$^{-1}$) of zooplankton and proportionate composition of zooplankton and phytoplankton. Zooplankton means are weighted by the volume of water taken for each sample.

| Station | September 18–20, 1984 | | | | | | | |
|---|---|---|---|---|---|---|---|---|
| | 1 | 2 | 3 | 4 | 6 | 9 | 10 | Mean |
| **ZOOPLANKTON** | | | | | | | | |
| Copepod nauplii | 11.3 | 7.7 | 8.8 | 4.5 | 1.6 | 2.7 | 9.6 | 6.3 |
| Rotifers | 41.2 | 85.0 | 182.5 | 74.3 | 7.2 | 13.9 | 24.9 | 59.3 |
| Daphniids | 0. | 1.4 | 0. | 0.4 | 0.2 | 9.3 | 3.1 | 2.3 |
| Chydorids | 0. | 0. | 0. | 0. | 0. | 0. | 0. | 0. |
| *Holopedium* | 0. | 0. | 0. | 0. | 0. | 0. | 0. | 0. |
| Bosminids | 30.5 | 162.2 | 124.6 | 47.7 | 37.1 | 63.7 | 178.3 | 94.9 |
| *Diaptomus* | 29.7 | 27.1 | 18.9 | 9.6 | 8.5 | 39.7 | 49.6 | 26.7 |
| Cyclopoids | 47.8 | 47.6 | 38.8 | 39.4 | 3.7 | 27.7 | 40.0 | 33.5 |
| *Limnocalanus/Epischura* | 0. | 5.8 | 6.2 | 0. | 0. | 15.1 | 2.6 | 4.6 |
| *Leptodora* | 0. | 7.9 | 3.5 | 2.0 | 0.1 | 0. | 4.4 | 2.6 |
| Biomass ($\mu$g l$^{-1}$) | 160.5 | 344.6 | 383.3 | 177.9 | 58.5 | 171.6 | 312.5 | 230.2 |
| Cladocerans (%) | 19.0 | 47.5 | 32.5 | 27.0 | 63.8 | 42.3 | 58.0 | 42.2 |
| Calanoids (%) | 20.6 | 10.3 | 7.2 | 5.8 | 15.9 | 32.8 | 18.0 | 15.4 |
| Cyclopoids (%) | 34.7 | 15.2 | 11.7 | 24.2 | 7.6 | 16.9 | 14.6 | 16.4 |
| Rotifers (%) | 25.7 | 24.7 | 47.6 | 41.8 | 12.4 | 8.1 | 8.0 | 25.8 |
| **PHYTOPLANKTON (%)** | | | | | | | | |
| Cyanophyta | 0.3 | 1.2 | 0.5 | 1.6 | 0. | 0. | 0.1 | 0.5 |
| Chlorophyta | 3.3 | 17.3 | 1.1 | 0. | 0.5 | 0.5 | 5.8 | 4.1 |
| Chrysophyceae | 36.4 | 14.1 | 25.8 | 64.9 | 70.3 | 65.4 | 13.7 | 41.5 |
| Diatomeae | 33.0 | 7.8 | 36.9 | 18.8 | 13.6 | 20.3 | 69.5 | 28.6 |
| Cryptophyceae | 27.0 | 53.5 | 32.2 | 12.6 | 7.8 | 13.7 | 5.8 | 21.8 |
| Dinophyceae | 0. | 6.0 | 3.5 | 2.0 | 7.8 | 0. | 5.1 | 3.5 |
| Biomass ($\mu$g l$^{-1}$) | 218.7 | 534.5 | 412.4 | 293.8 | 236.2 | 193.3 | 161.2 | 292.9 |

percent of this total respectively. Bricker *et al.* (1976) collected in the U.S. waters of the lake (our Stations 6 and 8, Fig. 1) in July, August and September, 1973 and found a seasonal mean abundance of 45.4 l$^{-1}$ comprising 92, 2.4, and 5.7 percent rotifers, copepods, and cladocerans respectively. The higher abundance in Ontario waters is presumably due to the rich southeastern water mass that includes much of this part of the lake. Averaging the seasonal means of Leach (1973) and Bricker *et al.* (1976) provides a crude estimate of lakewide zooplankton abundance of 165.7 l$^{-1}$ with rotifers, copepods, and cladocerans comprising 75, 18.1, and 6.5 percent respectively of this total. This compares with our mean seasonal zooplankton abundance for 1984 of 69.6 l$^{-1}$ made up of 23.5 percent rotifers, 30.3

percent copepods, and 46.0 percent cladocerans. Compared to 1972/1973, abundances in 1984 are lower, rotifers are relatively less abundant, and cladocerans are relatively more abundant. The reduction in rotifer biomass may be due to the increase in cladoceran competitors, but it is unclear what has caused the unusually high abundance of bosminids in Lake St. Clair. Decreases in fish planktivory could be involved although nutrient abatement and increases in industrial contaminants since the early 1970's may also be important.

Bricker *et al.* (1976) support the conclusion of Reighard (1894) that Lake St. Clair is very low in zooplankton abundance. This perception, however, undoubtedly arises because they sampled primarily in the U.S. and central channel waters

*Table 6.* Mean zooplankton biomass ($\mu g\,l^{-1}$) for various Great Lakes based on samples taken at different sites and on varying days from June to September. Numbers of samples are shown. Large invertebrates comprise *Mysis* and *Leptodora*. Dash indicates less than 0.01.

| | St. Clair 1984 $n = 30$ | Huron 1988 $n = 3$ | Erie 1984 $n = 6$ | Ontario 1984 $n = 11$ | Superior 1983 $n = 14$ |
|---|---|---|---|---|---|
| Copepod nauplii | 9.3 | 4.5 | 42.0 | 8.5 | – |
| Rotifers | 42.1 | 34.4 | 223.1 | 30.3 | 0.2 |
| Daphniids | 91.7 | 98.5 | 227.3 | 32.1 | 5.3 |
| Chydorids | 0.1 | 0. | 0. | 0. | 0. |
| *Holopedium* | 1.2 | 0. | 10.4 | 0.6 | 0.2 |
| Bosminids | 194.9 | 60.4 | 65.1 | 94.3 | 0.3 |
| *Diaptomus* | 93.5 | 197.1 | 296.8 | 84.4 | 8.5 |
| *Senecella* | 0. | 0.5 | 0. | 0. | 1.3 |
| Cyclopoids | 142.9 | 26.9 | 188.7 | 298.5 | 0.3 |
| *Limnocalanus/Epischura* | 81.6 | 54.3 | 14.1 | 30.0 | 13.5 |
| *Mysis* | 0. | 0.[1] | 0. | 22.1 | 3.2 |
| *Leptodora* | 5.6 | 0. | 2.8 | – | 0.06 |
| Biomass ($\mu g\,l^{-1}$) | 662.9 | 484.5[2] | 1070.4 | 601.0 | 32.9 |
| Cladocerans (%) | 43.4 | 32.8 | 28.6 | 21.1 | 17.8 |
| Calanoids (%) | 27.2 | 51.9 | 31.5 | 19.4 | 70.9 |
| Cyclopoids (%) | 22.2 | 5.6 | 19.1 | 50.7 | 0.8 |
| Rotifers (%) | 6.4 | 7.1 | 20.8 | 5.1 | 0.6 |
| Large invertebrates (%) | – | 1.6[2] | 0.3 | 3.7 | 9.8 |

[1] Underestimated because sampling not done at night.

[2] Includes $7.9\ \mu g \cdot l^{-1}$ of *Bythotrephes cederstroemii*, a predatory cladoceran that has recently invaded the Great Lakes.

of the lake where flow-through is very high. Compared with other St. Lawrence Great Lakes, Lake St. Clair actually has a high total biomass of zooplankton (only Lake Erie is higher) with the very high relative abundance of cladocerans, especially bosminids, already noted (Table 6). Phytoplankton biomass, which in our study averaged 586 $\mu g\,l^{-1}$ for the season, is more similar in this regard to Lake Huron than any of the other lakes (Munawar & Munawar, 1981). The high zooplankton biomass in Lake St. Clair may exist because, at least partly, bosminid cladocerans are efficient and relatively unselective grazers (DeMott, 1982). In comparison with all Great Lakes except Ontario, Lake St. Clair has a relatively low biomass of calanoid copepods and a relatively high one of cyclopoid copepods (Table 6). The reverse is true in Lake Ontario, which also has the highest absolute biomass of *Mysis relicta* of all the lakes. Lake Ontario also has a relatively high biomass of rotifers compared to the other Great Lakes (Table 6).

## 4. Conclusions

With a very high turnover rate of about nine days (Bricker *et al.*, 1976), one would expect Lake St. Clair to have a zooplankton community similar to that of Lake Huron from which it receives about 98 percent of its water (Leach, 1972). Nevertheless, major zooplankton groups differ in relative importance (Table 6), although our Lake Huron data were collected four years later from only three stations. In particular, cladocerans, principally bosminids, are more common in Lake St. Clair whereas calanoid copepods predominate in Lake Huron. These shifts in community structure, as well as the high biomass of zooplankton in Lake St. Clair, is probably due to the large, productive southeastern water mass already mentioned. This mass forms a large eddy in the southeast which is isolated from the fast currents of the central channel (Leach, 1980; Bricker *et al.*, 1976). It is likely that zooplankters flowing in from Lake Huron get entrained in this water mass where there is a selection for species typical of more productive habitats. This high level of production combined with rapid flushing and possible contaminant impacts in Lake St. Clair has resulted in a plankton community predominated by small-bodied organisms of reduced taxonomic diversity, particularly for zooplankton.

## 5. Acknowledgements

We are grateful to Eddy Jin for processing the zooplankton samples and data. Uwe Borgmann and Bill Taylor made helpful comments on the manuscript. This work was supported by operating grants from the Natural Sciences and Engineering Research Council of Canada and from the Donner Canadian Foundation.

## References

Bricker, K. S., F. J. Bricker & J. E. Gannon, 1976. Distribution and abundance of zooplankton in the U.S. waters of Lake St. Clair, 1973. Proc. 19th Conf. Great Lakes Res.: 256–271.

DeMott, W. R., 1982. Feeding selectivities and relative ingestion rates of *Daphnia* and *Bosmina*. Limnol. Oceanogr. 27: 518–527.

Edsall, T. A., B. A. Manny & C. N. Raphael, 1988. The St. Clair River and Lake St. Clair, Michigan: an ecological profile. U.S. Fish Wildl. Serv. Biol. Rep. 85(7.3). 130 pp.

Leach, J. H., 1972. Distribution of chlorophyll *a* and related variables in Ontario waters of Lake St. Clair. Proc. 15th Conf. Great Lakes Res.: 80–86.

Leach, J. H., 1973. Seasonal distribution, composition and abundance of zooplankton in Ontario waters of Lake St. Clair. Proc. 16th Conf. Great lakes Res.: 54–64.

Leach, J. H., 1980. Limnological sampling intensity in Lake St. Clair in relation to distribution of water masses. J. Great Lakes Res. 6: 141–145.

Munawar, M. & I. F. Munawar, 1981. A general comparison of the taxonomic composition and size analysis of the phytoplankton of the North American Great Lakes. Verh. int. Ver. Limnol. 21: 1695–1716.

Munawar, M., I. F. Munawar & W. G. Sprules, 1991. The plankton ecology of Lake St. Clair, 1984. In M. Munawar & T. Edsall (eds), Environmental Assessment and Habitat Evaluation of the Upper Great Lakes Connecting Channels. Hydrobiologia 219: 203–227.

Reighard, J. E., 1894. A biological examination of Lake St. Clair. Bull. Mich. Fish. Commiss. 4, 61 pp.

Schroeder, R., 1969. Ein summierender Wasserschopfer. Arch. Hydrobiol. 66: 241–243.

Sprules, W. G., 1984. Towards an optimal classification of zooplankton for lake ecosystem studies. Verh. int. Ver. Limnol. 22: 320–325.

Sprules, W. G. & M. Munawar, 1986. Plankton size spectra in relation to ecosystem productivity, size and perturbation. Can. J. Fish. aquat. Sci. 43: 1789–1794.

Sprules, W. G., L. B. Holtby & G. Griggs, 1981. A microcomputer-based measuring device for biological research. Can. J. Zool. 59: 1611–1614.

Sprules, W. G., J. M. Casselman & B. J. Shuter, 1983. Size distribution of pelagic particles in lakes. Can. J. Fish. aquat. Sci. 40: 1761–1769.

Winner, J. M., A. J. Oud & R. Ferguson, 1970. Plankton productivity studies in Lake St. Clair. Proc. 13th Conf. Great Lakes Res., pp. 640–650.

*Hydrobiologia* **219**: 239–250, 1991.
*M. Munawar & T. Edsall (eds),*
*Environmental Assessment and Habitat Evaluation of the Upper Great Lakes Connecting Channels.*
© *1991 Kluwer Academic Publishers.*

# Phosphorus cycling by mussels (Unionidae : Bivalvia) in Lake St. Clair

T.F. Nalepa, W.S. Gardner & J.M. Malczyk
*Great Lakes Environmental Research Laboratory, 2205 Commonwealth Blvd., Ann Arbor, MI 48105, USA*

*Key words:* nutrients, excretion, biodeposition, *Lampsilis*, phosphorus budget

### Abstract

The role of mussels in cycling phosphorus in Lake St. Clair during the May–October period was examined by measuring concentrations in the water column and in mussel tissue, and by measuring rates of biodeposition and excretion. Mean rates of biodeposition and excretion for *Lampsilis radiata siliquoidea*, the most abundant species, were 6.3 $\mu$g P (g shell-free dry wt)$^{-1}$ h$^{-1}$ and 1.3 $\mu$g P (g shell-free dry wt)$^{-1}$ h$^{-1}$, respectively; body tissue phosphorus content was 2.7 percent of dry wt. Seasonal changes in excretion rates appeared to be related to the gametogenic cycle of the organism, but seasonal changes in biodeposition rates were not apparent. Phosphorus assimilation efficiency for this species was about 40 percent. Overall, the mussel population in Lake St. Clair filtered about 210 MT of phosphorus, or about 13.5 percent of the total phosphorus load for the May–October study period. Of this amount, about 134 MT was sedimented to the bottom via biodeposition. Mussel biodeposition may be an important source of nutrients to other biotic components in the lake such as macrophytes and invertebrate deposit-feeders.

## 1. Introduction

Through their feeding and burrowing activities, benthic invertebrates can play an important role in the cycling of nutrients. In addition to mixing nutrient-rich pore waters with overlying waters, these organisms ingest organic material and excrete remineralized nutrients in forms readily available to phytoplankton. For example, in nearshore Lake Michigan, phosphorus excretion by benthic invertebrates was comparable to amounts released from the sediments (Gardner *et al.*, 1981). Of the various invertebrate groups, unionid bivalves (mussels) in particular can have a significant impact on nutrient cycling (Lewandowski & Stanczykowska, 1975; Walz, 1978; Stanczykowska & Planter, 1985; Kasprzak, 1986; James, 1987). These large filter-feeders have the capacity

to remove great amounts of particulate material from the water column. Some of this material is assimilated and used for growth, metabolism, and the production of offspring, but a large portion is voided through pseudofeces, feces, and inorganic excretion.

The purpose of this study was to examine the role of mussels in the cycling of phosphorus in Lake St. Clair. Mussels are the dominant macroinvertebrate in the lake, with standing stocks about four times greater than that of all other benthic invertebrates combined (Nalepa & Gauvin, 1988). A phosphorus budget through the mussel population was estimated from measurements of concentrations in the water column and in mussel tissue, and from rate measurements of biodeposition and excretion. Since material deposited on the bottom by mussels along with

240

naturally-sedimented material may play a role in the cycling of phosphorus, rates of phosphorus release from the sediments were also determined.

## 2. Description of study site

Lake St. Clair lies at the center of the 125 km long waterway between Lake Huron and Lake Erie. The main inflow is from the St. Clair River, which has a flow rate of $5\,100\,\mathrm{m^3\,s^{-1}}$ and contributes 98 percent of the flow into the lake. The only outflow is through the Detroit River. Hydraulic retention time in the lake is about 9 days. The lake has an area of $1\,110\,\mathrm{km^2}$, a volume of $3.4\,\mathrm{km^3}$, and a mean depth of 3 m. Because of its high flow-through rate and shallow depth, the lake is well-mixed; thermal stratification does not occur and oxygen concentrations remain close to saturation. Two distinct water masses have been distinguished within the lake: a northwestern mass, which consists primarily of low-nutrient water flowing into the lake from Lake Huron via the St. Clair River, and a southeastern mass, which consists of more stable water enriched by nutrient loadings from Ontario tributaries (Leach, 1980). As a result of these two distinct water masses, biological (chlorophyll *a* and zooplankton) and chemical (carbon, phosphorus, nitrogen, chloride, and total alkalinity) features of the water column tend to increase on a gradient from northwest to southeast (Leach, 1972, 1973). Because of the high inflow of low-nutrient water from Lake Huron, the water quality of Lake St. Clair has remained relatively good despite extensive shore-line development and agricultural land use in the watershed. The lake has been classified as meso-trophic, with overall nutrient concentrations lying between the low values of the upper lakes and the high values found in western Lake Erie (Herdendorf *et al.*, 1986).

## 3. Methods and materials

Mussels were collected at one site (Station 73) on five different dates in 1985 and two sites

*Fig. 1.* Location of sampling stations in Lake St. Clair. Mussels were collected at Stations 73 and 24 and intact sediment cores were collected at Stations 84, 71, 24, 4, and 14. Station designations are the same as used by Nalepa & Gauvin (1988).

(Stations 73 and 24) on six dates in 1986 (Fig. 1). The two collection sites in 1986 had different substrate types; sandy silt was dominant at Station 73, while silt was dominant at Station 24. Also, submergent macrophytes were present only at the former site during the summer months. Mussels were collected with an epibenthic sled, except in May and September 1985 when mussels were collected by divers. In the first year, excretion rates were determined on the first five to seven individuals collected, regardless of species. However, in the second year excretion was measured on only one species, *Lampsilis radiata siliquoidea*. This species is the dominant mussel in Lake St. Clair; it is distributed throughout the lake and accounts for 45 percent of the entire mussel population (Nalepa & Gauvin, 1988).

After collection, individual mussels were gently scrubbed and immediately placed in polyethylene containers having 2 liters of low-nutrient, particle-free, culture water (Lehman, 1980). The containers were placed in large coolers and the culture water was maintained at the *in situ* temperature. The incubation period lasted 4 hours with 1 ml samples for phosphorus and ammonia determinations drawn at 0, 2, and 4 h. Mussels began to filter water within 30 min after being placed in the containers. Phosphate ($PO_4$–P) concentrations in the culture water were determined with an autoanalyzer (Gardner & Malczyk, 1983) and ammonium ($NH_4$–N) was measured after reaction with o-phthalaldehyde (Gardner, 1978). In 1986, the biodeposition rate was determined by drying and weighing the fecal and pseudofecal material produced during the 4-h incubation period. Phosphorus content of this material was determined with an autoanalyzer after block digestion with mercuric acid, sulfuric acid, and potassium sulfate (Malczyk & Eadie, 1980). Dry weights of the mussels (soft tissue) were determined after drying at 60 °C for at least 48 h. After weighing, the soft tissue was ground to a fine powder and phosphorus content was measured as for the biodeposited material. The nutrient excretion rate of each individual was determined from the slope of a regression line between time and concentration during the 4-h time series. Since both excretion and biodeposition rates were determined on individuals placed in a non-food medium, these rates might be considered conservative estimates.

On most sampling dates in 1986, water samples were taken about 1 meter above the bottom with a Van Dorn water bottle. Total phosphorus, total particulate phosphorus, and total dissolved phosphorus were individually determined in these samples; the particulate fraction was separated by filtering the water through a glass-fiber filter (Malczyk & Eadie, 1980).

Intact cores for measurements of phosphorus flux out of the sediments were collected by divers at five sites in May and September 1985 (Fig. 1). The core tubes (4.2 cm dia. and 10 cm long) were inserted into the sediment about 5 cm, stoppered at both ends, and carefully brought to the surface. The core samples, which contained sediments along with bottom waters immediately overlying the sediments, were kept upright in a cooler during transport to the laboratory and were then placed in an incubator set at the *in situ* temperature. Aeration lines were placed through the top stopper and air was slowly bubbled into the overlying waters. This procedure kept the water well-mixed and also kept dissolved oxygen concentrations at near-saturation levels. All core tubes and aeration lines were made of high-density linear polyethylene to minimize phosphorus adsorption. Samples for phosphorus determinations were taken every 3 to 4 days by drawing out 1 ml of overlying water through a sampling port in the top stopper. Phosphorus levels in lake-water controls were also measured on each sampling day. The volume of overlying water was kept constant by adding 1 ml of lake water after each sample was drawn. The incubation period lasted between 65 and 70 days.

A total of 6 to 8 sediment cores were collected at each site on each sampling date. Since Lake St. Clair is shallow and bottom sediments are easily resuspended, the impact of resuspension on sediment phosphorus release was estimated by mixing the top 1 cm of sediment of one-half of the replicates at the beginning of the incubation period. This created a sediment slurry with the overlying waters. The sediments were mixed again every 10 days until the end of the incubation period. Phosphorus release rates in these mixed cores were compared to release rates in cores that were left undisturbed.

The rate of phosphorus release from the sediments was determined from the slope of a regression line between sampling day and phosphorus concentration in the overlying waters during the incubation period. If the regression was not significant at the 0.05 level, the release rate was considered to be zero. In many cases, phosphorus concentrations increased rapidly at the beginning of the incubation period and then remained relatively constant thereafter. Consequently, a maximal phosphorus release rate was also determined for each replicate core. This rate was calculated using concentrations at the beginning and end of the time period (within each incubation period) when phosphorus concentrations increased most rapidly.

*Table 1.* Phosphorus content of near-bottom waters at each of the two stations sampled in 1986. TSM = total suspended matter (mg l$^{-1}$); TDP = total dissolved phosphorus ($\mu$g l$^{-1}$); TPP = total particulate phosphorus ($\mu$g l$^{-1}$); TP = total phosphorus ($\mu$g l$^{-1}$).

| Sampling Date | Station 73 | | | | Station 24 | | | |
|---|---|---|---|---|---|---|---|---|
| | TSM | TDP | TPP | TP | TSM | TDP | TPP | TP |
| Apr 30 | 4.8 | – | 5.9 | 8.5 | – | – | – | – |
| May 19 | 4.6 | 3.9 | 5.9 | 10.0 | 8.2 | 7.0 | 14.8 | 21.9 |
| Jul 10 | 6.5 | 6.0 | 11.0 | 13.8 | 5.2 | 6.7 | 9.9 | 17.0 |
| Aug 4 | 3.6 | 4.2 | 3.2 | 6.4 | – | – | – | – |
| Sep 16 | 10.5 | 6.3 | 9.7 | 16.1 | 10.2 | 6.1 | 13.0 | 20.0 |
| Oct 15 | 8.8 | 5.6 | 10.0 | 15.2 | 25.5 | 5.1 | 24.2 | 28.1 |

## 4. Results

### 4.1. Phosphorus content of near-bottom waters

In general, more phosphorus was present at Station 24 than at Station 73. (Table 1). The former station was located farther south than the latter and, as mentioned, nutrient concentrations in the water column increase on a gradient from northwest to southeast. Also, the finer sediments present at Station 24 were more likely to be resuspended in the water column; the greatest difference in particulate phosphorus between the two stations occurred in October during the period of frequent storms. The percentage of particulate phosphorus in the seston ranged from 0.09 to 0.19 percent and was highest in spring/early summer at both stations (Fig. 2). This peak likely reflected a large biotic fraction in the water column at this time; chlorophyll *a* concentrations in Lake St. Clair are highest between April and June (Leach, 1972).

### 4.2. Mussel nutrient excretion

Mean rates of phosphorus and ammonia excretion by *Lampsilis radiata siliquoidea* in 1985 and 1986 are given in Table 2. Since excretion rates of individuals from Stations 73 and 24 were not significantly different (t-test; $P < 0.05$) on any of the sampling dates, only a mean value for the two stations is given. Seasonal trends in both phosphorus and ammonia excretion were similar for the two years. Phosphorus excretion in the summer declined from spring values to reach a minimum, and then increased to a peak in the fall;

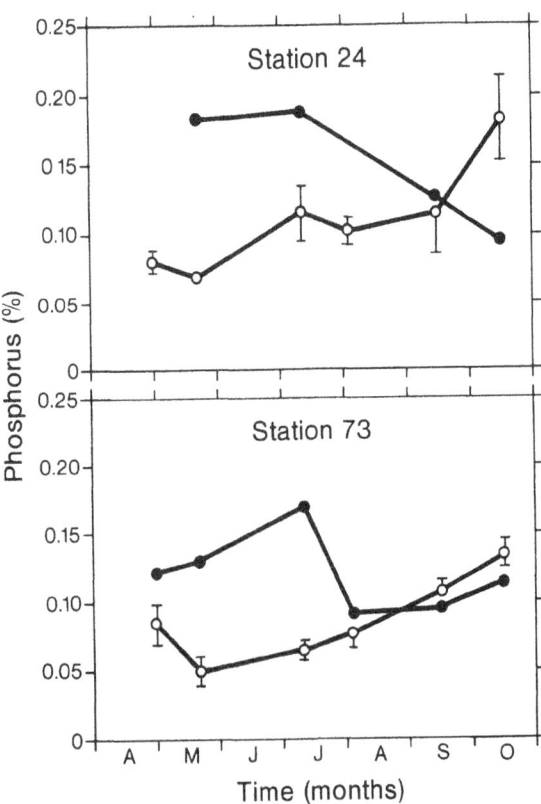

*Fig. 2.* Seasonal fluctuations in the phosphorus content (mean $\pm$ SE) of suspended material in the water column (•–•) and in mussel biodeposits (○–○) at the two sampling stations.

*Table 2.* Mean ($\pm$ SE) excretion rates of PO$_4$–P and NH$_4$–N by *Lampsilis radiata siliquoidea* in Lake St. Clair in 1985 and 1986. Rates given in $\mu$g (g dry wt)$^{-1}$ h$^{-1}$.

| Sampling Date | Temp (°C) | *n* | Mean Dry Wt (g) | Excretion Rate | |
|---|---|---|---|---|---|
| | | | | PO$_4$–P | NH$_4$–N |
| *1985* | | | | | |
| May 14 | 13.0 | 1 | 5.1 | 0.3 | 8.4 |
| Jul 16 | 22.5 | 5 | 3.3 | $<.1 \pm <.1$ | $22.3 \pm 1.4$ |
| Sep 3 | 21.0 | 2 | 1.6 | $1.9 \pm 0.1$ | $9.7 \pm 3.5$ |
| Sep 19 | – | 4 | 2.8 | $1.9 \pm 0.6$ | $12.3 \pm 1.6$ |
| *1986* | | | | | |
| Apr 30 | 10.0 | 9 | 1.7 | $1.1 \pm 0.3$ | $10.5 \pm 1.0$ |
| May 19 | 12.0 | 4 | 1.8 | $0.6 \pm 0.1$ | $9.4 \pm 1.8$ |
| Jul 10 | 22.0 | 9 | 2.3 | $0.5 \pm 0.1$ | $22.6 \pm 1.7$ |
| Aug 4 | 22.7 | 10 | 1.9 | $1.5 \pm 0.3$ | $23.6 \pm 3.8$ |
| Sep 16 | 17.0 | 10 | 1.9 | $1.9 \pm 0.4$ | $18.4 \pm 2.8$ |
| Oct 15 | 12.0 | 10 | 1.8 | $1.9 \pm 0.4$ | $9.6 \pm 1.0$ |

conversely, ammonia excretion was low in the spring and fall but reached a maximum in summer. Excretion rates of other mussel species were similar to those of *L. r. siliquoidea*, but seasonal trends were not as pronounced (Table 3). Reasons for the strong seasonal trends in excretion rates of *L. r. siliquoidea* are not clear, but are likely related to changes in the gametogenic cycle of the organism. *L. r. siliquoidea* is a long-term breeder (Clarke, 1981) in which gametes are formed in late spring/early summer, fertilized and deposited in

the brood pouch in late summer/early fall, and glochidia expelled the following spring/early summer (Coker *et al.*, 1921). Spawning condition affects phosphorus excretion, since a greater portion of assimilated material would be used for the production of gametes and not for metabolism. In the marine mussel *Modiolus*, sexually mature individuals retained more phosphorus during the spawning season (thereby excreting less, given the same food source) than individuals in the spent condition (Kuenzler, 1961). Dramatic

*Table 3.* Mean ($\pm$ SE) excretion rates of PO$_4$–P and NH$_4$–N by various species of mussels in Lake St. Clair in 1985. Rates given in $\mu$g (g dry wt)$^{-1}$ h$^{-1}$.

| Sampling Date | Temp (°C) | *n* | Mean Dry Wt (g) | Excretion Rate | |
|---|---|---|---|---|---|
| | | | | PO$_4$–P | NH$_4$–N |
| *1985* | | | | | |
| May 7 | 13.0 | 5[a] | 2.4 | $0.9 \pm 0.1$ | $11.1 \pm 1.7$ |
| May 14 | 13.0 | 6[b] | 8.0 | $1.0 \pm 0.4$ | $5.3 \pm 1.0$ |
| Jul 16 | 22.5 | 4[c] | 2.4 | $1.8 \pm 0.5$ | $16.4 \pm 1.1$ |
| Sep 3 | 21.0 | 5[d] | 2.3 | $4.7 \pm 1.0$ | $12.1 \pm 0.9$ |
| Sep 19 | – | 3[e] | 5.5 | $2.1 \pm 0.8$ | $12.6 \pm 4.9$ |

[a] *Leptodea fragilis*, 1 *Anodonta grandis*, 1 *Proptera alata*
[b] 2 *L. fragilis*, 1 *Elliptio complanata*, 1 *Lampsilis ovata*, 1 *Ligumea recta*, 1 *P. alata*
[c] 1 *Amblema plicata*, 1 *L. fragilis*, 1 *A. grandis*, 1 *P. alata*
[d] 3 *P. alata*, 2 *L. fragilis*
[e] 1 *A. grandis*, 1 *L. fragilis*, 1 *P. alata*

seasonal trends in the phosphorus content of *L. r. siliquoidea* were not apparent; yet, although differences were not significant, mean content was highest in late spring when gamete production supposedly occurs (Fig. 3). The increase in ammonia excretion in the summer also corresponds to the period of gamete production. Active protein catabolism (and hence increased ammonia excretion) occurs when the glycogen normally used for metabolism is used instead for gamete production (Gabbott & Bayne, 1973; Bayne & Scullard, 1977). The time lag between minimal phosphorus excretion (May) and maximal ammonia excretion (July) may have been a period of gradually increasing protein catabolism. Seasonal trends in food supplies tend to confirm increased phosphorus retention in mid-summer; the proportion of particulate phosphorus in the total seston was actually highest during the mid-summer period when phosphorus excretion rates were minimal (Fig. 2). Of course, this difference may also mean that the particulate phosphorus present in the water in late spring/early summer is

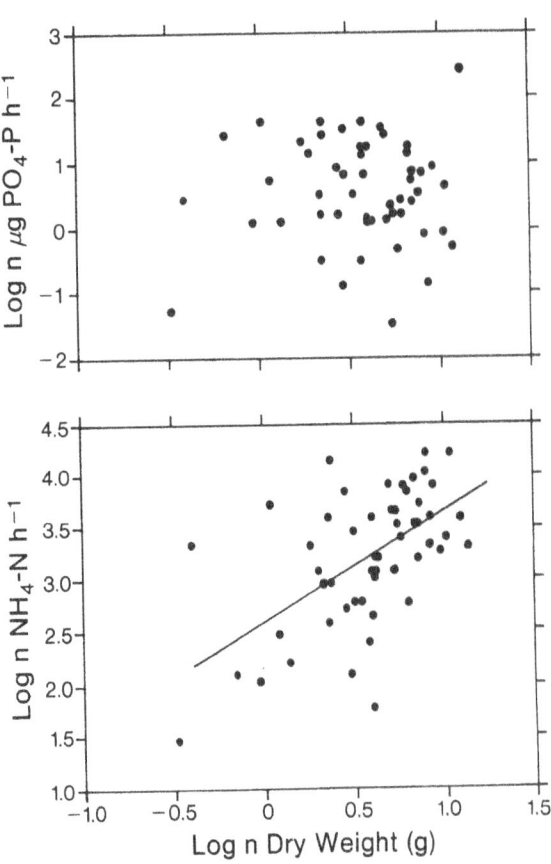

*Fig. 4.* Relationship between phosphorus (upper) and nitrogen (lower) excretion vs. dry-weight body tissue of *Lampsilis r. siliquoidea*. Regression coefficients (a, b) were 0.69 and $-0.42$ for phosphorus and 2.6 and 1.03 for nitrogen. Correlation coefficients ($r$) were 0.16 (non-significant; $P < 0.05$) and 0.63 (significant; $P < 0.05$) for the two nutrients, respectively.

more readily retained than that present at other times.

The relationship between ammonia excretion and tissue dry weight followed the general allometric equation $y = ax^b$, where $y$ = excretion, $x$ = shell-free dry weight, and $a$ and $b$ are regression coefficients, but a similar relationship for phosphorus excretion was not apparent (Fig. 4). This lack of a relationship for phosphorus contrasts to other studies of bivalve excretion which found both phosphorus and nitrogen excretion related to tissue dry weight (James, 1987; Lauritsen & Mozley, 1989). Overall, weight-specific excretion rates were somewhat lower than those found for other bivalves (Table 4).

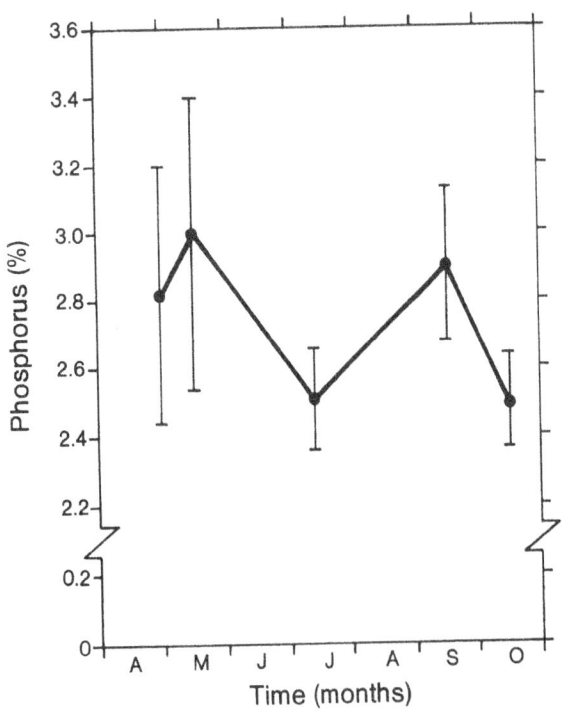

*Fig. 3.* Seasonal fluctuations in the phosphorus content (mean $\pm$ SE) of the body tissue of *Lampsilis r. siliquoidea*.

*Table 4.* Mean ($\pm$ SE) excretion rates of $PO_4$–P and $NH_4$–N by some bivalve species. Estimates are for an animal of about 1 gram. Rates given in $\mu g$ (g dry wt)$^{-1}$ h$^{-1}$.

| Benthic Organism | Excretion Rate | | Source |
|---|---|---|---|
| | $PO_4$–P | $NH_4$–N | |
| Freshwater: | | | |
| *Lampsilis r. siliquoidea* | 1.3 | 16.2 | This study |
| *Hyridella menziesi* | | 41.0 | James (1987) |
| Marine: | | | |
| *Mytilus californianus* | | 23.9 | Bayne *et al.* (1976) |
| *Mytilus edulis* | | 19.7 | Bayne & Scullard (1977) |
| *Geukensia demissa*\* | | 38.0 | Jordan & Valiela (1982) |
| *Modiolus demissus* | 2.7 | | Kuenzler (1961) |

\* *Geukensia demissa = Modiolus demissus.*

### 4.3. Mussel biodeposition

Mussel biodeposition is the sum of pseudofeces (material filtered but not ingested) and feces (material ingested but not assimilated) production. Generally, the phosphorus content of pseudofeces does not change relative to the material filtered (Stanczykowska & Planter, 1985). A comparison of phosphorus content of seston to the phosphorus content of biodeposited material showed that differences were most apparent in late spring/early summer (Fig. 2). At this time, phosphorus concentrations in the seston were high, while phosphorus concentrations in the biodeposited material were low. This further reflects the great efficiency of phosphorus retention during this period.

The mean rate of phosphorus biodeposition

was $6.3\ \mu g\ P\ g^{-1}\ h^{-1}$ and varied from 3.1 to $11.7\ \mu g\ P\ g^{-1}\ h^{-1}$ (Table 5). Differences between the two stations were not significant on any of the sampling dates (t-test, $P < 0.05$). The rates tended to be higher at both stations in the fall. This trend corresponded to the general tendency for rates to be positively related to amounts of particulate phosphorus in the water column (Fig. 5).

### 4.4. Phosphorus release from sediments

Phosphorus release from the sediments at each of the five stations on the two sampling dates is given

*Table 5.* Mean ($\pm$ SE) rates of phosphorus biodeposition by *Lampsilis r. siliquoidea* on each of the sampling dates in 1986. Rates given in $\mu g$ P (g dry wt)$^{-1}$ h$^{-1}$.

| Sampling Date | Phosphorus Biodeposition | |
|---|---|---|
| | Station 73 | Station 24 |
| Apr 30 | 5.8 $\pm$ 2.1 | 4.8 $\pm$ 1.8 |
| May 19 | 3.5 $\pm$ 0.5 | 3.1 |
| Jul 10 | 5.6 $\pm$ 1.7 | 7.2 $\pm$ 2.1 |
| Aug 4 | 5.4 $\pm$ 1.5 | 4.5 $\pm$ 0.7 |
| Sep 16 | 8.5 $\pm$ 1.9 | 11.7 $\pm$ 0.9 |
| Oct 15 | 6.2 $\pm$ 1.2 | 8.8 $\pm$ 3.2 |

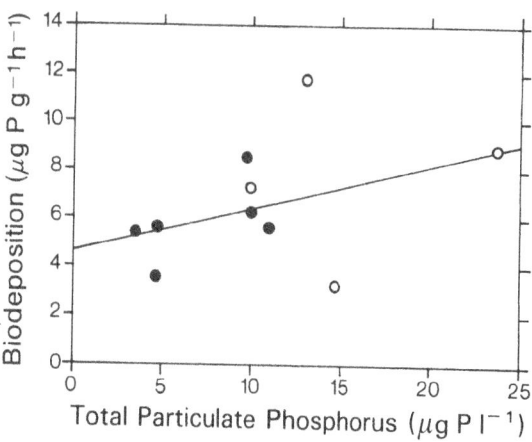

*Fig. 5.* Relationship between biodeposition rates of phosphorus and total particulate phosphorus in the water column ($r = 0.39$). Closed circles are Station 73 and open circles are Station 24.

*Table 6.* Mean ($\pm$ SE) rates of $PO_4$-P release from Lake St. Clair sediments at each of the stations on the two sampling dates in 1985. Maximum mean release rates are given in parentheses. Rates are given as $\mu g \, P \, m^{-2} \, d^{-1}$. Phosphorus release from sediments at Station 14 in September occurred early in the incubation period and then declined.

| Station | Sampling Date | |
|---|---|---|
| | May[1] | September[2] |
| 4 | 32.8 $\pm$ 4.7 (51.7) | 11.0 $\pm$ 6.3 (37.6) |
| 14 | 14.0 $\pm$ 7.4 (68.3) | 0.0 $\pm$ 0.0 (78.2) |
| 24 | 29.4 $\pm$ 9.9 (51.4) | 9.6 $\pm$ 6.2 (62.8) |
| 71 | 2.5 $\pm$ 0.9 (21.6) | 0.4 $\pm$ 0.4 (10.5) |
| 84 | 5.8 $\pm$ 2.7 (15.8) | 3.1 $\pm$ 3.1 (32.6) |

[1] Water temperature = 13 °C.
[2] Water temperature = 22 °C.

in Table 6. Mean release rates in this table include values from all replicates at a given station since there were no significant differences (t-test; $P < 0.05$) between release rates of mixed and unmixed cores at any of the five stations.

Release rates at the two sites in the northwestern portion of the lake (Stations 71 and 84) tended to be lower than rates at the three sites farther south (Stations 4, 14, and 24), which follows the trend of greater overall nutrient concentrations from northwest to southeast. Overall, release rates were higher in May than in September, despite the lower water temperatures during the former month. This difference can likely be attributed to the settling and subsequent mineralization of the spring phytoplankton bloom.

The net release of phosphorus from Lake St. Clair sediments was generally lower than sediment release rates in other areas of the Great Lakes. The mean release rate in this study was $11 \, \mu g \, P \, m^{-2} \, d^{-1}$ and the mean maximal release rate was $43 \, \mu g \, P \, m^{-2} \, d^{-1}$. In comparison, release rates of $170-570 \, \mu g \, P \, m^{-2} \, d^{-1}$ were reported from nearshore Lake Michigan (Quigley & Robbins, 1986) and $30-800 \, \mu g \, P \, m^{-2} \, d^{-1}$ were reported from Lake Ontario (Bannerman *et al.*, 1974). Because Lake St. Clair is relatively shallow, wave-induced resuspension occurs quite frequently (4 to 6 resuspension events per month in the spring and fall and 1 to 2 events per month

in the summer; N. Hawley, pers. com.). Resuspension diminishes the potential for phosphorus release from the sediments by keeping the upper sediments well oxidized; under such conditions phosphorus tends to remain strongly sorbed to the sediments (Syers *et al.*, 1973). In addition, frequent resuspension of the upper sediments reduces pore-water gradients and thus the potential for release through diffusive flux (Quigley & Robbins, 1986).

## 5. Discussion

The cycling of phosphorus through the mussel population of Lake St. Clair during the May–October period can be estimated from the determined values of concentrations and flux rates (Fig. 6). Values determined for *L. r. siliquoidea*, which was the dominant and most widespread species, were assumed to be representative of the entire mussel population. Phosphorus removal from the water column by mussels is the product of concentrations in the water, filtration rates, time spent filtering, and density. The mean concentration of particulate phosphorus 1 m above the bottom was $10.7 \, \mu g \, l^{-1}$ (Table 1), the filtration rate was $460 \, ml \, g^{-1} \, h^{-1}$ (Vanderploeg *et al.*, in prep.), and the mean biomass of the mussel population in the lake was 4.4 g (shell-free dry wt) $m^{-2}$ (Nalepa & Gauvin, 1988). Removal

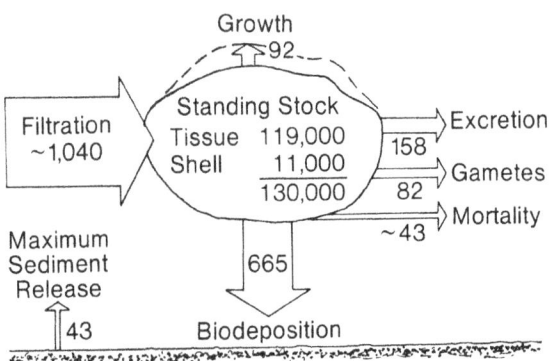

*Fig. 6.* Phosphorus flow through the mussel population in Lake St. Clair for the May–October 1986 period. Units are in $\mu g \, P \, m^{-2} \, d^{-1}$ except for standing stock which is in $\mu g \, P \, m^{-2}$.

rates of phosphate phosphorus and dissolved organic phosphorus were not considered since phosphate uptake comprises only a small fraction of total uptake (0.6% of particulate phosphorus uptake; Kuenzler, 1961) and the use of dissolved organic phosphorus by mussels has not been proved. The filtration rate of *L. r. siliquoidea* was determined in October 1987 and compares favorably to filtration rates of other mussel species of comparable size (Alimov, 1969; Lewandowski & Stanczykowska, 1975; Paterson & Cameron, 1985). The proportion of time spent filtering by individuals during the determination of their filtration rates was assumed to be equal to the proportion of time spent filtering by *in situ* populations. Given all the above values, the removal rate of phosphorus by mussels in Lake St. Clair was 520 $\mu$g P m$^{-2}$ d$^{-1}$. However, in examining other measured components of phosphorus flux through the population, the amount of phosphorus filtered should have been much higher. For instance, the mean excretion rate was 158 $\mu$g P m$^{-2}$ d$^{-1}$ and the biodeposition rate was 665 $\mu$g P m$^{-2}$ d$^{-1}$, giving a total of 823 $\mu$g P m$^{-2}$ d$^{-1}$ or an amount 1.6 times higher than that filtered only for these two budget components. A possible reason for this discrepency is that amounts of phosphorus in the water column at 1 m above the bottom are not the same as amounts occurring at a few centimeters above the bottom where the mussels are actually filtering. Indeed, although Lake St. Clair is well mixed, amounts of suspended material are about two times higher at 0.4 m above the bottom than at 1 m (Hamblin *et al.*, 1987). Therefore, a more accurate estimate of the amount of phosphorus actually filtered would be at least 1040 $\mu$g P m$^{-2}$ d$^{-1}$, or two times the amount originally calculated, assuming a similar proportional increase in the amount of particulate phosphorus. Since biodeposition equaled 665 $\mu$g P m$^{-2}$ d$^{-1}$, about 375 $\mu$g P m$^{-2}$ d$^{-1}$ or 36 percent of that assumed to be filtered is assimilated by the population. This proportion is generally comparable to that found for other bivalve species from a variety of habitats. For example, the freshwater mussel *Dreissena poly-*

*morpha* assimilated 49 percent of the phosphorus filtered in some Polish lakes (Stanczykowska & Planter, 1985) and *Hyridella menziesi* assimilated 80 percent in an oligotrophic lake in New Zealand (James, 1987). The estuarine species *Corbicula japonica* assimilated 41 percent in a poikilohaline lagoon (Fuji, 1979). In contrast, the marine mussel *Mytilus* effectively assimilated only 6 percent of the phosphorus filtered in a Georgia salt marsh (Kuenzler, 1961). This proportion is likely more a function of the quality and quantity of seston, and subsequently the amount of pseudofeces produced, rather than of great differences in assimilation efficiencies. Typically, as amounts of seston increase, the amount lost as pseudofeces also increases (Tenore & Dunstan, 1973; Fuji, 1979; Stanczykowska & Planter, 1985). Particulate phosphorus concentrations were ten times higher in the Georgia salt marsh (44 $\mu$g l$^{-1}$) than in the New Zealand lake (4 $\mu$g l$^{-1}$).

Of the phosphorus assimilated by the Lake St. Clair mussel population, 158 $\mu$g m$^{-2}$ d$^{-1}$ or 42 percent was excreted, with the remainder used by the population for production of biomass and generative elements, or lost through mortality during the period. The amount incorporated into standing stocks of the population can be estimated from growth rates and average content in the body and shell fractions. Production of *L. r. siliquoidea* in Lake St. Clair is 0.2 g m$^{-2}$ y$^{-1}$ and mean biomass is 1.5 g m$^{-2}$ (Nalepa & Gauvin, 1988). Extrapolated over the entire population, the mussel production rate in the lake is equal to 0.57 g m$^{-2}$ y$^{-1}$. Considering that most all growth in mussels occurs during the warmer months (Isely, 1914; Coker *et al.*, 1921), production during the May–October period was about 3.1 mg m$^{-2}$ d$^{-1}$. Mean phosphorus content of mussel tissue was 2.7 percent, so the amount stored in the body fraction equals 119,000 $\mu$g P m$^{-2}$ and the rate of incorporation into the body equals 84 $\mu$g P m$^{-2}$ d$^{-1}$. The standing stock of shells was 55.5 g m$^{-2}$ (Nalepa unpublished). Assuming that the phosphorus content of the shell fraction is equal to 0.02 percent (Kuenzler, 1961) and that phosphorus is incorporated into the shell fraction at the same rate as into the body fraction, phosphorus contained in the shells equaled 11 100 $\mu$g P m$^{-2}$ and incorporation into the shell fraction was 8 $\mu$g P m$^{-2}$ d$^{-1}$. Thus, the total amount of phosphorus incorporated into growth for the period was 92 $\mu$g P m$^{-2}$ d$^{-1}$.

The amount of phosphorus lost from the population in gametes can be inferred from the decrease in phosphorus content of the population over the study period (Kuenzler, 1961). Mean phosphorus content of the soft tissue was 2.82 percent on the first sampling date (April 30) and 2.48 percent on the last sampling date (October 15). Therefore, the net loss for the May–October period was 82 $\mu$g P m$^{-2}$ d$^{-1}$. Overall, the amount incorporated into biomass and generative elements was equal to 174 $\mu$g P m$^{-2}$ d$^{-1}$, or 17 percent of the amount filtered; this compares to the 12 percent estimated for the freshwater mussel *Dreissena polymorpha* (Stanczykowska & Planter, 1985).

The sum of biodeposition, growth, excretion, and gamete loss over the period, all measured independently, equaled 997 $\mu$g P m$^{-2}$ d$^{-1}$. This sum is very similar to the estimated filtration rate of 1040 $\mu$g P m$^{-2}$ d$^{-1}$. The difference of 43 $\mu$g P m$^{-2}$ d$^{-1}$ may be shown in Fig. 6 as mortality loss, which is the remaining pathway for phosphorus flux through the population during the period. This loss seems reasonable since mortality in adult mussels is generally low. The estimate of phosphorus loss due to mortality was 0.03 percent of the phosphorus contained in standing stocks (tissue and shell), whereas this loss was 0.06 percent in the marine mussel *Modiolus demissus* (Kuenzler, 1961).

As noted, total biodeposition by the mussels equaled 665 $\mu$g P m$^{-2}$ d$^{-1}$. This amount is potentially available for further use by deposit-feeding benthic invertebrates along with material that has been deposited on the bottom through natural sedimentation. Release from the sediments averaged 11 $\mu$g P m$^{-2}$ d$^{-1}$ with a mean maximal release rate of 43 $\mu$g P m$^{-2}$ d$^{-1}$. Because additional phosphorus was not released when the sediments were resuspended, the difference between sediment input and sediment release is phosphorus which either remains bound to sediment particles and is transported out of the system or buried, or which is incorporated into standing stocks of macrophytes or other benthic invertebrates besides the mussels. Accumulation in the sediments is not likely because wave-induced resuspension of the upper sediments is intense and phosphorus concentrations in the upper sediments are quite low (0.01% to 0.07%; Upper Great Lakes Connecting Channels Study – Sediment Workgroup, unpublished report). In addition, estimates of phosphorus loads and losses are not significantly different, indicating no apparent sources or sinks within the lake (Lang *et al.*, 1988).

The role of mussels in cycling phosphorus in Lake St. Clair can be evaluated by comparing rate measurements to total loading from outside sources. The total mean load of phosphorus into the lake is about 3100 MT (metric tons) y$^{-1}$ or 1550 MT for the May–October period (Lang *et al.*, 1988). Over this same period, mussels filtered 210 MT of phosphorus, or 13.5 percent of the total load. Part of the phosphorus is incorporated into biomass or excreted, but a large amount, 134 MT, is sedimented to the bottom via biodeposition. Thus, it appears that the primary role of mussels is to increase the retention of phosphorus in the lake by removing particles from the water column and depositing them on the bottom. Small suspended particles that may not easily settle to the bottom are made available as a food resource to other benthic components, thereby stimulating benthic productivity. Standing stocks of deposit-feeders are generally higher in areas with mussels than in areas without them (Sephton *et al.*, 1980; Radziejewska, 1986). Also, mussel biodeposits can stimulate the growth of aquatic macrophytes, presumably by enhancing nutrient levels in the sediments (Bertness, 1984). Phosphorus uptake by macrophytes in Lake St. Clair has been estimated at 219 MT y$^{-1}$ (Lang *et al.*, 1988). Mussel biodeposition may be especially important in enhancing nutrient availability in Lake St. Clair considering the high frequency of resuspension in the lake and the low rate of natural sedimentation.

# 6. Acknowledgements

We would like to thank M. Duquet of the R/V Bluewater for his outstanding support during field operations, J. Cavaletto for her help in making the nutrient determinations, and M. Quigley for his insightful comments on the manuscript.

## Note added in proof

This article was completed before the discovery and subsequent population increase of the zebra mussel, *Dreissena polymorpha* in Lake St. Clair. Since the phosphorus budget in this article is based on the activities of the unionid population only, all estimates of the role of the entire mussel population in the Lake must now be considered as minimum values.

## References

Alimov, A. F., 1969. Nekotorye obscie zakonomernosti processa filtracii u dvustvorcatych molljuskov. Z. Obsc. Biol. 30: 621–631.

Bannerman, R. T., D. E. Armstrong, G. C. Holdren & R. F. Harris, 1974. Phosphorus mobility in Lake Ontario sediments (IFYGL), Proc. 17th Conf. Great Lakes Res., pp. 158–178.

Bayne, B. L. & C. Scullard, 1977. Rates of nitrogen excretion by species of *Mytilus* (Bivalvia: Mollusca). J. Mar. Biol. Assoc. U.K. 57: 355–369.

Bayne, B. L., C. J. Bayne, T. C. Carefoot & R. J. Thompson, 1976. The physiological ecology of *Mytilus californianus* Conrad. 1. Metabolism and energy balance. Oecologia 22: 211–228.

Bertness, M. D., 1984. Ribbed mussels and *Spartina alterniflora* production in a New England salt marsh. Ecology 65: 1794–1807.

Clarke, A. H., 1981. The Freshwater Molluscs of Canada. Ntnl. Mus. Can., Ottawa.

Coker, R. E., A. F. Shira, M. W. Clarke & A. W. Howard, 1921. Natural history and propagation of fresh-water mussels. Bull. U.S. Bur. Fish. 37: 79–181.

Fuji, A., 1979. Phosphorus budget in natural population of *Corbicula japonica* Prime in poikilohaline lagoon, Zyusan-ko. Bull. Fac. Fish. Hokkaido Univ. 30: 34–49.

Gabbott, P. A. & B. L. Bayne, 1973. Biochemical effects of temperature and nutritive stress on *Mytilus edulis* L. J. Mar. Biol. Assoc. U.K. 53: 269–286.

Gardner, W. S., 1978. Microflurometric method to measure ammonium in natural waters. Limnol. Oceanogr. 23: 1069–1072.

Gardner, W. S. & J. M. Malczyk, 1983. Discrete injection flow analysis of nutrients in small-volume water samples. Anal. Chem. 55: 1645–1647.

Gardner, W. S., T. F. Nalepa, M. A. Quigley & J. M. Malczyk, 1981. Release of phosphorus by certain benthic invertebrates. Can. J. Fish. Aquat. Sci. 38: 978–981.

Hamblin, P. F., F. M. Boyce, J. Bull, F. Chiocchio & D. Robertson, 1987. Report to UGLCCS Work Groups. Ntnl. Wat. Res. Inst. Burlington, Ontario. Contrib. 87–87.

Herdendorf, C. E., C. N. Raphael & E. Jaworski, 1986. The ecology of Lake St. Clair wetlands: a community profile. U.S. Fish Wild. Serv. Biol. Rep. 85(7.7), Washington D.C.

Isely, F. B., 1914. Experimental study on the growth and migration of freshwater mussels. U.S. Bur. Fish. Doc. 7922, Washington, D.C.

James, M. R., 1987. Ecology of the freshwater mussel *Hyridella menziesi* (Gray) in a small oligotrophic lake. Arch. Hydrobiol. 3: 337–348.

Jordan, T. E. & I. Valiela, 1982. A nitrogen budget of the ribbed mussel, *Geukensia demissa*, and its significance in nitrogen flow in a New England salt marsh. Limnol. Oceanogr. 27: 75–90.

Kasprzak, K., 1986. Role of Unionidae and Sphaeriidae (Mollusca, Bivalvia) in the eutrophic Lake Zbechy and its outflow. Int. Rev. Ges. Hydrobiol. 71: 315–334.

Kuenzler, E. J., 1961. Phosphorus budget of a mussel population. Limnol. Oceanogr. 6: 400–415.

Lang, G. A., J. A. Morton & T. D. Fontaine, 1988. Total phosphorus budget for Lake St. Clair: 1975–80. J. Great Lakes Res. 14: 257–266.

Lauritsen, D. D. & S. C. Mozley, 1989. Nutrient excretion by the Asiatic clam *Corbicula fluminea*. J. North Amer. Benthol. Soc. 8: 134–139.

Leach, J. H., 1972. Distribution of chlorophyll *a* and related variables in Ontario waters of Lake St. Clair. Proc. 15th Conf. Great Lakes Res., pp. 80–86.

Leach, J. H., 1973. Seasonal distribution, composition and abundance of zooplankton in Ontario waters of Lake St. Clair. Proc. 16th Conf. Great Lakes Res., pp. 54–64.

Leach, J. H., 1980. Limnological sampling intensity in Lake St. Clair in relation to distribution of water masses. J. Great Lakes Res. 6: 141–145.

Lehman, J. H., 1980. Release and cycling of nutrients between planktonic algae and herbivores. Limnol. Oceanogr. 25: 620–632.

Lewandowski, K. & A. Stanczykowska, 1975. The occurrence and role of bivalves of the family Unionidae in Miklajskie Lake. Ekol. Pol. 23: 317–334.

Malczyk, J. M. & B. J. Eadie, 1980. Collection, preparation, and analysis procedures employed and precision achieved in the chemical field program, 1976–79. Open File Report No. 226. Great Lakes Env. Res. Lab., Ann Arbor, MI.

Nalepa, T. F. & J. M. Gauvin, 1988. Distribution, abundance, and biomass of freshwater mussels (Bivalvia:

Unionidae) in Lake St. Clair. J. Great Lakes Res. 14: 411–419.

Paterson, C. C. & I. F. Cameron, 1985. Comparative energetics of two populations of the unionid, *Anodonta cataracta* (Say). Freshwat. Invertebr. Biol. 4: 79–90.

Quigley, M. A. & J. A. Robbins, 1986. Phosphorus release processes in nearshore southern Lake Michigan. Can. J. Fish. Aquat. Sci. 43: 1201–1207.

Radziejewska, T., 1986. On the role of *Mytilus edulis* aggregations in enhancing meiofauna communities off the southern Baltic coast. Ophelia (Suppl. 4): 211–218.

Sephton, T. W., C. G. Paterson & C. H. Fernando, 1980. Spatial interrelationships of bivalves and nonbivalve benthos in a small reservoir in New Brunswick, Canada. Can. J. Zool. 58: 852–859.

Stanczykowska, A. & M. Planter, 1985. Factors affecting nutrient budget in lakes of the R. Jorka watershed (Masurian Lakeland, Poland) X. Role of the mussel *Dreissena polymorpha* (Pall.) in N and P cycles in a lake ecosystem. Ekol. Pol. 33: 345–356.

Syers, J. K., R. F. Harris & D. E. Armstrong, 1973. Phosphate chemistry in lake sediments. J. Env. Qual. 2: 1–14.

Tenore, K. R. & W. M. Dunstan, 1973. Comparison of feeding and biodeposition of three bivalves at different food levels. Mar. Biol. 21: 190–195.

Walz, V. N., 1978. Die Produktion der Dreissena-Population und deren Bedeutung im Stoffkreislauf des Bodensees. Arch. Hydrobiol. 82: 482–499.

*Hydrobiologia* **219**: 251–268, 1991.
*M. Munawar & T. Edsall (eds),*
*Environmental Assessment and Habitat Evaluation of the Upper Great Lakes Connecting Channels.*
© 1991 *Kluwer Academic Publishers.*

# Biology of the exotic zebra mussel, *Dreissena polymorpha*, in relation to native bivalves and its potential impact in Lake St. Clair

G.L. Mackie

*Department of Zoology, University of Guelph, Guelph, Ontario N1G 2W1, Canada*

*Key words:* Zebra mussel, *Dreissena*, biology, impact, exotic, Lake St. Clair, shell form, mode of life, reproduction, life cycle, population dynamics, dispersal, distribution, physiology, impact

## Abstract

The zebra mussel, *Dreissena polymorpha*, is a new exotic species that was introduced into the Great Lakes as early as the fall of 1985. It differs markedly from native species of bivalves in its: (i) shell form; (ii) mode of life; (iii) reproductive potential; (iv) larval life cycle; (v) population dynamics; (vi) distribution, (vii) dispersal mechanisms; (viii) physiology; (ix) potential impact on the ecosystem; and (x) impact on society and the economy. In body form, it has an anterior umbone, a flat ventral surface with permanent aperature for the byssal apparatus and a shape that together make the animal well adapted for life on a hard surface. The shell has a zebra-stripe pattern, a heteromyarian muscle condition and lacks hinge teeth which make it easily identifiable from native bivalves. The zebra mussel is strongly byssate and has an epifaunal mode of life not seen in native bivalves. The species is dioecious and has external fertilization, the eggs developing into pelagic veligers which remains planktonic for approximately 4 weeks. Gametogenesis begins in late winter to early spring, veligers appear in the water column in late May to early June and disappear in mid to late October in Lake St. Clair. Adults live for about 2 years and have very rapid growth rates. Maximum shell lengths average 2.3 to 2.5 cm. Standing crops as high as $200\,000\ \mathrm{m}^{-2}$ are present in the 1-m depths of the Ontario shores. Infestations may be interfering with the normal metabolism of native unionid clams and there is potential of the unionid clam populations being reduced or even eliminated from Lake St. Clair.

## 1. Introduction

The zebra mussel, *Dreissena polymorpha* (Pallas) was introduced into Lake St. Clair probably in 1985 (Hebert *et al.*, 1989). The species is native to Europe and appears to have been transported to Lake St. Clair in the freshwater ballast of a transoceanic ship. Until recently, ships have been discharging some of their ballast water into the Great Lakes and released either veliger larvae or young adults into Lake St. Clair. Adult shells about 2 cm

long were first found in August 1988, near Belle River in Lake St. Clair (Hebert *et al.*, 1989). Assuming it takes the species about 2 years to grow to 2 cm, it is most likely that the species was introduced in either the spring or summer of 1986 as young adults or in the summer or fall of 1985 as veliger larvae. Since adults are benthic and larvae are pelagic, and ships normally take on ballast water from the pelagic zone, the species was probably introduced as veliger larvae in the summer or fall of 1985. The species was not found in an extensive benthic survey in 1983 (Griffiths, 1987) or in an extensive unionid clam survey in 1986 (Tom Nalepa, U.S. Fish and Wildlife, pers. comm.). However, the numbers and sizes of adults of the pioneering population were probably so small that they could easily have been missed in their surveys.

*Dreissena polymorpha* differs from native species of bivalves in: (i) shell form; (ii) mode of life; (iii) reproductive potential; (iv) larval life cycle; (v) population dynamics; (vi) distribution; (vii) dispersal mechanisms; (viii) physiology, (ix) potential impact on the ecosystem; and (x) impact on society and the economy. This paper describes the most outstanding features in zebra mussels for each of these eleven attributes. The features are compared to native species of bivalves in the families Sphaeriidae and Unionidae in general, and to the Asian clam, *Corbicula fluminea*. The Asian clam is another exotic bivalve that was introduced to North America in the early 1930's in California (see Britton & Morton (1979) and McMahon (1982) for reviews). It has since spread throughout the United States but has not succeeded in invading the Great Lakes, although it has been found in warm water plumes of hydrogenerating plants or as small, somewhat benign populations in Lake Erie (Clarke, 1981; Scott-Wasilk *et al.*, 1983), Lake Michigan (White *et al.*, 1984) and the St. Clair River (French & Schloesser, 1991). Asiatic clams belong to the family Corbiculidae and are related to the native Sphaeriidae, both being of the superfamily Corbiculacea (Mackie, 1990). The comparisons between zebra mussels and other bivalves clearly demonstrate why the zebra mussel will probably become the dominant bivalve in Lake St. Clair, and in the Great Lakes.

## 2. Descriptions

### 2.1. Shell form

The shell of *D. polymorpha* is distinctive (Fig. 1a, b). The common name, zebra mussel, is derived from the pattern of zebra stripes on the shell and the scientific name, *polymorpha*, refers to the many morphs or forms that occur in the shell's colour pattern, including albino and solid black or brown. In spite of the variability in shell morphs, there are some significant functional, diagnostic features in external and internal shell morphology.

Externally, the umbone of the zebra mussel is acute and lies anterior in position (Fig. 1a). The ventral side of the shell is flattened and has a permanent opening through which the byssal apparatus extends. The shell is medium-sized, averaging 2.3 to 2.5 cm maximum length in Lake St. Clair. The shell shape is perfectly adapted to life on hard substrates: (1) The flat ventral surface allows the animal to be pulled tightly against the substrate by its byssal apparatus, making it difficult for predators to pry the shell from the substrate. (2) The umbone is lateral and adjacent to the substrate, giving the animal maximum upright stability at the surface of the substrate. (3) The shell is tapered dorsally (tent-shaped) making it difficult for predators to get a firm hold to pull the shell from the hard surface.

In other bivalves (see Clarke, 1973 for general reviews), the umbone is rounded and dorsal in position (Fig. 1c–h) and adapted for an infaunal life within sediments. The ventral margin is also rounded and lacks a permanent opening because adults do not have a functional byssal gland. *Corbicula* is medium-size, averaging 3 to 4 cm shell length (Fig. 1c, d) but has yet to be recorded from Lake St. Clair. The shell is glossy and has heavy, evenly-spaced ridges (Fig. 1c). Sphaeriids are small, averaging 5 to 8 mm shell length (Fig. 1e, f), and the shells are thin and fragile. Unionids are large, averaging about 10 cm shell length (Fig. 1g, h) and the shells are typically thick and robust.

Internally, the shell of the zebra mussel lacks cardinal and pseudocardinal teeth (Fig. 1b) but a vestigial lateral tooth may be present (Fig. 1b). The ligament is internal, anterior, and sunk in an elongated pit alongside the vestigial lateral tooth (when present). The umbonal end of the shell bears an apical septum or myophore plate (Fig. 1b) to which attaches the small anterior adductor muscle. The posterior end of the shell

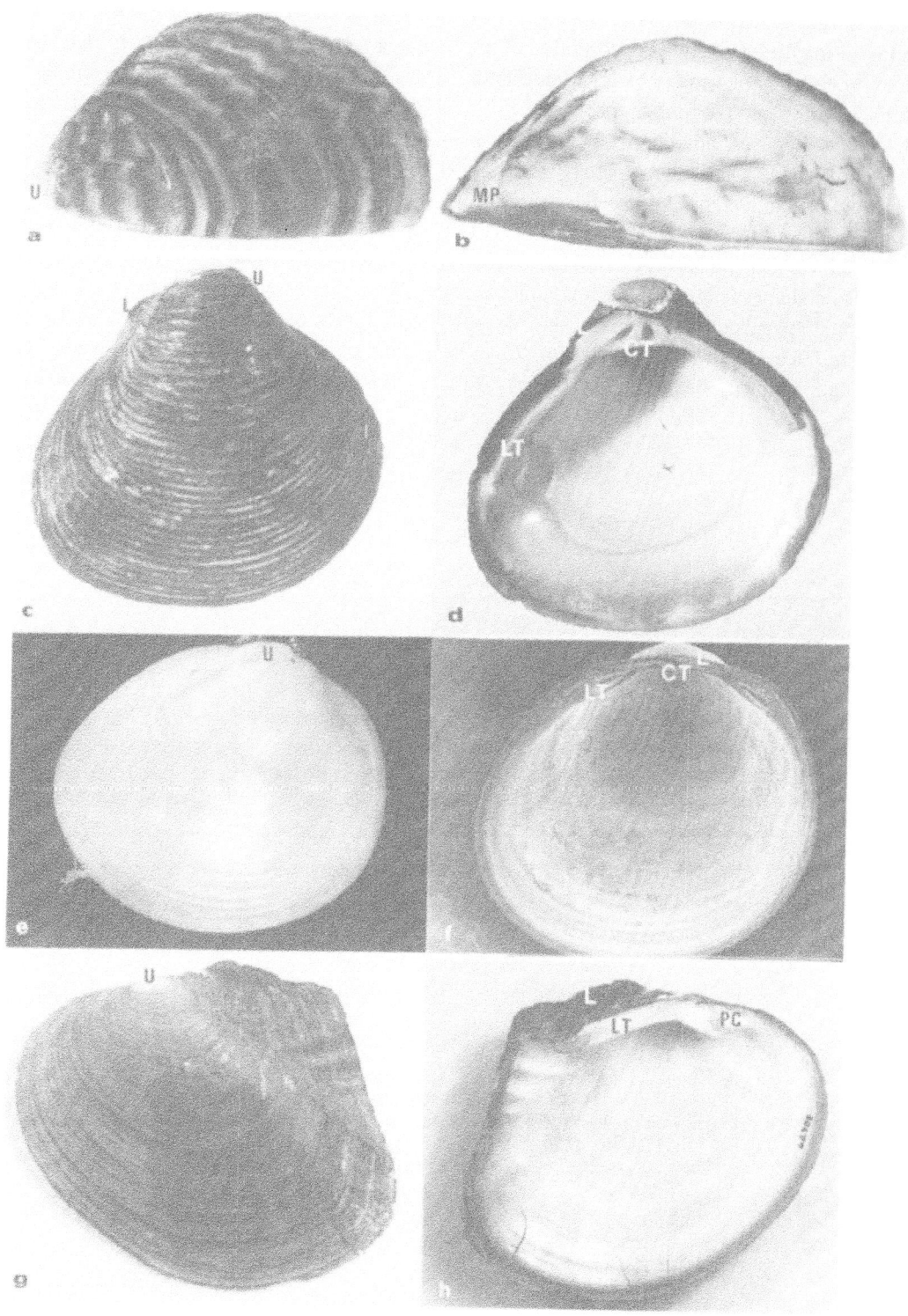

*Fig. 1.* External (a, c, e, g) and internal (b, d, f, h) features of shells of typical species in four families of bivalves in the Great Lakes. (a, b) Dreissenidae: *Dreissena polymorpha* from Lake St. Clair, shell 2 cm long. (c, d) Corbiculidae: *Corbicula fluminea*, another biofouling bivalve in North America, but not presently in Lake St. Clair, shell 3 cm long. (e, f) Sphaeriidae: *Pisidium lilljeborgi* shown, shell 2 mm long. (g, h) Unionidae: *Lasmigona complanata* shown, shell 9 cm long. CT = cardinal tooth; L = ligament; LT = lateral tooth; MP = myophore plate; PC = pseudocardinal tooth; U = umbone.

bears a large posterior adductor muscle scar, demonstrating the heteromyarian condition of the species (i.e. anterior and posterior adductor muscles of different size and shape).

In other bivalves, distinctive cardinal or pseudocardinal teeth and lateral teeth (Figs. 1f, h) are present; Asiatic clams have heavy triradiate cardinal teeth and serrated lateral teeth (Fig. 1d), sphaeriids have delicate cardinal and lateral teeth of diagnostic shapes (Fig. 1f), and most unionids have robust pseudocardinal teeth and large lamelliform lateral teeth (Fig. 1h) (some species, like *Anodonta grandis*, being an exception). All native bivalves and *Corbicula* have an external ligament, all lack a myophore plate and all are isomyarian, as indicated in the similar sizes and shapes of the scars for the anterior and posterior adductor muscles.

## 2.2 Mode of life

*Dreissena* is epifaunal, living byssally attached to all types of solid substrates, including rocks, floating and sunken logs, breakwalls, and various debris. They also attach to large, living inverte-

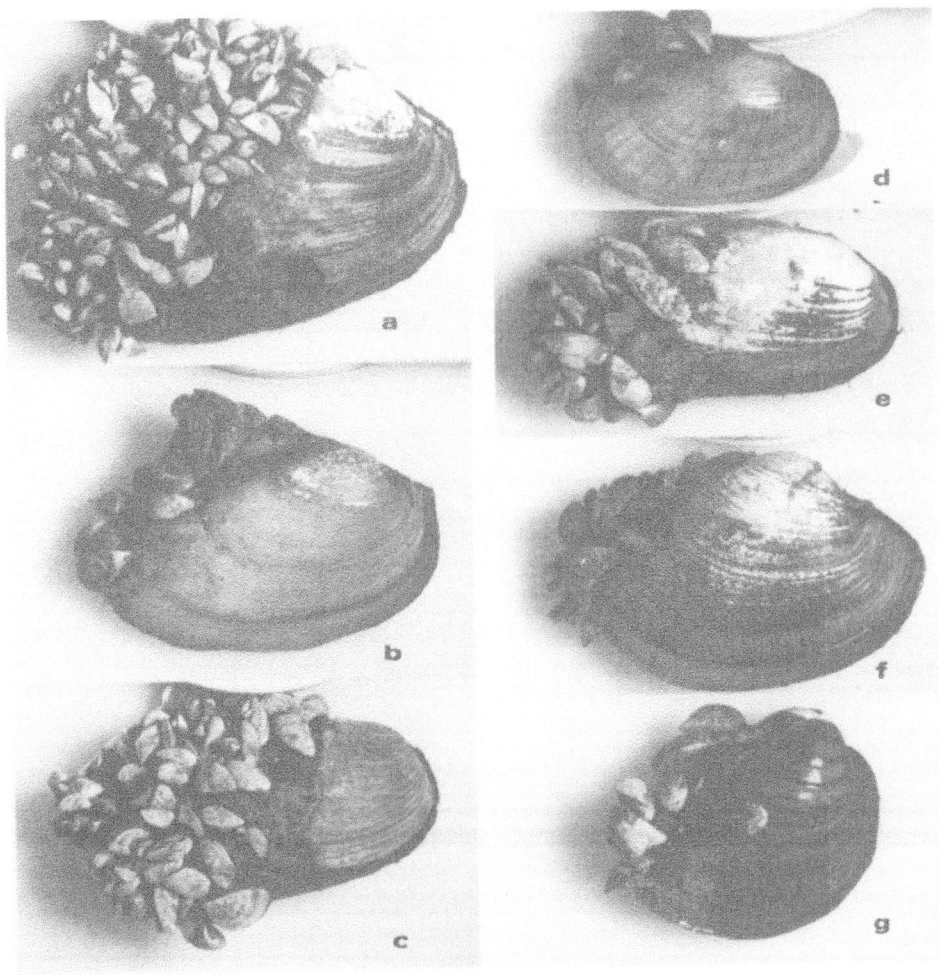

*Fig. 2.* Species of unionid clams from Lake St. Clair encrusted with zebra mussels. (a) *Proptera alata*; (b) *Lepodea fragilis*; (c) *Elliptio dilatata*; (d) *Lampsilis radiata siliquoidea*; (e) *Actinonais carinata*; (f) *Strophitus undulatus*; (g) *Pleurobema coccineum*.

brates such as unionid shells (Fig. 2) and crayfish. All other bivalves are infaunal, either living buried in the sediments or living at the mud-water interface partially buried in the sediments. Although unionid clams are infaunal and occupy an entirely different niche than zebra mussels, the unionids are ideal substrates for zebra mussels to colonize. Typically, only the posterior third to half of the unionid shell projects above the surface of the substrate while it is siphoning for respiratory and feeding purposes. Hence, only the posterior part of a unionid shell is usually colonized by zebra mussels (Fig. 2). The densities of zebra mussels on unionid shells is highly variable (Fig 12), but in 1989 averaged nearly 300 per unionid shell throughout Lake St. Clair. This is an order of magnitude higher than first reported in 1988 by Hebert *et al.* (1989). However, densities as high as 10 000 (D.W. Schloesser, U.S. Fish and Wildlife, per. comm.) and 15 000 (P.D.N. Hebert, University of Windsor) have been reported on some unionid shells in Lake St. Clair. All other freshwater bivalves, including *Corbicula*, are infaunal and typically live within the substrate or at the mud-water interface. Some species exhibit substrate preferences (e.g. *Musculium transversum* (Gale, 1969)), but most occur in a variety of substrate types, the particle sizes ranging from mud to coarse gravel. A few species (e.g. *Sphaerium fabale, Pisidium fallax*) also prefer to live in rivers and streams with some current. Currents may affect the shell shape of some species of bivalves (Bailey *et al.*, 1983; Mackie & Topping, 1988).

## 2.3. *Reproductive potential*

Zebra mussels are dioecious. Based on histological examinations of 40 specimens selected at random, the sex ratio of zebra mussels in Lake Erie is about 3 females: 2 males. Gonads are ripe at least by the first week of May (based on specimens first collected on May 2, 1988; Fig. 3a, b), indicating that gametogenesis begins in early spring, or perhaps even late winter. However, veligers do not appear in the water until the end of May to the first week of June (Fig. 3b, c show-

ing partially spent gonads), indicating that gametes are not released immediately. The water temperature was 15 °C when larvae were first seen in Lake St. Clair. The temperature threshold for reproduction in Polish populations of zebra mussels is 15 to 17 °C (Stanczykowska, 1977).

The eggs are 30 to 50 $\mu$ dia. According to Stanczykowska (1977), zebra mussels have exceedingly high fecundities, ranging from about 30 000 per female in their first year of sexual maturity to 40 000 in their third and fourth years. Oogenesis and spermiogenesis appear to occur at about the same time. In Lake St. Clair, most zebra mussels are sexually mature by the time they reach 8 to 10 mm shell length; a few mature at even 5.5 mm shell length (Gillis, 1989). Peak reproduction normally occurs in July and August, as indicated by the large proportion with partially spent gonads (Fig. 3e, f). Since most larvae are born in July and August and settle in August and September, most adults do not attain reproductive size until the following year. Only those born in June and July and grow quickly to reproductive size (8 to 10 mm) by August or September can reproduce in the same year of birth. Gametogenesis ceases in September, as indicated by the spent condition of the gonads in all length classes (Fig. 3g, h). Fertilization is external and occurs in the water (Franzen, 1983; Sprung, 1987).

The Corbiculacea are monoecious, the Sphaeriidae being simultaneous hermaphrodites (Mackie, 1990) and *C. fluminea* being dioecious or monoecious in lentic waters (Morton, 1983) and female or monoecious in lotic waters (Morton, 1986). The Unionidae are mostly dioecious, some species capable of developing hermaphroditism in a few individuals (Mackie, 1984).

Internal fertilization is used in all native species of bivalves and in *Corbicula*. However, in *Corbicula* the fertilized eggs are quickly released as trochophores and veligers. In native bivalves, the fertilized eggs are brooded for several months (see larval life cycles below). Since only a small number of larvae can be brooded by any one parent, the number of larvae that are produced is rather small. Hence, external fertilization and

256

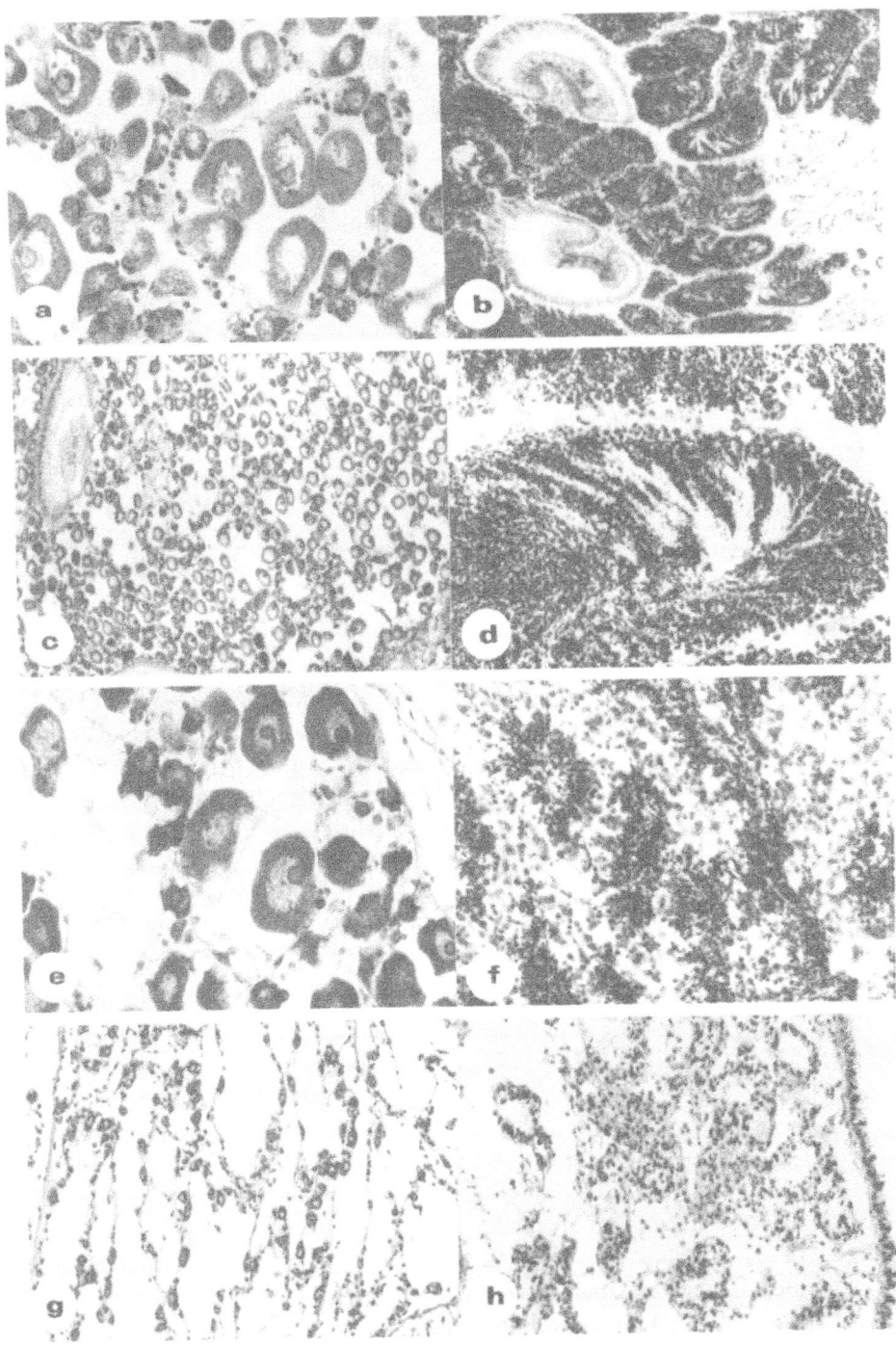

*Fig. 3.* Gametogenesis in *Dreissena polymorpha*. (a, b) May 2, 1989: (a) ripe female 10.0 mm shell length, 400 × ; (b) ripe male 12.0 mm shell length, 40 × . (c, d) May 28, 1989: (c) ripe female 10.8 mm shell length with some spent follicles, 200 × ; (d) ripe male 10.1 mm shell length, 100 × . (e, f) August 8, 1989: (e) partially spent female 13.0 mm shell length, 400 × ; (f) partially spent male 14.7 mm shell length, 200 × . (g, h) September 22, 1989: (g) spent female 17.8 mm shell length, 400 × ; (h) spent male 14.4 mm shell length, 100 × .

development partly explains why *Dreissena* will be much more prolific than native species of bivalves in North American surface waters.

*Corbicula* displays as great as, or even a greater, fecundity than *Dreissena*, with 25 000 to 75 000 veligers produced in the lifetime of a single clam (Aldridge & McMahon, 1978). Native species of Sphaeriidae, on the other hand, have very low fecundities (e.g. 1–40 eggs adult$^{-1}$) because they are brooders (Mackie, 1990). Unionidae have exceedingly high fecundities, with up to 2 000 000 glochidia produced by some females (Ellis, 1978), but mortalities are equally as high with only a few glochidia (1 in 10 000) developing and surviving to adulthood. The high reproductive capacities of the two exotic species also explains in part their success in North American surface waters.

## 2.4. *Larval life cycle*

The larval life cycle of *Dreissena* usually takes about four weeks to complete (Stanczykowska, 1964, 1977; Morton, 1969; Wiktor, 1969; Mackie *et al.*, 1989; Hopkins, 1990). Three *stages* are recognized (Fig. 4a) – a veliger stage; a post-veliger stage; and a settling stage. Each stage can be identified by certain *forms*. Veligers are identifiable by the presence of tuft of cilia, called the apical tuft. A 'D'-shaped, bivalved shell is

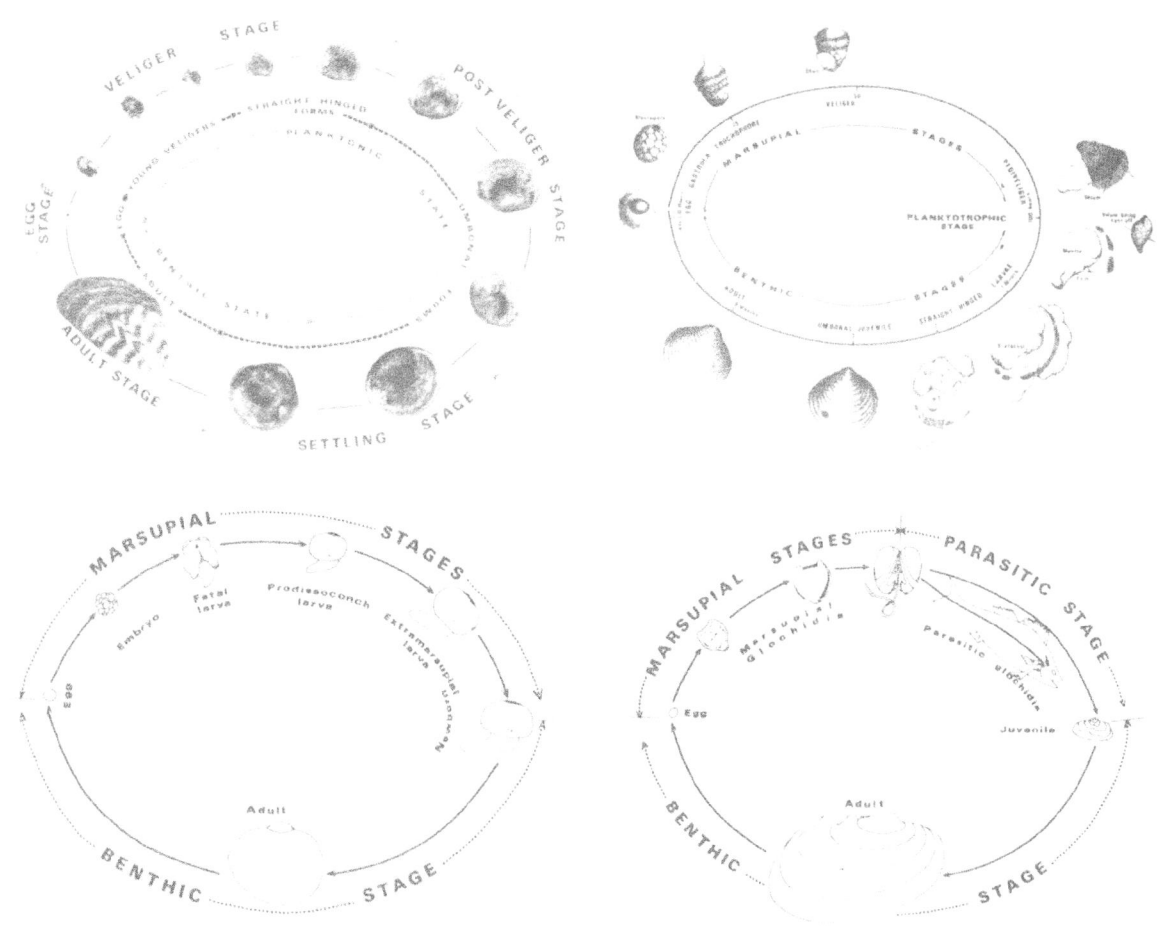

*Fig. 4.* Life cycles of four groups of bivalves in Lake St. Clair: (a) *Dreissena polymorpha*, times of development for each stage are unknown at present; (b) *Corbicula fluminea*; (c) Spaeriidae; (d) Unionidae.

258

formed as soon as the egg hatches. The veliger takes on a sphaerical shape as the shell developes (Fig. 4a). Straight-hinged forms are also part of the veliger stage because the velum is well developed and is not reduced in size or function until an umbone starts to form. The umbonal forms mark the beginning of several anatomical changes. Most notable are the reduction of the velum which becomes the siphons, lengthening of the foot and development of blood and some organ systems. With the loss of the velum the post-veliger larvae enter the settling stage. By the time the umbonal forms settle the byssal gland is completely functional and the young mussel attaches itself immediately to a firm substratum. All three stages occur in the *planktonic state*. Once the young mussel attaches itself to the substrate it becomes an adult and enters the *benthic state*.

The life cycle of *C. fluminea* (Fig. 4b) is similar to that of *D. polymorpha*, but the larval stages are planktonic for only a few days. After about 100 hrs, pediveligers lose the velum, the larvae begin to settle out as straight-hinged forms and take up a benthic existence (Kraemer & Galloway, 1986). A single byssal thread is secreted during the umbonal stage but is soon lost. The byssal gland becomes non-functional by the time the clam is a young adult.

The Sphaeriidae clams are ovoviviparous (Mackie, 1978) and brood their young in brood sacs on the inner gill of the parent (Fig. 4c). Four marsupial stages are recognized (Figs. 4c, 5): (i) Embryos first appear as gastrula in single-walled primary sacs; (ii) the embryos develop most of the organ systems and mature through the fetal larvae stage to the shelled prodissoconch stage within brood sacs. (iii) The prodissoconch larvae grow in size to become extramarsupial larvae which are so large they outgrow their brood sacs, tear the sac wall and eventually come to lie free in the inner gill space. The extramarsupial larvae are born through the excurrent siphon and begin a benthic existence as newborn. Complete larval development requires 1 to 3 months, depending on species. Details on larval development in both *Corbicula* and the Sphaeriidae can be found in Mackie (1990)

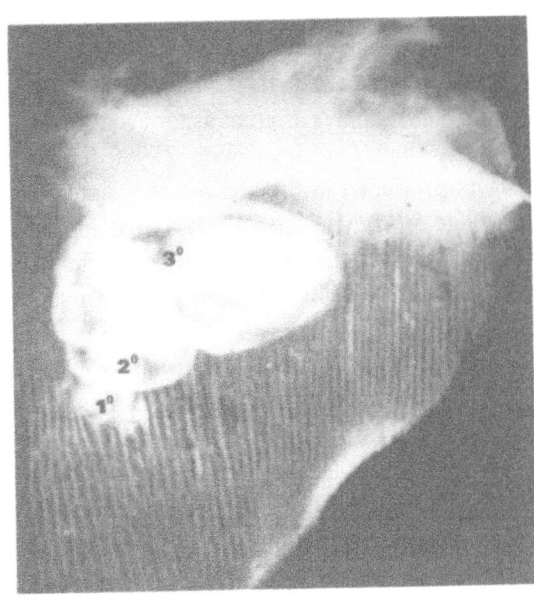

*Fig. 5.* Brood sac development on the inner gill of a typical sphaeriid, *Musculium securis*. Three stages are shown; the small primary sac (1°) contains embryos; the middle secondary sac (2°) contains fetal larvae; the large tertiary sac (3°) contains prodissoconch larvae. A fourth stage, the extramarsupial larvae which tear themselves free of tertiary brood sacs, is not shown.

The Unionidae clams have still another type of life cycle (Fig. 4d). They are the only freshwater bivalves that have a parasitic larval stage, called a glochidium. The glochidia develop in the gill marsupia for about 9 to 10 months, with 1 000 000 to 2 000 000 glochidia brooded (Ellis, 1978). The glochidia are released through the excurrent siphon in the spring or summer, depending on the species, to enter their parasitic stage. Clouds of glochidia are released when an appropriate fish species passes by. Opercular respiratory movements of the fish draw most of the glochidia into their gill cavity. The glochidia enter the gill tissue, live as parasites on fish blood, develop all the essential organ systems in about one month, enzymatically break free of the gill filaments and then drop to the bottom to begin a benthic existence.

The ovoviviparous and brooding habits of native sphaeriids is clearly less productive than the oviparous habit of zebra mussels. Although unionids produce millions of glochidia, the larval life cycle is not as efficient as that of the zebra mussel. The enormously high productivity of zebra mussels also partly explains why zebra mussels will become the dominant bivalve in Lake St. Clair.

## 2.5. Population dynamics

Adults of shell lengths exceeding 8 to 9 mm are reproductively mature by early May and contribute to the newly settled populations in June (Fig. 6, June 27). Larvae that were born late the

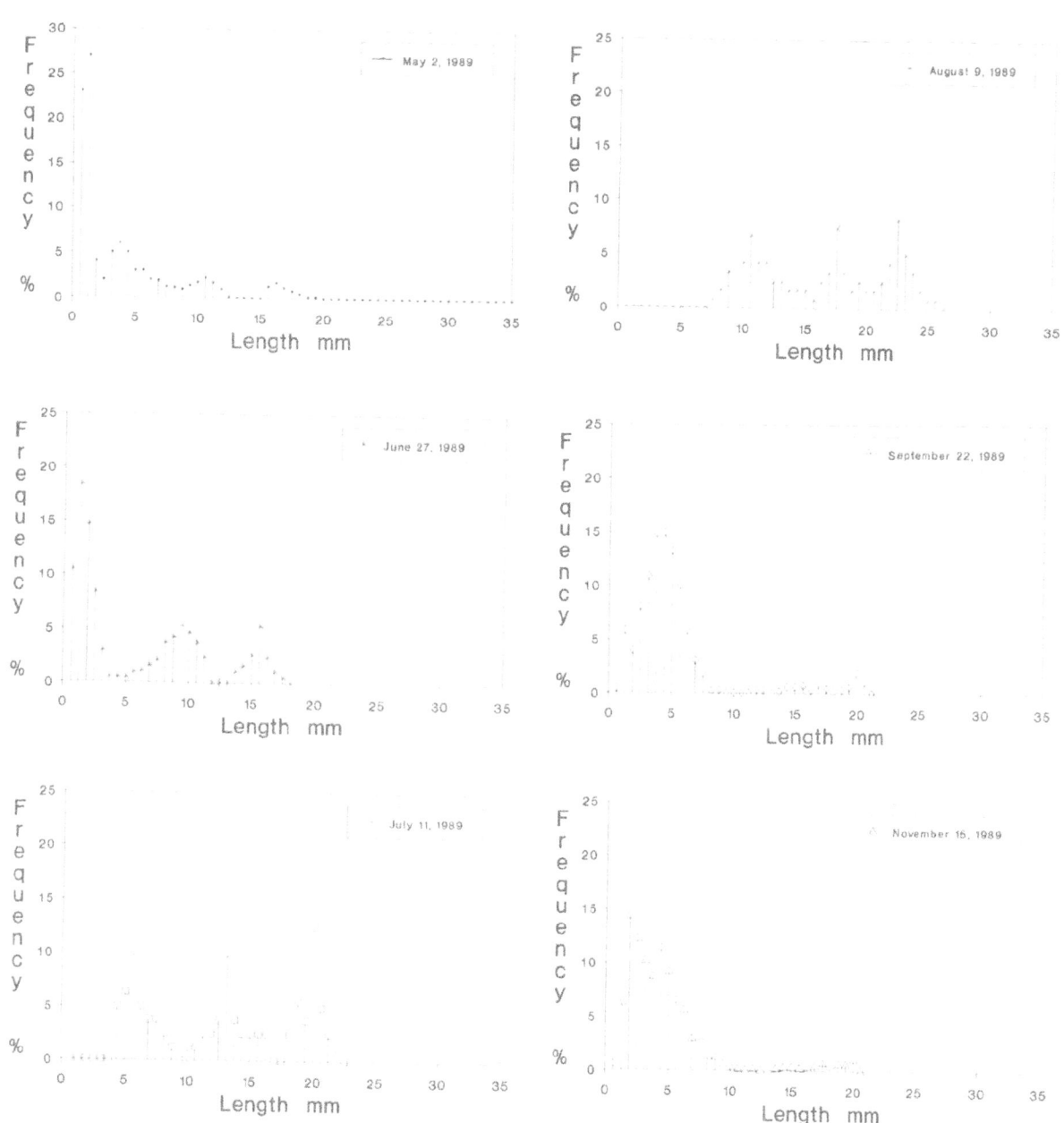

*Fig. 6.* Length frequency histograms depicting the length class distribution of *Dreissena polymorpha* in Lake St. Clair at Puce, Ontario in spring, summer and fall of 1989.

previous year (e.g. 1988) overwinter as young adults and vary in size between 1 and 4 mm (Fig. 6, May 2). These adults, and those born in late May to June (Fig. 6, June 27), attain a shell length of 15 to 20 mm by the end of the year, as shown by following the growth of mussels on cement blocks and by plotting mean sizes of each cohort (Fig. 7). By the summer or fall after their year of birth (e.g. 1989), all adults are reproductively active and contribute to the veliger population. Some adults that are born in late spring or early summer appear to grow and mature quickly during increasingly warm temperatures and contribute to the veliger population along with those adults born in the late fall of the previous year (Fig. 6, November 15). Hence, there are two periods of reproduction each year, as shown in the length-frequency distributions, once in the spring between May 2 and June 27 and again in the fall between September 22 and November 15 (Fig. 6). The growth rate of adults can be exceedingly fast, as much as 0.5 mm d$^{-1}$, as measure-

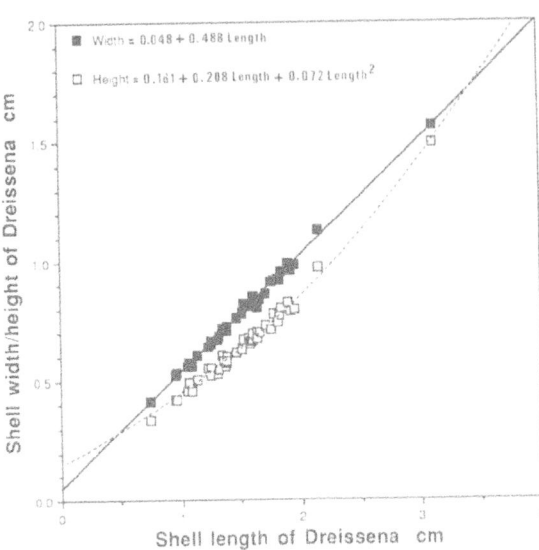

*Fig. 8.* Allometry of shell growth in *Dreissena polymorpha* in Lake St. Clair.

ments of adults on artificial substrates (cement blocks) in Lake St. Clair have shown. Typically, however, adults grow 1.5 to 2.0 cm y$^{-1}$. Most adults attain 1.5 cm shell length by the fall (Fig. 6). In the second year a few grow to exceed 3 cm in shell length. Most adults appear to die after 2 years of age (Fig. 6, September 22).

The allometric relationships between length, width and height of the shell of the zebra mussel in Lake St. Clair are shown in Fig. 8. The relationships between shell length and tissue dry weight and total dry weight are shown in Fig. 9. Allometry of shell growth and the length-weight relationships are similar to those reported for British (Morton, 1969) and Polish (Stanzcykowska, 1977) populations of *D. polymorpha*.

The population dynamics of *C. fluminea* is highly variable and will not be elaborated upon here because the species is not (yet) present in Lake St. Clair proper. Excellent overviews are provided in the proceedings of two *Corbicula* symposia, one edited by Britton (1979), and other appearing in the American Malacological Bulletin, Special Edition No. 2 (1986). The population dynamics of native species of bivalves in North American surface waters have been well studied and reported (see Mackie, 1990 for

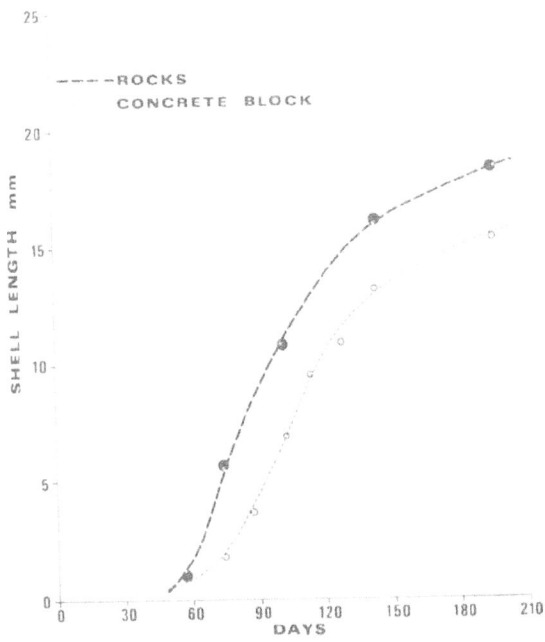

*Fig. 7.* Growth rates of *Dreissena polymorpha* as determined from plots of mean lengths of cohorts from Fig. 6 and from measurement of adults collected from cement blocks placed in Lake St. Clair in early May, 1989.

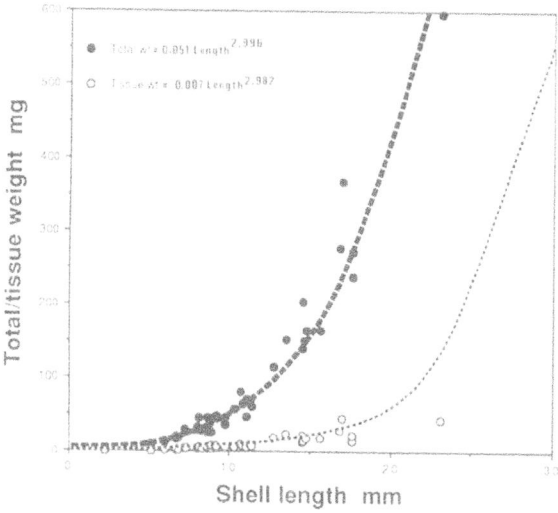

*Fig. 9.* Relationships between shell length and (a) tissue dry weight and (b) total dry weight of *Dreissena polymorpha* in Lake St. Clair.

Sphaeriidae and Clarke, 1973, for Unionidae). Nothing is known about the population dynamics of either sphaeriid or unionid species in Lake St. Clair. Knowledge of the natural variability in population dynamics of native species of bivalves is needed for a thorough understanding of the potential impact of zebra mussels on native bivalves, especially rare and endangered species. The exceedingly high reproductive capacity and rapid growth rate of the zebra mussel are two other features that will probably result in the dominance of zebra mussels over native bivalves in Lake St. Clair.

## 2.6. Distribution

In May, 1989, cement blocks (10.16 × 20.32 × 40.64 cm) were placed at Brights Grove in Lake Huron; at Corunna and Port Lambton in the St. Clair River; at Stoney Point, Belle River and Puce in Lake St. Clair; at Amherstburg in the Detroit River; at Wheatley, Port Stanley, Port Dover and Long Beach in Lake Erie; and at Niagara-on-the-Lake in Lake Ontario to determine the rate of spread throughout the Great Lakes. The blocks were examined every two weeks until September

and monthly in October, November, and December. Three vertical plankton hauls were taken with a 60 $\mu$ mesh student plankton sampler immediately before the blocks were examined. Only the presence of veliger larvae in the plankton samples were recorded for this part of the study.

In 1988 the zebra mussel was present at only a few isolated localities in the south-east end of Lake St. Clair and as far as Port Stanley in Lake Erie. Zebra mussels did not appear on the cement blocks or in plankton samples at Port Dover and Long Beach until September 1989. Living specimens of zebra mussels were also found at Port Colbourne in September, 1989 (Pers. Comm., Ron Dermott, Fisheries and Oceans, Canada Centre for Inland Waters) and at Port Weller in Lake Ontario in October 1989 (Pers. Comm., Larry King, Ontario Hydro). Based on these analyses, the zebra mussel has extended its range along the Ontario shorelines of Lake St. Clair and Lake Erie at a rate of approximately 250 km y$^{-1}$. As of December 1989, zebra mussels had not yet appeared on any of the cement blocks, nor in plankton samples taken at the two sites in the St. Clair River or at Brights Grove in Lake Huron.

Zebra mussels occur at all depths in the Ontario

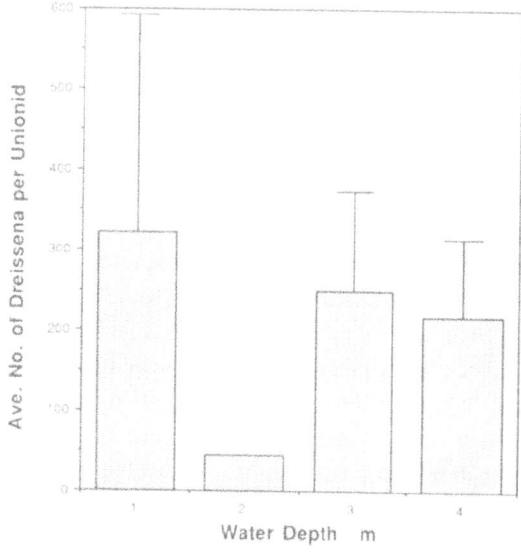

*Fig. 10.* Average number of *Dreissena polymorpha* per unionid shell in relation to depth in Lake St. Clair.

side of Lake St. Clair. The species occurs wherever there is suitable substrate present. Since the shells of unionid clams are the main suitable substrates in the 3 to 4 m depths, the distribution of zebra mussels in these depths (Fig. 10) is probably closely related to the distribution of unionid clams.

Unlike the zebra mussels, native species of bivalves are usually more common in soft sediments. They are present at all depths in Lake St. Clair but standing crops are relatively small, sphaeriids rarely exceeding 100 m$^{-2}$ (Griffiths, 1987) and unionids rarely exceeding 10 m$^{-2}$, as determined from SCUBA diving surveys of ten, 1 m$^2$ quadrats in each of the 1, 2, 3, and 4 m depths near Puce, Ontario; in the summer of 1989, densities of zebra mussels as high as 200 000 m$^{-2}$ were recorded in the shallow waters of Lake St. Clair (Mackie, unpublished data). This exceeds the maximum standing crops of sphaeriids and unionids by three and four orders of magnitude, respectively.

## 2.7. Dispersal mechanisms

The zebra mussel employs passive methods for dispersal during both the larval pelagic state and the adult benthic state. Primary dispersal occurs through the pelagic state by transport of the veligers and post-veligers by lake currents. As a result, the distribution of zebra mussels was initially concentrated toward the south-east end of Lake St. Clair. With the general eastward flow of water in the Great Lakes, the zebra mussel is distributed further eastward or northward than westward or southward. Attachment by means of its byssal apparatus to floating logs and debris has also helped to augment its eastward dispersal rate. Secondary dispersal may occur by the drifting of post-larvae and young adults using byssal and/or mucous threads. Enhanced transport by thread drifting has been demonstrated in several marine bivalve species (Sigurdson et al., 1976; Blok & Tan-Maas, 1977; Lane et al., 1982, 1985; Beukema & de Vlas, 1989) and in the freshwater C. fluminea (Prezant & Chalermwat, 1984).

Although similar behaviour has been observed in D. polymorpha held in laboratory aquaria (Ron Griffiths, pers. comm.), 'bysso-pelagic' (Lane et al., 1982) transport in zebra mussels in nature has yet to be verified. In such transport, the byssal or mucous threads are monofilaments, distinct in form and function from the attachment byssus threads (Lane et al., 1985). The threads are many times the length of the animal and increase the hydrodynamic drag to enable the young mussels to be transported in the water column by turbulence or currents. In Corbicula, small adults (7–14 mm shell length) secrete long mucous threads through their exhalent siphons and act as draglines to buoy the clam into the water column (Prezant & Chalermwat, 1984).

Currents exceeding 1 m s$^{-1}$ have prevented the veligers from invading the St. Clair River in the upstream direction. The dispersal of zebra mussels westward and northward will be accomplished by adults bysally attaching themselves to ships and by veligers in the ballast water of ships that travel from Lake Erie and release them in more northern lakes such as Lake Huron and Lake Superior. Commercial fishing boats and pleasure craft will be among the more common dispersal agents. Some veligers will probably be dispersed in bait buckets or in wet wells of some boats. In fact, the hull of a fishing vessel that had travelled from Lake Erie to Lake Michigan was observed to have been infested with zebra mussels when it was dry-docked at Green Bay, Lake Michigan in September 1989 (Pers. Comm., D.W. Schloesser, U.S. Fish and Wildlife, Ann Arbor, Michigan). Currents will now expedite the dispersal of zebra mussels in the connecting channels and lakes north of Lake St. Clair, including Lake Michigan, Lake Huron and the St. Clair River. Sphaeriids are dispersed mainly by passive mechanisms, such as by attaching to birds and insects (Mackie, 1990). Unionids rely mainly on their host fish species for dispersal during the parasitic glochidial stage. The rate of dispersal by such mechanisms is unknown but is probably relatively slow.

The rapid rate at which zebra mussels have been dispersed throughout Lake Erie and most of

Lake St. Clair is attributable to the planktonic veliger larvae, which are dispersed mainly by water currents, and the byssate feature of adults, which allows them to be transported upstream by attaching to hulls of boats. Clearly, these two features explain why the zebra mussel is becoming the dominant bivalve throughout the Great Lakes.

## 2.8. Physiology

The threshold temperatures for growth and reproduction apparently vary considerably among populations of zebra mussels. Preliminary analyses of data from Lake St. Clair populations suggest threshold temperatures of at least 10 °C for gametogenesis, 10 to 12 °C for growth and 14 to 16 °C for appearance of veligers. These temperatures are based on samples first taken on May 2 when the water temperature was 10 °C. These are similar to threshold temperatures of 11 to 12 °C for growth (Walz, 1978) and 15 to 17 °C for reproduction (appearance of veligers) reported by

Stanczykowska (1977) and Lewandowski and Ejsmont-Karabin (1983) for Polish populations and by Morton (1969) for British populations. Bij de Vaate (1989) reported threshold temperatures of 6 °C for growth and 12 °C for reproduction for populations in The Netherlands. Studies have yet to be made on specimens collected before May 2 and at temperatures less than 10 °C in Lake St. Clair. Nevertheless, the threshold temperatures for gametogenesis, growth, and larval release are well within the ranges of temperate species and suggest that the zebra mussel will not only succeed in the Great Lakes but many other lakes in North America.

The threshold temperatures for growth and reproduction of *Corbicula* are 16 °C and 22 °C, respectively. The threshold temperature for reproduction is rarely reached in most parts of Lake St. Clair and explains in part why the Asiatic clam will probably not succeed in colonizing the lake.

Collectively, zebra mussels are unrivaled in their capacity to clarify water. This was demonstrated by maintaining 50 mussels (shell length ranging from 2 mm to 15 mm in 10 l of 0, 1.5, 3.0,

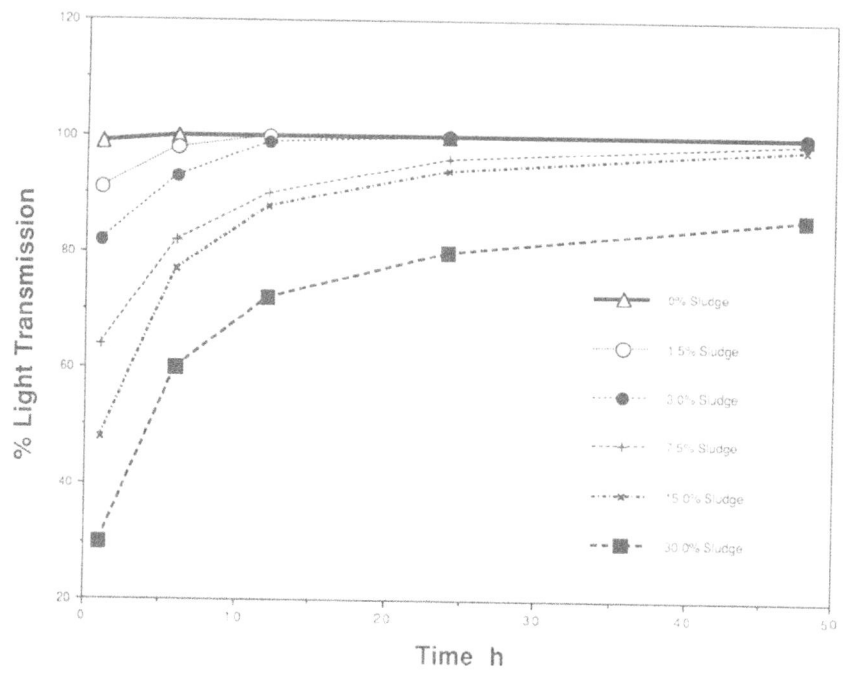

*Fig. 11.* Ability of *Dreissena polymorpha* to clarify different concentrations of activated sewage sludge.

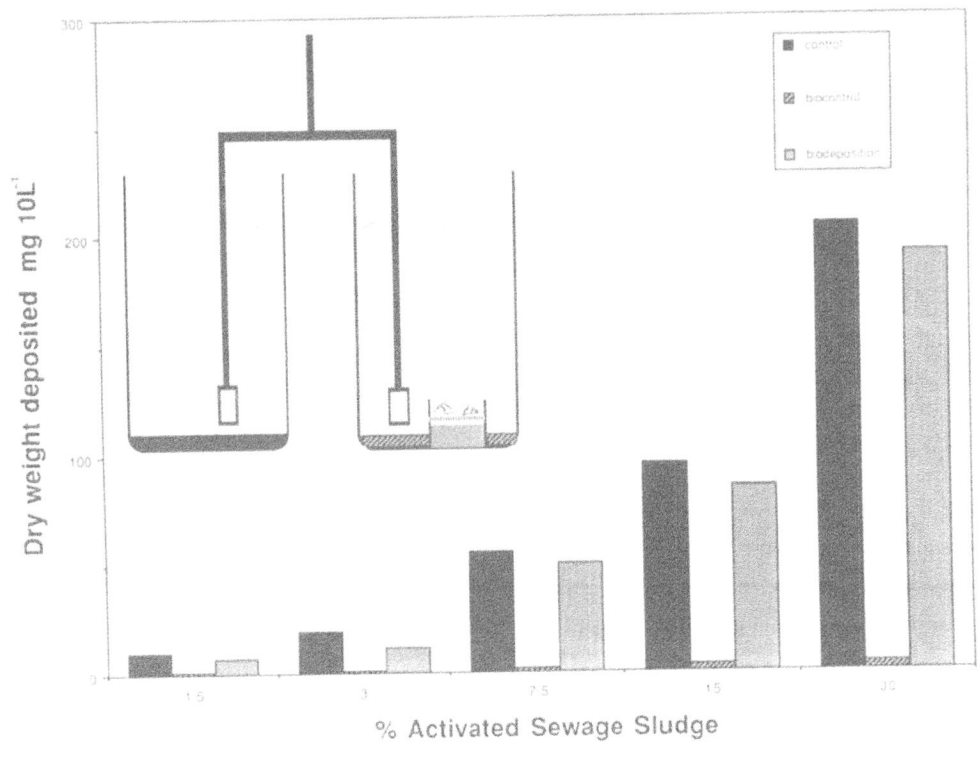

*Fig. 12.* Biodeposition of different concentrations of activated sewage sludge as pseudofaeces by *Dreissena polymorpha*.

7.5, 15 and 30 percent activated sewage sludge, with continuous aeration for 96 h. The mussels were suspensed 5 cm above the bottom of the bucket in 100 mm dia × 50 mm high acrylic tubes with a nylon mesh (1 mm openings) glued to the bottom of the tubes. Water clarity was restored to control levels (0 percent sludge) within 6 h for 1.5 percent sludge, 12 h for 3.0 percent sludge, 24 h for 7.5 percent sludge, 48 h for 15 percent sludge and 96 h for 30 percent sludge (Fig. 11). Nearly all of the suspended organic material was deposited on the bottom as pseudofaeces (Fig. 12). As freshwater bivalves, the zebra mussel's filtration rate (10 to 100 ml individual$^{-1}$ h$^{-1}$) is intermediate between Sphaeriidae (0.6 to 8.3 ml individual$^{-1}$ h$^{-1}$) and Unionidae (60 to 490 ml individual$^{-1}$ h$^{-1}$) (Stanczykowska *et al.*, 1976). *Corbicula* has a filtration rate that far exceeds these on an individual basis (60 to 800 ml h$^{-1}$). However, the well known ability of *Dreissena* to clarify lakes and watercourses (Morton, 1971; Stanczykowska, 1975; Piesik, 1983; Reeders

*et al.*, 1989) is a consequence of the enormous standing crops that usually prevail in aquatic systems.

## 3. Impacts

### 3.1. Ecosystem

One of the most obvious early impacts of zebra mussels is on native species of unionid clams. Zebra mussels are not selective and are colonizing all species of unionids in Lake St. Clair, including *Lampsilis radiata siliquoidea*, *L. ventricosa*, *Pleurobema coccineum*, *Proptera alata*, *Lasmigona complanata*, *L. compressa*, *Elliptio dilatata*, *Quadrula quadrula*, *Alasmidonta viridis*, *Anodonta grandis grandis*, *Obovaria subrotunda*, *Leptodea fragilis* and *Villosa iris*. There is an exponential increase in the numbers of zebra mussels on unionid shells in relation to the length and surface area of the host unionid shell (Fig. 13a, b). The infestations are

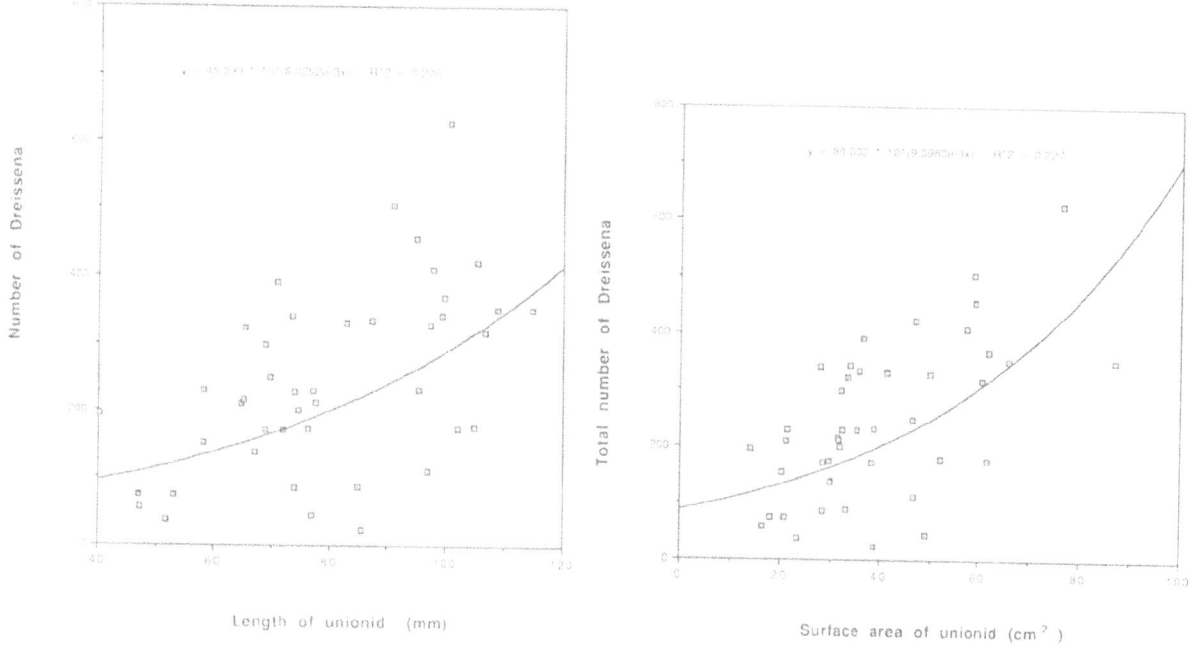

*Fig. 13.* Relationship between number of zebra mussels and (a) length and (b) surface area of unionid shells in Lake St. Clair.

so great that the unionids are not able to fully open their valves (Fig. 14a), or so invasive that the

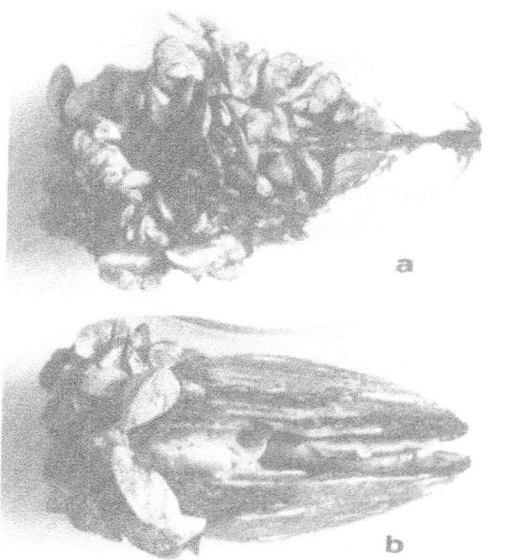

*Fig. 14.* Encrustations of *Dreissena polymorpha* so intense that the unionid is not able to fully open (a) or fully close (b) their valves.

valves cannot be closed (Fig. 14b). The massive infestations are hypothesized to have one or more of the following effects on unionid clams: (i) The biomass of zebra mussels is so great that normal locomotion and burrowing activities of the unionids are impaired. For most infestations, the biomass of zebra mussels is estimated to be as great as that of the unionid. For example, using the relationship, total dry weight = 0.051 shell length$^{2.996}$, for *D. polymorpha* in Lake St. Clair (Fig. 9) and total dry weight = 0.058 shell length$^{2.93}$ for *Lampsilis radiata* (Mackie & Flippance, 1983), a common unionid in Lake St. Clair, and assuming 1000 individuals at an average length of 10 mm for *D. polymorpha* on a single 100 mm long shell of *L. radiata*, the total biomass of zebra mussels (50 g) slightly exceeds that of the unionid (42 g). Unionids with 15 000 zebra mussels probably carry at least five times their own weight (i.e. 231 g of zebra mussels on a 42 g unionid), assuming an average length of 7 mm for zebra mussels. (ii) The concentration of zebra mussels at the posterior end of the unionid affects its balance and/or equilibrium during locomotion.

(iii) Invasive growth of zebra mussels between the valves of a unionid clam limits or even prevents valve closure and will expose the unionid to predators, parasites, disease, and noxious water quality. (iv) Invasive growth at the posterior end also interferes with the normal functioning of the siphons and processes associated with them, especially respiration and feeding. (v) Extensive infestations of zebra mussels limits or prevents valve opening which will impair normal metabolic functions for feeding, growth, reproduction, respiration and excretion. (vi) Enormous numbers of zebra mussels will strip the water of food and nutrients making little or none available to the unionid host which soon starves, loses weight, and dies. (vii) Infestations at the mantle margin of the shell will cause shell deformities, especially at the posterior end. (viii) Complete occlusion of the gape at the posterior end and perhaps of siphons which may cause death by smothering the unionid.

Large populations of zebra mussels have an enormous capacity to remove seston and clarify water. As shown above, they are able to remove suspended materials and nutrients from the water and biodeposit them on the bottom as pseudo-faeces. The following scenario can be hypothesized. The increase in water clarity will result in an increase in the size of the euphotic zone, but primary production can be expected to decline because most of the nutrients will have been biodeposited on the bottom and made available to the benthic community. Reductions in production and biomass of pelagic autotrophs, heterotrophs, herbivorous zooplankton and planktivores, including fish, can be expected to follow. Eventually, this single exotic bivalve species can potentially alter the entire pelagic-benthic energy balance of Lake St. Clair. An ecosystem that once was driven by energy derived from the pelagic zone will be driven by energy derived from the benthic zone. This may have severe socio-economic impacts, especially for commercial and sport fisheries.

## 3.2. Socio-economic

Impacts will not only be felt through changes in the energy balance of pelagic and benthic systems but socially and economically as well. The strongly byssate feature makes the zebra mussel one of the most tenacious and potent biofoulers of any exotic species ever introduced to North America. Navigational aids, such as fishing buoys and markers, have such heavy infestations that they sink below the surface (pers. comm., Joe Leach, Ontario Ministry of Natural Resources); commercial fishing gear, such as trap nets and gill nets, have accumulations of zebra mussels so large that the nets are rendered useless and are difficult to retrieve; hulls of boats and ships are so laden with zebra mussels that their sailing efficiency is impaired; beaches have accumulations of dead shells that are cutting the feet of swimmers; industrial and domestic intakes and pipelines drawing water from Lake St. Clair have such heavy infestations that flow rates have been reduced by as much of 50 percent (pers. comms., Joe Leach, Ontario Ministry of Natural Resources, Wheatley, Ontario and Ron Griffiths, Ontario Ministry of Environment, London, Ontario).

## 4. Acknowledgements

The information provided above are based on data collected during the first year of a three-year study funded by the Ontario Ministry of Environment to examine the biology and ecological impact of zebra mussels in the Great Lakes. The life history and growth data were collected by D. Pathy, P. Gillis and K. Wingerden and are the first to be reported on the population of zebra mussels in Lake St. Clair, a population that will be the epicentre of dispersal of zebra mussels on the North American continent.

## References

Aldridge, D. W. & R. F. McMahon, 1978. Growth, fecundity and bioenergetics in a natural population of the asiatic clam, *Corbicula manilensis* Philippi, from north central Texas. J. Moll. Stud. 44: 49–70.

American Malacological Bulletin, 1986. Proceedings, Second International *Corbicula* Symposium. Special Publication No. 2.

Bailey, R. ,C., E. H. Anthony & G. L. Mackie, 1983. Environmental and taxonomic variation in fingernail clams (Bivalvia: Pisidiidae) shell morphology. Can. J. Zool. 61: 2781–2788.

Beukema, J. J. & J. de Vlas, 1989. Tidal-current transport of thread-drifting postlarval juveniles of the bivalve *Macoma balthica* from the Wadden Sea to the North Sea. Mar. Ecol. Prog. Ser. 52: 193–200.

Bij de Vaate, A., 1989. Occurrence and population dynamics of the zebra mussel, *Dreissena polymorpha* (Pallas, 1771), in the Lake IJsselmeer area (the Netherlands). Manuscript of pap.pres. at New York Sea Grant Conference on Zebra Mussels in the Great Lakes, November 22, 23, 1989. 5 pp.

Blok, J. W., de & M. Tan-Maas, 1977. Function of byssus threads in young postlarval *Mytilus*. Nature 267: 558.

Britton, J. C. (Ed.), 1979. Proceedings, first international *Corbicula* symposium. Texas Christian Univ., October 1977. Texas Christian Univ. Res. Found. Publ., Fort Worth, Texas. 313 pp.

Britton, J. C. & B. Morton, 1979. *Corbicula* in North America: The evidence reviewed and evaluated. In; J. C. Britton (ed.), Proceedings, First International *Corbicula* Symposium, Texas Christian Univ., Fort Worth, Texas. Texas Christian Univ. Res. Found. Publ. pp. 249–287.

Clarke, A. H., Jr., 1973. The freshwater Mollusca of the Canadian interior basin. Malacologia 13: 1–509.

Clarke, A. H, 1981. *Corbicula fluminea* in Lake Erie. Nautilus 95: 83–84.

Ellis, A. E., 1978. British freshwater bivalve Mollusca. Keys and notes for the identification of the species. Synopses of the British fauna No. 11. Academic Press, London. 109 pp.

Franzen, A., 1983. Ultrastructural studies of the spermatozoa in 3 bivalve species with notes on evolution of elongated sperm nucleus in primitive spermatozoa. Gamete Res. 7: 199–214.

French, J. R. P. & Schloesser, D. W., 1991. Growth and overwinter survival of the Asiatic clam, *Corbicula fluminea*, in the St. Clair river, Michigan. In: M. Munawar & T. Edsall (eds), Environmental Assessment and Habitat Evaluation of the Upper Great Lakes Connecting Channels. Developments in Hydrobiology 65. Kluwer Publishers, Dordrecht: 165–170. Reprinted from Hydrobiologia 219

Gale, W. F., 1969. Bottom fauna of Pool 19, Mississippi River, with emphasis on the life history of the fingernail clam, *Musculium transversum*. Ph. D. Dissert., Iowa State Univ., Ames. University microfilm No. 69-20642, Ann Arbor, MI.

Gillis, P., 1989. Gametogenesis in *Dreissena polymorpha*. An unpublished report. Department of Zoology, University of Guelph, 19 pp.

Griffiths, R. W., 1987. Environmental quality assessment of Lake St. Clair in 1983 as reflected by the distribution of benthic macroinvertebrate communities. Rep. Prep. for the Ont. Min. Env. by Aquatic Ecosystems Limited. ISBN No. 0-7729-2339-6. 35 pp. + appendices.

Hebert, P. D. N., B. W. Muncaster & G. L. Mackie, 1989. Ecological and genetic studies on *Dreissena polymorpha* (Pallas): a new mollusc in the Great Lakes. Can. J. Fish. Aquat. Sci. 46: 1587–1591.

Hopkins, G. J., 1990. The zebra mussel, *Dreissena polymorpha*: a photographic guide to the identification of microscopic veligers. Available from Queen's Printer for Ontario, ISBN: 0-7729-6349-5.

Kraemer, L. R. & M. L. Galloway, 1986. Larval development of *Corbicula fluminea* (Muller) (Bivalvia: Corbiculacea): An appraisal of its heterochrony. Amer. Malacol. Bull. 4: 61–79.

Lane, D. J. W., J. A. Nott & D. J. Crisp, 1982. Enlarged stem glands in the foot of the post-larval mussel, *Mytilus edulis*: Adaptation for bysso-pelagic migration. J. Mar. Biol. Assoc. U.K. 62: 809–818.

Lane, D. J. W., A. R. Beaumont & J. R. Hunter, 1985. Byssus drifting and the drifting threads of young post-larval mussel *Mytilus edulis*. Mar. Biol. 84: 301–308.

Lewandowski, K. & J. Ejsmont-Karabin, 1983. Ecology of planktonic larvae of *Dreissena polymorpha* (Pall.) in lakes of different degrees of heating. Pol. Arch. Hydrobiol. 30: 89–102.

Mackie, G. L., 1978. Are sphaeriid clams ovoviviparous or viviparous? Nautilus 92: 145–147.

Mackie, G. L., 1984. 5. Bivalves. *In*: K. M. Wilbur (ed.). The Mollusca. Vol. 7. Reproduction (eds. A. S. Tompa, N. H. Verdonk & J. A. M. van de Biggelaar), Academic Press, NY. pp. 351–418.

Mackie, G. L., 1990. Biology of corbiculacean clams of North America. In press, Government Publications of Canada, Ottawa, Ontario, Canada. 1285 pp. manuscript.

Mackie, G. L. & L. A. Flippance, 1983. Intra- and interspecific variations in calcium content of freshwater Mollusca in relation to calcium content of the water. J. Moll. Stud. 49: 204–211.

Mackie, G. L., W. N. Gibbons, B. W. Muncaster & I. M. Gray, 1989. The zebra mussel, *Dreissena polymorpha*: a synthesis of European experiences and a preview for North America. Rep. prep. for Wat. Resour. Br., Great Lakes Sect. Available from Queen's Printer for Ontario, ISBN 0-7729-5647-2.

Mackie, G. L. & J. M. Topping, 1988. Historical changes in the unionid fauna of the Sydenham River watershed and downstream changes in shell morphometrics of three common species. Can. Field-Nat. 102: 617–626.

McMahon, R. F., 1982. The occurrence and spread of the introduced Asiatic freshwater clam, *Corbicula fluminea* (Muller), in North America: 1924–1982. Nautilus 96: 134–141.

Morton, B., 1969. Studies on the biology of *Dreissena poly-*

*morpha* Pall. III. Population dynamics. Proc. Malacol. Soc. Lond. 38: 471–482.

Morton, B., 1971. Studies on the biology of *Dreissena polymorpha* Pall. 5. Some aspects of filter feeding and the effect of micro-organisms upon the rate of filtration. Proc. Malacol. Soc. Lond. 39: 289–301.

Morton, B., 1983. The sexuality of *Corbicula fluminea* in lentic and lotic waters in Hong Kong. J. Moll. Stud. 49: 81–83.

Morton, B., 1986. *Corbicula* in Asia – An updated synthesis. Amer. Malacol. Bull., Spec. Edn No. 2 (1986): 113–124.

Piesik, Z., 1983. Biology of *Dreissena polymorpha* settling of stylon nets and the role of this mollusk in eliminating the seston and the nutrients from the water course. Pol. Arch. Hydrobiol. 30: 353–362.

Prezant, R. S. & K. Chalermwat, 1984. Flotation of the bivalve *Corbicula fluminea* as a means of dispersal. Science 225: 1491–1493.

Reeders, H. H., A. Bij de Vaate & F. J. Slim, 1989. The filtration rate of *Dreissena polymorpha* (Bivalvia) in three Dutch lakes with reference to biological water quality management. Freshw. Biol. 22: 133–141.

Scott-Wasilk, J., G. G. Downing & J. S. Lietzow, 1983. Occurrence of the Asiatic clam *Corbicula fluminea* in the Maumee River and Western Lake Erie. J. Great Lakes Res. 9: 9–13.

Sigurdson, J. B., C. W. Titman & P. A. Davies, 1976. The dispersal of young post-larval bivalve molluscs by byssus threads. Nature 262: 386–387.

Sprung, M., 1987. Ecological requirements of developing *Dreissena polymorpha* eggs. Arch. Hydrobiol. Suppl. 79: 69–86.

Stanczykowska, A., 1964. On the relationship between abundance, aggregations and 'condition' of *Dreissena polymorpha* Pall. in 36 Masurian lakes. Ekol. Pol. A12: 653–690.

Stanczykowska, A., 1975. Ecosystem of the Mikolajskie Lake Poland regularities of the *Dreissena polymorpha* Bivalvia occurrence and its function in the lake. Pol. Arch. Hydrobiol. 22: 73–78.

Stanczykowska, A., 1977. Ecology of *Dreissena polymorpha* (Pall.) (Bivalvia) in lakes. Pol. Arch. Hydrobiol. 24: 461–530.

Stanczykowska, A., W. Lawacz, J. Mattice & K. Lewandowski, 1976. Bivalves as a factor affecting the circulation of matter in Lake Mikolajskie (Poland). Limnologica 10: 347–352.

Walz, N., 1978. The energy balance of the freshwater mussel *Dreissena polymorpha* in laboratory experiments and in Lake Constance. II. Reproduction. Arch. Hydrobiol. Suppl. 55: 106–119.

White, D. S., M. H. Winnell & D. D. Jude, 1984. Discovery of the Asiatic clam, *Corbicula fluminea*, in Lake Michigan. J.Great Lakes Res. 10: 329–331.

Wiktor, L., 1969. The biology of *Dreissena polymorpha* (Pall.) and its ecological importance in the Firth of Szczecin. Stud. Mater. Morsk. Inst. Ryb. Gdynia, Ser. A, 5: 1–88. (cited in Stanczykowska, 1977).

*Hydrobiologia* **219**: 269–279, 1991.
*M. Munawar & T. Edsall (eds),*
*Environmental Assessment and Habitat Evaluation of the Upper Great Lakes Connecting Channels.*
© 1991 Kluwer Academic Publishers.

# The Detroit River: effects of contaminants and human activities on aquatic plants and animals and their habitats[1]

B.A. Manny[1] & D. Kenaga[2]
[1]*U.S. Fish and Wildlife Service, National Fisheries Research Center – Great Lakes, Ann Arbor, MI 48105, USA*; [2]*6191 Pollard, East Lansing, MI 48823, USA*

*Key words:* water, sediments, oil, heavy metals, PCBs, plankton, clams, fish, and birds

## Abstract

Despite extensive urbanization of its watershed, the Detroit River still supports diverse fish and wildlife populations. Conflicting uses of the river for waste disposal, water withdrawals, shipping, recreation, and fishing require innovative management. Chemicals added by man to the Detroit River have adversely affected the health and habitats of the river's plants and animals. In 1985, as part of an Upper Great Lakes Connecting Channels Study sponsored by Environment Canada and the U.S. Environmental Protection Agency, researchers exposed healthy bacteria, plankton, benthic macroinvertebrates, fish, and birds to Detroit River sediments and sediment porewater. Negative impacts included genetic mutations in bacteria; death of macroinvertebrates; accumulation of contaminants in insects, clams, fishes, and ducks; and tumor formation in fish. Field surveys showed areas of the river bottom that were otherwise suitable for habitation by a variety of plants and animals were contaminated with chlorinated hydrocarbons and heavy metals and occupied only by pollution-tolerant worms. Destruction of shoreline wetlands and disposal of sewage and toxic substances in the Detroit River have reduced habitat and conflict with basic biological processes, including the sustained production of fish and wildlife. Current regulations do not adequately control pollution loadings. However, remedial actions are being formulated by the U.S. and Canada to restore degraded benthic habitats and eliminate discharges of toxic contaminants into the Detroit River.

## 1. Introduction

For many years, the channels connecting the Great Lakes, including the Detroit River, have been used for the disposal of toxic wastes that impaired beneficial uses of these waters and their biological resources (EC & EPA, 1988; Hartig & Thomas, 1988). Since 1974, each of these channels has been recognized as a pollution problem area and more recently as an 'area of concern' by the International Joint Commission because pollution impaired their use as drinking water or prevented the consumption of fish from them (Hartig & Thomas, 1988). Part of each channel falls under the jurisdiction of both Canada and the United States.

In 1985, the U.S. and Canada conducted a study of the Upper Great Lakes Connecting Channels to integrate scientific information and develop recommendations for restoring beneficial

---

[1]Contribution 738 of the National Fisheries Research Center-Great Lakes, 1451 Green Road, Ann Arbor, MI 48105, U.S.A.

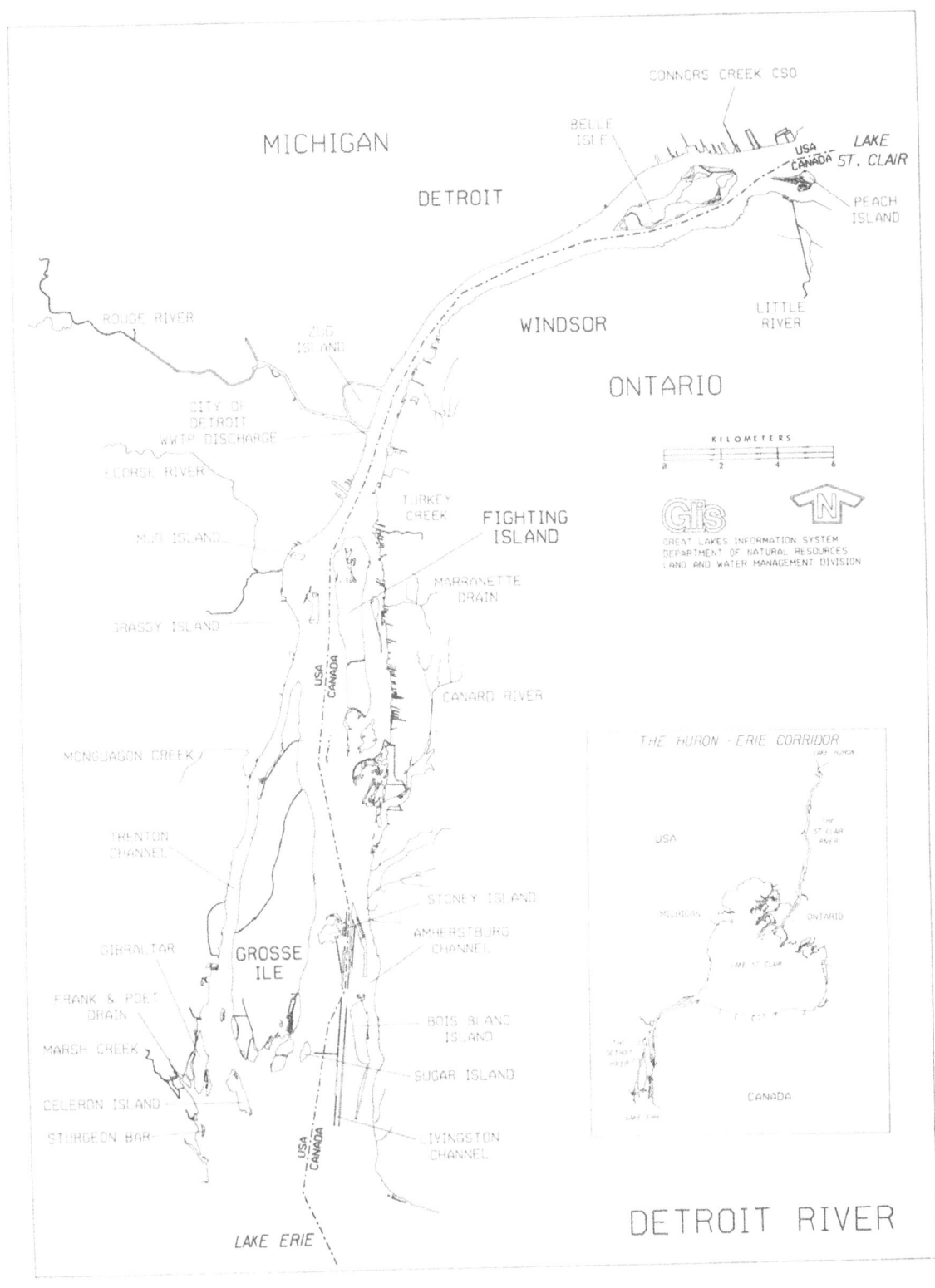

*Fig. 1.* The Detroit River, showing localities of special interest in this study.

uses in these areas (EC & EPA, 1988). Scientific studies were conducted to identify and measure sources of contaminants and their impacts on beneficial uses, to determine the adequacy of existing control measures, to recommend additional pollution controls, and to recommend surveillance needed to monitor the effects of restoration efforts. The findings of these and earlier studies, which focused on the Detroit River, its biota, and habitats, are summarized in two reports (EC & EPA, 1988; Manny *et al.*, 1988), upon which we drew freely in preparing this publication.

## 2. Description of the study area

The Detroit River forms the lower third of the strait or channel connecting Lakes Huron and Erie and is bisected by the border between Canada and the United States (Fig. 1). The Detroit River is 51 km long and falls only 0.9 m (Derecki, 1984). The upper 21 km is 700 to 1000 m wide, less than 15 m deep, and contains two islands. The lower 30 km is 1500 to 6000 m wide, less than 9 m deep, and contains 10 islands. The Trenton Channel lies between the U.S. shoreline and the largest island, Grosse Ile. Extensive excavation was required to construct channels for commercial navigation in the river and about $4 million is spent each year to maintain channel depth by dredging (USACE, 1981). Water velocities average 0.3 to 0.9 m s$^{-1}$ but exceed 1.7 m s$^{-1}$ in main channels. Water passes from the head to the mouth of the river in about 20 h. The mean annual river temperature is about 10 °C; monthly temperatures vary from 0.5 to 22 °C (Muth *et al.*, 1986).

Tributaries and wastewater discharges add much sediment to the river. The principal tributaries are the Rouge and Ecorse Rivers; the Frank & Poet drain, and Marsh and Monguagon Creeks in Michigan and the Little River, Turkey Creek, Marranette drain, and the Canard River in Ontario (Fig. 1). The total combined discharge of these tributaries (about 32 m$^3$ s$^{-1}$) is equal to the combined effluent of the eight Detroit-area waste-water treatment plants (EC & EPA, 1988), and is less than 1 percent of the river's average flow of 5240 m$^3$ s$^{-1}$. The Detroit River provides habitat for at least 82 species of phytoplankton, 31 species of aquatic macrophytes, 300 species of macrozoobenthos, 65 species of fish, and 27 species of waterfowl (Manny *et al.*, 1988). In colonial times, coastal wetlands bordered most of the Detroit River but now only 31 small, isolated wetlands covering 1380 ha remain in the river (Manny *et al.*, 1988). Since 1955, despite improvements in water quality, the abundance of the most common submersed plant, wild celery (*Vallisneria americana*), has declined by 72 percent in the lower Detroit River, coincident with declining use of the river by diving ducks, such as the canvasback duck (*Aythya vallisneria*); (Schloesser & Manny, 1990). Habitat is provided for coldwater fishes from September to June, but such fishes move out during the period of maximum water temperature from July to September (Manny *et al.*, 1988).

More than 80 political jurisdictions exist within the river's 1844 square km watershed, including the cities of Detroit, Michigan and Windsor, Ontario with a combined population of about 5 million people (EC & EPA, 1988). About 90 percent of the river's Canadian watershed is agricultural or undeveloped (Manny *et al.*, 1988). The river's U.S. watershed is 30 percent agricultural, 30 percent residential, 10 percent industrial, and the remainder is urban. About 46 km of the U.S. shoreline is privately owned and 87 percent of it has been filled and bulkheaded (Muth *et al.*, 1986). Most of the Canadian shoreline is also privately owned but less of it has been filled and bulkheaded (Manny *et al.*, 1988). Historically, ice formed across the river from December to March. Presently, the river seldom freezes over because commercial vessels ply the river throughout the winter and large volumes of heated effluent are added continuously to the river by waste discharges. The river is a major source of drinking water (five water intakes) and a source of process or cooling water for more than 30 industries and power plants. As the busiest waterway for commerce on the Great Lakes, the river transports 60 million metric tonnes of iron ore, coal, limestone, grains, and other cargoes worth over a billion dollars each year (USACE, 1984). Recreational boating, primarily for fishing, is well established on the Detroit River. In 1983–85, anglers spent 1.4 million hours on the river and caught over 1.4 million fish, mostly white bass (*Morone chrysops*), walleye (*Stizostedion vitreum*), and yellow perch (*Perca flavescens flavescens*) (Haas *et al.*, 1985).

## 3. Sources of contamination

Each day, eight municipal sewage treatment plants, utilizing pollution control measures costing nearly $500 million, discharge nearly 3 million m$^3$ of effluent into the river (EC & EPA, 1988). Power plants, steel mills, petroleum refineries, salt mines, and manufacturers of chemicals, automobiles, and plastics, primarily on the U.S. shore, add another 3 million m$^3$ of effluent per day. Since 1977, wastes from many industries have been diverted to municipal sewage treatment plants. As a result, the Detroit Waste Water Treatment Plant discharge in the Detroit River is now the principle source of 15 troublesome contaminants, including polychlorinated biphenyls (PCBs), hexachlorobenzene, cadmium, nickel, chromium, zinc, phenol, ammonia, phosphorus, oil and grease, and cyanide (EC & EPA, 1988). Sixty-six permitted industrial discharges along the U.S. shore contribute additional oil and grease, ammonia, iron, phosphorus, phenols, cyanide, copper, chromium, cadmium, cobalt, zinc, nickel, polyaromatic hydrocarbons, PCBs, and hexachlorobenzene (Table 1). The 214 combined sewer overflows on the U.S. shore are the primary source of lead and mercury (Table 1); 26 combined sewer overflows from the City of Windsor discharge smaller amounts of contaminants

*Table 1.* Estimated mean annual loadings of contaminants to the Detroit River from various sources, 1979–1986 (from EC & EPA, 1988).

| Contaminant | Total measured annual loading (thousands) of kilograms | Percentage of loading from each source | | | | | |
| --- | --- | --- | --- | --- | --- | --- | --- |
| | | Point sources | | Tributaries | | Combined sewer overflows | |
| | | Michigan | Ontario | Michigan | Ontario | Detroit | Windsor |
| Chloride | 623,207 | 21.6 | 63.9 | 12.3 | 1.2 | 0.2 | 0.8 |
| Suspended solids | 57,608 | 33.7 | | 50.0 | 1.9 | 14.4 | |
| Oil and grease | 15,720 | 77.8 | 0.4 | | | 21.0 | 0.8 |
| Ammonia | 11,693 | 77.6 | 5.3 | 6.2 | 5.4 | 5.3 | 0.2 |
| Iron | 1,561 | 75.7 | 7.8 | 2.7 | 0.1 | 5.1 | 8.6 |
| Phosphorus | 916 | 49.4 | 9.1 | 16.6 | 11.3 | 12.7 | 1.0 |
| Zinc | 316 | 54.4 | 19.5 | 17.0 | 0.6 | 6.2 | 2.3 |
| Nickel | 74 | 49.6 | 8.4 | 10.0 | 24.3 | 6.6 | 1.0 |
| Phenol | 55 | 64.0 | 34.6 | | | 1.2 | 0.2 |
| Lead | 49 | 16.3 | 22.0 | 19.4 | 1.7 | 31.7 | 8.8 |
| Cyanide | 44 | 97.9 | 1.9 | | | | 0.2 |
| Copper | 38 | 27.4 | 24.5 | 24.9 | 0.5 | 19.9 | 2.9 |
| Chromium | 16 | 71.3 | | | | 28.7 | |
| Cadmium | 6 | 52.6 | 5.9 | 13.0 | 0.1 | 25.3 | 3.1 |
| PAHs* | 2 | 82.9 | 13.3 | | | | 3.8 |
| Mercury | 2 | 2.4 | 0.1 | 1.1 | 0.1 | 96.2 | 0.1 |
| PCBs* | 0.20 | 37.8 | 5.6 | 22.1 | 0.1 | 33.9 | 0.6 |
| Cobalt | 0.02 | 99.0 | 0.5 | | | | 0.5 |
| Hexachlorobenzene | 0.001 | 96.7 | | | | 3.3 | |

* PAHs = polyaromatic hydrocarbons, and PCBs = polychlorinated biphenyls.

(Marsalek & Ng, 1987). Combined sewer over-flows are also estimated to add 14 percent of the phosphorus and suspended solids, 22 percent of the oil and grease, 28 percent of cadmium, 29 percent of the chromium, 23 percent of the copper, and 35 percent of the PCBs entering the Detroit River (Table 1). These estimates are conservative because inputs from combined sewer overflows occur during rainfall events and are therefore difficult to sample adequately (Pollman & Danek, 1988). Moreover, sediments in discharges from some combined sewer overflows contain high concentrations of suspended particles, PCBs, and other contaminants that accumulate in the river (Kenaga, 1986; Kenaga & Crum, 1987; Marsalek & Ng, 1987).

Accidental spills of hazardous substances can result in shock loadings that are equal to or greater than annual loadings from regulated discharges and are a major source of some contaminants elsewhere in the upper connecting channels (Edsall et al., 1988). From 1973 to 1979, there were 581 spills of petroleum products totaling over 700 m³ and 45 spills of other hazardous substances totaling over 334 m³ into U.S. waters of the Detroit River, primarily from land-based facilities (Manny et al., 1988). The amounts of contaminants spilled into Ontario waters of the river were probably smaller, but records are not available. Additional toxic substances probably enter the river from the 110 sites of ground water contamination located within 3 km of the river and on islands in the river (EC & EPA, 1988).

Trace metals and organic contaminants adsorbed to fine-grained particles that enter the river settle in depositional zones adjacent to islands and shorelines (Fallon & Horvath, 1985). Zinc, nickel, chromium, cobalt, copper, and lead accumulate in the fine clay fraction ($< 13$ $\mu$m dia.) and in a large-sized silt fraction of 48 to 63 $\mu$m (Mudroch, 1985). Sediments in many areas of the lower Detroit River are heavily contaminated with hazardous and toxic substances, including PCBs and heavy metals (Hamdy & Post, 1985; EC & EPA, 1988; Nichols et al., 1990). Such substances are loosely bound to sediments by adsorption and cation-exchange processes and

may be easily released if sediments are disturbed (DePinto et al., 1987). To prevent the escape of contaminants and contamination of the food chain, polluted sediments dredged from the Detroit River are confined in sealed enclosures (IJC, 1982).

## 4. Levels of contamination in river sediments

The large volume of clean water that enters the Detroit River from Lake St. Clair maintains river water quality in an acceptable range for aquatic life (Manny et al., 1988). However, river sediments are seriously contaminated with a variety of toxic organic substances and heavy metals (Table 2). Many of these contaminants, which are only slightly soluble in water, are present in the sediments in concentrations that are greatly in excess of the Canadian guideline for open water disposal of dredged sediments. In 1981, levels of PCBs were ten times higher in sediments along the U.S. shore than along the Canadian shore (Thornley & Hamdy 1984; Kauss & Hamdy, 1985). In 1986, the highest PCB concentration yet found in Detroit River sediments (40 mg kg⁻¹) was found along the Michigan shore about 5 km

Table 2. Contaminant levels (mg kg⁻¹ dry wt) in Detroit River sediments and Ontario pollution guideline for each (compiled from IJC, 1982; Limno-Tech Inc., 1985; Lum & Gammon, 1985; Bertram et al., 1991).

| Contaminant | Level (range) | Guideline |
|---|---|---|
| Volatile solids | 11 000–379 000 | 60 000 |
| Oil and grease | 100–29 000 | 1 500 |
| Polychlorinated biphenyls | 0.02–3.8 | 0.05 |
| Cyanide | 0.5–0.8 | 0.1 |
| Mercury | 0.04–56 | 0.3 |
| Lead | 4.8–960 | 50 |
| Zinc | 21–5 300 | 100 |
| Iron | 15 800–3 710 000 | 10 000 |
| Chromium | 4–330 | 25 |
| Copper | 0.5–380 | 25 |
| Cadmium | 0.3–17 | 1 |
| Nickel | 5–293 | 25 |
| Hexachlorobenzene | 0.0031–0.36 | none |
| Octachlorostyrene | 0.001–0.01 | none |

downstream from Belle Isle (Kenaga, 1986; Kenaga & Crum, 1987). Mercury levels in sediment declined in the Detroit River between 1968 and 1980 but cadmium, chromium, copper, lead, nickel, and zinc concentrations in sediments increased significantly during that period, especially around the mouth of the Rouge River and in the Trenton Channel (Thornley & Hamdy 1984; Nichols *et al.*, 1991).

## 5. Effects of contaminants on Biota

### 5.1. General

The demonstrated effects of contaminants on Great Lakes biota include mutagenicity, carcinogenicity, phototoxicity, body deformities, and reproductive dysfunction (Henry *et al.*, 1989). The effects of measurable sublethal contaminant concentrations on individual organisms or their population, such as formation of external tumors on fish, are largely unknown. Mortality of burrowing mayflies, (*Hexagenia limbata*), reduces the productivity of their populations in contaminated areas of the connecting channels (Edsall *et al.*, 1991). Reproductive failure in populations of herring gulls (*Larus argentatus*), bald eagles (*Haliaeetus leucocephalus*), and double-crested cormorants (*Phalacrocorax auritus*) reduced the entire affected population (Gilbertson, 1988). Finally, there are human health concerns caused by bioaccumulation of contaminants in aquatic animals (Humphrey, 1983, 1988). The following biological effects have been demonstrated at various levels of the Detroit River food chain.

### 5.2. Bacteria

A bacterial luminescence assay using sediment pore water from 136 locations in the lower Detroit River showed that sediments were acutely toxic to *Photobacterium phosphoreum* at 25 locations, mostly in an area along the western shore of the Trenton Channel; sediments in this area also supported no benthic macroinvertebrates (Giesy *et al.*, 1988). Extracts of sediments from this area also contained toxic, synthetic, organic substances that caused mutations in *Salmonella microsomes* (Maccubbin, 1987; DePinto *et al.*, 1987; Furlong *et al.*, 1988).

### 5.3. Phytoplankton

Bioassessment of sediment toxicity to phytoplankton was determined by carbon-14 algal fractionation bioassays that were conducted with various dilutions of standard and chelator-treated elutriates (Munawar *et al.*, 1985). Toxicity, measured as a decrease in carbon assimilation by small ($< 20 \, \mu$m) phytoplankton, was directly related to the concentration of water soluble metals, such as zinc, manganese, cadmium, and lead. The results confirmed that sediment toxicity should not be evaluated as it presently is on the basis of sediment chemical measurements (IJC, 1982).

### 5.4. Zooplankton

Feeding and reproduction of zooplankton (*Daphnia pulicaria* and *Ceriodaphnia* sp.) were reduced 50–100 percent in 7-d bioassays in sediment elutriates (White *et al.*, 1987). Likewise, sediment porewater collected at 10 of 30 stations in the Trenton Channel was acutely toxic to *Daphnia magna* in 4-d bioassays (Giesy *et al.*, 1987).

### 5.5. Benthic macroinvertebrates

Contaminated sediments negatively affect benthic macroinvertebrates in the river. Because the quality and productivity of benthic habitats throughout large areas of the river are impaired (Thornley, 1985; EC & EPA, 1988), the river was designated an Area of Concern for remedial action (Hartig & Thomas, 1988). Sediment avoidance was measured in 2-d tests with the aquatic worm, *Stylodrilus* sp. (White *et al.*, 1987). In uncontaminated sediments, all worms burrowed, all remained buried, and none died; in sediments collected near Monguagon Creek in the Trenton Channel, only 10 percent of the worms remained buried and 53 percent died. Growth of larval midges (*Chironomus tentans*) in contaminated sediments from the Trenton Channel (0.02 to 0.08 mg d$^{-1}$) was slower than in uncontaminated sediments collected elsewhere in the river (0.48 to 0.53 mg d$^{-1}$) (Giesy *et al.*, 1987).

The production of burrowing mayflies (*Hexagenia limbata*) in April to October 1986 was significantly lower in the Detroit River (708 to 1 035 mg dry wt m$^{-2}$), where sediment concentrations of oil and metals exceeded established guidelines for disposal of polluted sediments, than in other areas of the upper connecting channels (980 to 3 481 mg dry wt m$^{-2}$), where sediment concentrations did not exceed the guidelines (Edsall *et al.*, 1990).

Unionid bivalves are used to monitor environmental contaminant concentrations because they accumulate many contaminants present in the environment (Nalepa & Landrum, 1988). Native Detroit River clams (*Lampsilis radiata siliquoidea*) contained concentrations of lead, cadmium, PCBs, and octachlorostyrene that were up to 59 times higher than concentrations of these contaminants in surficial sediments (0 to 10 cm) from which the clams were collected (Great Lakes Institute, 1984; Pugsley *et al.*, 1985). Perhaps coincidentally, the distribution and abundance of this clam, particularly young individuals, has decreased markedly in Lake St. Clair and western Lake Erie during the past 25 years (Nalepa & Gauvin, 1988; Thomas F. Nalepa, NOAA Great Lakes Environ. Res. Lab., Ann Arbor, Mich., pers. comm.). Caged, non-native clams (*Elliptio compalanta*) placed in the Detroit River for 18 months accumulated hexachlorobenzene, octachlorostyrene, PCBs, and polyaromatic hydrocarbons (Kauss & Hamdy, 1985). High tissue concentrations of these contaminants were found in clams along the U.S. shore near the Conners Creek combined sewer overflow, the Rouge River, and in the lower Trenton Channel. Lower tissue concentrations of these organochlorine residues were found in clams along the Ontario shore,

indicating that sources of these substances were likely on the U.S. shore.

### 5.6. Fishes

Contaminated sediments also negatively affect fishes in the river. Larval channel catfish (*Ictalurus punctatus*) fed significantly more slowly when exposed to contaminated sediments from the Trenton Channel than when exposed to uncontaminated sediments (White *et al.*, 1987). Injection of 'eyed' eggs of rainbow trout (*Oncorhynchus mykiss*) with dilute extracts from Detroit River sediments increased mortality of embryos 2- to 3-fold, compared to that of control eggs injected only with solvent carrier. One year after injection, 3 percent of the surviving fish, injected as eggs with extracts from sediments collected near Mongaugon Creek, had liver neoplasms. Neoplasms and pre-neoplastic lesions were also found on brown bullhead (*Ictalurus nebulosus*), walleye, redhorse sucker (*Moxostoma* sp.), white sucker (*Catostomus commersoni*), and bowfin (*Amia calva*) collected in the lower Detroit River (Maccubbin, 1987). Dermal or oral neoplasms were found on 14.4 percent of the bullhead and on 4.8 percent of the walleye. Liver neoplasms were found in 15.4 percent of the bowfin. Spottail shiners (*Notropis hudsonius*) collected near Gibraltar, Michigan contained PCB concentrations (912 to 2 997 ng g$^{-1}$) that were significantly higher than those in these fish near the Canadian shore (153 to 316 ng g$^{-1}$) (Suns *et al.*, 1985).

The lower Detroit River is a major spawning ground for fishes that inhabit the river and western Lake Erie. All but one of the 39 fish species that spawn in or near the mouth of the Detroit River deposit their eggs on the bottom in contaminated sediments (Manny *et al.*, 1988). Heavy metals, such as chromium, that are present in Detroit River sediments can kill eggs and larvae of several of the fish species (Eisler, 1986) that use these historically important spawning grounds in the Detroit River; the eggs and larvae of other fishes that spawn there may also be threatened by these contaminants.

## 5.7. Birds

Eggs of the herring gull (*Larus argentatus*) collected near an industrial waste dump on Fighting Island from 1978 to 1982 contained the highest PCB and hexachlorobenzene concentrations measured anywhere in the Great Lakes basin (Struger *et al.*, 1985). Because these gulls eat fish, lay eggs containing organochlorine contaminants present near the nesting colony, and suffer high rates of reproductive failure (Peakall, 1988) they are monitored as part of a contaminant surveillance plan of the International Joint Commission (Gilbertson, 1988).

Carcasses of 13 diving ducks that foraged during winter on contaminated sediments near Mud Island in the Detroit River contained higher concentrations of more toxic and persistent forms of PCBs than did common carp (*Cyprinus carpio*), aquatic worms, and sediments collected at the same time and place (Smith *et al.*, 1985). Fifteen young-of-the-year diving ducks collected at the same time and place also contained high PCB concentrations (Kreis, 1988).

## 6. Discussion

In this paper we have attempted to describe contaminant concentrations in the Detroit River aquatic environment, relate that information to laboratory and site-specific field studies, and show existing contaminant concentrations produce adverse effects on aquatic life present in the river. Although the picture that emerges is incomplete, the evidence shows that large areas of river bottom habitat have been degraded by contaminants and that many animal populations in the river have been impacted by contaminants.

Levels of toxic substances in Detroit River sediments, such as oil, PCBs, and heavy metals often exceed standards and guidelines designed to protect aquatic life, particularly near urban industrial discharges. Point sources add the most contaminants to the river even though most discharges are regulated. Spills and combined sewer overflows are large unregulated sources of contaminants. Loadings of 'conventional pollutants' (e.g. oil, phenol, phosphorus) have decreased substantially since 1970, but accumulations of oil and heavy metals persist in the sediments in areas where benthic animal populations are greatly altered or eliminated. Fine-grained sediments in the lower river are heavily contaminated from historic, unregulated discharges and continue to expose aquatic biota to toxic substances.

There is a serious lack of information on the effects of specific toxic contaminants or complex mixtures of these contaminants on the aquatic food web, waterfowl, and aquatic mammals. However, it is readily apparent that plant and animal populations throughout large areas of the river have been disrupted and altered by destruction of habitat and chemical contamination of the water and sediments. In these respects, the Detroit River resembles many large rivers of the world that have been used for waste disposal by developed nations (Hynes, 1966; Oglesby *et al.*, 1972; Ajmal *et al.*, 1987; Fremling *et al.*, 1989; Lelek, 1989).

The primary environmental objectives set forth in the bi-national Great Lakes Water Quality Agreement (IJC, 1988) are restoring and maintaining the chemical, physical, and biological integrity of the Great Lakes, including the Detroit River, and eliminating the impacts of toxic substances on man's uses of their aquatic resources. The Agreement seeks to reduce or eliminate the discharge of any or all persistent toxic substances. Criteria for permissible quantities of contaminants in sediments are now being developed and remedial action plans are being formulated to restore all beneficial uses of the Detroit River (Hartig & Thomas, 1988). The draft plan for the Detroit River calls for monitoring tissue contaminant concentrations in river biota to measure remedial progress. Guidelines for such monitoring are being developed (IJC, 1986, 1987; Evans, 1988) and specific measures to reduce pollutant loadings to the Detroit River, including better monitoring of contaminant loadings by combined sewer overflows, have been proposed (EC & EPA, 1988). Compliance schedules in discharge permits for controlling combined sewer overflows are now required for most wastewater treatment plants that discharge into the Detroit River (Michigan Department of Natural Resources, Open files).

An overall reduction in the amount of wastes and toxic substances added to the Detroit River is needed to meet existing water quality objectives (IJC, 1988). By using less-toxic or non-toxic materials in manufacturing processes and by recycling usable materials from waste effluents, progress toward these objectives could be made now with available technology (Lawrence, 1988; Calvin et al., 1988). Many uses of Detroit River water could be met by recycling heated effluents or by reclaiming wastewaters. Sewage may ultimately be disposed of on land rather than in rivers (Hynes, 1966). If so, the present practice of mixing toxic substances with sewage then discharging it into rivers may change to some form of centralized hazardous waste treatment, such as that used in Canada (Hrudey & Simpson, 1988). Habitat improvements (Gore & Petts, 1989) and protection of remaining island habitat in the river (Manny et al., 1988) would enhance the survival of desirable plants and animals in the river. In the long run, such measures may be the most economical way to meet the water needs of modern society.

The potential risks to human health of consuming PCB-laden fish or waterfowl from the river have been noted (Smith et al., 1985; Humphrey, 1988; Hebert et al., 1990). Human exposure to ubiquitous contaminants, such as PCBs, is potentially higher from eating Great Lakes fish, than from direct exposure to terrestrial, atmospheric, or drinking water sources (Swain, 1983; Humphrey, 1983; Davies, 1988). Because Detroit River fish contain PCBs and mercury, these contaminants may be consumed by Detroit River anglers and their families. To protect human health, authorities in Michigan and Ontario have issued consumption advice for large carp, walleye, rock bass (Ambloplites rupestris), and freshwater drum (Aplodinotus grunniens) from the Detroit River (Ontario Ministry of Natural Resources, 1985; Michigan Department of Natural Resources, 1989). No con-

sumption advisory has yet been issued for Detroit River waterfowl.

## 7. Acknowledgements

We thank T.A. Edsall, G.D. Haffner, R.G. Kreis, Jr., and J. Marsalek for bibliographic assistance and F.J. Horvath of the Michigan Department of Natural Resources for Fig. 1. This work was performed as part of a cooperative project between the U.S. Fish and Wildlife Service and the U.S. Environmental Protection Agency, under Interagency Agreement No. DW 14931214-01-0.

## References

Ajmal, M., R. Khan & A. U. Khan, 1987. Heavy metals in water, sediments, fish and plants of river Hindon, U. P., India. Hydrobiologia 148: 151–157.

Bertram, P., T. A. Edsall, B. A. Manny, S. J. Nichols & D. W. Schloesser, 1991. Chemical contamination and physical characteristics of sediments in the upper Great Lakes connecting channels, 1985. U.S. Env. Protect. Agency, Great Lakes Ntnl. Program Off., 230 South Dearborn St., Chicago, IL. 60604. USA In Press.

Calvin, D. W., J. M. Rio & B. L. Haviland, 1988. Integrated waste management: An industrial perspective. In; N. W. Schmidtke (ed.), Toxic contaminants in large lakes. Vol. III. Sources, fate, and controls of toxic contaminants. Lewis Publishers, Chelsea, Mich.: 165–175.

Davies, K., 1988. Human exposure routes to persistent toxic chemicals in the Great Lakes basin: A case study. In; N. W. Schmidtke (ed.), Toxic contamination in large lakes. Vol. I Chronic effects of toxic contaminants in large lakes. Lewis Publishers, Chelsea, Mich.: 195–226.

DePinto, J. V., T. L. Theis, T. C. Young, D. Vanetti, M. Waltman & S. Leach, 1987. Exposure and biological effects of in-place pollutants: sediment exposure potential and particle-contaminant interactions. Clarkson Univ., Env. Eng., Rep. No. 87–9. 75 pp. + appendices.

Derecki, J. A., 1984. Detroit River, physical and hydraulic characteristics. Ntnl. Oceanogr. Atmos. Admin., Great Lakes Env. Res. Lab., Ann Arbor, Mich. GLERL Contrib. No. 417.

EC & EPA (Environment Canada & U.S. Environmental Protection Agency). 1988. Upper Great Lakes connecting channels study. Final report, Volume II. U.S. Env. Protect Agency, Chicago, Ill. 591 pp.

Edsall, T. A., B. A. Manny & C. N. Raphael, 1988. The St. Clair River and Lake St. Clair, Michigan: An ecological

278

profile. U.S. Fish Wildl. Serv. Washington, D. C. Biol. Rep. 85(7.3). 130 pp.

Edsall, T. A., B. A. Manny, D. W. Schloesser, S. J. Nichols & A. M. Frank, 1991. Production of *Hexagenia limbata* nymphs in contaminated sediments in the Upper Great Lakes connecting Channels. Hydrobiologia 219: 353–361.

Eisler, R., 1986. Chromium hazards to fish, wildlife, and invertebrates: A synoptic review. U.S. Fish Wildl. Serv., Patuxent Wildl. Res. Center, Laurel, MD 20708. Biol. Rep. 85(1.6).

Evans, M. S. (ed.), 1988. Toxic contaminants and ecosystem health: a Great Lakes focus. John Wiley & Sons, New York. 602 pp. ISBN 0-471-85556-1.

Fallon, M. E. & F. J. Horvath, 1985. Preliminary assessment of contaminants in soft sediments of the Detroit River. J. Great Lakes Res. 11: 373–378.

Fremling, C. R., J. L. Rasmussen, R. E. Sparks, S. P. Cobb, C. F. Bryan & T. O. Claflin. 1989. Mississippi River fisheries: A case history. Large Rivers Symposium. Can J. Fish. Aquat. Sci. Spec. Pub. 106: 309–351.

Furlong, E. T., D. S. Carter & R. A. Hites, 1988. Organic contaminants in sediments from the Trenton Channel of the Detroit River, Michigan. J. Great Lakes Res. 14: 489–501.

Giesy, J. P., R. L. Graney, J. L. Newsted & C. J. Rosiu, 1987. Toxicity of in-place pollutants to benthic invertebrates. Interim Rep. to U.S. Env. Protect. Agency, Large Lakes Res. Sta., Grosse Ile, Mich. & Mich. State Univ., East Lansing, Mich.

Giesy, J. P., C. J. Rosiu, R. L. Graney, J. L. Newsted, A. Benda, R. G. Kreis, Jr. & F. J. Horvath, 1988. Toxicity of Detroit River sediment interstitial water to the bacterium *Photobacterium phosphoreum*. J. Great Lakes Res. 14: 502–513.

Gilbertson, M., 1988. Epidemics in birds and mammals caused by chemicals in the Great Lakes. In; M. S. Evans, (ed.) Toxic contaminants and ecosystem health: A Great Lakes focus. John Wiley & Sons, New York: 133–152.

Gore, J. A. & G. E. Petts, 1989. Alternatives in regulated river management. CRC Press. Boca Raton, Florida. 352 pp. ISBN 0-849-34877-3.

Great Lakes Institute, 1984. A case study of selected trace contaminants in the Essex region. Univ. Windsor, Ann. Rep. UP-175, Vol. 1: Physical Sciences. March 31. Windsor, Ontario.

Haas, R. C., W. C. Bryant, K. D. Smith & A. J. Nuhfer, 1985. Movement and harvest of fish in Lake St. Clair, the St. Clair River, and the Detroit River. U.S. Army Corps Engin., Detroit, Michigan. 141 pp.

Hamdy, Y. & L. Post, 1985. Distribution of mercury, trace organics, and other heavy metals in Detroit River sediments. J. Great Lakes Res. 11: 353–365.

Hartig, J. H. & R. L. Thomas, 1988. Development of plans to restore degraded areas in the Great Lakes. Env. Managem. 12: 327–347.

Hebert, C. E., G. D. Haffner, I. M. Weis, R. Lazar & L. Montour, 1990. Organochlorine contaminants in duck populations of Walpole Island. J. Great Lakes Res. 16: 21–26.

Henry, M., M. Mac, B. Manny, T. Kubiak & J. Gannon, 1989. Contaminants in the Great Lakes: a threat to fish and wildlife. U.S. Fish Wildl. Serv., Ntnl. Fish. Res. Center-Great Lakes, Ann Arbor, Mich. Unpubl. Rep.

Hrudey, S. E. & K. J. Simpson, 1988. Management of residues from centralized hazardous waste treatment facilities. In; N. W. Schmidtke, (ed.), Toxic contamination in large lakes. Vol. III. Sources, fate, and controls of toxic contaminants. Lewis Publishers, Chelsea, Mich.: 177–190.

Humphrey, H. E. B., 1983. Populations studies of PCBs in Michigan residents. In; F. M. D'Itri & M. A. Kamrin (eds), PCBs: Human and Environmental Hazards. Ann Arbor Sci. Publishers, Ann Arbor, Mich.: 299–310.

Humphrey, H. E. B., 1988. Chemical contaminants in the Great Lakes: The human health aspect. In; M. S. Evans (ed.), Toxic contaminants and ecosystem health: a Great Lakes focus. John Wiley & Sons, New York.: pp. 153–165.

Hynes, H. B. N., 1966. The biology of polluted waters. Liverpool Univ. Press. Great Britain. 202 pp. ISBN 0-820-16901.

IJC (International Joint Commission), 1982. Guidelines and register for evaluation of Great Lakes dredging projects. Rep. Dredging Subcomm. to Great Lakes Wat. Qual. Bd., Windsor, Ontario. 365 pp.

IJC. (International Joint Commission), 1986. Evaluation of sediment bioassessment techniques. Rep. Dredging Subcomm. to Great Lakes Wat. Qual. Bd, Windsor, Ontario. 123 pp.

IJC. (International Joint Commission), 1987. Guidance on characterization of toxic substances problems in areas of concern in the Great Lakes basin. Windsor, Ontario. 179 pp.

IJC. (International Joint Commission), 1988. Revised Great Lakes Water Quality Agreement of 1978, as amended by protocol signed November 18, 1987. Windsor, Ontario.

Kauss, P. B. & Y. Hamdy, 1985. Biological monitoring and organochlorine contaminants in the St. Clair and Detroit Rivers using introduced clams, *Elliptio complanatus*. J. Great Lakes Res. 11: 247–263.

Kenaga, D., 1986. Concentrations of PCBs in sediment depositional zones in the upper Detroit River along the United States shore, July 8 and 9, 1986 and the 18th Street sewer system, Detroit Michigan, July 14 and 15, 1986. Mich. Dept. Ntrl. Resources, Surf. Wat. Qual. Div., Lansing, Mich., Staff Rep. 13 pp.

Kenaga, D. & J. Crum, 1987. Sediment PCB concentrations along the U.S. shoreline of the upper Detroit River, and sediment and water PCB concentrations in two City of Detroit combined sewers with overflows to the Detroit River, July and October, 1986 and January 1987. Mich. Dept. Ntrl. Resources, Surf. Wat. Qual. Div., Lansing, Mich. Staff Rep. 6 pp.

Kreis, R. G., Jr. (ed.), 1988. Integrated study of exposure and biological effects of in-place sediment pollutants in the Detroit River, Michigan: An upper Great Lakes connecting channel. U.S. Env. Protect. Agency, Large Lakes Res. Lab., 9311 Groh Road, Grosse Ile, MI 48138, USA.

Lawrence, J. R., 1988. Wealth generator or environmental protector. The approach of a chemical company to process and product development under increasing environmental pressures. In; N. W. Schmidtke, (ed.), Toxic contamination in large lakes. Vol. III. Sources, fate, and controls of toxic contaminants. Lewis Publishers, Chelsea, Mich.: 121–149.

Lelek, A., 1989. The Rhine River and some of its tributaries under human impact in the last two centuries. Large Rivers Symposium. Can. J. Fish. Aquat. Sci. Spec. Publ. 106: 469–487.

Limno-Tech Inc., 1985. Summary of the existing status of the upper Great Lakes connecting channels data. 2395 Huron Parkway, Ann Arbor, Mich. 48104. 156 pp.

Lum, K. R. & K. L. Gammon, 1985. Geochemical availability of some trace and major elements in surficial sediments of the Detroit River and western Lake Erie. J. Great Lakes Res. 11: 328–338.

Maccubbin, A. E., 1987. Biological effects of in-place pollutants: Neoplasia in fish and related causal factors in the Detroit River system. Interim Rep. U.S. Env. Protect. Agency Large Lakes Res. Sta., 9311 Groh Road, Grosse Ile, Mich.

Manny, B. A., T. A. Edsall & E. Jaworski, 1988. The Detroit River, Michigan: an ecological profile. U.S. Fish. Wildl. Serv. Biol. Rep. 85(7.17). 86 pp.

Marsalek, J. & H. Y. F. Ng, 1987. Contaminants in urban runoff in the upper Great Lakes connecting channels area. Canada Centre for Inland Waters, Ntnl. Wat. Res. Inst., Burlington, Ontario, Canada. Rep. No. RRB-87-27.

Michigan Department of Natural Resources, 1989. Michigan fishing guide. East Lansing, Mich.

Mudroch, A., 1985. Geochemistry of Detroit River sediments. J. Great Lakes Res. 11: 193–200.

Munawar, M., R. L. Thomas, W. Norwood & A. Mudroch, 1985. Toxicity of Detroit River sediment-bound contaminants to ultraplankton. J. Great Lakes Res. 11: 264–274.

Muth, K. M., D. R. Wolfert & M. T. Bur, 1986. Environmental study of fish spawning and nursery areas in the St. Clair-Detroit River system. U.S. Fish Wildl. Serv., Ntnl. Fish. Res. Center – Great Lakes, Ann Arbor, Mich. Admin. Rep. 85–6. 53 pp.

Nalepa, T. F. & J. M. Gauvin, 1988. Distribution, abundance, and biomass of freshwater mussels (Bivalvia: Unionidae) in Lake St. Clair. J. Great Lakes Res. 14: 411–419.

Nalepa, T. F. & P. F. Landrum, 1988. Benthic invertebrates and contaminant levels in the Great Lakes: effects, fates, and role in cycling. In; M. S. Evans (ed.), Toxic contaminants and ecosystem health; A Great Lakes focus. John Wiley & Sons. New York.: 77–102.

Nichols, S. J., B. A. Manny, D. W. Schloesser & T. A. Edsall, 1991. Heavy metal contamination of sediments in the upper connecting channels of the Great Lakes. Hydrobiologia 219: 307–315.

Oglesby, R. T., C. A. Carlson & J. A. McCann, (eds.), 1972. River ecology and man. Proc. Internat. Symp., Univ. Massachusetts, June 20–23, 1971. Academic Press, New York. 465 pp. ISBN 0-12-524450-9.

Ontario Ministry of the Environment, 1985. Guide to eating Ontario sport fish. 135 St. Clair Ave, West, Toronto, Ontario, Canada, M4V 1P5. 254 pp.

Peakall, D. B., 1988. Known effects of pollutants on fish-eating birds in the Great Lakes of North America. In; N. W. Schmidtke, (ed.), Toxic contamination in large lakes. Vol. I, Chronic effects of toxic contaminants in large lakes. Lewis Publishers, Chelsea, Mich.: 39–54.

Pollman, C. D. & L. J. Danek, 1988. Contribution of urban activities to toxic contamination of large lakes. In; N. W. Schmidtke, (ed.), Toxic contamination of large lakes. Vol. III, Sources, fate, and controls of toxic contaminants. Lewis Publishers, Chelsea, Michigan.: 25–40.

Pugsley, C. W., P. D. N. Hebert, G. W. Wood, G. Brotea & T. W. Obal, 1985. Distribution of contaminants in clams and sediments from the Huron-Erie corridor. I – PCBs and Octachlorostyrene. J. Great Lakes Res. 11: 275–289.

Schloesser, D. W. & B. A. Manny, 1990. Decline of wild celery buds in the lower Detroit River, 1950–1985. J. Wildl. Managem. 54: 72–76.

Smith, V. E., J. M. Spurr, J. C. Filkins & J. J. Jones, 1985. Organochlorine contaminants of wintering ducks foraging on Detroit River sediments. J. Great Lakes Res. 11: 231–246.

Struger, J., D. V. Weseloh, D. J. Hallette & P. Mineau, 1985. Organochlorine contaminants in herring gull eggs from the Detroit and Niagara Rivers and Saginaw Bay (1978–1982): contaminant discriminants. J. Great Lakes Res. 11: 223–230.

Suns, K., G. Crawford & D. Russell, 1985. Organochlorine and mercury residues in young-of-the-year spottail shiners from the Detroit River, Lake St. Clair, and Lake Erie. J. Great Lakes Res. 11: 347–352.

Swain, W. R., 1983. An overview of the scientific basis for concern with polychlorinated biphenyls in the Great Lakes. In; F. M. D'Itri & M. A. Kamrin (eds), PCBs: Human and Environmental Hazards. Ann Arbor Sci. Publ., Ann Arbor, Mich.: pp. 11–48.

Thornley, S., 1985. Macrozoobenthos of the Detroit and St. Clair Rivers with comparisons to neighbouring waters. J. Great Lakes Res. 11: 290–296.

Thornley, S. & Y. Hamdy, 1984. An assessment of the bottom fauna and sediments of the Detroit River. Ontario Min. Environ., Toronto. 48 pp.

USACE (U.S. Army Corps of Engineers), 1981. Essayons. A history of the Detroit District. Detroit, Mich. 215 pp.

USACE (U.S. Army Corps of Engineers), 1984. Waterborne commerce of the United States, calendar year 1983. Part 3 – Waterways and harbors of the Great Lakes. U.S. Government Printing Office, Washington, D.C. 103 pp.

White, D. S., J. Bowers, D. Jude, R. Moll, S. Hendricks, P. Mansfield & M. Flexner, 1987. Exposure and biological effects of in-place pollutants (bioassays). Interim Rep. to the U.S. Env. Protect. Agency, Large Lakes Res. Sta., 9311 Groh Road, Grosse Ile, MI 48138, USA.

*Hydrobiologia* **219**: 281–299, 1991.
*M. Munawar & T. Edsall (eds),*
*Environmental Assessment and Habitat Evaluation of the Upper Great Lakes Connecting Channels.*
© *1991 Kluwer Academic Publishers.*

# Response of bacteria and phytoplankton to contaminated sediments from Trenton Channel, Detroit River[1]

Russell A. Moll and Pamela J. Mansfield
*Center for Great Lakes and Aquatic Sciences, The University of Michigan, 2200 Bonisteel Blvd., Ann Arbor, Michigan 48109-2099 USA*

*Key words:* contaminated sediment bioassay, phytoplankton, bacteria, primary productivity, heterotrophic activity, Detroit River

## Abstract

Several types of bioassays were used in 1986 and 1987 to investigate the effect of contaminated sediments on natural populations of bacteria and phytoplankton from the Trenton Channel, Detroit River. The approach included the measurement of uptake of $^3$H-glucose or $^3$H-adenine by bacteria and $^{14}$C-bicarbonate by phytoplankton in the presence of different amounts of Trenton Channel and Lake Michigan (control) sediments. Trenton Channel sediments are contaminated by high levels of toxic organic compounds and metals, especially zinc, lead, and copper. Because levels of biomass of bacteria and phytoplankton varied widely among the different bioassays, it was necessary to adjust uptake rates for biomass. Biomass adjustments were made using acridine orange counts for bacteria and chlorophyll measurements for phytoplankton. The results show a statistically significant suppression of uptake of substrates for both bacteria and phytoplankton with increasing amounts of sediment. Uptake was suppressed as much as 90 percent for bacteria and 93 percent for phytoplankton at 1200 mg l$^{-1}$ of Trenton Channel sediments compared to bioassays without sediment. Uncontaminated Lake Michigan sediment suppressed uptake much less than Detroit River sediment; the difference in suppression of uptake between the two types of sediment was statistically significant for both bacteria and phytoplankton.

## 1. Introduction

A large number of studies have attempted to quantify the degree to which various components of the food web are affected by contaminated sediment (Struger *et al.*, 1985; Smith *et al.*, 1985; Pugsley *et al.*, 1985). Some of these have employed bioassay techniques which have evolved into a protocol for the investigation of the effects of contaminated sediment on biota (Mount & Norberg, 1984). Results of these studies lead to the general conclusion that a large number of factors are determinants in the outcomes of the experiments (Munawar *et al.*, 1983). Some of these factors are: the type and concentration of contaminants, the location of the contaminant in the sediment (bound to particle, pore water, etc.), the amount of organic matter in the sediment, the pH of the sediment and the overlying water, the Eh of the sediment, the temperature, and the intensity of mixing of sediment and water (Munawar *et al.*, 1985). Further, the response of the organisms toward the contaminant depends upon the species and its location in the water column or sediment (Wong *et al.*, 1978).

[1] Contribution No. 518 of the Center for Great Lakes and Aquatic Sciences of the University of Michigan.

In effect, the bioassay of sediment has been species, site, and date specific.

Experience from the Great Lakes Basin supports the same general conclusions discussed above. Results from the Detroit River, Niagara River, and Lakes Erie and Ontario could not be readily extended to other locations within the basin. In effect, these conclusions predicate that a generic bioassay be developed to answer questions from specific locations (U.S. Environmental Protection Agency/U.S. Army Corp of Engineers, 1977). The generic nature of the bioassay allows a common yardstick upon which to compare the degree of toxicity at many locations; this common yardstick can then be used for management decisions.

These concerns were the rationale behind this study which had as its focus contaminant problems associated with sediment in connecting rivers and water bodies. Realistic appraisals of the potential exposure of organisms to toxicants and the response to those exposures was the primary objective of the study. This paper deals with a subset of that problem, the response of phytoplankton and bacteria to contaminated sediment from the Trenton Channel.

The objective of the phytoplankton and bacteria studies was to estimate the suppression of bacterial metabolism and phytoplankton productivity from exposure to Detroit River sediments. From this objective a set of experiments was developed which attempted to determine biological effects with several types of bioassays. These experiments, carried out during the ice-free season in 1986 and 1987, focused on the use of bioassays incorporating the uptake of several different types of radioactive-labelled organic and inorganic compounds. The different bioassays were designed to investigate these hypotheses: (1) the response of microorganisms to contaminated sediment is a combination of a particle effect and a toxic effect, (2) toxicity is primarily from the sediment rather than the pore water, and (3) toxicity varies seasonally and spatially.

The study location was the Trenton Channel which is a branch of the southern portion of the Detroit River west of Grosse Ile. The Trenton

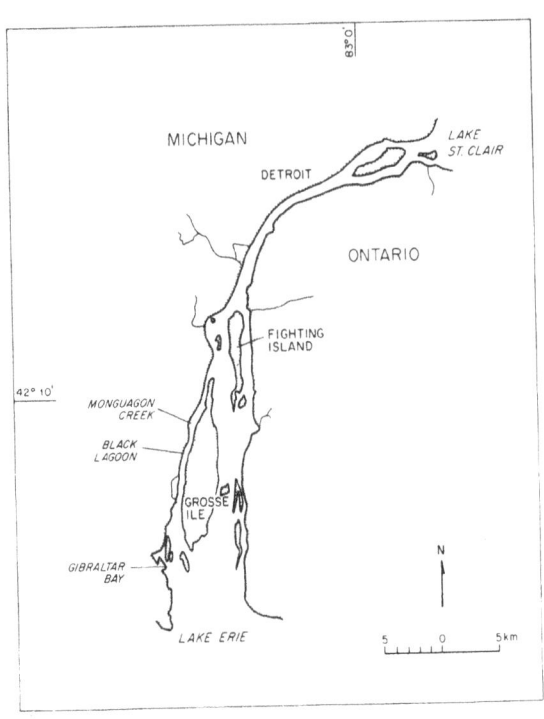

Fig. 1. Sampling locations in the Detroit River.

Channel is analogous to the main portion of the Detroit River in that it is dredged for 2/3 of its length resulting in a hard gravel and rock bottom. Streamflows average $2\,100 \text{ m}^3 \text{ s}^{-1}$ in the Trenton Channel which is approximately 1/3 of the total flow of the Detroit River. The dredging activity and the large streamflows keep the bottom of the channel scoured. There are numerous small creeks and embayments which have accumulations of soft sediment along the length of the Trenton Channel. Three of these areas served as the primary sampling locations for this study (Fig. 1).

Much of the western shore of the Trenton Channel is heavily industrialized including steel manufacturing, chemical processing plants, automotive parts manufacturing, and electrical generation facilities (coal fired). In conflict with these industrial processes, the Trenton Channel has many waterfront homes, parks, and boat launching facilities which provide ready access for recreation. The banks of the Trenton Channel also hold a hazardous waste storage site which shows evidence of leaking of toxic materials into the river (Furlong *et al.*, 1988). Finally, the

Trenton Channel receives extensive input of treated sewage effluent including some of the City of Detroit treatment plant discharge. These numerous environmental insults have resulted in designating the Trenton Channel and the Detroit River as one of the 42 Areas of Concern in the Laurentian Great Lakes.

## 2. Methods

### 2.1. Field procedures

Samples were collected three times during the 1986 field season, spring (May–June), summer (August), and autumn (October), and seven times during 1987, approximately once per month from April through October. Sediment was collected with a Ponar grab from three sites, Monguagon Creek, Black Lagoon, and Gibraltar Bay (Fig. 1). Sediment was stored in 250-ml wide-mouth polyethylene bottles which had been acid-washed and thoroughly rinsed. River water was collected in 20-l carboys from just below the surface at the center of the Trenton Channel, near each station in 1986 and near Monguagon Creek in 1987. This water was the source of bacteria and phytoplankton populations used in the bioassays. Water and sediment samples were kept cool until they reached Ann Arbor. All samples were then refrigerated in darkness, in sealed containers, until bioassays were performed, usually within 48 hours of collection.

### 2.2. Bacterial bioassays

Bioassays with bacteria were conducted during the 1986 and 1987 field seasons with sediment from the three Trenton Channel stations. These bioassays were conducted with three different protocols: clean versus contaminated sediment, elutriate, and pore water. Metabolism of bacteria was measured by uptake of $^3$H-glucose or $^3$H-adenine (New England Nuclear).

Sediment-water solutions for the clean versus contaminated and elutriate bioassays were pre-

*Table 1.* Dry weight of sediment as percent of wet weight; mean of two replicates.

| Lake Michigan | 52 |
|---|---|
| Monguagon Creek | 37 |
| Black Lagoon | 32 |
| Gibraltar Bay | 26 |

pared as follows: The prescribed amount (see below) of unprocessed sediment (unscreened) was weighed on an electronic balance to the nearest 0.1 mg. Leaves, small twigs and other debris were excluded from the sediment. Sediment was weighed as wet weight in order to keep handling to a minimum after collection. Laboratory analysis showed that wet weight of sediment varied between two and four times oven-dried weight (Table 1). The sediment was then washed into 1000 ml of bioassay water, thoroughly mixed and rapidly dispensed into the bioassay incubation vessels. Incubation vessels consisted of Pyrex-type hard glass 15-ml screw-top test tubes which were sterilized just before use.

The clean versus contaminated sediment bioassays consisted of incubation of four matched pairs of samples per bioassay. One member of each pair received clean (central Lake Michigan) sediment and the other member contaminated sediment from one of the Detroit River stations. One bioassay was conducted for each site for each time period. Bacterial bioassays were completed for all sites (Lake Michigan plus the three Trenton Channel stations) in one day for a given time period. Sediment concentrations were 0, 12, 120, and 1200 mg l$^{-1}$ wet weight in 1986; in 1987 a 600 mg l$^{-1}$ weight was added, and the control was not assayed with a 0 mg l$^{-1}$ 'contaminated' matched pair.

The sterile test tubes were filled with 15 ml of sediment-water solution. These samples were inoculated with 2 $\mu$Ci of carrier-free, high-specific activity (30 Ci/mmole) $^3$H-isotope in 0.2 ml of sterile water. The charge of isotope contained 0.012 $\mu$g glucose (New England Nuclear – NET 100) or 0.016–0.018 $\mu$g adenine (New England Nuclear – NET 350). Glucose was only used in the first experiments (May–June 1986), with all

284

other bioassays in 1986 and 1987 employing $^3$H-adenine. After inoculation, the tubes were immediately tightly capped and wrapped in aluminum foil. Tubes were then placed in an incubator and rotated, at 1 rpm on a 46-cm-diameter vertical disk-shaped rotator, to keep sediment in suspension. Samples were incubated for two hours in darkness with the incubator set to the river water temperature.

The elutriate bioassays followed the above protocol except that the isotope was not added immediately after the sediment-water mixture was placed in the test tubes. Rather, the bioassay samples were rotated for 0.5 hour, allowed to settle for one hour, then inoculated with isotope and incubated for two hours in the dark without rotation. The clean sediment from the bioassays described above served as the other half of the matched pair for the elutriate bioassays. The elutriate bioassays were not conducted in 1987.

Pore water was extracted from sediment by centrifuging 250 ml of wet sediment at 2000 rpm for 45 min and then drawing off the supernatant. Pore water was diluted to concentrations of 0.5, 1.0 and 10.0 ml of pore water (corresponding to approximately 1.9, 3.7, and 37 g of sediment, wet weight) per liter bioassay water.

Pore water bioassays followed the same protocol as the clean versus contaminated bioassays after the pore water-river water solution was mixed. The solution was dispensed into the sterile test tubes and the tubes inoculated with 2 $\mu$Ci of $^3$H isotope (glucose or adenine). Incubations were conducted in the dark for two hours without rotation. The pore water bioassays were not conducted in 1987.

All bacterial bioassays were conducted in triplicate, three incubation tubes per treatment. The average of the three tubes was considered the response of the bioassay at one sediment concentration at one location. Bacterial bioassays were corrected for instantaneous and abiotic uptake of isotope by subtracting a time zero blank from each uptake value. Time zero blanks were determined by inoculating three test tubes with the 120 or 1200 mg l$^{-1}$ sediment-water mixture and immediately filtering the samples without any in-

cubation period. The amount of radioactive isotope injected into each test tube (standards) was determined by injecting 0.2 ml of isotope and sterile water directly into a scintillation vial.

Immediately after incubation the contents of the test tubes were filtered onto Millipore SA filters (0.2-$\mu$ pore size) with a low vacuum. Each filter was rinsed with five 1-ml aliquots of ice-cold trichloroacetic acid (5% v/v) except for the glucose experiments (May–June 1986) where each filter was rinsed with 5 ml of sterile water. The damp filters were placed into scintillation vials containing 10 ml of Safety-Solve, a water-dispersing cocktail. All samples, standards, and blanks from each bioassay were assayed with a liquid scintillation counter at the same time.

### 2.3. Phytoplankton bioassays

The phytoplankton bioassays used the same overall protocol as the bacterial bioassay. The sediment-water and pore water assay solutions were prepared with the same technique as the bacterial bioassays. The differences between the two bioassays were the use of $^{14}$C-sodium bicarbonate (New England Nuclear – NEC 086H), larger incubation vessel, incubations in the light and different sediment concentrations in 1987. Results from 1986 showed that sediment concentrations approaching 1200 mg l$^{-1}$ suppressed uptake of isotope to such an extent that differences between matched pairs were difficult to determine. In 1987 sediment concentrations for phytoplankton bioassays were changed to 75, 150, 300, and 600 mg l$^{-1}$ wet weight.

Phytoplankton bioassays consisted of three 125-ml bottles per sediment weight and type (the bottles actually held 145 ml). Each triplicate was comprised of two light bottles and one dark bottle. After the bottles were filled with the appropriate solution, 2 ml of incubation water was removed to provide space for the charge of isotope. Two $\mu$Ci of $^{14}$C-sodium bicarbonate in 2 ml water was added to each bottle which was immediately capped. Incubations were conducted in a well-lit incubator (300–500 $\mu$E m$^{-2}$ s$^{-1}$) which was maintained at river water temperature. Samples were rotated for the 4-hour incubation. Elutriate bioassays were conducted by rotating the samples for 0.5 hour, then letting them settle for one hour followed by injecting with the $^{14}C$ and a 4-hour incubation without rotation. Pore water bioassays were incubated for four hours without rotation. Similar to the bacterial bioassays, only clean versus contaminated bioassays were conducted in 1987.

The $^{14}C$ isotope was prepared by using conventional methods in 1986 and ultra-clean metal-free methods in 1987 (Fitzwater et al., 1982). The conventional methods consisted of purchasing 1.0 mCi of bicarbonate $^{14}C$, diluting the 1.0 ml to 1000 ml with sterile water with the pH adjusted to approximately 8.5, and dispensing this solution into glass ampoules which were immediately sealed with a torch. Sealed ampoules were sterilized and stored in a refrigerator until use. The ultra-clean modification of the inoculation procedure consisted of the following: The initial charge of $^{14}C$ was cut to 100 ml with sterile water which had a pH of about 8.5. This 100 ml was then stored in a Teflon bottle in the dark in a refrigerator until the day of bioassay. In the morning before the bioassays began, the stock was cut to working strength by diluting 12 ml of the stored $^{14}C$ solution to 60 ml with sterile water. This working solution was kept in another Teflon bottle and dispensed with non-metallic pipettes, 1 ml (2 $\mu$Ci) per incubation bottle.

Bioassays were terminated by removing the bottles from the incubator and placing them in the dark until filtration. Samples were filtered through Millipore HA filters (0.45 $\mu$ pore size) and the damp filters placed in a scintillation vial with 1.0 ml of 10 percent HCl. After a 2-hour soak in the dilute acid to remove carbonates on the filter, 10 ml of water-dispersing scintillation cocktail was added.

The dark bottle of the triplicate incubation was taken to represent non-photosynthetic uptake of $^{14}C$. The dark uptake was subtracted from the uptake of the two light bottles. The amount of $^{14}C$ reaching the samples was determined by injecting one charge (1.0 ml) into a scintillation vial containing 10 ml of cocktail and 1.0 ml of scintillation-grade phenethylamine. The latter substance was used to keep the $^{14}C$ trapped in the vial by preventing any $^{14}CO_2$ from escaping from solution.

The uptake of the samples was converted to units of uptake of $\mu$g C l$^{-1}$ h$^{-1}$ which required the determination of alkalinity in the Detroit River water. The alkalinity determination was made by mixing 20.0 ml of sample with 5.0 ml of 0.010 N HCl and measuring the pH of the resultant mixture.

## 2.4. Isotope assays

Activity of samples with $^3H$ and $^{14}C$ was determined with a Beckman LS 7500 liquid scintillation counter. Each type of isotope was measured with a separate program which optimized counting efficiency by an external standard and Compton Edge. Samples were counted for up to 10 minutes or until the standard error of counting efficiency fell below 2 percent of the mean counts per minute. Counts per minute for each sample were converted to disintegrations per minute by external standards, a quench curve, and H numbers. The quench curve was determined by counting external quenched standards of known specific activity.

## 2.5. Biomass estimates

At the time of each bacterial bioassay, 5-ml samples were taken from all sediment-water concentrations, preserved with one drop of glutaraldehyde, and stored in a refrigerator. Within a few weeks bacteria samples were processed for acridine orange direct counts using a modification (Moll & Brahce, 1986) of the technique of Hobbie et al. (1977). A Leitz Wetzlar Dialux 20 microscope was used to count bacteria, at a magnification of 787 × or 1250 ×. Bacteria were counted in 6 to 10 fields of view, depending on variability of the counts.

During each algal bioassay, water from each sediment-water type and concentration was filtered for chlorophyll measurement, using methods based on those in Davis & Simmons (1979). Chlorophyll measurements involved collection of algae on a Whatman GF/C filter, initial extraction in 90 percent acetone buffered with magnesium carbonate, and further extraction by grinding. The amount of chlorophyll in the sample was estimated with a Turner Designs Model 10 fluorometer. The fluorometer was calibrated by taking measurements of known dilutions of 1 g chlorophyll $a$. These same dilutions were also measured with a Beckman spectrophotometer. Chlorophyll equations were obtained from Strickland & Parsons (1972).

## 2.6. Statistical analyses

All statistical techniques described below were conducted with the Michigan Interactive Data Analysis System (MIDAS) and BMDP Statistical Software (Fox & Guire, 1976; Dixon et al., 1981). Clean versus contaminated sediment bioassays were analyzed using two-factor Analysis of Variance (ANOVA) where the main effects were sediment type (clean or contaminated, or site from which sediment was collected) and concentration of sediment. Elutriate bioassay results were contrasted with clean and contaminated sediment results using two-factor ANOVA, where the main effects were sediment concentration and experimental treatment. A one-factor ANOVA contrasted the four pore water concentrations. Levene's Test for equality of variances was conducted for all two-factor ANOVAs and for pore water assays (Brown & Forsythe, 1974). The Brown-Forsythe Test, similar to ANOVA but less sensitive to violations of the assumptions of parametric analysis, was also used for all bioassay results. In those instances where the Levene's Test indicated that intergroup variances were not homogeneous, the Brown-Forsythe Test was considered the more reliable form of analysis. Otherwise, the ANOVA results were viewed as the definitive statistical analysis.

# 3. Results

## 3.1. Bacterial bioassays

### 3.1.1. Clean versus contaminated sediment bioassays

The clean versus contaminated sediment bioassays comprised the major portion of the bacterial studies in 1986 and 1987. This strategy was adopted on the basis of 1986 studies where these types of bioassays were found to yield reproducible results. Consistent results were observed despite a change in substrates from glucose to adenine during 1986 and adjustment to techniques between 1986 and 1987.

During 1986, the uptake of substrate for the clean versus contaminated sediment bioassay showed a pattern with relatively constant uptake among the four different clean sediment concentrations. The average of three sets of bioassays showed a reduction in uptake of about 20 percent between the 0 and 1200 mg $l^{-1}$ clean sediment concentrations. In contrast, bioassays with contaminated sediments showed a large suppression of uptake, typically about 50 percent between the 0 and 1200 mg $l^{-1}$ sediment concentrations. These trends were generally observed throughout 1986 but differed among the three sampling locations.

The results of the bacterial bioassays were transformed into units of uptake adjusted for biomass. The rationale for this transformation was that the sediment additions significantly increased bacterial biomass. The resultant increase in bacterial numbers caused a subsequent increase in uptake which could mask the suppression of isotope uptake by the contaminated materials. The transformation yielded units of picograms per hour per million cells (pg $h^{-1}$ $M^{-1}$ cells), which are referred to as adjusted uptakes.

Adjusted uptake results from the 1986 bioassays showed a decline in uptake with both clean and contaminated sediment additions. Suppression of uptake of glucose and adenine with clean sediment ranged from 34.3 to 52.0 percent (Table 2). Suppression of uptake with contaminated sediment ranged from 65.2 to 79.2 percent

*Table 2*. Bacteria uptake adjusted for biomass (pg h$^{-1}$ M$^{-1}$ cells) at various concentrations of sediment from Lake Michigan (Clean), Monguagon Creek (MC), Black Lagoon (BL), and Gibraltar Bay (GB), pooled over all months, 1986.

| Site | Sediment concentration (mg l$^{-1}$) | | | |
|---|---|---|---|---|
| | 0(control) | 12 | 120 | 1200 |
| Clean | 17.2 | 21.1 | 22.8 | 11.3 |
| Contam: MC | 12.3 | 15.7 | 16.1 | 4.28 |
| Elutriate: MC | 24.6 | 17.3 | 16.2 | 5.53 |
| Clean | 27.7 | 21.9 | 33.2 | 13.3 |
| Contam: BL | 27.6 | 30.6 | 10.6 | 5.73 |
| Elutriate: BL | 20.1 | 22.8 | 8.02 | 2.16 |
| Clean | 20.0 | 20.9 | 17.3 | 13.5 |
| Contam: GB | 23.7 | 26.7 | 29.6 | 5.83 |
| Elutriate: GB | 33.7 | 36.3 | 25.5 | 11.8 |

*Table 3*. Bacteria uptake adjusted for biomass (pg h$^{-1}$ M$^{-1}$ cells) at various concentrations of sediment from Lake Michigan and Trenton Channel sites, pooled over all months, 1987.

| Site | Sediment concentration (mg l$^{-1}$) | | | | |
|---|---|---|---|---|---|
| | 0(control) | 12 | 120 | 600 | 1200 |
| Lake Michigan | 9.81 | 6.11 | 7.02 | 5.68 | 4.75 |
| Monguagon Creek | 9.81 | 7.12 | 5.14 | 4.55 | 3.76 |
| Black Lagoon | 9.81 | 7.02 | 5.52 | 2.46 | 2.27 |
| Gibraltar Bay | 9.81 | 6.31 | 6.83 | 3.42 | 3.27 |

(Table 2). This pattern of suppression of adjusted uptake with increasing sediment concentration was observed in every set of bioassays except the May–June experiments using Monguagon Creek sediment and August using clean (Lake Michigan) sediment.

The 1987 studies were modified based on the results of the 1986 experiments. The clean versus contaminated sediment bioassay became the primary mechanism of investigating the effects of Trenton Channel sediments on bacteria in 1987. Bioassays were conducted once per month from April through October inclusive. The basic format of the bioassay remained unchanged although an additional sediment concentration of 600 mg l$^{-1}$ was added while the 0 mg l$^{-1}$ concentration was viewed as the no-sediment control for both the clean and contaminated bioassays.

The uptake of adenine by bacteria followed a somewhat different pattern in 1987 compared to the 1986 results. Both years had similar trends for bioassays with clean sediment; uptake remained relatively constant among all sediment concentrations. Bioassays with contaminated sediment showed an increase in uptake with seasonal averages of 9.31 and 17.21 ng l$^{-1}$ h$^{-1}$ for 0 and 1200 mg l$^{-1}$ of sediment respectively. The trend in uptake of adenine was relatively uniform

among the three sampling sites. Uptake was slightly lower at Gibraltar Bay than the other two locations. These patterns of uptake for both clean and contaminated sediment were observed for all 1987 bioassays.

Despite the increase in uptake with contaminated sediment from the 1987 bioassays, when uptake was adjusted for bacterial biomass, the patterns followed those observed in 1986 (Table 3). Average adjusted uptake for 1987 for the clean sediment bioassays declined from 9.81 to 4.75 pg h$^{-1}$ M$^{-1}$ cells, as clean sediment concentrations increased from 0 to 1200 mg l$^{-1}$ (Table 3). For a similar increase in contaminated sediments, average adjusted uptake declined from 9.81 to 3.10 pg h$^{-1}$ M$^{-1}$ cells. These trends of decline in adjusted uptake with increasing sediment concentrations were observed at all three locations for most bioassays. Also, in all but 2 of 21 sets of experiments, the suppression of adjusted uptake was greater for the contaminated sediments than the clean sediments. Figures 2 and 3 show examples of the 1987 experiments with adjusted uptake for increasing sediment additions and differences between adjusted and unadjusted uptake results for one location, respectively.

The similarities of results from 1986 and 1987 were striking considering that the method changed between years and the experiments were conducted on natural bacteria populations which were not preconditioned to the substrate. Experiments conducted in 1986 showed a decline in the average adjusted uptake of 41 and 75 percent

288

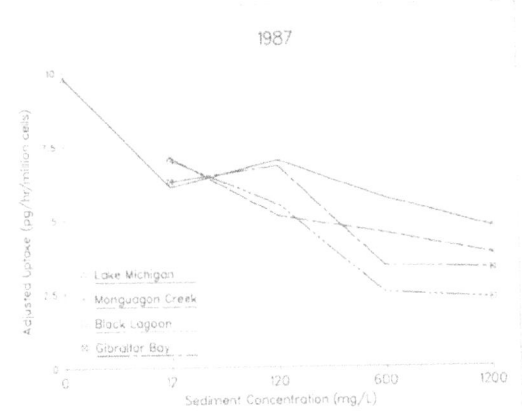

*Fig. 2*. Average 1987 ³H uptake adjusted for bacteria biomass. Each observation is the mean of 21 incubations.

from 0 to 1200 mg l⁻¹ for clean and contaminated sediment, respectively. Likewise, the same averages from 1987 showed a decline of 51 and 67 percent for clean and contaminated sediment, respectively. Further, these patterns were repeated for most bioassays in both years. The key to reproducing and understanding the bioassays appears to arise from transforming the uptake values to adjusted uptake values, i.e., correcting for biomass.

While the results from the clean versus contaminated experiments provided many useful inferences, the overall experimental design also con-

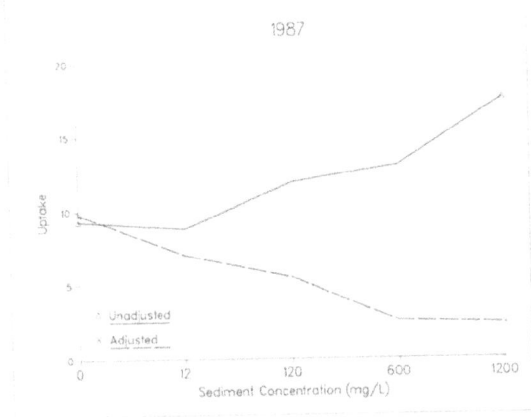

*Fig. 3*. Black Lagoon average 1987 ³H uptake, adjusted and unadjusted for bacteria biomass. Each observation is the mean of 21 incubations.

formed to a specific statistical model. A two-factor ANOVA was used to test two hypotheses: that uptake levels were significantly affected by sediment concentration (from 0 to 1200 mg l⁻¹), and that uptake levels were significantly affected by sediment type (clean or contaminated). Two different types of ANOVAs are presented in each table, the conventional parametric analysis and a modified version called the Brown-Forsythe ANOVA. The latter form has less restrictive assumptions because the equality of within-group variances is not required (Dixon *et al.*, 1981). Generally, the Brown-Forsythe ANOVAs are more conservative than their parametric counterparts. Because more often than not the assumption of equality of variances was met (tested with Levene's Statistic), the results discussed below refer only to the regular parametric ANOVA. The Brown-Forsythe results are presented in every case as a comparison to the regular analysis.

The ANOVAs from 1986 were conducted on each of the three sites for each of three months for a total of nine analyses (Table 4). Two out of nine

*Table 4*. Summary of ANOVAs for 1986 bacteria clean versus contaminated sediment bioassays, uptake rates adjusted for biomass.

| Site and Date | Standard ANOVA | | | Brown-Forsythe | | |
|---|---|---|---|---|---|---|
| | Treat | Conc | Int | Treat | Conc | Int |
| MC, May | ** | * | ns | ns | ns | ns |
| BL, June | ns | ** | ** | ns | ns | ns |
| GB, June | ns | ** | ns | ns | ** | ns |
| MC, August | ns | ns | ns | ns | ns | ns |
| BL, August | ns | ** | ** | ns | * | ** |
| GB, August | ns | ** | ns | ns | * | ns |
| MC, October | ns | ns | ns | ns | ns | ns |
| BL, October | ns | ** | ** | ns | ** | ** |
| GB, October | ** | ** | ** | ** | ** | ** |

| | |
|---|---|
| * | = significant at 0.01 |
| ** | = significant at 0.001 |
| ns | = not significant |
| Treat | = experimental treatment, Conc = concentration, Int = interaction. |
| MC | = Monguagon Creek, BL = Black Lagoon, GB = Gibraltar Bay. |

*Table 5.* Summary of ANOVAs for 1987 bacteria clean versus contaminated sediment bioassays, uptake rates adjusted for biomass.

| Date | Standard ANOVA | | | Brown-Forsythe | | |
|---|---|---|---|---|---|---|
| | Treat | Conc | Int | Treat | Conc | Int |
| April | ns | ns | ns | ns | ns | ns |
| May | ** | ** | ** | ns | ns | ns |
| June | ** | ** | * | * | ** | * |
| July | ** | ** | ns | * | ** | ns |
| August | ** | ** | ** | ** | ** | * |
| September | ** | ** | * | ** | ** | ns |
| October | ns | ** | ns | ns | * | ns |

| | |
|---|---|
| * | = significant at 0.01 |
| ** | = significant at 0.001 |
| ns | = not significant |
| Treat | = experimental treatment, Conc = concentration, Int = interaction. |

analyses showed significant effects due to substrate type and seven out of nine were significant for sediment concentration. Four interactions were significant for the adjusted uptake results.

The experimental design shifted slightly in 1987 with one ANOVA used to analyze data for all three sites plus the clean sediment. This resulted in seven ANOVAs, one for each set of experiments each month (Table 5). Results from the ANOVAs with adjusted uptake results showed significant effects from sediment concentration for six out of seven analyses and four significant interactions. Five out of seven ANOVAs showed significant effects due to sediment type. The increase in the number of statistically significant treatment effects for sediment type for 1987 appeared to arise from a reduction in the within-group variance (among replicate incubations) between the two years.

### 3.1.2. Elutriate bioassays

Patterns of uptake for the elutriate bioassays matched those of the clean versus contaminated experiments. Average unadjusted uptakes declined almost 60 percent from 48.3 to 20.2 ng $l^{-1}$ $h^{-1}$ as Black Lagoon sediment concentrations increased from 0 to 1200 mg $l^{-1}$. Likewise,

Monguagon Creek unadjusted uptakes declined 51 percent and Gibraltar Bay uptakes 54 percent. When adjusted for bacterial biomass, reductions in 1986 average uptakes were 77 percent for Monguagon Creek, 89 percent for Black Lagoon, and 65 percent for Gibraltar Bay (Table 2). These same trends were observed for almost every elutriate experiment conducted in 1986.

The statistical significance of sediment concentration and type for the elutriate experiments was tested with two-factor ANOVA. As with the clean versus contaminated experiments, nine ANOVAs were conducted, one for each of three sampling locations and each of three field trips. In these analyses, sediment type was actually a contrast between the contaminated elutriate bioassay and the Lake Michigan sediment from the clean versus contaminated experiments. The ANOVAs showed that for adjusted uptake, six of nine experiments had significant effects due to sediment type and six of nine for sediment concentration (Table 6). Five of the nine ANOVAs had significant interaction terms.

*Table 6.* Summary of ANOVAs for 1986 bacteria elutriate versus clean sediment bio assays, uptake rates adjusted for biomass.

| Site and Date | Standard ANOVA | | | Brown-Forsythe | | |
|---|---|---|---|---|---|---|
| | Treat | Conc | Int | Treat | Conc | Int |
| MC, May | ns | ** | ns | ns | * | ns |
| BL, June | ** | ** | ** | ns | ns | ns |
| GB, June | ** | ** | * | ** | * | ns |
| MC, August | * | ns | ns | ns | ns | ns |
| BL, August | ** | ns | ** | ** | ns | * |
| GB, August | * | ns | ns | * | ns | ns |
| MC, October | ns | ** | ns | ns | ns | ns |
| BL, October | ns | ** | ** | ns | * | * |
| GB, October | ** | ** | * | ** | * | ns |

| | |
|---|---|
| * | = significant at 0.01 |
| ** | = significant at 0.001 |
| ns | = not significant |
| Treat | = experimental treatment, Conc = concentration, Int = interaction. |
| MC | = Monguagon Creek, BL = Black Lagoon, GB = Gibraltar Bay. |

290

### 3.1.3. Pore water bioassays

Bioassays using pore water were intended to determine if the locus of toxicity was the pore water alone or bulk sediment. The results showed that the pore water was much less toxic than bulk sediment. The results of the pore water bioassays were uneven with some suppression, enhancement, and unchanged rates of uptake (Table 7). The levels of bacterial biomass remained relatively constant with the addition of pore water; at two locations the amount of pore water added to the river water had little effect on bacterial abundances, while at Gibraltar Bay the biomass almost doubled.

There was no consistent trend in the uptake of substrate in bioassays using pore water. At Monguagon Creek, the addition of pore water stimulated uptake, while at Black Lagoon and Gibraltar Bay the pore water suppressed uptake (Table 7). The patterns were relatively unchanged after the uptake rates were adjusted for biomass. The bioassays from three different locations yielded three different results. Uptake at Monguagon Creek was highly stimulated by the addition of pore water, remained unchanged at

Black Lagoon, and was suppressed at Gibraltar Bay.

### 3.2. Phytoplankton bioassays

#### 3.2.1. Clean versus contaminated sediment bioassays

Phytoplankton were extremely sensitive to the addition of sediment to the bioassay water. Uptake of $^{14}C$ by phytoplankton, unadjusted for biomass, showed approximately a linear decrease with increasing concentration of sediment.

Increasing chlorophyll concentrations with addition of sediment indicated Trenton Channel sediment contained algae. Because higher biomass could increase uptake and thus mask the effect of contaminated sediment, uptake rates were adjusted by dividing by chlorophyll concentrations. Uptake adjusted for biomass showed a somewhat more complex pattern than unadjusted uptake. Averaged over all months and sites, adjusted uptake also declined with concentration, but in a less linear fashion than unadjusted uptake. During 1986, two sites had approximately the same average adjusted uptake, while Gibraltar Bay uptakes were about 30 percent higher (Table 8). The results show a reduction in uptake

*Table 7.* Bacteria biomass (million cells ml$^{-1}$), uptake (ng l$^{-1}$ h$^{-1}$), and uptake adjusted for biomass (pg h$^{-1}$ M$^{-1}$ cells) at various concentrations of pore water from Trenton Channel sites, pooled over all months, 1986.

| Site | Pore water concentration (ml l$^{-1}$) | | | |
|---|---|---|---|---|
| | 0(control) | 0.5 | 1.0 | 10.0 |
| *Biomass* | | | | |
| Monguagon Creek | 2.20 | 2.15 | 2.33 | 2.15 |
| Black Lagoon | 1.73 | 1.98 | 2.78 | 2.09 |
| Gibraltar Bay | 2.58 | 4.00 | 3.09 | 4.57 |
| *Uptake* | | | | |
| Monguagon Creek | 27.4 | 24.7 | 17.3 | 29.1 |
| Black Lagoon | 44.0 | 34.9 | 35.1 | 30.1 |
| Gibraltar Bay | 58.6 | 32.4 | 28.8 | 27.8 |
| *Adjusted uptake* | | | | |
| Monguagon Creek | 12.3 | 20.7 | 13.0 | 33.2 |
| Black Lagoon | 24.8 | 19.7 | 14.4 | 20.6 |
| Gibraltar Bay | 22.5 | 20.1 | 20.0 | 9.27 |

*Table 8.* Phytoplankton uptake adjusted for biomass ($\mu$g $^{14}C$ h$^{-1}$ $\mu$g$^{-1}$ chlorophyll-$\alpha$) at various concentrations of sediment from Lake Michigan (Clean), Monguagon Creek (MC), Black Lagoon (BL), and Gibraltar Bay (GB), pooled over all months, 1986.

| Site | Sediment concentration (mg l$^{-1}$) | | | |
|---|---|---|---|---|
| | 0(control) | 12 | 120 | 1200 |
| Clean | 15.1 | 14.2 | 8.3 | 2.2 |
| Contam: MC | 14.1 | 12.9 | 8.4 | 1.7 |
| Elutriate: MC | 15.6 | 10.8 | 6.9 | 0.7 |
| Clean | 22.5 | 15.8 | 12.9 | 1.8 |
| Contam: BL | 16.1 | 16.4 | 5.2 | 0.4 |
| Elutriate: BL | 14.4 | 10.2 | 4.6 | 0.5 |
| Clean | 19.7 | 17.8 | 13.8 | 2.2 |
| Contam: GB | 23.2 | 16.5 | 9.7 | 1.7 |
| Elutriate: GB | 8.2 | 14.8 | 7.1 | 0.9 |

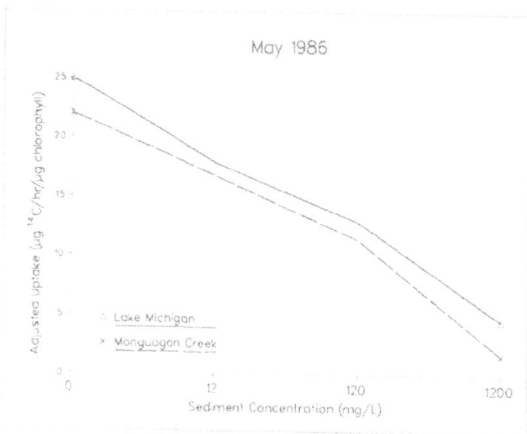

*Fig. 4.* Phytoplankton uptake adjusted for biomass, May and June 1986. Each observation is the mean of 2 incubations.

at all concentrations of sediment. Uptake was reduced approximately 15 percent for clean and contaminated sediment at low concentrations (12 mg l$^{-1}$ – Table 8). The reduction was about 40 percent for the clean sediment and 56 percent for the contaminated sediment at 120 mg l$^{-1}$, and 89 and 93 percent at 1200 mg l$^{-1}$ for clean and contaminated sediment, respectively (Table 8). Figure 4 shows the trend in uptake with increasing sediment for two locations for the May 1986 bioassay.

Because of the extreme suppression of uptake at the sediment concentration of 1200 mg l$^{-1}$, in 1987 the highest concentration was changed to 600 mg l$^{-1}$ and intermediate concentrations were adjusted to 75, 150, and 300 mg l$^{-1}$. For 1987, uptake adjusted for biomass dropped approximately 50 percent from 0 to 600 mg l$^{-1}$ (Table 9). The biggest drop in adjusted uptake occurred between 75 and 150 mg l$^{-1}$ added sediment (Table 9).

During both 1986 and 1987, there was some variation among, and sometimes within, bioassays in both rates of unadjusted uptake and uptake adjusted for biomass. Some individual treatment groups in the bioassays showed results contrary to the general trends of reduced uptake with increasing sediment concentration, while other bioassays expressed the trends intensely.

*Table 9.* Phytoplankton uptake adjusted for biomass ($\mu$g $^{14}$C h$^{-1}$ $\mu$g$^{-1}$ chlorophyll-$\alpha$) at various concentrations of sediment from Lake Michigan and Trenton Channel sites, pooled over all months, 1987.

| Site | Sediment concentration (mg l$^{-1}$) | | | | |
|---|---|---|---|---|---|
| | 0 (control) | 75 | 150 | 300 | 600 |
| Lake Michigan | 21.0 | 23.4 | 19.3 | 17.5 | 14.6 |
| Monguagon Creek | 21.0 | 31.1 | 21.6 | 14.9 | 8.7 |
| Black Lagoon | 21.0 | 19.2 | 15.5 | 8.8 | 5.2 |
| Gibraltar Bay | 21.0 | 34.7 | 22.7 | 18.4 | 14.1 |

Two-factor ANOVA was used to test the hypothesis that uptake rates were not significantly affected by sediment concentration or sediment type (site from which sediment came). In 1986 bioassays, uptake adjusted for biomass was always significantly different among sediment concentrations and was significantly different between sediment types for four out of nine bioassays (Table 10). Interactions were significant for five out of the nine ANOVAs. In both 1986

*Table 10.* Summary of ANOVAs for 1986 phytoplankton clean versus contaminated sediment biossays, uptake rates adjusted for biomass.

| Site and Date | Standard ANOVA | | | Brown-Forsythe | | |
|---|---|---|---|---|---|---|
| | Treat | Conc | Int | Treat | Conc | Int |
| MC, May | ** | ** | ns | ns | * | ns |
| BL, June | ** | ** | ** | * | * | ns |
| GB, June | ns | ** | ** | ns | ns | ns |
| MC, August | ns | ** | ns | ns | ** | ns |
| BL, August | ** | ** | ** | ns | * | ns |
| GB, August | * | ** | ** | ns | ** | ns |
| MC, October | ns | ** | ns | ns | ns | ns |
| BL, October | ns | ** | ** | ns | ** | * |
| GB, October | ns | * | ns | ns | ns | ns |

| | |
|---|---|
| * | = significant at 0.01 |
| ** | = significant at 0.001 |
| ns | = not significant |
| Treat | = experimental treatment, Conc = concentration, Int = interaction. |
| MC | = Monguagon Creek, BL = Black Lagoon, GB = Gibraltar Bay. |

and 1987, significant interactions were usually due to large uptake differences between sediment types at one or two concentrations and not at other concentrations.

The experimental design used in the 1987 studies permitted direct comparison among all three Trenton Channel sites and Lake Michigan sediment. Pairwise t-tests showed that Gibraltar Bay uptake rates were often higher, and Black Lagoon uptake often lower, than uptake at other sites for a given concentration of sediment. Uptake not adjusted for biomass was always significantly different (ANOVA) among sediment concentrations, and was significantly different among sites in four of seven months. Uptake adjusted for biomass similarly was always significantly different among sediment concentrations, and was significantly different among sites in five of seven months (Table 11). Four interactions were significant for adjusted uptake (Table 11). For the June and August bioassays, adjusted uptake was notably different in one cell (Lake Michigan 300 mg l$^{-1}$ and Monguagon Creek 75 mg l$^{-1}$, respectively), causing significant interactions. Brown-Forsythe tests revealed a pattern similar to the standard ANOVA, but with a somewhat lower level of significance (Table 11).

### 3.2.2. Elutriate bioassays

The elutriate bioassays yielded results similar to the clean versus contaminated sediment bioassays. Uptake of $^{14}C$ adjusted for biomass was slightly reduced (about 6%) at the 12 mg l$^{-1}$ sediment concentration, substantially reduced (51%) at the 120 mg l$^{-1}$ concentration, and severely reduced (94%) at the 1200 mg l$^{-1}$ concentration (Table 8). These results show the extreme sensitivity of the algae to the incorporation of sediment in the incubation even if that sediment is allowed to settle out of the incubation water.

Two-factor ANOVA was used to test the hypothesis that uptake was not significantly different among concentrations or treatments, where treatment refers to rotated clean sediment, rotated contaminated sediment, or non-rotated contaminated sediment (elutriate). All of the nine standard ANOVAs showed significant differences in both unadjusted and adjusted uptake among sediment concentrations (Table 12). Most Brown-Forsythe tests (seven of nine for unadjusted uptake, six of nine for adjusted uptake) also showed uptakes were significantly different among sediment concentrations. However, only four Brown-Forsythe ANOVAs out of nine showed significant

Table 11. Summary of ANOVAs for 1987 phytoplankton clean versus contaminated sediment bioassays, uptake rates adjusted for biomass.

| Site and Date | Standard ANOVA | | | Brown-Forsythe | | |
|---|---|---|---|---|---|---|
| | Treat | Conc | Int | Treat | Conc | Int |
| April | ** | ** | ns | ns | ns | ns |
| May | ** | ** | ** | ** | ** | ** |
| June | ** | ** | ** | * | ** | * |
| July | ns | * | ns | ns | ns | ns |
| August | ** | ** | * | ** | ** | ns |
| September | ** | ** | * | ns | * | ns |
| October | ns | ** | ns | ns | * | ns |

* = significant at 0.01
** = significant at 0.001
ns = not significant
Treat = experimental treatment, Conc = concentration, Int = interaction.

Table 12. Summary of ANOVAs for 1986 phytoplankton elutriate versus clean versus contaminated sediment bioassays, uptake rates adjusted for biomass.

| Site and Date | Standard ANOVA | | | Brown-Forsythe | | |
|---|---|---|---|---|---|---|
| | Treat | Conc | Int | Treat | Conc | Int |
| MC, May | ** | ** | * | ** | ** | ns |
| BL, June | ** | ** | ** | ** | ** | * |
| GB, June | ** | ** | ** | * | * | ns |
| MC, August | ns | ** | ns | ns | ** | ns |
| BL, August | ** | ** | ** | * | ** | ns |
| GB, August | ns | ** | * | ns | ns | ns |
| MC, October | ns | ** | ns | ns | ns | ns |
| BL, October | ** | ** | ** | ns | * | ns |
| GB, October | ns | ** | ns | ns | ns | ns |

* = significant at 0.01
** = significant at 0.001
ns = not significant
Treat = experimental treatment, Conc = concentration, Int = interaction.

differences in adjusted uptake between experimental treatments (Table 12). Five standard ANOVAs were likewise significant. Six of the nine standard ANOVAs had significant interaction terms.

### 3.2.3. Pore water bioassays

The addition of up to 1 percent v/v pore water had little effect on the uptake of $^{14}C$ by phytoplankton (Table 13), in contrast to the incubations using sediment, which showed a very large reduction in the amount of uptake with increasing amounts of sediment. The only pore water bioassay that showed significant differences in uptake among concentrations was Black Lagoon in October. Pairwise comparisons showed that unadjusted uptake was significantly higher for controls than for treatments, and that 0.5 and 1 ml l$^{-1}$ samples had higher adjusted uptake than 0 or 10 ml l$^{-1}$.

### 3.3. Sensitivity, accuracy, and precision of methods

Bioassays reported in this study were relatively sensitive measures of toxicity, as a small amount of sediment yielded a large reduction in uptake.

Table 13. Phytoplankton biomass ($\mu g$ chlorophyll-$\alpha$ l$^{-1}$), uptake ($\mu g$ $^{14}C$ l$^{-1}$ h$^{-1}$), and uptake adjusted for biomass ($\mu g$ $^{14}C$ h$^{-1}$ $\mu g^{-1}$ chlorophyll-$\alpha$) at various concentrations of pore water from Trenton Channel sites, pooled over all months, 1986.

| Site | Pore water concentration (ml l$^{-1}$) | | | |
|---|---|---|---|---|
| | 0(control) | 0.5 | 1.0 | 10.0 |
| *Biomass* | | | | |
| Monguagon Creek | 1.88 | 1.80 | 1.95 | 2.04 |
| Black Lagoon | 1.06 | 0.60 | 0.60 | 0.67 |
| Gibraltar Bay | 0.67 | 0.72 | 0.64 | 0.65 |
| *Uptake* | | | | |
| Monguagon Creek | 30.5 | 29.0 | 24.7 | 27.3 |
| Black Lagoon | 17.1 | 11.8 | 13.3 | 13.0 |
| Gibraltar Bay | 13.5 | 14.1 | 13.9 | 14.8 |
| *Adjusted uptake* | | | | |
| Monguagon Creek | 14.1 | 14.1 | 11.7 | 12.5 |
| Black Lagoon | 18.4 | 18.4 | 22.0 | 19.9 |
| Gibraltar Bay | 17.8 | 16.7 | 18.5 | 18.7 |

Sediment concentrations were chosen to include those likely in a moderate resuspension event (12 to 150 mg l$^{-1}$ wet wt), and concentrations probable under extreme mixing conditions (600 to 1200 mg l$^{-1}$). The lowest concentrations, 12 to 75 mg l$^{-1}$, had little effect on or enhanced uptake, while medium concentrations (120 to 150 mg l$^{-1}$) began to show inhibition of uptake. Thus a relatively severe resuspension event could cause a decrease in productivity of phytoplankton or metabolism of bacteria, while a less severe event might have little effect as shading and toxicity would be offset by nutrient or substrate availability.

Accuracy of bioassays in terms of the real world was difficult to determine. No doubt, the bioassays simplified the situation, as one small, well-mixed water mass under constant conditions was used for all treatments, whereas field conditions would be heterogeneous, complex, and changing. However, baseline data on biological effects under controlled conditions are needed before hypotheses can be formulated for design of field research. Absolute uptake values probably reflected higher phytoplankton productivity than most field conditions, because of the bright light in the incubator. In contrast, bacterial metabolism was probably underestimated because only one substrate was used in the bioassays.

Bioassay results were reasonably precise. The standard error of uptake for phytoplankton bioassays for a given date and treatment was generally an order of magnitude less than the mean uptake. The results were not quite as good for bacteria as for algae, but the standard error was seldom more than half of the mean. The precision of the results was reflected in the similarity among replicates and contributed to significant differences between treatments as shown by ANOVAs.

## 4. Discussion

### 4.1. Bacterial bioassays

Bacteria were chosen as one of the two functional groups of organisms for this investigation because they constitute a major link in the food web that comes in contact with toxic materials in the sediments. Further, some bacteria are known to thrive on certain toxic materials, while others respond negatively to the presence of the same materials (Colwell & Grimes, 1986).

A large variety of metabolic effects could lead to the suppression of uptake of substrates by bacteria in the presence of both synthetic organic compounds and metals. Some of these effects include: alteration to the chemistry of cell membranes, suppression of the uptake of certain amino acids, interference in the synthesis of nucleic acids, and the disruption in certain catabolic pathways (Lal & Saxena, 1982). Because these effects change due to both the concentration of the toxic substance and the type of bacteria present, *a priori* it is difficult to predict the response of the community to the presence of the materials. Combinations of certain substances appear to have a synergistic effect that is as yet not fully understood (Wong *et al.*, 1978). For these reasons, bioassays become site- and time-dependent and must be repeated frequently.

The use of adenine as a substrate is somewhat surrounded in controversy with some authors (Karl & Winn, 1984, 1986; Karl, 1981) attributing uptake to both algae and bacteria. Other authors (Fuhrman *et al.*, 1986a, 1986b; Peele *et al.*, 1985) have found either equivocal results in the offshore marine environment or unequivocal results nearshore that only bacteria take up adenine. Following the findings of Fuhrman *et al.* (1986a), this study assumed that, in enriched nearshore environments, adenine uptake was mediated almost exclusively by bacteria. In contrast, glucose uptake may be substantially affected by algal heterotrophy (Moll, 1984). Thus the rationale for shifting to adenine from glucose after the first set of experiments in late May and early June, 1986.

Bioassays with sediments are more complicated than with little or no suspended materials. The sediment can mask the effect of the toxic material due to its binding nature and shifts in redox potential. For these reasons, paired analyses comparing non-toxic sediments and contaminated sediments appear the most reliable bio-assays. But, this additional constraint adds considerable complexity to the experimental design in that the number of controls is greatly increased. The results from this study encourage use of such controls.

The results of the bioassays must be interpreted with caution because in some cases the sediment actually stimulated rather than suppressed uptake. This was often true at the lower sediment concentrations, and almost always when uptake was not corrected for biomass. Munawar *et al.* (1988) occasionally noticed a similar effect in phytoplankton bioassays during investigation of the effects of Toronto Harbour sediment on algae.

Isotope dilution in all experiments was possible by the introduction of non-radioactive substrate (adenine) with the sediment additions. This dilution could have the effect of reducing isotope uptake at high sediment concentrations thus confounding the experimental results. While the effect of isotope dilution was not investigated *per se*, three lines of reasoning suggest this was not a major problem. First, in many experiments substrate uptake did not decrease with increasing sediment concentrations. Only after the uptakes were adjusted for biomass did the suppressed uptakes change. If isotope dilution was important, unadjusted uptake rates would decline significantly given the two orders of magnitude increase in sediment concentration with the various additions for each set of bioassays. Second, most of the pore water incubations showed no decline in either adjusted or unadjusted uptake rates with increasing amounts of pore water additions. Clearly, isotope dilution was not operating in these experiments. Third, the effect of contaminated sediment was determined against the clean sediment control. Isotope dilution could mask the results only by substantial differences in dilution between the two sediment types. While this condition may have been true in a few cases, it is most unlikely to occur on a consistent basis.

The salient conclusion from this study was that when uptake of substrate was adjusted for bacterial biomass, a significant effect was observed due to the amount and type of sediment. In 1986 the reduction in uptake between the control

(0 mg l$^{-1}$ sediment addition) and the highest concentration (1200 mg l$^{-1}$ sediment addition) was 65.3, 79.2, and 75.4 percent for Monguagon Creek, Black Lagoon, and Gibraltar Bay, respectively. This compares to a reduction of 39.6 percent for the clean Lake Michigan sediment addition. The pattern was similar for 1987 with the reductions of 61.7, 76.9, and 66.7 percent for Monguagon Creek, Black Lagoon, and Gibraltar Bay, respectively, and a reduction of 51.6 percent for the Lake Michigan sediment. These results imply a high level of sensitivity by the bacteria to the addition of the contaminated sediment.

A similar level of sensitivity was not observed by Ribo et al. (1985) using the Microtox bioassay with Detroit River samples. They found that of a total of 76 stations investigated, only four showed acute levels of toxicity. Two of these four sites were from the Trenton Channel, with one at the mouth of Monguagon Creek. These bioassays used Photobacterium phosphoreum and Detroit River water, but no sediment. Most of the Ribo et al. (1985) Microtox assays showed responses less than 20 percent different from the controls.

There have been a number of studies investigating the effect of specific toxic materials on microorganisms. Many of these studies have been summarized by Lal & Saxena (1982). The results are highly variable with some compounds such as heptachlor proving extremely toxic to bacteria, while DDT and its derivatives were relatively innocuous to bacteria. Lal & Saxena (1982) suggested that the variable nature of the results was attributed in part to the degree of susceptibility of the various bacteria species to the different toxic materials. Cunningham et al. (1986) came to the same conclusion using the Microtox bioassay from a set of experiments with samples from New Jersey and Pennsylvania.

Given the ephemeral nature of the results from previous investigations, the consistent results from this study are unusual. This consistency is especially interesting given the change in substrates over the two years, the various locations investigated, and the large range in seasons. Apparently, the experimental protocol used in this study was sufficiently different from the previous

studies to generate a novel type of result. Perhaps the exposure of mid-river water samples, containing relatively unaffected microbes, to the toxic materials found in the sediments of the backwaters of the Trenton Channel elicited the significant response in the bioassays. In addition, Lal & Saxena (1982) indicate that in many highly contaminated environments a natural bacterial population develops which actually metabolizes some of the toxicants; in some instances these bacteria become dependent on the toxicant as an energy source. Thus the bioassay response could be considerably different if conducted with populations from the backwater versus mid-river. Finally, a large number of the toxic substances in the backwaters of the Trenton Channel are metals (Fallon & Horvath, 1985; Theis et al., 1988). These compounds have proven more toxic to most microorganisms than synthetic organic compounds (Wong et al., 1978). Further studies are needed on the types of responses observed in bioassays using a mixture of toxic materials versus a single class of compounds.

## 4.2. Phytoplankton bioassays

As autotrophs, phytoplankton are an important base to the food web, thus adverse effects on their growth or health may have repercussions on productivity of higher trophic levels. Phytoplankton are believed relatively sensitive organisms as subjects for bioassay investigations (Munawar & Thomas, 1986).

Effects of contaminants can be grouped into three categories: changes in species composition and size structure, metabolic and physiological response, and ultrastructural response (Munawar et al., 1988). Because the first effect is long-term, only the second and third effects may be operating in these Trenton Channel bioassays. Broadly, decreased uptake with increasing sediment is caused by several factors: shading, toxicity, physical damage to cells, adsorption of nutrients to the sediment (decreasing their bioavailability), and other mechanisms. Lake Michigan sediment was believed to inhibit uptake primarily by shading and possible cell damage, while an additional toxic response was observed from Trenton Channel sediment exposure. Toxicity includes a number of physiological responses which are different among species and contaminants and are not well understood, but include inhibition of photosynthesis from interference with electron transport, cell membranes, development and structure of chloroplasts, and the light and dark reactions (Lal & Saxena, 1982; Munawar et al., 1988). The multiple metals and organic contaminants in Trenton Channel sediment may have unknown synergistic or antagonistic effects, of which only the net sum is observed. Combinations of metals have been shown to be more toxic to certain algae than were the same metals presented singly (Wong et al., 1978).

By expressing results of the Trenton Channel bioassays as percent reduction of uptake from controls to high sediment concentration, they can be compared with results from other sites and experimental protocols. Uptake adjusted for biomass decreased an average of over 90 percent between 0 and 1200 mg l$^{-1}$ sediment in 1986 (Table 8). In 1987, from 75 to 600 mg l$^{-1}$ the decrease in uptake averaged about 65 percent (Table 9). To observe only the toxic effect, not the particle effect, one may compare uptake rates in clean versus contaminated sediment for only one concentration. Monguagon Creek sediment reduced uptake 40 percent at 600 mg l$^{-1}$ and 23 percent at 1200 mg l$^{-1}$ compared to equal concentrations of Lake Michigan sediment. Similarly, Black Lagoon sediment reduced uptake 64 percent at 600 mg l$^{-1}$ and 78 percent at 1200 mg l$^{-1}$ compared to 600 and 1200 mg l$^{-1}$ of Lake Michigan sediment.

The previous research with phytoplankton sediment bioassays most comparable to this study was that of Munawar & Thomas (1986) and Munawar et al. (1983, 1985), who conducted elutriate bioassays using $^{14}$C uptake, natural phytoplankton assemblages, and sediment from polluted sites around the Great Lakes. The following percentages of uptake reduction from Munawar's several studies were all from solutions of 20 percent elutriate. Because of water addition to the

sediment, 20 percent elutriate corresponds to approximately 3 percent liquid actually derived from the sediment, while the phytoplankton bioassays from this study using 1200 mg l$^{-1}$ sediment contained about 0.08 percent water. Sites closest to the three Trenton Channel sites of this study – just upstream of Fighting Island, and the mouth of the Detroit River – showed no trend in reduction of uptake with addition of elutriate (Munawar et al., 1985). Elutriate from just downstream of Belle Isle reduced uptake 17 to 20 percent, and that from northwest Lake Erie reduced uptake 15 to 25 percent, but the difference was nonsignificant for Lake Erie. Toronto Harbour sediment elutriate reduced uptake 20 to 30 percent, Toledo Harbor elutriate about 27 percent (Munawar & Thomas, 1986), Niagara River elutriate 45 to 85 percent and Lake Ontario elutriate 28 to 48 percent (Munawar et al., 1983); ranges refer to different size fractions of phytoplankton. Overall, these uptake reductions were small compared to this study, considering the actual concentrations of sediment-derived material. The Detroit River sites studied by Munawar et al. (1985) may not have been as highly contaminated as those of this study.

Other studies investigated dealt with specific toxicants rather than field-collected sediment alone. Powers et al. (1982) adsorbed Aroclor, a PCB, into the sediment. They then subjected a natural phytoplankton assemblage to the particles, from which Aroclor desorbed at a final concentration of 50 $\mu$g l$^{-1}$. Compared to the control with clay only, Aroclor with sediment decreased uptake about 80 percent after one day (Powers et al., 1982). Suppression of uptake continued throughout the 3-day incubation. The high reduction in uptake, presence of sediment particles, and inclusion of the adsorption/desorption process are similar to the present study.

Wong et al. (1978) exposed Scenedesmus quadricauda cultures to a combination of ten metals at recommended water quality objective concentrations, which were generally determined from bioassays using a single metal. In culture media, productivity was inhibited 68 percent, and one-tenth the metals concentrations reduced uptake 40 percent, demonstrating the synergistic toxic effects of metals on phytoplankton. Water quality objective concentrations of metals added to algal cultures in Hamilton Harbour water, which itself contains metals, reduced uptake 78 percent, and one-tenth the concentrations reduced uptake 21 percent (Wong *et al.*, 1978). These reductions in uptake are relatively large, and suggest that uptake reductions in the present study may be related to high concentrations of metals in Trenton Channel sediment.

## 4.3. Relevance of bioassays

All sediment and pore water bioassays were conducted in the laboratory. But every effort was made to maintain conditions as close to the natural environment as possible. Bioassays were usually conducted within 48 hours of collection, and never more than a few days after sample collection. Both sediment and incubation water were handled as little as possible before the bioassays and stored in dark cold rooms until analysis. All sediment containers were kept completely full, cold, and tightly stoppered until opened for bioassays. The operational principle was that an important component of the sediment was the volatile organic fraction which could be readily lost by casual treatment of the sediment. Finally, bacteria and phytoplankton populations used in the experiments were from the Trenton Channel and not preadapted to the environmental condition of the bioassays. This approach was used in order to simulate the probable circumstances surrounding an event which would cause resuspension of contaminated sediment in the Detroit River.

The 'freshness' of the natural populations of microorganisms was critical to the outcome of the experiments. An unplanned set of experiments demonstrated this fact. River water that was allowed to sit one extra day in the cold room was used as a second control for several phytoplankton bioassays. After four sets of bioassays, the results clearly showed that this was not a control because that extra day had the effect of signifi-

cantly changing the uptake of substrate in comparison to the bioassays conducted the day before. When the experimental protocol was altered to run the controls on the same day, this 'control' effect disappeared.

The standard elutriate bioassay used by other investigators, where sediment is mixed with extra water, allowed to settle, and the water used after filtering, was modified for this study to make the protocol more ecologically relevant. Chapman *et al.* (1986) found that aged sediment-water mixtures and elutriates were not as toxic to *Daphnia* as freshly mixed sediment and water. Potential mechanisms for this were loss of volatile toxic material, sequential solution and binding or precipitation of the toxic component, or the rapid solution of a toxic component followed by slower solution of an antagonist (Chapman *et al.*, 1986). When a resuspension event occurs in the field, the organisms present are instantly exposed to the whole sediment, and the bioassay methods chosen here reflect that process.

Pore water bioassays showed that toxicity was associated more with particles than the pore water. Some portions of the toxic contaminants were apparently bound or adsorbed to particles and did not enter the pore water during its preparation. The highest concentration of pore water $(10 \text{ ml l}^{-1})$ was derived from a quantity of sediment much higher than the highest concentration in whole-sediment bioassays (about 40 g versus 1.2 g). Thus pore water was clearly much less toxic than an equivalent amount of whole sediment. However, these results do not suggest that pore water was not toxic. To the contrary, when pore water concentrations were raised to high enough levels, a significant response was induced (Giesy *et al.*, 1988).

The conclusions from this study suggest that the sediments in the backwaters of the Trenton Channel are highly contaminated and pose a risk to the food web. Although those sediments do reside in backwaters, the contaminated material is found on the surface of the sediments implying that relatively minor mixing events can mobilize the toxic materials. Further, location of the contaminated material on the top of the sediment suggests a continuing input to these regions. Additional research should be conducted to track the movement of these toxicants through the food web.

## 5. Acknowledgements

This research was supported by the Large Lakes Research Station of the U.S. Environmental Protection Agency under Grant No. EPA-CR-812569. The authors gratefully acknowledge the many useful suggestions made by Drs. K. Burnison and P. Wong during the review process.

## References

Brown, M. B. & A. B. Forsythe, 1974. Robust tests for the equality of variances. J. Am. Stat. Ass. 69: 364–367.

Chapman, G., M. Cairns, D. Krawczyk, K. Mulueg, A. Nebeker & G. Schuytema, 1986. Report on the toxicity and chemistry of sediments from Toronto and Toledo harbors. In; Rep. Dredging Subcommittee Great Lakes Wat. Qual. Bd. Internat. Joint Commiss., Windsor, Ontario, Canada: 91–118.

Colwell, R. R. & D. J. Grimes, 1986. Evidence for genetic modification of microorganisms occurring in natural aquatic environments. In T. M. Poston & R. Purdy (eds), Aquatic Toxicology and Environmental Fate: Ninth Volume. ASTM Spec. Tech. Pub. 921: 222–232.

Cunningham, V. L., M. S. Morgan & R. E. Hannah, 1986. Effect of natural water source on the toxicity of chemicals to aquatic microorganisms. In T. M. Poston & R. Purdy (eds), Aquatic Toxicology and Environmental Fate: Ninth Volume. ASTM Spec. Tech. Pub. 921: 436–449.

Davis, C. O. & M. S. Simmons, 1979. Water chemistry and phytoplankton field and laboratory procedures. Great Lakes Res. Div. Spec. Rep. No. 70. Univ. Mich., Ann Arbor, Mi. 88 pp.

Dixon, W. J., M. B. Brown, L. Engelman, J. W. Frane, M. A. Hill, R. I. Jennrich & J. D. Toporek, 1981. BMDP Statistical Software. Univ. California Press, Los Angeles. 726 pp.

Fallon, M. E. & F. J. Horvath, 1985. Preliminary assessment of contaminants in soft sediments of the Detroit River. J. Great Lakes Res. 11: 373–378.

Fitzwater, S. E., G. A. Knauer & J. H. Martin, 1982. Metal contamination and its effect on primary production measurements. Limnol. Oceanogr. 27: 544–551.

Fox, D. J. & K. E. Guire, 1976. Documentation for MIDAS. Statist. Res. Lab., Univ. Michigan, Ann Arbor, Mi. 203 pp.

Fuhrman, J. A., H. W. Ducklow, D. L. Kirchman, J. Hudak, G. B. McManus & J. Kramer, 1986a. Does adenine incorporation into nucleic acids measure total microbial production? Limnol. Oceanogr. 31: 627–636.

Fuhrman, J. A., H. W. Ducklow, D. L. Kirchman, J. Hudak & G. B. McManus, 1986b. Adenine and total microbial production: a reply. Limnol. Oceanogr. 31: 1395–1400.

Furlong, E. T., D. S. Carter & R. A. Hites, 1988. Organic contaminants in sediments from the Trenton Channel of the Detroit River, Michigan. J. Great Lakes Res. 14: 489–501.

Giesy, J. P., R. L. Graney, J. L. Newsted, C. J. Rosiu, A. Benda, R. G. Kreis & F. J. Horvath, 1988. A comparison of three sediment bioassay methods for Detroit River sediments. Env. Toxicol. Chem. 7: 483–498.

Hobbie, J. E., R. J. Daley & S. Jasper, 1977. Use of Nuclepore filters for counting bacteria by fluorescence microscopy. Appl. Env. Microbiol. 33: 1225–1228.

Karl, D. M., 1981. Simultaneous rates of ribonucleic acid and deoxyribonucleic acid synthesis for estimating growth and cell division of aquatic microbial communities. Appl. Env. Microbiol. 42: 802–810.

Karl, D. M. & C. D. Winn, 1984. Adenine metabolism and nucleic acid synthesis: applications to microbiological oceanography, In; J. E. Hobbie & P. J. Williams (eds), Heterotrophic Activity in the Sea: 197–215. Plenum, New York.

Karl, D. M. & C. D. Winn, 1986. Does adenine incorporation into nucleic acids measure total microbial production?: a response to comments by Fuhrman et al. Limnol. Oceanogr. 31: 1384–1394.

Lal, R. & D. M. Saxena, 1982. Accumulation, metabolism and effects of organochlorine insecticides on microorganisms. Microbiol. Rev. 46: 95–127.

Moll, R. A., 1984. Heterotrophy by phytoplankton and bacteria in Lake Michigan. Verh. Int. Ver. Limnol. 22: 431–434.

Moll, R. A. & M. Brahce, 1986. Seasonal and spatial distribution of bacteria, chlorophyll, and nutrients in nearshore Lake Michigan. J. Great Lakes Res. 12: 52–62.

Mount, D. I. & T. J. Norberg, 1984. A seven-day life-cycle cladoceran toxicity test. Soc. Env. Tox. Chem. 3: 425–434.

Munawar, M. & R. L. Thomas, 1986. Bioassessment of Toronto-Toledo sediments. In; Rep. Dredging Subcommittee Great Lakes Wat. Qual. Bd. Internat. Joint Commiss., Windsor, Ontario, Canada: 9–50.

Munawar, M., A. Mudroch, I. F. Munawar & R. L. Thomas, 1983. The impact of sediment-associated contaminants from the Niagara River mouth on various size assemblages of phytoplankton. J. Great Lakes Res. 9: 303–313.

Munawar, M., R. L. Thomas, W. Norwood & A. Mudroch, 1985. Toxicity of Detroit River sediment-bound contaminants to ultraplankton. J. Great Lakes Res. 11: 264–274.

Munawar, M., P. T. S. Wong & G-Y. Rhee, 1988. The effects of contaminants on algae: an overview. In; N. W.

Schmidtke (ed), Toxic Contamination in Large Lakes, Vol. I: Chronic Effects of Toxic Contaminants in Large Lakes. pp. 113–160. Lewis Publishers, Inc., Chelsea, Michigan.

Peele, E. R., R. E. Murray, R. B. Hanson, L. R. Pomeroy & R. E. Hodson, 1985. Distribution of microbial biomass and secondary production in a warm-core Gulf Stream ring. Deep-Sea Res. 32: 1393–1403.

Powers, C. D., G. M. Nau-Ritter, R. G. Rowland & C. F. Wurster, 1982. Field and laboratory studies of the toxicity to phytoplankton of polychlorinated biphenyls (PCBs) desorbed from fine clays and natural suspended particulates. J. Great Lakes Res. 8: 350–357.

Pugsley, C. W., P. D. N. Hebert, G. W. Wood, G. Brotea & T. W. Obal, 1985. Distribution of contaminants in clams and sediments from the Huron-Erie corridor. I. PCBs and octachlorostyrene. J. Great Lakes Res. 11: 275–289.

Ribo, J. M., B. M. Zaruk, H. Hunter & K. E. Kaiser, 1985. Microtox toxicity test results for water samples from the Detroit River. J. Great Lakes Res. 11: 297–304.

Smith, V. E., J. M. Spurr, J. C. Filkins & J. J. Jones, 1985. Organochlorine contaminants of wintering ducks foraging on Detroit River sediments. J. Great Lakes Res. 11: 231–246.

Strickland, J. D. H. & T. R. Parsons, 1972. A practical handbook of seawater analysis. Fish. Res. Bd. Can. Bull. No. 167. 310 pp.

Struger, J., D. V. Weseloch, D. J. Hallett & P. Mineau, 1985. Organochlorine contaminants in herring gull eggs from the Detroit River and Saginaw Bay (1978–1982): contaminant discriminants. J. Great Lakes Res. 11: 223–230.

Theis, T. L., T. C. Young & J. V DePinto, 1988. Factors affecting metal partitioning during resuspension of sediments from the Detroit River. J. Great Lakes Res. 14: 216–226.

U.S. Environmental Protection Agency/U.S. Army Corps of Engineers, 1977. Ecological evaluation of proposed discharge of dredged material into ocean waters. Env. Effects Lab. U.S. Army Corps Engin. Waterways Effects Lab., Vicksburg, Miss.

Wong, P. T. S., Y. K. Chau & P. L. Luxon, 1978. Toxicity of a mixture of metals on freshwater algae. J. Fish. Res. Bd. Can. 35: 479–481.

*Hydrobiologia* **219**: 301–306, 1991.
*M. Munawar & T. Edsall (eds),*
*Environmental Assessment and Habitat Evaluation of the Upper Great Lakes Connecting Channels.*
© 1991 *Kluwer Academic Publishers.*

# Tumors in fish from the Detroit River

Alexander E. Maccubbin & Noreen Ersing
*Grace Cancer Drug Center, Roswell Park Memorial Institute, 666 Elm Street, Buffalo, NY 14263, USA*

*Key words:* neoplasia, sediment contaminants, in-place pollutants, Great Lakes

## Abstract

A fish tumor survey was conducted in the Lower Detroit River during 1985 to 1987. Five species of fish were collected from sediment deposition zones and were examined for neoplasia. Neoplasms and related lesions were found in bullhead (*Ictalurus nebulosus*), walleye (*Stizostedion vitreum*), redhorse sucker (*Moxostoma, sp*), white sucker (*Catostomus commersoni*), and bowfin (*Amia calva*). Overall, 8.2 percent of the fish examined had oral/dermal lesions and 10.1 percent had liver lesions. Liver and skin tumors were found to be age-related in bullheads and size-related in walleye. Based on the fish surveyed, the incidence of tumors in fish from the Detroit River was similar to that observed in other chemically contaminated waterways.

## 1. Introduction

The Detroit River, one of the 42 Great Lakes Areas of Concern of the International Joint Commission, has a long history of pollution from agricultural, municipal and industrial sources. Along with other Upper Great Lakes Connecting Channels – the St. Marys River and St. Clair River – the Detroit River has been classified as an Area of Concern because of contaminated sediments, also known as in-place pollutants (Great Lakes Water Quality Board 1985). Previous studies of contaminants in the Detroit River have documented the presence of at least 200 chemical pollutants in the sediments (Fallon & Horvath, 1985; Hamdy & Post, 1985; Pranckevicius, 1987). Determining the relative importance of in-place pollutants with respect to their biological effects has become a focus of research as a part of establishing plans for reducing environmental degradation and rehabilitating polluted areas.

In recent years, several studies have suggested that toxic chemicals in sediments and neoplasia

observed in certain fish species may be related (Black *et al.*, 1982; Black, 1983; Malins *et al.*, 1984, 1987; Baumann *et al.*, 1987). Tumors have been observed in several fish species collected from the Great Lakes ecosystem (Sonstegard, 1977; Black *et al.*, 1982; Baumann *et al.*, 1982, 1987; Black, 1983; Maccubbin *et al.*, 1987; Cairns & Fitzsimons, 1988). In general, fish with tumors have been captured in waterways that are chemically contaminated. As part of the United States EPA's in-place pollutant project, we conducted a fish tumor survey in the Detroit River to determine the extent of neoplasia in selected fish species. We report here preliminary results of that survey describing the types of tumors found and their incidence in the selected fish species.

## 2. Materials and methods

Fish were collected by electro-shocking during ice-free months of 1985, 1986, and 1987. Sampling sites were in areas of sediment depo-

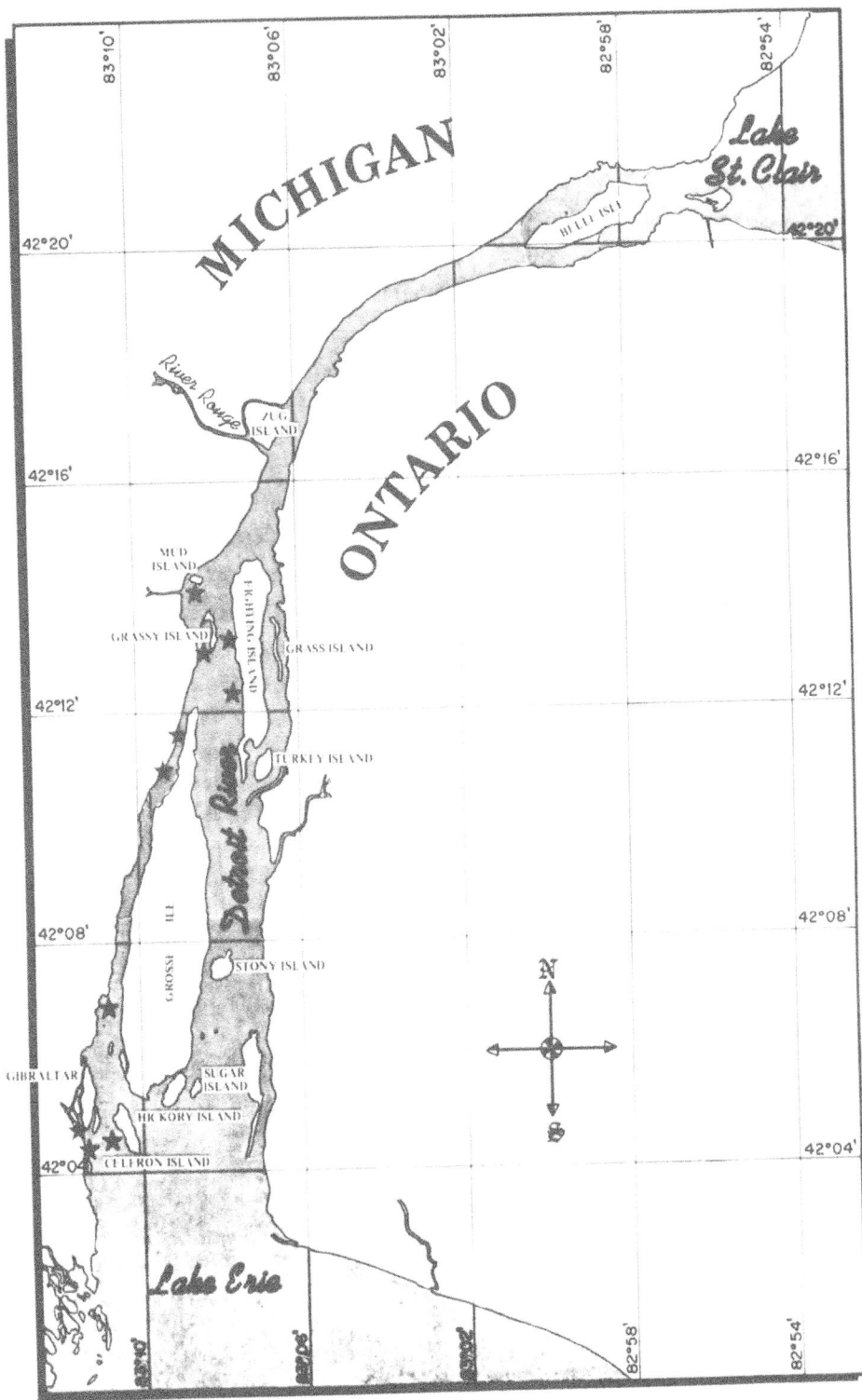

*Fig. 1.* The Detroit River System. Fish were collected from several sites (★) within the Lower Detroit River, Trenton Channel.

sition in the lower Detroit River (Fig. 1). Five species were targeted for examination: bullhead (*Ictalurus nebulosus*), white sucker (*Catostomus commersoni*), redhorse sucker (*Moxostoma sp.*), bowfin (*Amia calva*), and walleye (*Stizostedion vitreum*). Fish were netted from the water after electro-shocking and were placed on ice for transport to the laboratory where they were examined.

Fish were killed by cervical dislocation and were weighed and measured. External and internal examinations were conducted to document any grossly visible lesions. Any suspected tumors and adjacent normal tissue were fixed in 10 percent neutral buffered formalin for histological preparation. In addition, tissues were taken from normal appearing individuals. Pectoral spines were removed from bullheads to determine age.

Tissues that had been fixed in 10 percent neutral buffered formalin were dehydrated through a graded series of ethyl alcohol. Dehydrated tissues were then embedded in paraffin blocks and 6 μm thick sections were made. Sections were mounted on glass slides, deparaffinized, stained with hematoxylin and eosin, and examined by light microscopy. All suspected tumors were confirmed by histological evaluation.

## 3. Results

A number of neoplasms and related lesions were observed (Table 1). The classification of these lesions is somewhat arbitrary because there is no comprehensive guide for the classification of tumors in wild fish. We have used terminology derived from two reports on the histology of tumors in fish (Hendricks *et al.*, 1984; Myers *et al.*, 1987) and from standard terminology for histopathology (Thomas & Richter, 1984). In general, lesions could be classified into two groups based on anatomic location, namely oral/dermal and liver. The incidence of these lesions varied with species. However, based on the total number of fish examined; 8.2 percent had oral/dermal lesions and 10.1 percent had liver lesions. Dermal/oral tumors were observed in 10.2 percent of bullheads and 4.5 percent of walleye (Table 2). Among bullheads the majority of dermal/oral lesions were epidermal papillomas found on the exterior of the upper or lower jaws and occassionally involved both jaws. The lesions ranged from small transluscent raised nodules (< 0.5 cm dia.) to large cauliflower-like gelatinous masses of 2 to 3 cm diameter. In a few individuals these masses extended into the mouth covering the teeth and involving the oral cavity. In general, these lesions were non-invasive; however, in a few cases, the lesion was observed to be invading muscle tissue. Epidermal papillomas were also observed on the dorsal aspect of the head and body and on the flanks of some bullheads. These tumors were usually small, slightly raised nodules (< 1 cm dia.)

*Table 1.* Neoplasms and related lesions observed in fish from the Detroit River

| Species | Neoplasms/Lesions | Anatomic location |
|---|---|---|
| Bowfin | Hepatoneoplastic Nodule | Liver |
| Bullhead | Eosinophilic Focus | Liver |
| | Clear Cell Focus | Liver |
| | Hepatocellular Carcinoma | Liver |
| | Cholangioma | Liver |
| | Epidermal Papilloma | Mouth, Skin – Various Sites |
| | Neuroepithelioma | Barbel |
| | Epidermal Hyperplasia | Mouth, Skin, Barbel |
| Redhorse Sucker | Clear Cell Focus | Liver |
| Walleye | Eosinophilic Focus | Liver |
| | Clear Cell Focus | Liver |
| | Hepatocellular Carcinoma | Liver |
| | Fibroma | Skin – Various Sites |
| White Sucker | Clear Cell Focus | Liver |

304

*Table 2.* Incidence of neoplasia in fish species from the Detroit River.

| Species | Tumor Type | |
|---|---|---|
| | Dermal/Oral[a] | Liver[b] |
| Bullhead | 46/449 (10.2) | 27/306 (8.8) |
| Walleye | 3/ 66 (4.5) | 6/ 60 (10.0) |
| Redhorse Sucker | 0/ 46 (0.0) | 5/ 37 (13.5) |
| White Sucker | 0/ 13 (0.0) | 2/ 11 (18.2) |
| Bowfin | 0/ 22 (0.0) | 4/ 22 (18.2) |

Numbers in parentheses are % incidence.

[a] Dermal/oral tumors included epidermal papilloma and neuroepithelioma in bullheads and fibroma in walleye.

[b] Liver tumors included foci of altered hepatocytes, hepatocellular carcinoma and cholangioma.

and often were pigmented with melanin. Epidermal papillomas were characterized by replacement of the normal stratified structure of the epidermis by enlarged papillary formations of epidermis. In addition, the epidermis was nearly devoid of alarm substance cells and had increased mitotic activity. In addition to the epidermal papillomas, a single neuroepithelioma on the barbel of one bullhead was observed. Dermal lesions observed in walleye were always found on the flanks or gill covers and were visible as raised white or pinkish-white masses. When examined by light microscopy, lesions were observed to be uncapsulated fibromas that often had foci of ossification and were similar to those described by Black *et al.* (1982).

Liver neoplasms in bullheads were of both hepatocellular and bile duct origin. Hepatocellular lesions ranged from small foci of altered hepatocytes, either eosinophilic or clear cell, to hepatocellular carcinoma. Tumors arising from bile ducts were comprised of cells in duct- or gland-like formations, often with abnormal morphology and abundant mitotic activity. These lesions, when plainly visible, were usually small (1 to 2 mm dia.) white or cream colored foci on the liver surface or within the liver substance. Liver neoplasms in walleye were usually visible only upon microscopic examination. Lesions ranged from small foci of altered hepatocytes, both eosinophilic and clear cell type, to large hepatocellular carcinoma. Liver lesions observed in

suckers were always in the form of clear cell foci of altered hepatocytes and were sometimes visible during autopsy as white or pale gray foci within the liver mass. The most dramatic liver lesions were observed in bowfin. These lesions were neoplasms of hepatocellular origin and were visible during autopsy as tan or white nodules within the liver mass. Lesions were comprised of pale-staining hepatocytes, filled with an unknown storage product (possibly fat), and a few cords of more basophilic hepatocytes. A detailed description of this lesion has been previously published (Maccubbin *et al.*, 1987).

As part of our study, we hoped to establish age-specific tumor incidence in bullheads. Analysis of the fish collected in 1986 demonstrated a definite trend for increased tumor incidence with increasing age (Fig. 2). In 1986, we collected bullheads that were from 1 to 7 years old with the majority being from 1 to 4 years old. The tumor incidence in fish aged less than 4 years was less than 5 percent whereas fish of 4 years and older had incidence values greater than 10 percent and approaching 50 percent in 5-year-old fish. We are continuing the age analysis on the 1985 and 1987 collections.

## 4. Discussion

Most of the types of neoplasms observed in the target species from the Detroit River were similar

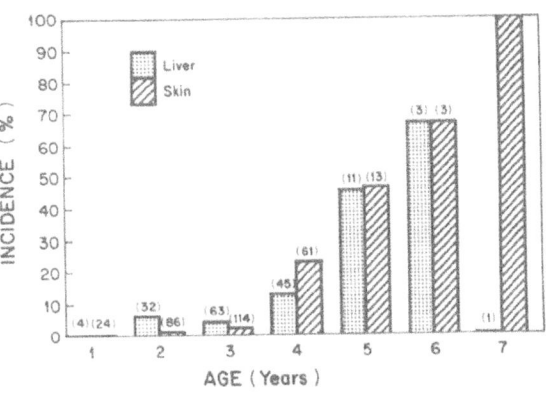

*Fig. 2.* The incidence of liver and dermal/oral (skin) tumors in bullheads from the Detroit River collected in 1986. The percentage of fish with tumors was determined for each age class. The numbers in parenthesis are the number of fish examined.

to those described in other studies. Dermal/oral tumors have been described in bullheads and walleye from the Niagara River and other Lake Erie tributaries (Black, 1983), in bullheads from a final oxidation pond (Grizzle et al., 1981; 1984) and walleye from Torch Lake (Black et al., 1982). In addition, liver neoplasms have previously been described in bullheads (Dawe et al., 1964; Brown et al., 1973; Baumann et al., 1982, 1987; Black, 1983), white suckers (Dawe et al., 1964, 1976; Black, 1983), redhorse suckers (Black, 1983) and walleye (Black et al., 1982). Although oral tumors have been described in white suckers in the Great Lakes (Sonstegard, 1977; Black, 1983; Cairns & Fitzsimons, 1988) and intestinal tumors have been described in redhorse suckers (Black, 1983), these lesions were not observed in fish from the lower Detroit River. The observation of liver neoplasia in bowfin is the first such report. Whether bowfin from other areas have similar liver neoplasia remains to be determined. In addition, a larger sample size of bowfin is needed to determine extent and significance of neoplasia.

There have been relatively few comprehensive studies of fish tumors in other freshwater systems and the study areas have characteristics that may not be completely comparable to the Detroit River. However, the incidence of tumors in fish from the Detroit River was similar to that observed in studies of other industrially-impacted waterways. For example, the incidence of liver neoplasia in brown bullheads has been found to be about 10 percent in the Buffalo River (J.J. Black, personal communication), 12.2 percent in the Fox River (Brown et al., 1973) and 33.0 percent in the Black River (Baumann et al., 1982). The overall liver tumor incidence in bullheads from the Detroit River (8.8%) was lower. However, a recent report (Baumann et al., 1987) demonstrated a significant positive correlation between age and tumor incidence. Our preliminary analysis of tumor incidence data on age-specific basis indicates this is also true in bullheads from the Detroit River. When adjusted for age, the tumor incidence is similar to that found in the Black River (Baumann et al., 1987). The incidence of liver neoplasia in walleye and white suckers was also somewhat lower than that found

in other studies. Dawe et al. (1964) observed a 25 percent incidence of liver tumors in white suckers from Deep Creek Lake, Maryland, whereas we observed only a 18.2 percent incidence in a smaller sample size. However, because of the small sample size and lack of age data, the significance of these lesions remains to be determined. Black et al. (1982) reported a 30 to 35 percent incidence of liver neoplasia in walleye from Torch Lake, Michigan as compared to 10.0 percent for the Detroit River. However, walleye described by Black et al. (1982) were all greater than 7 years old. The ages of walleye collected in this study were not determined; however, the incidence of neoplasms was correlated with the length of the fish. Tumors were only observed in walleye greater than 50 cm. With respect to tumor type, 14.2 percent ($3/21$) had dermal/oral tumors and 23.8 percent ($5/21$) had liver neoplasms. Assuming that older fish are larger, after adjustment for age, the incidence in the Detroit River may approach that reported by Black et al. (1982).

Many of the above cited studies have suggested a pollutant etiology for the observed neoplasia in fish. At this time, we cannot make any conclusive statements as to the precise cause of neoplasms in the Detroit River fish. All of the fish were collected in sediment deposition zones which may have been contaminated with organic pollutants some of which may be carcinogenic. At present, sediment chemistry analyses are being conducted by other researchers in the EPA in-place pollutants project. Relationships between sediment chemical profiles and tumors are still to be determined. However, preliminary results of the analysis of bile demonstrated that the fish collected had been exposed to multi-ringed aromatic hydrocarbons (Maccubbin, unpublished observations). Studies in the Puget Sound have shown correlations between aromatic hydrocarbon metabolites and liver lesions in the English sole (Krahn et al., 1986). It is possible that similar correlations exist in the fish collected in this study.

The results presented in this report document that at least five fish species found in the Detroit River have liver tumors and two species have oral/dermal tumors. The cause of these tumors has yet to be determined. Four of the target species with tumors are bottom dwelling/feeding species and it is possible that they have been exposed to carcinogens in contaminated sediments. Studies are currently under way to determine the mutagenic and carcinogenic potential of sediments in the Detroit River. In addition, studies have been conducted to measure DNA adduct levels in liver tissue to determine the extent of exposure to genotoxic agents. The results of these studies and analyses of the chemical contaminants found in sediment may help to determine possible causes of the observed neoplasia.

## Acknowledgements

This work was supported by a cooperative agreement (CR-812575) with the United States Environmental Protection Agency. Although the USEPA provided financial support, this report has not been peer reviewed by the agency.

We would like to thank Dr. J. J. Black for his assistance in the diagnosis of lesions and Mrs. R. Weaver for histologic preparations. We also thank Mr. W. Richardson and the staff of the USEPA Large Lakes Research Station, Grosse Ile, Michigan for laboratory space and logistical support and Dr. M. Mac and Mr. S. Smith and the U.S. Fish and Wildlife staff, Ann Arbor, Michigan for help in collecting fish and aging the bullheads. In addition, we thank Karen Marie Schrader for assistance in the preparation of this manuscript.

## References

Baumann, P. C., W. D. Smith & M. Ribick, 1982. Hepatic tumor rates and polynuclear aromatic hydrocarbon levels in two populations of brown bullheads (*Ictalurus nebulosus*), In; M. J. Cooke, M. J. Dennis & G. L. Fischer (Eds.) *Polynuclear Aromatic Hydrocarbons: Physical and Biological Chemistry*, pp. 93–102. Battelle Press, Columbus, Ohio.

Baumann, P. C., W. D. Smith & W. K. Parland, 1987. Tumor frequencies and contaminant concentrations in brown bullheads from an industrialized river and a recreational river. Trans. Amer. Fish. Soc. 116: 79–86.

Black, J. J. 1983. Field and laboratory studies of environmental carcinogenesis in Niagara River fish. J. Great Lakes Res. 9: 326–334.

Black, J. J., E. D. Evans, J. C. Harshbarger & R. F. Zeigel, 1982. Epizootic neoplasms in fishes from a lake polluted by copper mining wastes. J. Ntnl. Cancer Inst. 69: 915–926.

Brown, E. R., J. J. Hazdra, L. Keith, I. Greenspan & J. B. G. Kwapinski, 1973. Frequency of fish tumors found in a polluted watershed as compared to nonpolluted Canadian waters. Cancer Res. 24: 189–198.

Cairns, V. W. & J. D. Fitzsimons, 1988. The occurrence of epidermal papillomas and liver neoplasia in white suckers (*Catostomus commersoni*) from Lake Ontario. In; A. J. Niimi & K. R. Solomon (eds.), Proc. 14th Ann. Aquat. Toxicol. Workshop, Toronto, Canada. Can. Tech. Rep. Fish. Aquat. Sci. 1607: 151–152.

Dawe, C. J., M. F. Stanton & F. J. Schwartz, 1964. Hepatic neoplasms in native bottom-feeding fish of Deep Creek Lake, Maryland. Cancer Res. 24: 1194–1201.

Dawe, C. J., R. Sonstegard, M. F. Stanton, D. E. Woronecke, & R. T. Reppert, 1976. Intrahepatic biliary neoplasms in *Catostomus commersoni*. Prog. Exper. Tumor Res. 20: 195–204.

Fallon, M. E. & F. J. Horvath, 1985. Preliminary assessment of contaminants in soft sediments of the Detroit River. J. Great Lakes Res. 11: 373–378.

Great Lakes Water Quality Board, 1985. 1985 Report on Great Lakes Water Quality. Great Lakes Water Quality Board Report to the International Joint Commission. Presented June 1985. Kingston, Ontario, p. 212.

Grizzle, J. M., T. E. Schwedler & A. L. Scott, 1981. Papillomas of black bullheads, *Ictalurus melas* (Rafinesque), living in a chlorinated sewage pond. J. Fish Dis. 4: 345–351.

Grizzle, J. M., P. Melius & D. R. Strength, 1984. Papillomas on fish exposed to chlorinated wastewater effluent. J. Ntnl. Cancer Inst. 73: 1133–1142.

Hamdy, Y. & L. Post, 1985. Distribution of mercury, trace organics, and other heavy metals in Detroit River sediments. J. Great Lakes Res. 11: 353–365.

Hendricks, J. D., T. R. Meyers & D. W. Shelton, 1984. Histological progression of hepatic neoplasia in rainbow trout (*Salmo gairdneri*) Ntnl. Cancer Inst. Monogr. 65: 321–336.

Krahn, M. M., L. D. Rhodes, M. S. Myers, L. K. Moore, W. D. MacLeod & D. C. Malins, 1986. Associations between metabolites of aromatic compounds in bile and the occurrence of hepatic lesions in English sole (*Parophrys vetulus*) from Puget Sound, Washington. Arch. Env. Contam. Toxicol. 15: 61–67.

Maccubbin, A. E., J. J. Black & J. C. Harshbarger, 1987. A case report of hepatocellular neoplasia in bowfin, *Amia calva* L. J. Fish Dis. 10: 329–331.

Malins, D. C., B. B. McCain, D. W. Brown, S. L. Chan, M. S. Myers, J. T. Landahl, P. G. Prohaska, A. J. Friedman, L. D. Rhodes, D. G. Burrows, W. O. Gronlund & H. O. Hodgins, 1984. Chemical pollutants in sediments and diseases of bottom-dwelling fish in Puget Sound, Washington. Env. Sci. Technol. 18: 705–713.

Malins, D. C., B. B. McCain, M. S. Myers, D. W. Brown, M. M. Krahn, W. T. Roubal, M. H. Schiewe, J. T. Landahl & S.-L. Chan, 1987. Field and laboratory studies of the etiology of liver neoplasms in marine fish from Puget Sound. Env. Health Perspect. 71: 5–16.

Myers, M. S., L. D. Rhodes & B. B. McCain, 1987. Pathologic anatomy and patterns of occurrence of hepatic neoplasms, putative preneoplasic lesions, and other idiopathic hepatic conditions in English sole (*Parophrys vetulus*) from Puget Sound, Washington. J. Ntnl. Cancer Inst. 78: 333–363.

Pranckevicius, P. E. 1987. 1982 Detroit Michigan area sediment survey. EPA 905/4-87-003, Great Lakes Ntnl. Program Off. Rep. 87–11.

Sonstegard, R. A. 1977. Environmental carcinogenesis studies in fishes of the Great Lakes of North America. Ann. New York Acad. Sci. 298: 261–269.

Thomas, C. & G. W. Richter, 1984. Sandritter's Color Atlas and Textbook of Histopathology, 7th Edition. Year Book Medical Publishers, Inc., Chicago, IL.

*Hydrobiologia* **219**: 307–315, 1991.
*M. Munawar & T. Edsall (eds),*
*Environmental Assessment and Habitat Evaluation of the Upper Great Lakes Connecting Channels.*
© 1991 *Kluwer Academic Publishers.*

# Heavy metal contamination of sediments in the Upper Connecting Channels of the Great Lakes[1]

S.J. Nichols, B.A. Manny, D.W. Schloesser & T.A. Edsall
*National Fisheries Research Center-Great Lakes, U.S. Fish and Wildlife Service, 1451 Green Road, Ann Arbor, Michigan 48105, USA*

*Key words:* metals, sediment contamination, deposition zones, rivers

## Abstract

In 1985, sampling at 250 stations throughout the St. Marys, St. Clair, and Detroit rivers and Lake St. Clair – the connecting channels of the upper Great Lakes – revealed widespread metal contamination of the sediments. Concentrations of cadmium, chromium, copper, lead, mercury, nickel, and zinc each exceeded U.S. Environmental Protection Agency sediment pollution guidelines at one or more stations throughout the study area. Sediments were polluted more frequently by copper, nickel, zinc, and lead than by cadmium, chromium, or mercury. Sediments with the highest concentrations of metals were found (in descending order) in the Detroit River, the St. Marys River, the St. Clair River, and Lake St. Clair. Although metal contamination of sediments was most common and sediment concentrations of metals were generally highest near industrial areas, substantial contamination of sediments by metals was present in sediment deposition areas up to 60 km from any known source of pollution.

## 1. Introduction

Areas within each of the connecting channels of the upper Great Lakes – the St. Marys River, the St. Clair River, Lake St. Clair, and the Detroit River – have been designated 'Areas of Concern' by the International Joint Commission. Elevated sediment concentrations of iron and zinc in the St. Marys River and of mercury in the St. Clair and Detroit rivers contributed to the designation of Areas of Concern in those waters (GLWQB, 1983). Efforts to restore these degraded waters pointed to gaps in our knowledge of pollution problems throughout the connecting channels. The Upper Great Lakes Connecting Channels Study of the United States and Canada was established to integrate all existing data, perform studies to fill gaps in the data base, and develop recommendations for restoring ecosystem quality (EC & USEPA, 1988).

As a part of this study, in 1985, we conducted a comprehensive survey of heavy metal contaminants in the sediments within each of the connecting channels, from the head of the St. Marys River to the mouth of the Detroit River. The purpose of our survey was to determine the distribution of cadmium, chromium, copper, lead, mercury, nickel, and zinc in the connecting channels of the upper Great Lakes and to identify locations where metal concentrations exceeded the criteria of the U.S. Environmental Protection

---
[1] Contribution 735 of the National Fisheries Research Center-Great Lakes, U.S. Fish and Wildlife Service, 1451 Green Road, Ann Arbor, MI 48105.

308

*Table 1.* Criteria for sediment concentration of metals established by the Ontario Ministry of the Environment (OME) and the U.S. Environmental Protection Agency (USEPA). All units in $\mu g\,g^{-1}$ (International Joint Commission, 1982).

| Metal | OME | USEPA | | |
|---|---|---|---|---|
| | | Non-polluted | Moderate pollution | Heavy pollution |
| Mercury | 0.3 | < 1.0 | – | > 1.0 |
| Cadmium | 1.0 | – | – | > 6 |
| Chromium | 25.0 | < 25 | 25–75 | > 75 |
| Copper | 25.0 | < 25 | 25–50 | > 50 |
| Nickel | 25.0 | < 20 | 20–50 | > 50 |
| Lead | 50.0 | < 40 | 40–60 | > 60 |
| Zinc | 100.0 | < 90 | 90–200 | > 200 |

Agency (USEPA) for open-water disposal of polluted sediments (Table 1).

## 2. Methods

Sediments were collected in May and June 1985 from 250 stations established within a grid covering the St. Marys River, the St. Clair River, Lake St. Clair, and the Detroit River (Fig. 1). Each grid cell was 2.2 × 2.2 km in the St. Marys and St. Clair rivers, 4.8 × 6.9 km in Lake St. Clair, and 2.8 × 1.8 km in the Detroit River. Within each grid cell one station was established at the first location where sampling with a standard Ponar grab (0.05 m²) yielded fine-grained, silty sediment, in which the clay fraction was high. Such sediments bind metals more readily than other sediments (Mudroch, 1985). All stations were referenced to readily distinguishable landmarks, including navigation buoys, on National Oceanic and Atmospheric Administration navigation charts (chart pack series 14884 for the St. Marys River and 14853 for the St. Clair and Detroit rivers and Lake St. Clair). Of the 250 stations, 96 were in Canadian waters and 154 were in U.S. waters.

At each station a sample of sediment for metals analysis was collected with the Ponar grab. When the Ponar containing the sediments was brought up from the river, it was allowed to drain, placed in a large metal tub and opened carefully, so that the sediments were deposited gently in the tub and the sediment profile was undisturbed. About 500 g of surficial sediment from the top 3 cm thick layer of the Ponar sample were removed with a stainless steel spoon, placed in an acetone-washed glass jar, and refrigerated at 4 °C in darkness. This subsampling procedure ensured that none of the sediments that came in contact with the surface of the Ponar, or the tub were taken for analysis. We also thoroughly rinsed the tub and Ponar with clean water before collecting each sample to prevent contamination of one sample by another.

All of the glass jars, lid-liners, and lids were washed with detergent (Alconox 1), rinsed with distilled water and acetone, and heated to 135 °C for five days. Ten jars, lids, and liners were randomly selected after heat treatment and rinsed with 10 percent nitric acid. The rinses were combined (150 ml total volume) and compared to a 10 percent nitric acid blank on a plasma emission spectrophotometer to check for metal contamination. Only chromium and copper were detected in the rinse solution and these ocurred at low concentrations ($<0.1\,\mu g\,ml^{-1}$) that would not contaminate any of the sediment samples.

Sixteen bottle blanks, consisting of heat-treated jars, were opened and resealed in the field during regular sampling at 16 stations chosen at random. After being resealed these jars were handled like the sediment samples, and later analyzed for metal contamination.

Sediment chemistry services were provided by the Bionetics Corporation at the U.S. Environmental Protection Agency, Central Regional Laboratory in Chicago, Illinois. Each sample was analyzed for mercury, cadmium, chromium, copper, lead, nickel, and zinc. Heavy metals, with the exception of mercury, were analyzed using Inductively Coupled Argon Plasma (ICAP) methodology (Jarrell-Ash, 1979; USEPA, 1979). Mercury was analyzed using a tin chloride reduction and cold vapor trapping technique (Plumb, 1981).

Detailed information on the quality assurance

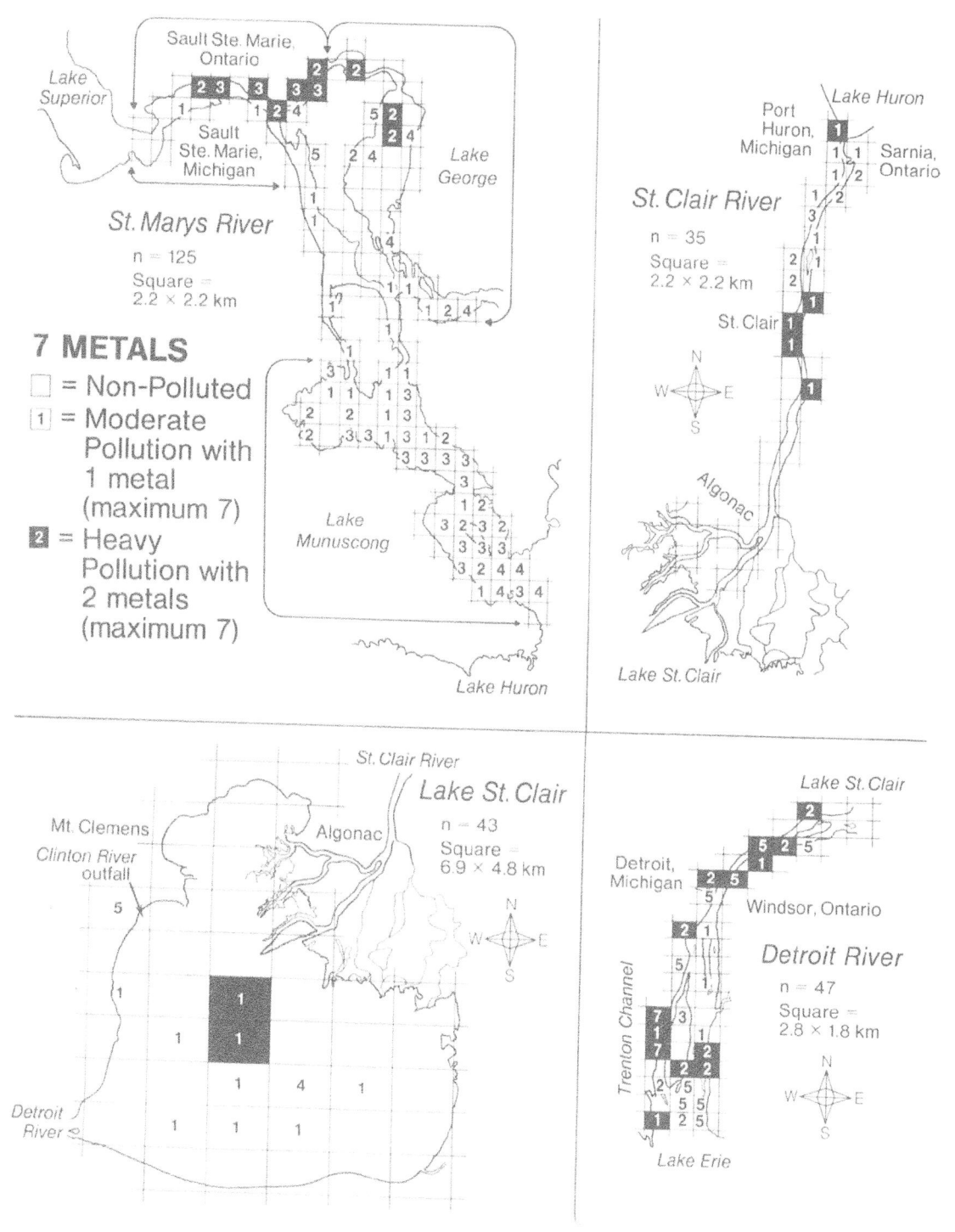

*Fig. 1.* Metals pollution of sediments in the connecting channels of the upper Great Lakes in 1985. U.S. Environmental Protection Agency (USEPA) criteria for dredged sediments (International Joint Commission, 1982) were used to distinguish non-polluted sediments from those with moderate or heavy pollution. The numerals in each grid square indicate the number of metals (1–7) that exceeded USEPA criteria at the station sampled in that grid cell.

program used during analysis can be found in Bertram *et al.* (In Press). In general, the 250 samples were analyzed in units of ten. After each set of ten samples were run through the ICAP or tin chloride procedure a duplicate sediment sample was analyzed, followed by a National Bureau of Standards sample (Standard Reference Material 1645) and three in-house standards. The limits of detection of the procedure were (in mg $kg^{-1}$): mercury, 0.1; cadmium, 0.2; chromium, 0.8; copper, 0.6; lead, 7.0; nickel, 2.0; zinc, 4.0. Approximate precision was $\pm 10$ percent or $\pm 0.1$ mg $kg^{-1}$ for mercury, and $\pm 10$ percent or $\pm 1 \times$ the detection limit for all other metals. The approximate accuracy was plus 5 percent for mercury and plus 20 percent for all other metals. Recovery of metal spikes added to the sediment samples was 93 to 103 percent.

## 3. Results

At 128 (51%) of the 250 stations sampled, sediment concentrations of at least one of the seven metals we measured exceeded USEPA criteria for moderate or heavy pollution (Fig. 1). In the Detroit River, sediment pollution was moderate or heavy at 29 (62%) of the stations and at stations 230 and 236 in the Trenton Channel the sediment was heavily polluted with all seven metals. In the St. Marys River, sediments were moderately or heavily polluted at 71 (57%) of the stations, most of which were in the Sault Ste. Marie area. The percentage of moderately polluted stations was similar in the St. Marys (57%) and Detroit rivers (53%), but the percentage of heavily contaminated sites was significantly higher in the Detroit River (30%) than in the St. Marys River (7%).

In the St. Clair River, sediments were moderately or heavily polluted by at least one metal at 16 (46%) of the stations (Fig. 1). In Lake St. Clair, sediments at 11 (26%) of the stations were polluted, and mercury exceeded the criterion for heavy pollution at two of them. Most of the contaminated stations in the St. Clair River were clustered near towns along the middle and upper regions of the river. In Lake St. Clair, the contaminated stations were offshore in the deeper, central basin and at the mouth of the Clinton River cut-off channel.

Of the seven metals measured, chromium, nickel, copper, zinc, and lead were the most common sediment contaminants throughout the study area (Table 2). Chromium levels exceeded USEPA pollution criteria in only the St. Marys and Detroit rivers and Lake St. Clair (Table 2; Fig. 2). Although moderate chromium contamination was common to all areas of the St. Marys and Detroit rivers, heavily contaminated sediments were limited mostly to the lower Detroit River.

*Table 2.* Number of stations in the Upper Great Lakes Connecting Channels where sediment concentrations of metals exceeded moderate (M) or heavy (H) pollution criteria of the U.S. Environmental Protection Agency. Number of stations in each channel is shown in parentheses.

| Metal | St. Marys River (125) | | St. Clair River (35) | | Lake St. Clair (43) | | Detroit River (47) | | All channels combined (250) | |
|---|---|---|---|---|---|---|---|---|---|---|
| | M | H | M | H | M | H | M | H | M | H |
| Mercury | 0 | 0 | 0 | 0 | 0 | 2 | 0 | 4 | 0 | 6 |
| Cadmium | 0 | 0 | 0 | 0 | 0 | 0 | 0 | 2 | 0 | 2 |
| Chromium | 69 | 1 | 0 | 0 | 2 | 0 | 19 | 4 | 90 | 5 |
| Copper | 38 | 4 | 14 | 1 | 5 | 0 | 11 | 12 | 68 | 17 |
| Lead | 6 | 8 | 2 | 3 | 1 | 0 | 10 | 13 | 19 | 24 |
| Nickel | 42 | 0 | 5 | 0 | 9 | 0 | 18 | 6 | 74 | 6 |
| Zinc | 14 | 9 | 4 | 1 | 2 | 0 | 17 | 11 | 37 | 21 |

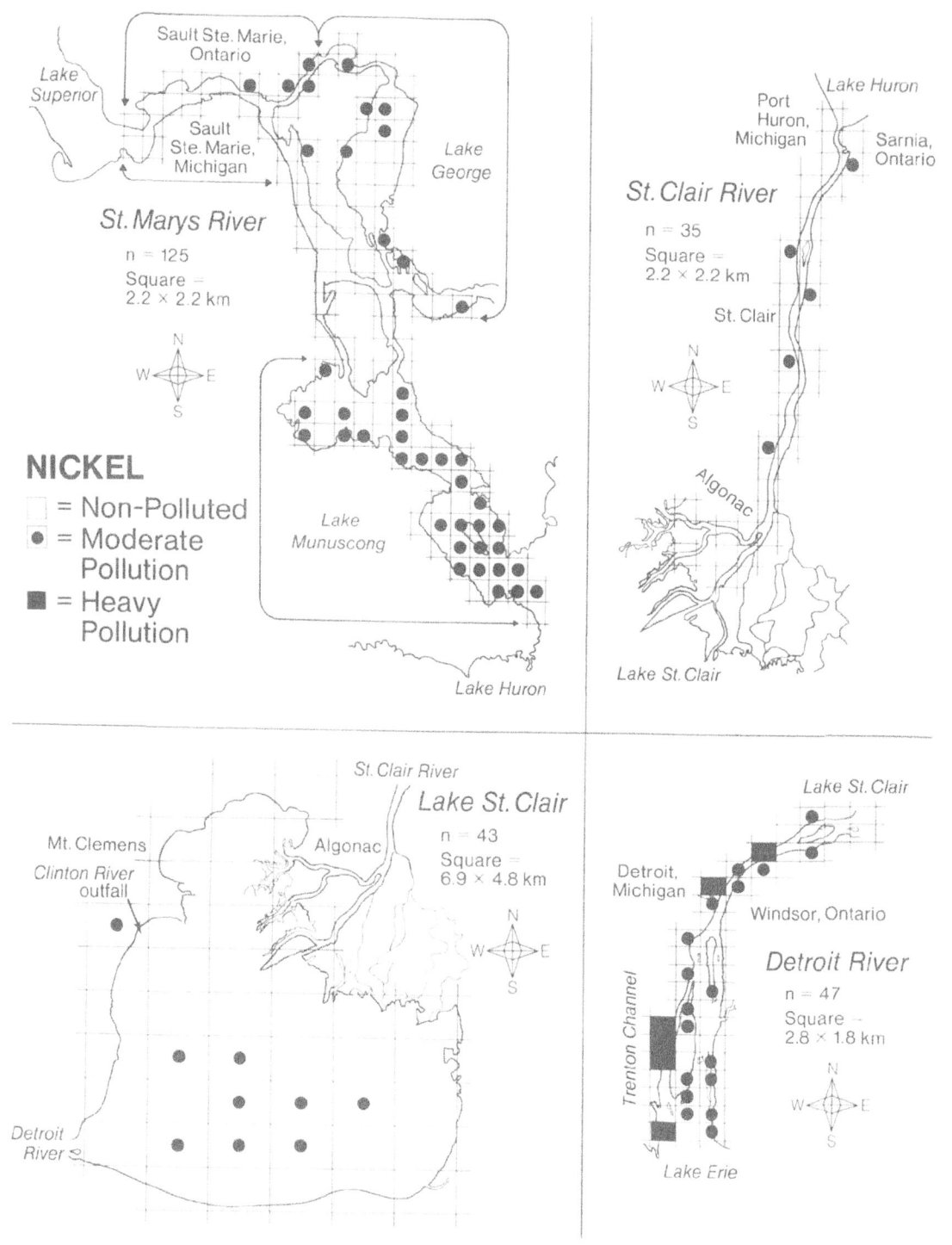

*Fig. 2.* Nickel pollution of sediments in the connecting channels of the upper Great Lakes in 1985. Pollution criteria are U.S. Environmental Protection Agency guidelines for dredged sediments (International Joint Commission, 1982).

Nickel contamination occurred at 80 (32%) of the 250 stations (Fig. 2). Moderate nickel contamination occurred in all three areas of the St. Marys River, the upper St. Clair River, the center and Clinton River areas of Lake St. Clair, and both the upper and lower reaches of the Detroit River. Heavy nickel contamination was restricted to the Detroit River.

The pattern of copper and zinc contamination was similar to that of nickel. Copper exceeded the USEPA criterion at 85 (34%) of the 250 stations, and only 17 of these were heavily polluted (Table 2). Zinc concentrations exceeded criteria for moderate or heavy pollution at 23 percent of the stations. Heavy contamination by zinc was slightly more widespread than heavy copper or nickel contamination.

Lead was the only metal tested that exceeded the sediment criterion for heavy pollution at more stations than it did for moderate pollution – 24 versus 19 stations (Table 2). Lead contamination was localized near industrial areas and was less widespread than nickel, copper, or zinc contamination.

Sediment concentrations of mercury and cadmium rarely exceeded the pollution criteria in Table 1. Mercury pollution occurred at 6 and cadmium at 2 of the 250 stations sampled. Mercury was found in central Lake St. Clair and the upper and lower ends of the Detroit River. Cadmium pollution was limited to the lower Detroit River.

The mean and maximum concentrations of each of the seven metals tested were highest in the Detroit River and, except for mercury, were lowest in Lake St. Clair (Table 3). The mean concentrations of copper, nickel, lead, and zinc were similar in the St. Marys and St. Clair rivers, and intermediate to those in the Detroit River and Lake St. Clair.

## 4. Discussion

In this study we found that metal contamination was widespread in the connecting channels of the upper Great Lakes. The highest concentrations of metals occurred near heavily industrialized areas of the Detroit River, the Sault Ste. Marie portion

*Table 3.* Mean ($\mu g\ g^{-1}$), standard error (SE), and range of concentrations of seven metals in the sediments of the connecting channels of the upper Great Lakes in 1985. Number of stations in each channel is shown in parentheses.

| | St. Marys River (125) | | | St. Clair River (35) | | | Lake St. Clair (43) | | | Detroit River (47) | | |
|---|---|---|---|---|---|---|---|---|---|---|---|---|
| | Mean | SE | Range | Mean | SE | Range | Mean | SE | Range | Mean | SE | Range |
| Mercury | 0.05 | 0.005 | 0.00 0.31 | 0.15 | 0.035 | 0.00 0.60 | 0.25 | 0.075 | 0.00 2.71 | 1.61 | 1.180 | 0.04 55.80 |
| Cadmium | 0.31 | 0.024 | 0.20 1.80 | 0.44 | 0.040 | 0.20 1.20 | 0.28 | 0.035 | 0.20 1.40 | 1.99 | 0.716 | 0.20 33.00 |
| Chromium | 30.72 | 1.718 | 2.70 78.00 | 10.23 | 0.534 | 5.20 15.00 | 11.73 | 0.880 | 2.80 26.00 | 37.01 | 6.806 | 4.80 260.00 |
| Copper | 20.57 | 1.370 | 1.80 110.00 | 24.68 | 2.376 | 7.10 55.00 | 12.69 | 1.238 | 1.40 30.00 | 38.23 | 5.413 | 3.30 150.00 |
| Nickel | 15.31 | 0.886 | 1.80 44.00 | 13.29 | 0.879 | 5.90 24.00 | 13.62 | 1.033 | 3.80 33.00 | 27.24 | 3.099 | 6.10 130.00 |
| Lead | 20.95 | 1.752 | 7.00 130.00 | 29.07 | 3.444 | 7.00 92.00 | 16.95 | 1.649 | 7.00 50.00 | 65.57 | 11.866 | 7.20 360.00 |
| Zinc | 69.34 | 6.374 | 6.00 470.00 | 67.85 | 8.333 | 28.00 310.00 | 48.44 | 3.369 | 11.00 130.00 | 272.70 | 112.349 | 21.00 5,300.00 |

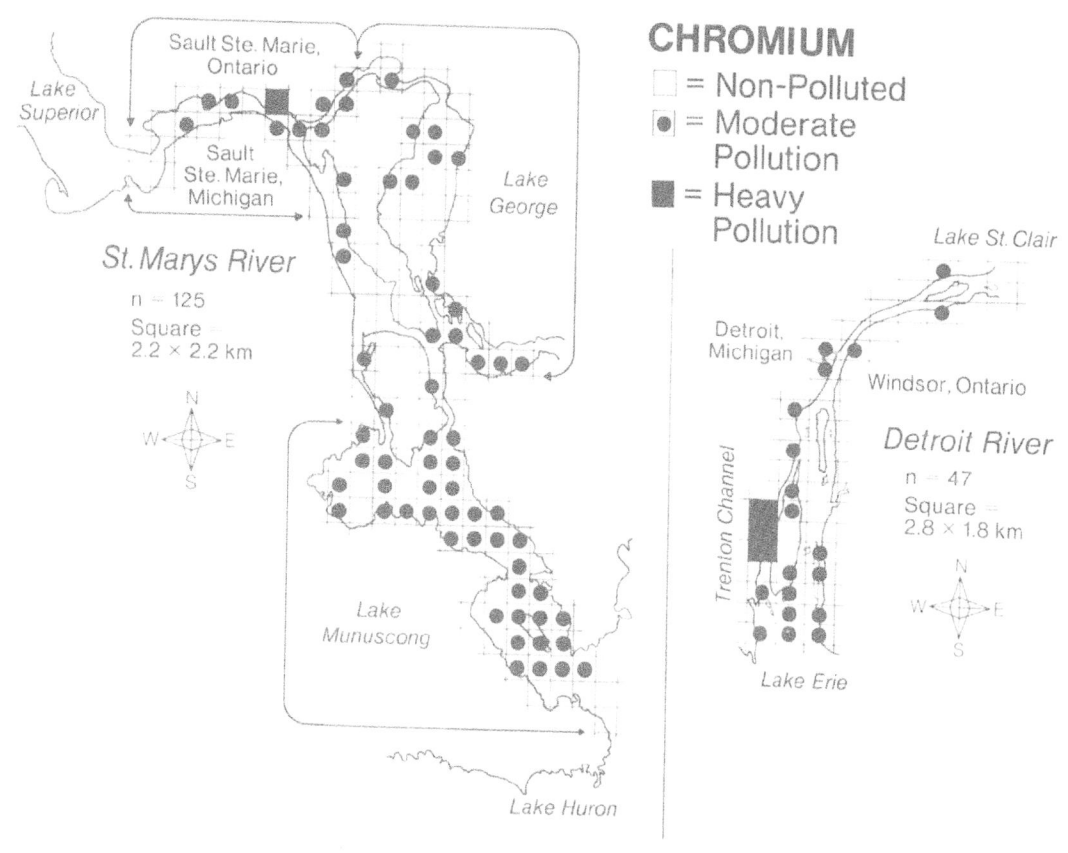

*Fig. 3.* Chromium pollution of sediments in the connecting channels of the upper Great Lakes in 1985. Pollution criteria are U.S. Environmental Protection Agency guidelines for dredged sediments (International Joint Commission, 1982).

of the St. Marys River, and the Port Huron-Sarnia-St. Clair areas of the St. Clair River, as reported in earlier studies by Environmental Control Technology Corporation (1974), Ontario Ministry of the Environment (1979), Thornley & Hamdy (1984), Fallon and Horvath (1985), Handy and Post (1985), Limno-Tech (1985), Mudroch (1985), and Mudroch *et al.* (1988). We also found elevated levels of metals in sediment deposition areas in Lake George and Lake Munuscong in the St. Marys River and in Lake St. Clair, but not in the stretches of river between the sources and the sediment deposition areas. Our data do not indicate how these metals moved from sources to deposition areas, but we speculate that metals were absorbed onto sediment particles and carried downstream by river currents. Support for this interpretation is provided by what is

now perhaps a classic example of the movement of metals from a point source in the St. Clair River to a remote sediment deposition area in Lake St. Clair. In the 1950's and 1960's two chloro-alkali plants discharged over 100 000 kg of mercury into the St. Clair River near Sarnia, Ontario (Limno-Tech, 1985). By 1970, elevated levels of mercury were found immediately below the point source and also 60 km farther downstream in the center of Lake St. Clair, but not in the intervening portion of the river where currents are swift and fine sediments do not accumulate (Thomas *et al.,* 1975; Edsall *et al.,* 1988; EC & USEPA, 1988).

Of the seven metals we measured, only copper and cadmium were among those described by Limno-Tech (1985) as 'parameters of concern' throughout the connecting channels of the upper Great Lakes. Our data show that sediment contamination by nickel, zinc, and lead was almost as widespread as contamination by copper. Chromium contamination has been considered a problem in the St. Clair and Detroit rivers for some years (Limno-Tech, 1985), but the widespread chromium pollution we found in the St. Marys River was not documented in earlier work. Tissue samples from brown bullheads (*Ictalurus nebulosus*) collected in 1985 from the lower St. Marys River contained high levels of chromium (National Fisheries Research Center-Great Lakes, Unpubl. data), indicating this metal can enter the aquatic food chain.

Cadmium and mercury are well documented as contaminants in the connecting channels of the upper Great Lakes (Ontario Ministry of the Environment, 1979; Limno-Tech, 1985; EC & USEPA, 1988). One difficulty in describing sediment contamination by cadmium and mercury stems from the fact that the pollution guidelines used by USEPA and the Ontario Ministry of the Environment (OME) are not in agreement (Table 1). As a result, cadmium exceeded the OME criterion at 28 of the 250 stations, but exceeded USEPA criterion at only 2 stations, and mercury concentrations exceeded the OME criterion at 45 out of 250 stations and the USEPA criterion at only 6 stations.

In conclusion, metal contamination of sediments is widespread in the connecting channels of the upper Great Lakes. Sediment concentrations of the seven metals we studied were above USEPA criteria for moderate or heavy pollution at more than half of the sampling sites. Although pollution was heaviest near industrial areas, metal contaminants from these areas were also concentrated in sediment deposition zones as far as 60 km downstream from pollution sources. As metal loadings from point sources are reduced throughout the connecting channels of the upper Great Lakes, the historical accumulations of metals in these sediments will become relatively more important. Future pollution surveys in these waters should include sampling near industrial point sources and in downstream deposition zones, to permit more accurate estimates of impacts of metals on the aquatic ecosystem.

## 5. Acknowledgements

This work was performed as a cooperative project between the U.S. Fish and Wildlife Service, National Fisheries Research Center-Great Lakes, and the U.S. Environmental Protection Agency, Great Lakes National Program Office, under the terms of Interagency Agreement No. DW 14931214-01-0. Sediment chemistry services were provided by USEPA under the supervision of Project Officer Paul Bertram.

## References

Bertram, P., T. Edsall, B. Manny, S. Nichols & D. Schloesser, 1985. Physical and chemical characteristics of sediments in the upper Great Lakes connecting channels, 1985. U.S. Env. Protect. Agency Tech. Rep., Chicago, Illinois. In Press.

Edsall, T. A., B. A. Manny & C. N. Raphael, 1988. The St. Clair River and Lake St. Clair, Michigan: an ecological profile. U.S. Fish Wildl. Serv. Biol. Rep. 85(7.3). 130 pp.

EC & USEPA (Environment Canada & U.S. Environmental Protection Agency), 1988. Upper Great Lakes connecting channels study. Final report. Volume II. U.S. Env. Protect. Agency, Chicago, Illinois. 626 pp.

Environmental Control Technology Corporation, 1974. Water Pollution Investigation: Detroit and St. Clair Rivers. U.S. Env. Protect. Agency, Rep. No. EPA/905/9-74-013), Ann Arbor, Michigan. 361 pp.

Fallon, M. & F. Horvath, 1985. Preliminary assessment of contaminants in soft sediments of the Detroit River. J. Great Lakes Res. 11(3): 373–378.

GLWQB (Great Lakes Water Quality Board), 1983. Report on Great Lakes Water Quality; Appendix A: Areas of Concern in the Great Lakes Basin; 1983 Update of Class A areas. Rep. to Internat. Joint Commiss. 113 pp.

Hamdy, Y. & L. Post, 1985. Distribution of mercury, trace organics, and other heavy metals in Detroit River sediments. J. Great Lakes Res. 11(3): 353–365.

International Joint Commission, 1982. Guidelines and register for evaluation of Great Lakes dredging projects. 365 pp.

Jarrell-Ash, 1979. Mark III Atomcomp interim operator's manual. Jarrell-Ash Division, Fisher Scientific Co., Waltham, Massachusetts. March 1979, M 79–1.

Limno-Tech, 1985. Summary of the existing status of the upper Great Lakes connecting channels data. Prepared for: Upper Great Lakes Connecting Channels Study. Limno-Tech, Inc., Ann Arbor, Michigan. 159 pp.

Mudroch A., 1985. Geochemistry of the Detroit River sediment. J. Great Lakes Res. 11(3): 193–200.

Mudroch, A., L. Sarazin & T. Lomas, 1988. Summary of surface and background concentrations of selected elements in the Great Lakes sediments. J. Great Lakes Res. 14(2): 241–251.

Ontario Ministry of the Environment, 1979. St. Clair River organics study; biological surveys, 1968 and 1977. Ont. Min. Nat. Resour., Wat. Resour. Assess. Unit, Toronto, Ontario. 90 pp.

Plumb, R. H. Jr., (ed.), 1981. Procedures for handling chemical analyses of sediment and water samples. Technical Report EPA/CE-81-1. Prepared by Great Lakes Laboratory, State University College at Buffalo, Buffalo, New York, for USEPA/CORPS Technical Committee on Criteria for Dredged or Fill Material. U.S. Army Corps of Engineers, Waterways Experiment Station, Vicksburg, Mississippi. 386 pp.

Thomas, R., J. Jaquet & A. Mudroch, 1975. Sedimentation processes and associated changes in surface sediment trace metal concentrations in Lake St. Clair, 1970–1974. Internat. Conf. Heavy Metals Env., Toronto, Ontario, October 27–31, 1975. pp. 691–708.

Thornley, S. & Y. Hamdy, 1984. An assessment of the bottom fauna and sediments of the Detroit River. Ont. Min. Env. Tech. Rep. ISBN. 0-7743-8474-3. February. 48 pp.

USEPA (U.S. Environmental Protection Agency), 1979. Simultaneous analysis of liquid samples for metals by inductively coupled argon plasma atomic emission spectroscopy (ICAP-AES). U.S. Env. Protect. Agency, Region V, Chicago, Illinois. Unpublished.

*Hydrobiologia* **219**: 317–324, 1991.
*M. Munawar & T. Edsall (eds),*
*Environmental Assessment and Habitat Evaluation of the Upper Great Lakes Connecting Channels.*
© *1991 Kluwer Academic Publishers.*

# Application of a microcomputer-based algal fluorescence technique for assessing toxicity: Lake St. Clair and St. Clair River examples

M. Munawar[1], S.T. Severn[2] & C.I. Mayfield[2]
[1]*Fisheries & Oceans Canada, Canada Centre for Inland Waters, P.O. Box 5050, Burlington, Ontario, L7R 4A6 Canada*; [2]*Biology Department, University of Waterloo, Waterloo, Ontario, N2L 3G1, Canada*

*Key words:* algae, fluorescence, toxicity, microcomputer, Great Lakes, rivers

## Abstract

The use of a multi-trophic assay strategy is now being encouraged in toxicological investigations which provides for rapid and sensitive tests. Such a strategy, a microcomputer-based algal fluorescence technique, was applied for the bioassessment of Lake St. Clair and St. Clair River ecosystems. The technique was found to be rapid, sensitive, and relatively inexpensive. In addition, it permitted microscopic examination of the impact of contaminants on individual cells/organisms, a feature which is not possible by other tests using radioisotopes and enzymes. The algal fluorescence technique appears to have a considerable potential for fast screening of large numbers of environmental samples.

## Introduction

The alarming increase of toxic substances in our environment is a serious threat to the integrity and health of aquatic ecosystems. It is now well-established that environmental evaluation and hazard assessment based solely on chemical analysis is neither desirable nor useful for management purposes. Consequently, the use of bioassays is receiving much attention as a tool for environmental assessment and habitat evaluation. A large variety of bioassays is increasingly becoming available and a multi-trophic assay strategy is gaining support (Munawar *et al.*, 1989). The development of new techniques and assays has been facilitated by the availability of inexpensive microcomputers and associated software.

The availability of inexpensive interfaces between microcomputers and video cameras has made possible the development of image-analysis systems which can measure the relative brightness of many individual points in a microscope image. (Mayfield, 1984; Severn *et al.*, 1986; Mayfield & Munawar, 1988; Severn *et al.*, 1989).

In photosynthetic organisms, when all the reaction centres of the photosystem are open, the Q quinone is completely oxidized and chlorophyll fluorescence is at a minimum (Schreiber, 1983). When exposed to ultraviolet light, the reaction centres become saturated and the electron acceptors become fully reduced. The chlorophyll molecules are unable to transfer excitation to the reaction centres at the rate it is absorbed and dissipate excess energy as fluorescence (Clayton, 1980; Foyer, 1984). Toxicants are thought to inhibit photosynthesis and chlorophyll fluorescence by interference with the reproduction, development, structure, and integrity of the chloroplasts, by affecting the processes of the light reaction involved in the conversion of light energy to chemical energy, and by affecting the biosynthetic pathways involved in the production of photosynthetic products (Moreland & Hilton, 1976; Ashton & Crafts, 1981).

Interference with the chloroplast structure or integrity by a toxicant such as a heavy metal results in a decrease in the fluorescence intensity (Mayfield & Munawar, 1988; Severn et al., 1989). Inhibition of the photosynthetic electron transport pathway by certain organic pesticides such as atrazine enhance chlorophyll fluorescence (Rafii et al., 1979; Richard et al., 1983; Moody et al., 1983; Severn et al., 1986). In both of these examples a decrease in $^{14}C$ fixation would be observed (Ashton & Crafts, 1981; Forstner & Wittmann, 1981). The major advantage in using the algal fluorescence procedure is that after the initial expense for the hardware and software there is little operational cost for routine experimental use of the equipment. Furthermore, the data from the video-fluorescence assay can be processed very rapidly with minimal expense.

This paper deals with the following objectives:
(1) To determine the feasibility of using a computerized algal fluorescence technique as a rapid, sensitive, and inexpensive screening test for toxicity evaluation and inclusion in a multi-trophic battery of tests strategy.
(2) To compare the effects of the sediment elutriates from Lake St. Clair and the St. Clair River on natural phytoplankton (from mid-lake Lake Ontario) and laboratory grown algal cultures (*Ankistrodesmus braunii* and *Chlorella vulgaris*) and to determine if data based on laboratory populations could be extrapolated to the natural environment.
(3) To determine the effects of the elutriate after prolonged exposure (96 h) to contaminants.

## Materials and methods

### Microcomputer-microscope assembly

An Apple II microcomputer with a Digisector-65 video digitization interface card (Micro Works, Del Mar, CA) accepted input for a black and white video camera or video cassette recorder and presented the image as a series of points on the Apple II high resolution graphics screen.

There are different methods to digitize the image received; the simplest is to present only those points on the screen that are greater than a predetermined brightness value ranging from 0 to 63 (i.e., 64 brightness values). Software or controls on the interface card permit the brightness and contrast of the image to be varied.

Each point (called a pixel) in the video image can be individually addressed, and the current brightness value measured by the computer. With appropriate software, a large number of points in a predetermined pattern can be addressed and evaluated. In this manner any video image can be converted to a series of brightness values and these values may be stored in a file on a disk drive attached to the microcomputer. The information contained in this file can then be further processed by an analysis program to extract and collate various types of information.

A Nikon Labophot microscope with a 100-watt mercury lamp providing epi-illumination through fluorescence objectives was equipped with a violet dichroic mirror and barrier filter system (Nikon Canada). The images of the chlorophyll fluorescence of the algal cells were relayed through a Panasonic WV-1050A high sensitivity black and white video camera to a video cassette recorder. A still frame image from the recorder was used as input for the digitization process. Still frame images are not essential for the digitization process but their use allows many more images to be stored on a tape because of the short taping duration required for each image.

The program used for digitization was written partly in Applesoft BASIC and partly in assembly language. This assembly language programming was required because 10000 individual points (pixels) on a square of $100 \times 100$ pixels were individually digitized and their brightness values measured. This process was too slow in BASIC (about 5 min) but was of acceptable duration (about 80 sec) in assembly language. As each point was addressed, its brightness was measured and the value stored in the memory of the computer. Upon completion of the scanning process, the entire 10000 brightness values were transferred to a disk file.

Another program written in assembly language was used to analyze the information stored in the disk files. The program read in the 10 000 brightness values and categorized them into groups of 0–7, 8–15, 16–22, etc., up to 55–63. The average brightness of all points greater than 7 was then calculated. Further analysis calculated the average brightness of all points that were not black (i.e. all points greater than brightness level 0). The percentage of points with brightness levels greater than 0 was also calculated.

### Elutriate preparation

The sediments from St. Clair River (Station 79) and Lake St. Clair (Station 54) were collected (Fig. 1) and elutriates were prepared according to the methodology described in Munawar & Munawar (1987) and U.S.E.P.A. (1977). A control with filtered lake water was also included to which no elutriate was added.

### Experimental procedure

#### Natural phytoplankton experiment
Natural phytoplankton from Lake Ontario was used as a test assemblage. The effect of a 20 percent dosage of sediment elutriates was examined at 0, 4, 24, 48, and 96 hours time intervals by the algal fluorescence technique. The phytoplankton was size fractionated in a pre-filtration step, into $< 20 \mu m$ and $> 20 \mu m$ mesh size assemblages (Munawar & Munawar, 1987). A control consisted of pre-fractionated samples of $< 20 \mu m$ and $> 20 \mu m$ size assemblages to which no elutriate was added. The algal cells were collected on black membrane filters (Sartorius 0.45 $\mu m$). The filters were placed on the stage of the epifluorescence microscope. Triplicate microscope image fields were recorded and 12 cells digitized for each treatment (Mayfield, 1984).

#### Algal culture experiment
In a second experiment, laboratory grown cultures of *Ankistrodesmus braunii* and *Chlorella vulgaris* were mixed and inoculated into filtered lake water with 20 percent elutriate dosage additions and control of filtered lake water at a concentration of approximately $10^4$ cells ml$^{-1}$. Triplicate microscope images were recorded and five cells were digitized for each species and treatment.

## Results and discussion

### Natural phytoplankton
The data acquired with the algal fluorescence assays are presented in Figs. 2 and 3 for natural phytoplankton. The graph was constructed to depict the change in emitted fluorescence of the algal chlorophyll during a 100 s exposure to ultraviolet light. The initial fluorescence describes the intensity in the first few seconds of exposure to the ultraviolet light. Fluorescence decay (decay) represents the rate at which emitted fluorescence declines due to exposure to the ultraviolet light. Final fluorescence is that value when no further decay is observed. In most cases this is taken to be the fluorescence at 100 s.

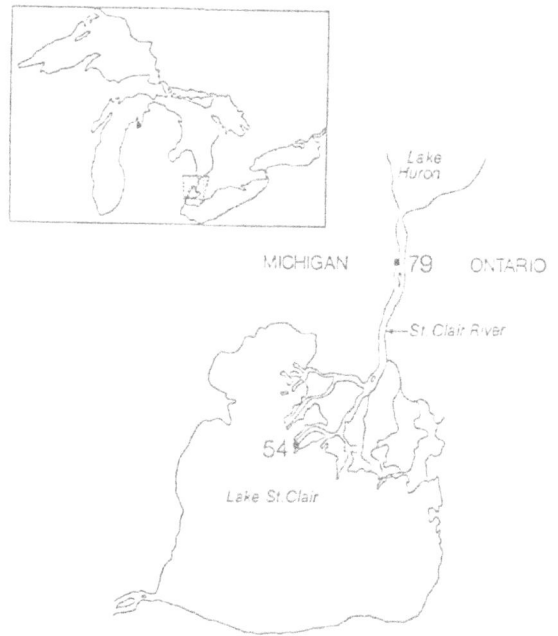

*Fig. 1.* Sampling locations.

320

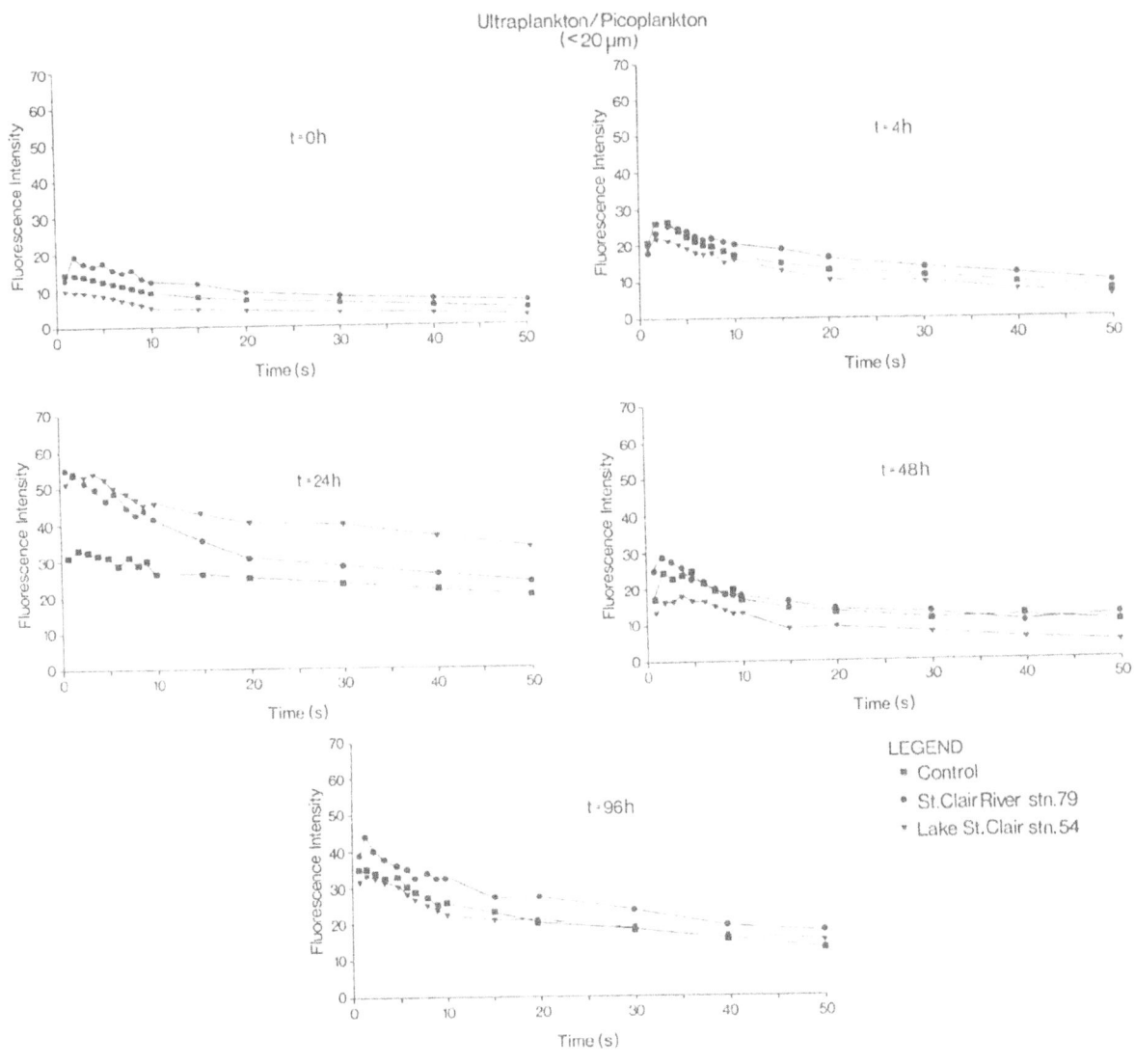

*Fig. 2.* Algal Fluorescence assay with ultraplankton/picoplankton ($< 20 \ \mu$m).

The addition of St. Clair River sediment elutriate (20%) caused slight enhancement of fluorescence intensity for ultraplankton/picoplankton ($< 20 \ \mu$m) at $t = 0$ and $t = 4$ h. On the other hand, addition of Lake St. Clair elutriate was responsible for a slight reduction in fluorescence. At $t = 24$ both the elutriates enhanced fluorescence. There was a decline in the initial fluorescence intensities and rates of decay from 24 to 48 h. Exposure to elutriates for both 48 and 96 h showed results similar to those at $t = 0$ h.

The impact of sediment elutriate on microplankton/netplankton ($> 20 \ \mu$m) is shown in Fig. 3. Initially the fluorescence intensities due to the spiking of both the elutriates (Stn 79 and Stn 54) were higher than the control at $t = 0$ h. Slightly higher fluorescence was observed due to the addition of Lake St. Clair elutriate at $t = 4$ h whereas higher fluorescence for St. Clair River was observed at $t = 24$ h. The fluorescence intensity was consistently low for Lake St. Clair elutriate at $t = 48$ h. Both the elutriates showed lower

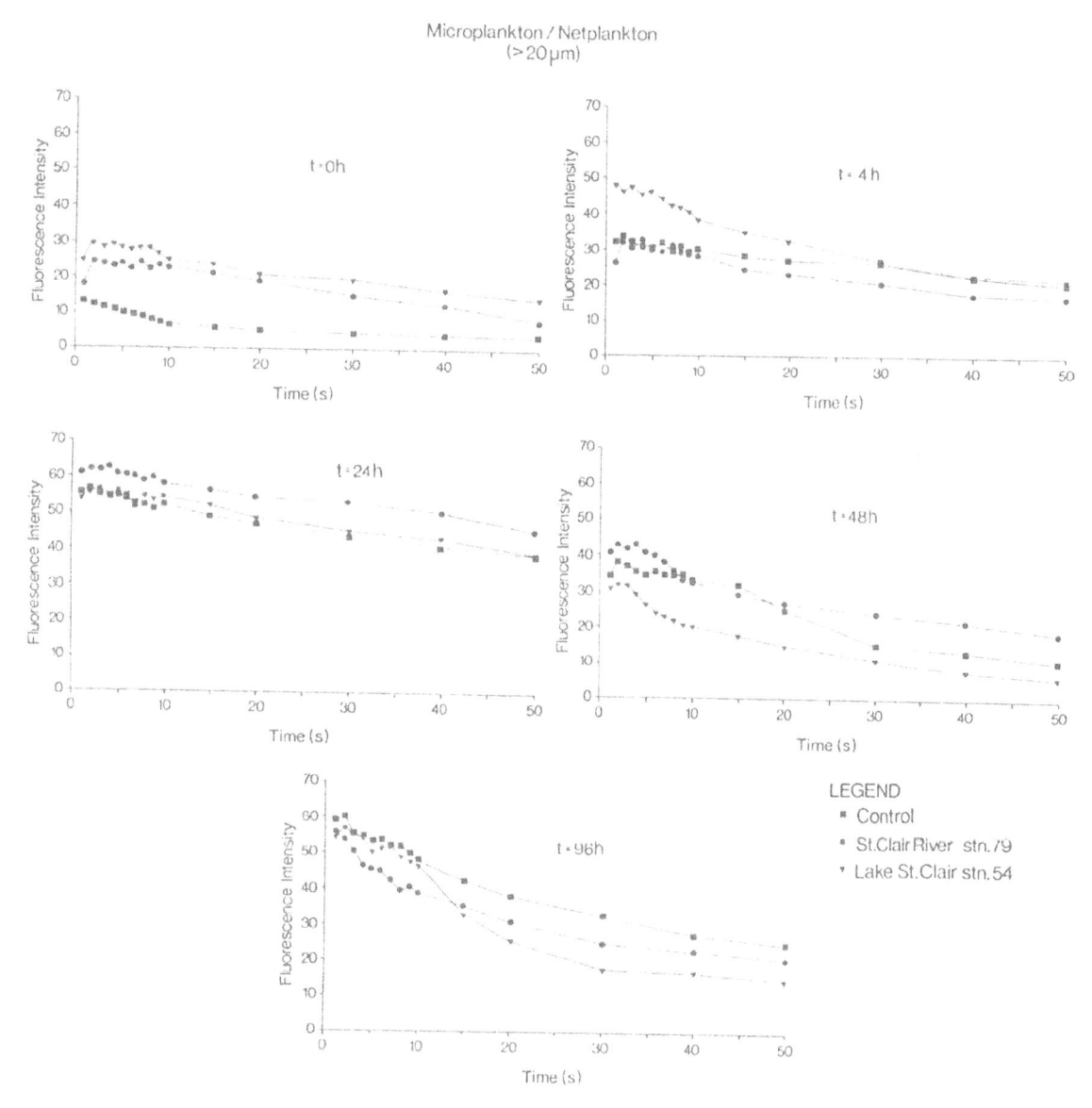

*Fig. 3.* Algal Fluorescence assay with microplankton/netplankton (> 20 μm).

fluorescence intensities than the control at *t* = 96 h.

Fluorescence enhancement is considered to be an indicator of electron blockage. The observed fluorescence enhancement for ultraplankton/picoplankton resulting from the spiking of both the elutriates (Stn 79 and Stn 54) is indicative of stress. Similarly the enhancement shown by the microplankton/netplankton demonstrates elutriate-induced stress at *t* = 0 h by the St. Clair River and Lake St. Clair sediments. On the other hand, inhibition of fluorescence emission shown by the microplankton/netplankton at *t* = 48 h (Lake St. Clair) and at *t* = 96 h (St. Clair River) is suggestive of a reduction in the chloroplast structural integrity of the test species resulting in a higher sensitivity to the ultraviolet light. This is in agreement with the bioassay results with heavy metals by Mayfield & Munawar (1983) and Severn *et al.* (1989).

322

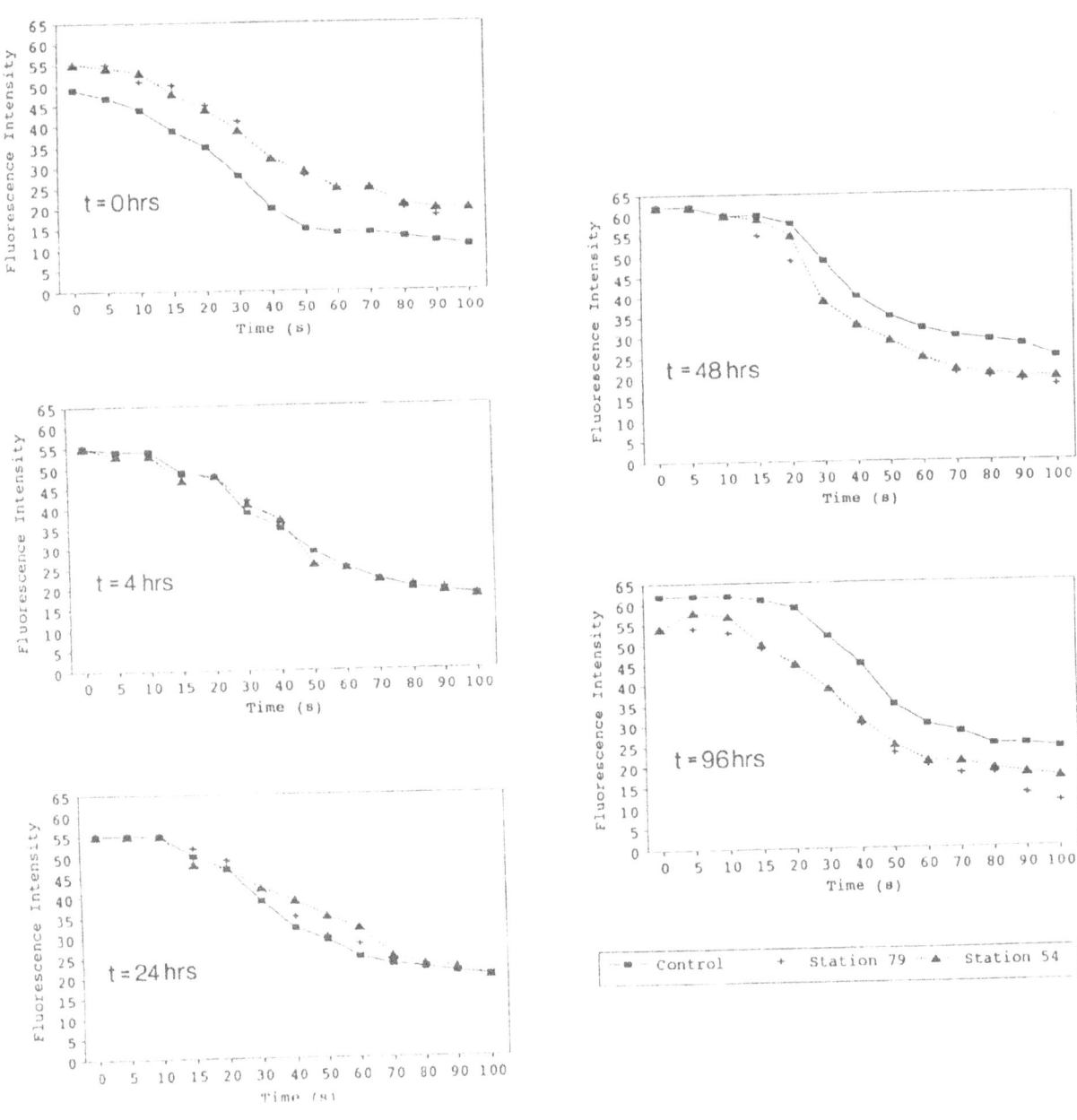

*Fig. 4.* Algal Fluorescence assay with the laboratory grown culture of *Ankistrodesmus braunii* and *Chlorella vulgaris.*

The above experimental observations reveal some interesting and useful information concerning the physiological status of the algal cells. For example, both the size assemblages of natural phytoplankton appear to have been stressed by the addition of sediment elutriates from Station 79 (St. Clair River) and Station 54 (Lake St. Clair). However, the results suggest that the natural phytoplankton adapted physiologically to the presence of sediment-bound contaminants after a period of long exposure. For instance, the ultraplankton/picoplankton seem to have recovered after $t = 48$ h (Fig. 2) for both elutriates. The microplankton/netplankton appear to have responded variably to the addition of both elutriates. However, they adapted to elutriate presence by showing slightly lower fluorescence than the control at $t = 24$ h. Long exposure resulted in damage to the chloroplast integrity as evidenced by the $t = 96$ h experimental fluorescence data of the microplankton/netplankton (Fig. 3)

*Algal culture experiment*

The experimental results with the mixed culture are presented in Fig. 4. Fluorescence enhancement was observed at $t = 0$ h. Within 48 h of exposure to the elutriates, an increase in the rate of fluorescence decay was observed. Generally, compared to the control, the trend from $t = 0$ h to $t = 96$ h was initially higher fluorescence intensities through equivalent results at $t = 4$ and eventual lower fluorescence intensities at $t = 48$ h and, particularly at $t = 96$ h, a marked decrease in fluorescence intensity over the exposure period.

A comparison of the experimental data collected from natural phytoplankton with that of the laboratory grown culture indicates that the mixed pure cultures of algae were more sensitive to the addition of elutriates than the natural phytoplankton. Therefore, conclusions based on laboratory grown cultures should not be extrapolated without regard to the natural environment.

## Conclusions

The microcomputer-based algal fluorescence test was found to be rapid, sensitive, and relatively inexpensive. The test has several advantages over other procedures being used in toxicity testing by means of radioisotopes and enzymes. The test permits visual examination of the impact of contaminants on individual cells/organisms under the microscope which is not possible by other techniques. After the initial and modest capital expense the fluorescence technique is extremely cost effective particularly when the computers are available at extremely low cost. Recent advances in microcomputer technology make the procedure more accurate and precise; greater detail (up to

1024 by 1024 pixels per field) can now be acquired and processed by microcomputers and the increased processing speed allows faster analysis of acquired data. We believe the test could be very useful for fast screening of large numbers of environmental samples.

With the growing increase in environmental problems in both fresh and marine waters, it is imperative that new techniques and approaches be developed for toxicity testing to assist management decisions. In the light of the growing support being given to the multi-trophic assay strategy by the scientific community (Munawar *et al.*, 1989), the microcomputer-based algal fluorescence technique seems to have considerable potential as a rapid, sensitive, and inexpensive procedure for hazard assessment and environmental evaluation.

## Acknowledgements

We would like to thank Iftekhar F. Munawar, Warren Norwood, Lynda McCarthy, Debra Myles, A. Aujla, S. Nielsen, and J. Wotherspoon for their assistance in various aspects of the work and the preparation of this manuscript. We are grateful to W.E. Inniss (University of Waterloo) and H.F. Nicholson for useful suggestions and technical editing.

## References

Ashton, F. M. & A. S. Crafts, 1981. Mode of Action of Herbicide (2nd edition). A Wiley-Interscience Publication, John Wiley and Sons, New York.

Clayton, R. K., 1980. Photosynthesis: Physical mechanisms and chemical patterns. Cambridge University Press, Cambridge.

Forstner, U. & G. T. W. Wittman, 1981. Metal Pollution in the Aquatic Environment. Springer-Verlag, Berlin.

Foyer, C. H., 1984. Photosynthesis. John Wiley and Sons, Inc. New York.

Mayfield, C. I., 1984. A simple computer-based video image analysis system and potential applications to microbiology. J. Microbiol. Meth. 3: 61–67.

Mayfield, C. I. & M. Munawar, 1983. Preliminary study of the effects of contaminants from sediments on algal membranes. J. Great Lakes Res. 9(2): 314–316.

324

Mayfield, C. I. & M. Munawar, 1988. Micro-computer based measurement of algal fluorescence as a potential indicator of environmental contamination. Bull. Envir. Contam. Toxicol. 41: 261–266.

Moody, R. P., P. Weinberger & R. Greenhalgh, 1983. Algal fluorometric determination of the potential phytotoxicity of environmental pollutants. In: J.O. Nriagu, (ed.). Aquatic Toxicology, Advances in Environmental Science and Technology John Wiley & Sons, New York.

Moreland, D. E. & J. L. Hilton, 1976. Actions on photo-systems. In: L. J. A. Audus (ed.), Herbicides, Vol. 1. pp. 427–461. Academic Press, New York.

Munawar, M. & I. F. Munawar, 1987. Phytoplankton bio-assays for evaluating toxicity of in situ sediment con-taminants. In: R. Thomas, R. Evans, A. Hamilton, M. Munawar, T. Reynoldson & H. Sadar (eds), Ecological Effects of in situ Sediment Contaminants. Hydrobiologia 149: 87–105.

Munawar, M., I. F. Munawar, C. I. Mayfield & L. H. McCarthy, 1989. Probing ecosystem health: a multi-disciplinary and multi-trophic assay strategy. In: M. Munawar, G. Dixon, C. I. Mayfield, T. Reynoldson & M. H. Sadar (eds), Environmental Bioassay Techniques and their Application. Hydrobiologia 188/189: 93–116.

Rafii, Z. E., F. M. Ashton & R. K. Glenn, 1979. Metabolic sites of action of fluridone in isolated mesophyll cells. Weed Sci. 27: 422–426.

Richard, E. P. Jr., J. R. Gross, C. J. Arntzen & F. W. Slife, 1983. Determination of herbicide inhibition of electron transport by fluorescence. Weed Sci. 31: 361–367.

Severn, S. T., C. I. Mayfield & W. E. Inniss, 1986. Evaluation of a computer-assisted video data system for toxicological testing. J. Microbiol. Methods 5: 23–31.

Severn, S. R. T., M. Munawar & C. I. Mayfield, 1989. Measurement of sediment toxicity of autotrophic and heterotrophic picoplankton by epifluorescence micro-scopy. Hydrobiologia 176/177: 525–530.

USEPA (U.S. Environmental Protection Agency), 1977. Technical Committee on Criteria for Dredged and Fill Material. Ecological evaluation of proposed discharge of dredged material into ocean waters. Environ. Effects Lab., U.S. Army Corp of Engineers. Waterways Exp. Station, Vicksburg, Miss. 24 pp.

*Hydrobiologia* **219**: 325–332, 1991.
*M. Munawar & T. Edsall (eds),*
*Environmental Assessment and Habitat Evaluation of the Upper Great Lakes Connecting Channels.*
© 1991 *Kluwer Academic Publishers.*

# A method for evaluating the impact of navigationally induced suspended sediments from the Upper Great Lakes Connecting Channels on the primary productivity

M. Munawar, W. P. Norwood & L. H. McCarthy
*Fisheries and Oceans Canada, Great Lakes Laboratory for Fisheries and Aquatic Sciences, Canada Centre for Inland Waters, P.O. Box. 5050, Burlington, Ontario, L7R 4A6, Canada*

*Key words:* navigation, sediments, primary productivity, toxicity, Great Lakes

## Abstract

The impact of navigationally induced suspended sediments from the Upper Great Lakes Connecting Channels on the size-fractionated primary productivity was evaluated by the Carbon-14 technique. The method applied was on-site, rapid, sensitive, inexpensive, and provided dynamic-toxicological information essential for hazard assessment. Both enhancement and inhibition of the primary productivity was observed in various parts of the Upper Great Lakes Connecting Channels. These responses appear to depend on the type of natural plankton and their exposure to various contaminant/nutrient complexes generated by the disturbance of the bottom sediments during the passage of ships. Traditionally, only the inhibition of primary productivity has been monitored from a toxicity point of view, but it is important also to evaluate the implications of enhancement since it may result in increased eutrophication, propagation of nuisance blooms, and change of intricate food-web interactions. The procedure adopted in this study for the first time appears to possess considerable potential for a simple and rapid screening of environmental perturbations resulting from navigational activities.

## Introduction

Routine dredging of navigation channels is an important segment in the maintenance of Great Lakes harbours, since it assures safe access for national and international shipping. This procedure, together with disposal, and navigational activities, causes considerable disturbance of bottom and suspended sediments with which contaminants may be associated. The impact of such a disturbance of sediments on the aquatic biota of the Great Lakes' ecosystem is largely unknown.

Evaluation of in-place pollutants must be based on experimental toxicological testing and biological impact assessment instead of the traditional chemical characterization. The importance of these bioassessment studies has further increased due to the recent identification of 'Great Lakes Areas of Concern' by the International Joint Commission (IJC, 1987), which includes several harbours where contaminated sediments pose a serious threat to the aquatic ecosystem health and its integrity.

During the past five years, *in situ* monitoring techniques have been developed and adapted which allow on-site evaluation of environmental perturbation (Munawar *et al.*, 1986, 1989a; Munawar & Thomas, 1989). Using these techniques, an attempt has been made to evaluate the impact of navigation in the Upper Great Lakes

Connecting Channels (St. Mary's River, St. Clair River, Lake St. Clair, and Detroit River). A similar bioassessment of navigational impact was successfully conducted in Toronto Harbour earlier in our laboratory.

## Methods and materials

### Monitoring and experimental protocol

Sampling of the Upper Great Lakes Connecting Channels was conducted during the late spring and summer of 1987. The experimental test sites (Fig. 1) included Lake Munuscong on St. Mary's River, mid-way down the St. Clair River outside the Lambton generating station, the delta of the St. Clair River as it flowed into Lake St. Clair (St. Clair Flats), mid-way down Detroit River (Fighting Island), and at the mouth of the Detroit River including western Lake Erie. Primary productivity has been extensively used as an indicator of environmental stress, in several *in situ* studies of

the Great Lakes (Munawar *et al.*, 1986, 1989a; Munawar & Thomas, 1989). However, in the present study, this parameter has been used for the first time for evaluating the impact of navigation. Just prior to the passage of a tanker or other large ship, a water sample was collected for size-fractionated primary productivity experiments, which has been used in this investigation as a physiological indicator of biological (phytoplankton) activity. Immediately after the passage of the vessel from the test site, another water sample was collected. The frequency of vessel traffic through the test sites determined the time of post-navigation sampling. (i.e.: 30 minutes, 60 minutes after the passage of the vessel).

### Size-fractionated primary productivity experiments

Size-fractionated primary productivity of natural phytoplankton was determined for samples collected before and after passage of ships using the well-known Carbon-14 uptake technique of Vollenweider *et al.*, (1974) and Munawar *et al.* (1989b). A portion of thoroughly mixed water sample containing natural phytoplankton was inoculated with $NaH^{14}CO_3$ and incubated for four hours under constant light intensity on a shaker in a temperature controlled incubator. After incubation, the entire contents were size-fractionated through a 20 $\mu$m Nitex screen for the determination of microplankton/netplankton productivity ($>20\,\mu$m). The portion of the sample that passed through the 20 $\mu$m Nitex screen was filtered directly onto a 0.45 $\mu$m Millipore membrane for the determination of ultraplankton/picoplankton ($<20\,\mu$m) productivity. The membrane filters were then kept in a Phase Combining System for liquid scintillation counting (see Munawar *et al.*, 1989b for further details). The primary productivity is expressed as a rate (mg C m$^{-3}$ h$^{-1}$).

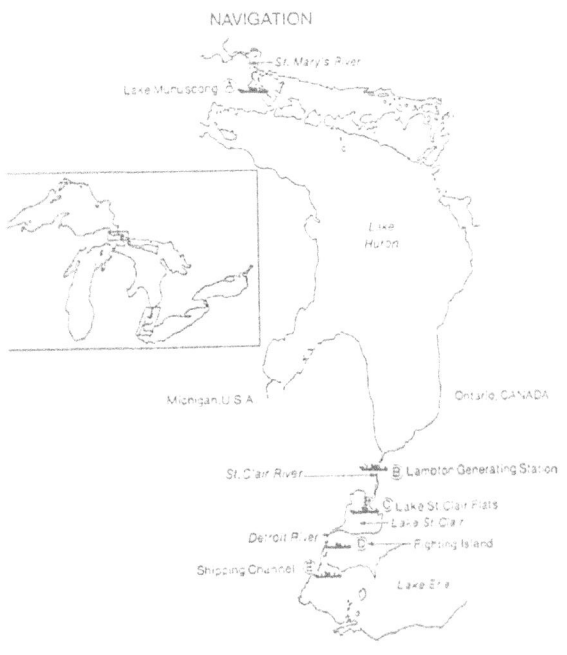

*Fig. 1.* Sampling sites in the Upper Great Lakes Connecting Channels.

## Results – Chemical

A cursory inspection of the chemical results (Tables 1 to 5) from all sites tested indicate some chemical changes resulting in either an increase or decrease of various variables between pre-navigation and post-navigation episodes. Also, none of the chemical variables appear to significantly violate the guidelines (Table 1) of the Great Lakes Water Quality Agreement (IJC, 1988). Consequently, based only on these observed chemical variables it is difficult to evaluate the impact of navigation on the ecosystem. It is apparent then that some bioassessment approach is essential for this purpose.

## Results – Biological

### Lake Munuscong (St. Mary's River)

The monitoring of primary productivity in Lake Munuscong (St. Mary's River) immediately after tanker passage revealed significant increase ($P < 0.01$) in microplankton/netplankton productivity by 33 percent (Fig. 2). Similarly, the ultraplankton/picoplankton also experienced a 37 percent enhancement ($P < 0.01$) of productivity.

### Lake St. Clair Flats

At Lake St. Clair Flats only significant effects on the productivity ($P < 0.05$) were shown by the

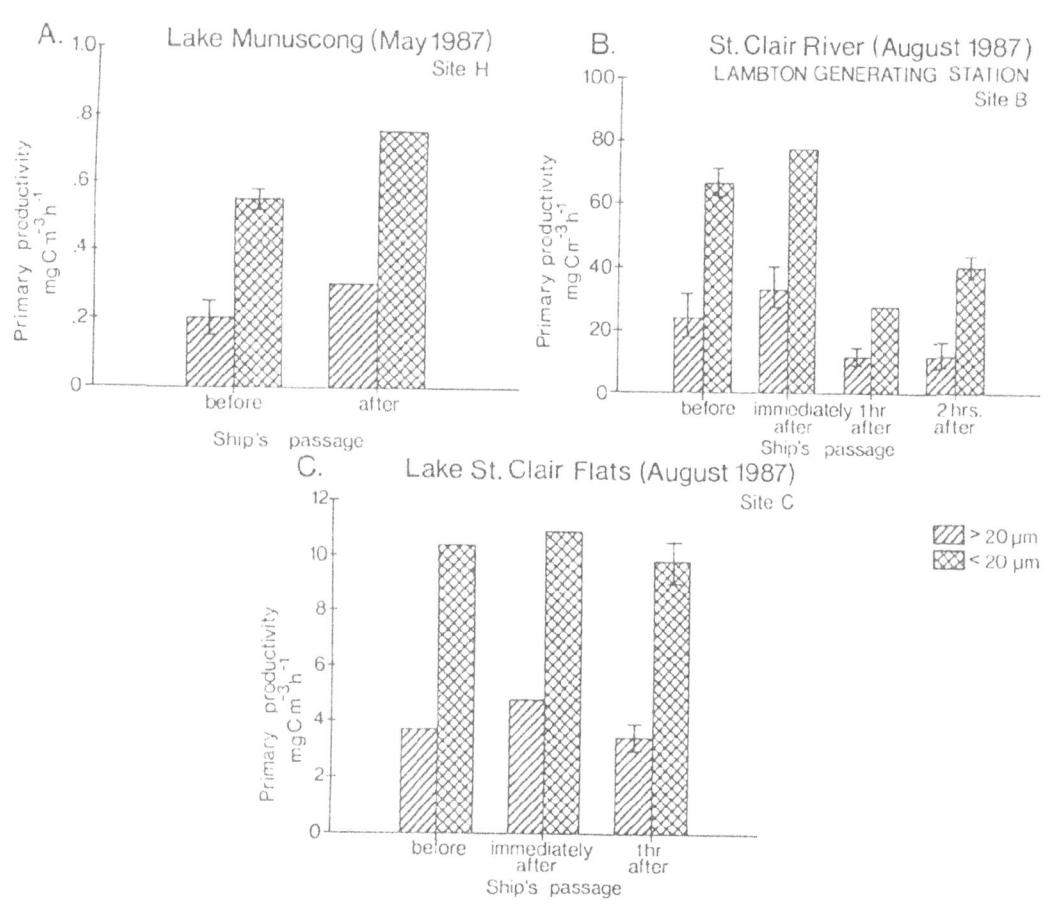

*Fig. 2.* Size-fractionated primary productivity in Lake Munuscong (A); St. Clair River (B); and Lake St. Clair Flats (C).

microplankton/netplankton assemblage which indicated an increase of 21 percent ($P < 0.05$) immediately after ship's passage compared to pre-navigation (Fig. 2).

*Lambton Generating Station (St. Clair River)*

At the Lambton Generating Station (St. Clair River) neither microplankton/netplankton nor ultraplankton/picoplankton demonstrated any significant change immediately after navigation. However, statistically significant inhibition ($P < 0.1$) was measured one hour after the passage of the ship in both size assemblages, showing 60 percent and 53 percent reductions of productivity respectively. Significant reduction of productivity ($P < 0.05$) was still detectable two hours after navigation in both the size assemblages, with declines of 39 percent compared to pre-navigation levels.

*Detroit River (Fighting Island)*

Navigational activity in the Detroit River (Fighting Island) produced no significant change in microplankton/netplankton productivity in samples collected immediately after ship's passage (Fig. 3). The ultraplankton/picoplankton

productivity, on the other hand, was significantly inhibited ($P < 0.01$) by 19 percent. Subsequent sampling after both one and two hours of ship's passage revealed that there was a recovery of the ultraplankton/picoplankton productivity compared to pre-navigation.

*Shipping Channel (Western Lake Erie)*

The impact of navigation in the shipping channel (Western Lake Erie) presented some interesting results. No significant differences were measured in samples collected both immediately after and one hour after ship's perturbation (Fig. 3). However, the microplankton/netplankton productivity showed a slight, yet significant inhibition ($P < 0.05$) in samples collected approximately two hours after navigation. The ultraplankton/picoplankton productivity rate showed no significant changes in the samples collected during the episode.

**Discussion**

As indicated earlier the chemistry results showed some fluctuations in their concentration when pre-navigation and post-navigation episodes were compared (Tables 1 to 5). However, such expen-

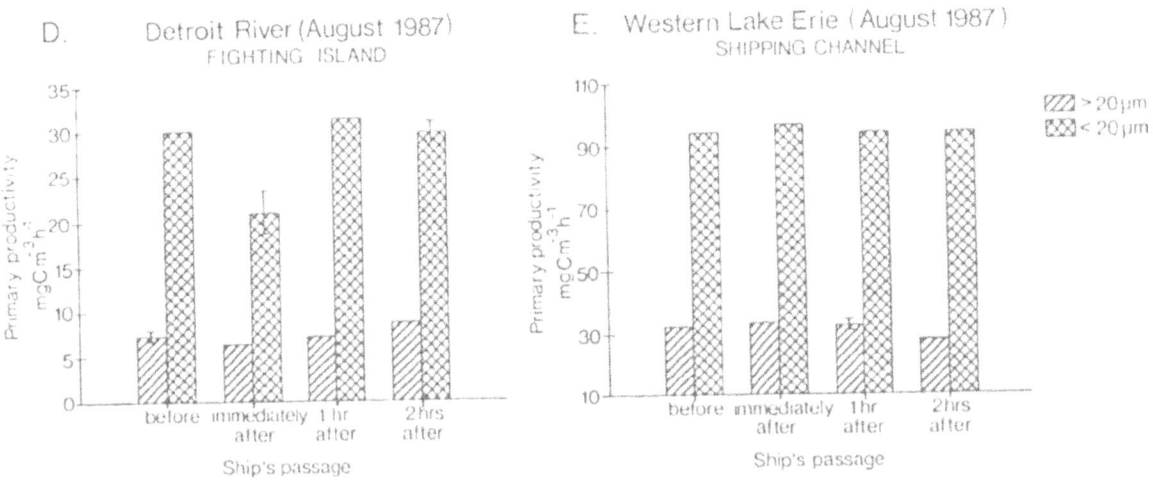

*Fig. 3.* Size-fractionated primary productivity in Detroit River (D) and Western Lake Erie (E).

*Table 1.* Results of chemical analysis of water during pre-navigation and post-navigation episodes in Lake Munuscong (A: St. Mary's River).

| | pre-navigation | post-navigation* |
|---|---|---|
| Organic ($\mu$g l$^{-1}$) | | |
| A-BHC | 0.0054 | 0.0045 |
| G-BHC | 0.0010 | 0.0009 |
| Dieldrin | 0.0004 | 0.0006 |
| Total PCB | 0.0772 | 0.2924 |
| Inorganic (mg l$^{-1}$) | | |
| DOC | 1.1 | 1.1 |
| DIC | 9.6 | 9.8 |
| NO$_3$NO$_2$ | 0.364 | 0.369 |
| NH$_3$ | <0.005 | <0.005 |
| SRP | 0.0012 | 0.0011 |
| Cd | 0.0005 | <0.0001 |
| Co | <0.0005 | <0.0005 |
| Cu | 0.0010 | 0.0011 |
| Fe | 0.0753 | 0.0772 |
| Mn | 0.004 | 0.003 |
| Ni | <0.0005 | <0.0005 |
| Pb | 0.0013 | 0.0012 |
| Zn | 0.0013 | 0.0009 |

* Post-navigation sampling after ship's passage.
Ontario Ministry of Environment Open Water Dredge Disposal Guidelines (concentrations in mg/kg dry weight): Arsenic 8, Cadmium 1, Chromium 25, Copper 25, Cyanide 0.1, Iron 10,000, Lead 50, Mercury 0.3, Nickel 25, Oil and grease 1500, Phosphorus 1000, Polychlorinated biphenyls 0.05, total Kjeldahl nitrogen 2000, volatile solids 60,000, Zinc 100.

sive data were not useful for a realistic evaluation of the impact of navigation on the various ecosystems investigated. Therefore, the utility of a biological assessment strategy as adopted in the present study (by applying the size-fractionated primary productivity as a sensitive indicator of perturbation) is apparent. For example, in Lake Munuscong (St. Mary's River) and St. Clair Flats, the navigationally induced suspended sediments appear (as indicated by the turbid nature of the samples collected) to enhance the productivity of both size assemblages of phytoplankton.

At Lambton Generating Station (St. Clair River) significant inhibition of productivity was recorded after one and two hours of navigational

impact ($P$, 0.5 to 0.01). Similarly the results of the experiment in Detroit River (Fighting Island) exhibit immediate inhibition of ultraplankton/picoplankton but recovery to pre-navigation levels as shown by the hourly and 2-hourly rates. (Fig. 3).

On the other hand, in the shipping channel of Western Lake Erie, a combination of no response (immediately and an hour after ship's passage) to significant inhibition of productivity ($P$, 0.05) of microplankton/netplankton was observed after two hours of perturbation (Fig. 3).

These experimental results indicate firstly that the primary productivity varied greatly from station to station in various components of the Upper Great Lakes Connecting Channels. Secondly, that stations respond differently to navigational perturbation, which is not surprising since each of the sites harboured different types of plankton assemblages and exposure to various

*Table 2.* Results of chemical analysis of water during pre-navigation and post-navigation episodes at Lambton Generating Station (B: St. Clair River).

| | pre-navigation | immediately* | after 1 hour* | after 2 hours* |
|---|---|---|---|---|
| Organic ($\mu$g l$^{-1}$) | | | | |
| A-BHC | 0.0026 | 0.0029 | 0.0025 | 0.0024 |
| G-BHC | 0.0005 | 0.0006 | 0.0005 | 0.0006 |
| Dieldrin | 0.0006 | 0.0006 | 0.0004 | 0.0005 |
| Total PCB | 0.0451 | 0.0578 | 0.0413 | 0.0436 |
| Inorganic (mg l$^{-1}$) | | | | |
| DOC | 1.5 | 1.2 | 1.3 | 1.3 |
| DIC | 19.1 | 19.1 | 18.9 | 19.4 |
| NO$_3$NO$_2$ | 0.458 | 0.258 | 0.48 | 1.288 |
| NH$_3$ | 0.088 | 0.03 | 0.058 | 0.233 |
| SRP | 0.0021 | 0.001 | 0.0022 | 0.001 |
| TP | 0.0105 | 0.0117 | 0.0136 | 0.0158 |
| Alk | 84.9 | 82.3 | 81.5 | 83.9 |
| SiO$_2$ | 0.93 | 0.99 | 1.05 | 1.08 |
| Cd | 0.0003 | 0.0001 | <0.0001 | <0.0001 |
| Co | <0.0005 | <0.0005 | <0.0005 | 0.0006 |
| Cu | 0.0018 | 0.0005 | 0.0007 | 0.0013 |
| Fe | 0.104 | 0.104 | 0.143 | 0.150 |
| Mn | 0.013 | 0.01 | 0.006 | 0.01 |
| Ni | 0.0007 | 0.0008 | 0.0009 | 0.0008 |
| Pb | <0.0007 | <0.0007 | 0.0008 | <0.0007 |
| Zn | 0.0045 | 0.001 | 0.0016 | 0.0029 |

* Post-navigation sampling after ship's passage.

*Table 3.* Results of chemical analysis of water during pre-navigation and post-navigation episodes at Lake St. Clair Flats (F).

| | pre-navigation | after 1 hour* | after 2 hours* |
|---|---|---|---|
| **Organic ($\mu g\,l^{-1}$)** | | | |
| A-BHC | 0.0024 | 0.0024 | 0.0022 |
| G-BHC | 0.0007 | 0.0006 | 0.0006 |
| p, p'-DDE | 0.0005 | 0.0004 | |
| Dieldrin | 0.0007 | 0.0004 | 0.0005 |
| Total PCB | 0.0887 | 0.0638 | 0.0796 |
| **Inorganic ($mg\,l^{-1}$)** | | | |
| DOC | 5.0 | 2.4 | 1.2 |
| DIC | 18.9 | 22.8 | 18.8 |
| $NO_3NO_2$ | 0.14 | 0.575 | 0.276 |
| $NH_3$ | <0.005 | <0.091 | 0.058 |
| SRP | 0.0014 | 0.021 | 0.0019 |
| TP | 0.0088 | 0.0164 | 0.0130 |
| Alk | 79.4 | 84.0 | 82.0 |
| $SiO_2$ | 1.08 | 1.08 | 1.13 |
| Cd | <0.0001 | 0.0001 | <0.0001 |
| Co | 0.0005 | <0.0005 | <0.0005 |
| Cu | 0.002 | 0.0007 | 0.0009 |
| Fe | 0.118 | 0.194 | 0.162 |
| Mn | 0.003 | 0.005 | 0.005 |
| Ni | 0.0005 | 0.0019 | 0.0006 |
| Pb | <0.0007 | <0.0007 | <0.0007 |
| Zn | 0.001 | 0.0011 | 0.0026 |

* Post-navigation sampling after ship's passage.

types of contaminant/nutrient complexes. For example, the phytoplankton assemblage in Lake Munuscong originated from oligotrophic Lake Superior, and that of St. Clair Flats from oligotrophic Lake Huron, both of which are phosphorus limited. Therefore, the enhancement of their productivity rate by exposure to the suspended sediments resulting from the passage of ships is not surprising. The observed inhibition of primary productivity in various experiments in the St. Clair and Detroit Rivers is an extremely complex puzzle resulting from intricate nutrient/contaminant interactions, and an evaluation is beyond the scope of this paper.

As observed in this study, the net result of navigationally induced stress is either the enhancement or inhibition of primary produc-

tivity. If evaluated carefully and systematically, the inhibition is usually considered as an indicator of environmental stress. On the other hand, the enhancement is traditionally ignored. This may be misleading since we should be concerned with both types of community responses, since the enhancement response may result either in the growth of certain size assemblages or species which may not be desirable from the eutrophication point of view, or which might change the food web dynamics of an ecosystem (Rhee, 1982; Munawar *et al.*, 1983).

## Conclusions

Our results demonstrate the effectiveness and simplicity of physiological tests adopted such as size-fractionated productivity, for navigational impact assessment. The test applied in this investigation is on-site, rapid, inexpensive, and pro-

*Table 4.* Results of chemical analysis of water during pre-navigation and post-navigation episodes at Fighting Island (D: Detroit River)

| | pre-navigation | immediately* | after 1 hour* | after 2 hours* |
|---|---|---|---|---|
| **Organic ($\mu g\,l^{-1}$)** | | | | |
| A-BHC | 0.0021 | 0.0019 | 0.0019 | 0.0019 |
| G-BHC | 0.0005 | 0.0006 | 0.0005 | 0.0018 |
| Dieldrin | 0.0005 | | | |
| Total PCB | 0.0678 | 0.0706 | 0.0543 | 0.0717 |
| **Inorganic ($mg\,l^{-1}$)** | | | | |
| DOC | 1.3 | 1.4 | 1.2 | 1.3 |
| DIC | 19.4 | 19.0 | 18.7 | 19.0 |
| $NO_3NO_2$ | 0.237 | 0.433 | 0.367 | 0.459 |
| $NH_3$ | 0.038 | 0.076 | 0.047 | 0.064 |
| SRP | 0.0025 | 0.0061 | 0.0031 | 0.0084 |
| TP | 0.0156 | 0.0142 | 0.0127 | 0.0182 |
| Alk | 83.4 | 83.4 | 82.5 | 83.4 |
| $SiO_2$ | 1.14 | 1.13 | 1.11 | 1.04 |
| Cd | <0.0001 | <0.0001 | <0.0001 | 0.0001 |
| Co | <0.0005 | <0.0005 | <0.0005 | <0.0005 |
| Cu | 0.0009 | 0.0006 | 0.0007 | 0.0009 |
| Fe | 0.162 | 0.152 | 0.178 | 0.208 |
| Mn | 0.017 | 0.008 | 0.011 | 0.017 |
| Ni | 0.0011 | 0.0012 | 0.0012 | 0.0013 |
| Pb | 0.001 | <0.0007 | <0.0007 | <0.0007 |
| Zn | 0.0017 | 0.0013 | 0.0016 | 0.0024 |

* Post-navigation sampling after ship's passage.

*Table 5.* Results of chemical analysis of water during pre-navigation and post-navigation episodes at the mouth of Western Lake Erie (E).

| | pre-navigation | immediately* | after 1 hour* | after 2 hours* |
|---|---|---|---|---|
| **Organic ($\mu$g l$^{-1}$)** | | | | |
| A-BHC | | 0.0019 | 0.0017 | 0.0019 |
| G-BHC | | 0.0005 | 0.0004 | 0.0005 |
| Total PCB | | 0.0806 | 0.0517 | 0.0562 |
| **Inorganic (mg l$^{-1}$)** | | | | |
| DOC | 1.3 | 1.3 | 1.2 | 1.2 |
| DIC | 18.5 | 19.0 | 18.9 | 18.9 |
| NO$_3$NO$_2$ | 0.93 | 0.328 | 0.411 | 0.509 |
| NH$_3$ | 0.056 | 0.021 | 0.043 | 0.058 |
| SRP | 0.0041 | 0.0028 | 0.003 | 0.0049 |
| TP | 0.0171 | 0.0158 | 0.0124 | 0.0125 |
| Alk | 80.8 | 83.4 | 82.5 | 84.4 |
| SiO$_2$ | 0.99 | 1.06 | 1.11 | 1.09 |
| Cd | 0.0001 | <0.0001 | <0.0001 | 0.0001 |
| Co | <0.0005 | <0.0005 | <0.0005 | <0.0005 |
| Cu | 0.0021 | 0.0006 | 0.0022 | 0.0006 |
| Fe | 0.171 | 0.178 | 0.217 | 0.178 |
| Mn | 0.003 | 0.005 | 0.006 | 0.007 |
| Ni | 0.0009 | 0.0014 | 0.0015 | <0.0005 |
| Pb | 0.001 | <0.0007 | 0.0008 | <0.0007 |
| Zn | 0.0071 | 0.0034 | 0.0101 | 0.0026 |

* Post-navigation sampling after ship's passage.

vides much more dynamic-toxicological information based on physiological response compared to routine chemistry results. Furthermore, these on-site experiments provide an overall picture of the community's response to nutrient/contaminant interactions/complexes (Munawar *et al.*, 1986). A similar approach was successfully adopted for the *in situ* bioassessment of dredging/disposal of contaminated sediments in Toronto Harbour (Munawar & Thomas, 1989; Munawar *et al.*, 1989a). Since routine dredging, disposal, and navigational activities are essential and expensive for the maintenance of Great Lakes harbours, the on-site techniques utilized in our investigation appear to have considerable potential for a rapid and sensitive assessment of navigational and other environmental perturbations, and may be useful for an efficient bioassessment of contaminants and better management of the Great Lakes ecosystem.

## Acknowledgements

We are grateful to Drs. R.L. Thomas (University of Windsor), C.I. Mayfield (University of Waterloo), and Dr. G.G. Leppard (National Water Research Institute) for their advice, suggestions and comments. We would like to thank D. Myles, P. Desroches, H.F. Nicholson, I. Orchard, and Y. Hamdy for their assistance. We greatly appreciate the support and encouragement of the Sediment Group of the Upper Great Lakes Connecting Channels; and the Environmental Protection Service, Ontario Region, in the development and implementation of this project. The assistance of the Captain and crew of C.S.S. Limnos is greatly appreciated for the field work.

## References

IJC (International Joint Commission), 1987. Guidance on characterization of toxic substrates problems in Areas of Concern in the Great Lakes. Windsor, Ontario, 179 pp.

IJC, 1988. Revised Great Lakes Water Quality Agreement of 1978 as amended by Protocol signed November 18, 1987. Agreement with annexes and terms of reference, between the United States of America and Canada signed at Ottawa November 22, 1978 and Phosphorus Load Reduction Supplement signed October 7, 1983. Windsor, Ontario.

Munawar, M. & R. L. Thomas, 1989. Sediment toxicity testing in two areas of concern of the Laurentian Great Lakes: Toronto (Ontario) and Toledo (Ohio) harbours. In: P. G. Sly & B. T. Hart (eds), Sediment/Water Interaction. Hydrobiologia 176/177: 397–409.

Munawar, M., A. Mudroch, I. F. Munawar & R. L. Thomas, 1983. The impact of sediment-associated contaminants from the Niagara River mouth in various size assemblages of phytoplankton. J. Great Lakes Res. 9(2): 303–313.

Munawar, M., R. L. Thomas, W. P. Norwood & S. A. Daniels, 1986. Sediment toxicity and production-biomass relationships of size-fractionated phytoplankton during on-site simulated dredging experiments in a contaminated pond. In: P. G. Sly (ed), Sediment and Water Interactions, pp. 407–426 Springer Verlag New York Inc.

Munawar, M., W. P. Norwood, L. H. McCarthy & C. I. Mayfield, 1989a. *In situ* bioassessment of dredging and disposal activities in a contaminated ecosystem: Toronto Harbour. In: M. Munawar, G. Dixon, C. Mayfield, T. Reynoldson & M. Sadar, Environmental Bioassay Techniques and their Application. Hydrobiologia. 188/189: 601–618.

332

Munawar, M., I. F. Munawar, C. I. Mayfield & L. H. McCarthy, 1989b. Probing ecosystem health: a multi-disciplinary and multi-trophic assay strategy. In: M. Munawar, G. Dixon, C. Mayfield, T. Reynoldson & M. Sadar (eds), Environmental Bioassay Techniques and their Application. Hydrobiologia. 188/189: 93–116.

Rhee, G.-Yull, 1982. Overview of phytoplankton contaminant problems. J. Great Lakes Res. 8: 326–327.

Vollenweider, R. A., M. Munawar & P. Stadelmann, 1974. A comparative review of phytoplankton and primary productivity in the Laurentian Great Lakes. J. Fish. Res. Bd Can. 31(5): 739–762.

*Hydrobiologia* **219**: 333–344, 1991.
*M. Munawar & T. Edsall (eds),*
*Environmental Assessment and Habitat Evaluation of the Upper Great Lakes Connecting Channels.*
© *1991 Kluwer Academic Publishers.*

# Heavy metals in aquatic macrophytes drifting in a large river[1]

B. A. Manny, S. J. Nichols & D. W. Schloesser
*National Fisheries Research Center-Great Lakes, U.S. Fish and Wildlife Service, Ann Arbor, MI 48105,
USA*

*Key words:* vascular plants, cadmium, nickel, copper, zinc, lead, Detroit River, Lake Erie

## Abstract

Macrophytes drifting throughout the water column in the Detroit River were collected monthly from May to October 1985 to estimate the quantities of heavy metals being transported to Lake Erie by the plants. Most macrophytes (80–92% by weight) drifted at the water surface. Live submersed macrophytes made up the bulk of each sample. The most widely distributed submersed macrophyte in the river, American wildcelery (*Vallisneria americana*), occurred most frequently in the drift. A total of 151 tonnes (ash-free dry weight) of macrophytes drifted out of the Detroit River from May to October. The drift was greatest (37 tonnes) in May. Concentrations of heavy metals were significantly higher in macrophytes drifting in the river than in those growing elsewhere in unpolluted waters. Annually, a maximum of 2 796 kg (eight heavy metals combined) were transported into Lake Erie by drifting macrophytes. The enrichment of all metals was remarkably high (range: $4 000 \times$ to $161 000 \times$) in macrophytes, relative to their concentration in water of the Detroit River. Detroit River macrophytes are thus a source of contaminated food for animals in the river and in Lake Erie.

## 1. Introduction

Because aquatic macrophytes contain contaminants it has been suggested that they be used to monitor contaminant loadings to rivers (Abo-Rady, 1980; Harding & Whitton, 1981; Say & Whitton, 1983; Edwards *et al.*, 1989). As river macrophytes die, they usually break loose and drift downstream (Haslam, 1978). Inasmuch as the movement of contaminants in aquatic macrophytes in large rivers has not been investigated, we undertook the present study because the Detroit River is contaminated with heavy metals (Manny

& Kenega, 1991) and adds large amounts of nutrients and contaminants to Lake Erie (Beeton, 1961; Mudroch, 1985; Manny *et al.*, 1988). Macrophytes are abundant and widely distributed throughout the littoral areas of the river (Schloesser & Manny, 1986; Hudson *et al.*, 1986). Drifting macrophytes have been noted in the Detroit River and in other channels connecting the Laurentian Great Lakes (Hunt, 1962; Michigan Water Resources Commission, 1967; Bryant & Nuhfer, 1984), but their contribution to the movement of contaminants has not been investigated. All previous studies of drift in large rivers, including the Great Lakes channels, have either ignored drifting macrophytes (Galtsoff, 1924; Wefring & Hopwood, 1981; Pocklington & Tan, 1987; Admiraal & Van Zanten, 1988; Lau

---

[1] Contribution 734 of the National Fisheries Research Center-Great Lakes, U.S. Fish and Wildlife Service, 1451 Green Road, Ann Arbor, MI 48105.

et al., 1989) or presented only qualitative information about them, not including their contaminant content (Poe & Edsall, 1982; Jude *et al.*, 1986; Edwards *et al.*, 1989). Because we suspected that aquatic plants contained contaminants and knew they were the major energy source for secondary production in this river (Edwards *et al.*, 1989) we investigated the heavy metal content of macrophytes drifting down the Detroit River into Lake Erie. We emphasized submersed macrophytes i.e., those adapted for growth beneath the water (Sculthorpe, 1967) because shoots of such plants predominated in the samples and they accumulate higher concentrations of metals than emergent macrophytes (Mudroch & Capobianco, 1979; Welsh & Denny, 1980; Mudroch, 1981; Denny, 1987).

## 2. Methods

Plant drift was sampled monthly from a boat at three cross-river channel transects in the Detroit River, from May to October 1985 (Fig. 1). Transect I was about 8 km downstream from Lake St. Clair directly opposite the city of Detroit, Michigan, at 83° 02′ 25″. Transects II and III were about 25 km downstream from transect I on either side of Grosse Ile at 42° 08′ 30″. We established three stations along each transect at points of maximum channel depth: station 1 was adjacent to and north or west of the shipping channel, station 2 was in the middle of the shipping channel, and station 3 was south or east of the shipping channel (Fig. 1). Vessel position on station was determined by reference to navigation buoys and objects on shore. The length of each transect was determined by reference to navigation charts (National Oceanographic and Atmospheric Administration, 1979). On June 25, 1985, using an echo sounder, we determined water depth at 30 evenly spaced intervals along transect I, 16 along II, and 46 along III. These measurements were used to prepare Fig. 1. Water depths at stations 1, 2, and 3 were, respectively, 8.2, 13.7, and 13.7 m along transect I, 5.2, 8.2, and 6.1 m along transect II, and 6.7, 11.0, and 7.0 m along transect III.

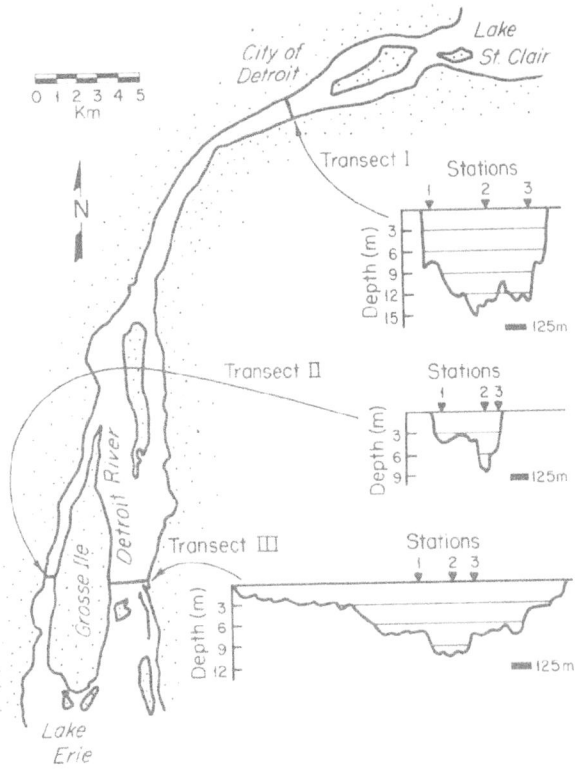

Fig. 1. Transects and sampling stations where drifting plants were collected in the Detroit River in 1985.

Drifting macrophytes were collected with a 50 cm, cylinder-on-cone plankton tow net made of 3 mm mesh that was towed from a derrick attached to a boat that moved at constant speed against the current (Schloesser *et al.*, 1989). Flowmeters were mounted inside and outside the net to monitor net filtration efficiency. There was no closing device on the net, but we believe there was little contamination from other than intentionally sampled depths because at the end of each tow we allowed the boat to drift with the current, causing the net to collapse during retrieval.

We made a total of 126 tows during the study. Plants drifting at the surface were collected in a 10 min surface tow made in a transverse-upstream direction across the entire transect; about one-third of the net diameter was above the water surface. In rough water, we towed the net along the lee side of the boat to prevent it from pulling out of the water or being entirely submersed. These surface samples were used to estimate the amounts of plants drifting near the water surface (0–0.34 m) at each transect. Plants drifting below the surface were sampled at 3-m depth intervals. At each station, we made one to four 10-min tows of the net at the following depth intervals (meters): 0.34 to 3, 3 to 6, 6 to 9, and (transect I only) 9 to 12. We did not sample the depth interval of 12 to 15 m at transect I, 6 to 9 m at transect II, or 9 to 12 m at transect III because large debris on the bottom snagged our tow net the first time we tried to sample at these depths.

Within each depth interval, the net was operated continuously for 2.5 min in a stepwise manner. For example, within the first interval, the net was towed for 2.5 min at each of four depths: 0.5, 1.2, 2.1, and 3.0 m. About 160 m³ of water passed through the net during each tow. After each tow, we pulled the net into the boat, discarded vegetation hanging outside the net, removed all macrophyte material from the net, and rinsed it over a 2 mm screen with river water. The filtering efficiency of our tow net dropped below 87 percent during only 4 of 126 tows, when vegetation became entangled in the propellers of the flowmeter. Macrophytes collected in each tow were wrapped in aluminum foil, placed on ice, taken to the laboratory, and refrigerated (4 °C) in the dark; they remained fresh for at least seven days.

In the laboratory, we separated living submersed macrophytes from detritus and recorded their taxonomic composition. Materials contributed by emergent and terrestrial plants were treated as detritus. Wet weights of plants and detritus were determined to the nearest 0.01 g.

The metals content (cadmium, cobalt, copper, chromium, lead, mercury, nickel, and zinc) of six samples of live, drifting macrophytes and a control sample of National Bureau of Standards spinach leaves were determined by digestion in strong acid, followed by analysis by inductively coupled plasma, mass spectroscopy instrumentation. Mercury in macrophytes was analyzed by atomic adsorption after tin-chloride reduction by a cold vapor-trapping procedure.

To convert wet weights to ash-free dry weights (AFDW), we used a factor of 0.05 derived for submersed macrophytes in the St. Clair-Detroit River system from the data of Hudson *et al.* (1986). To calculate total water discharge at each transect in the Detroit River, we used a computer model that integrated daily water velocity measurements at several points in the river (Quinn & Hagman, 1977). At each transect, we divided the river cross-section into a surface depth interval, 0 to 0.34 m, and the 3 m depth intervals previously mentioned (see Fig. 1). The proportion of the total river cross-section represented by each depth interval at each transect was determined from Fig. 1 by electronic planimetry (Brown & Manny, 1985). The water discharge through each depth interval at each transect was calculated as the product of total water discharge through the transect and the proportion of cross-sectional area represented by each depth interval; uniform linear variation in water discharge with depth was assumed between stations. This assumption is probably valid because the river is well mixed at each of these transects by turbulence. Having measured the concentration of drifting plants in each depth interval on each sampling date at each sampling station, and knowing the water discharge through each depth interval, we obtained the volume-weighted transport (drift) of macrophytes through each depth interval. The sum of those values equalled the total macrophyte transport through each transect on the monthly sampling day. The biomass of drifting plants, in tonnes per month (AFDW), was calculated as the product of the total average volume-weighted macrophyte biomass on the day of measurement for that month at transects II and III combined, and the river discharge during that month.

We calculated the mass of each heavy metal moved out of the Detroit River per year by plants as twice the product of the average concentration of that metal in the six samples of live plant drift and the tonnes of live plant biomass at transects II and III combined, during the half-year period of measurement from May to October. Our estimate of each heavy metal transported out of the

Detroit River into Lake Erie by all submersed macrophytes produced in the Detroit River was calculated as the product of the mean concentration of that metal in the six live plant samples and the estimated total net production of submersed macrophytes in the Detroit River (12 380 tonnes; Edwards *et al.*, 1989). Statistical significance of results was evaluated using *t*-tests at $P \leq 0.05$.

## 3. Results and discussion

### 3.1. Composition of the drift

Throughout the study, 81 percent of the drift consisted of live submersed macrophytes, most of which floated at the surface; the rest consisted of detritus, most of which drifted in the upper 3 m of the water column (Table 1). In May, most plants

Table 1. Mean monthly biomass (in grams wet weight 1000 m$^{-3}$ of water $\pm$ one SE; $n = 6$) of macrophytes drifting at various depths in the Detroit River from May to October 1985. Percentages of total plant biomass, all depths combined, are shown in parentheses.

| Transect and Depth (m) | Biomass (group) | |
|---|---|---|
| | Live | Dead |
| **I** | | |
| 0–0.34 | $307 \pm 178$ | $66 \pm 30$ |
| 0.34–3 | $11 \pm 4$ | $8 \pm 5$ |
| 3–6 | $6 \pm 2$ | $3 \pm 1$ |
| 6–9 | $4 \pm 2$ | $4 \pm 2$ |
| 9–12 | $6 \pm 2$ | $3 \pm 1$ |
| Combined | $335(80) \pm 4$ | $84(20) \pm 9$ |
| **II** | | |
| 0–0.34 | $185 \pm 78$ | $36 \pm 20$ |
| 0.34–3 | $22 \pm 7$ | $19 \pm 13$ |
| 3–6 | $8 \pm 4$ | $11 \pm 5$ |
| Combined | $215(76) \pm 31$ | $66(24) \pm 8$ |
| **III** | | |
| 0–0.34 | $94 \pm 39$ | $12 \pm 7$ |
| 0.34–3 | $9 \pm 3$ | $4 \pm 1$ |
| 3–6 | $14 \pm 5$ | $2 \pm 1$ |
| Combined | $117(87) \pm 16$ | $18(13) \pm 2$ |

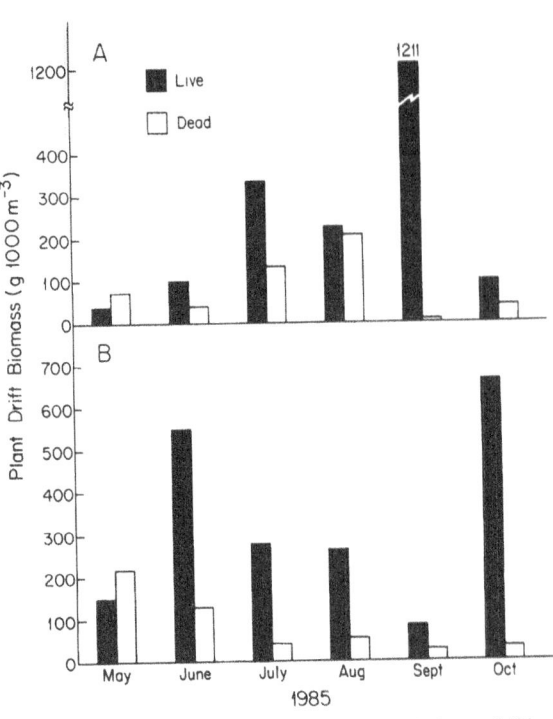

*Fig. 2.* Wet weight biomass of live and dead plants drifting monthly at (*A*) Transect 1 and (*B*) Transects II and III (combined) in the Detroit River from May to October 1985.

we collected below the water surface were fragmented and partly decomposed, possibly representing growth from the previous year (Schloesser *et al.*, 1985). We believe that the live macrophytes floated because oxygen was generated and trapped in them during photosynthesis (Sculthorpe, 1967: 120). At transects II and III combined, live drift was greatest in October and dead plant drift in May; from June to October, live drift exceeded dead drift (Fig. 2).

Although 21 species of submersed macrophytes and 11 species of emergent macrophytes are known to live in the Detroit River (Manny *et al.*, 1988), we collected only seven taxa of submersed and one taxon of emergent macrophytes in the drift (Table 2). The most abundant and widely distributed submersed plant in the river, American wildcelery (*Vallisneria americana*), also occurred most frequently in the drift. For the 126 tows, it was taken alive in 49 tows and dead in 21 tows, mostly in September and October. Other macrophytes (*Myriophyllum spicatum*, narrow-leafed *Potamogeton* spp., and *Elodea canadensis*)

Table 2. Frequency of occurrence[a] of live and dead plants drifting in the Detroit River in 1985 (all dates and transects combined).

| Taxa | Total | |
|---|---|---|
| | Live | Dead |
| **Submersed** | | |
| *Vallisneria americana* | 49 | 21 |
| *Myriophyllum spicatum* | 23 | 11 |
| Narrow-leafed *Potamogetons* | 27 | 5 |
| *Elodea canadensis* | 8 | |
| *Potamogeton richardsonii* | 2 | 1 |
| *Najas flexilis* | 4 | |
| *Potamogeton crispus* | | 1 |
| **Emergent** | | |
| *Typha* | | 6 |
| **Other** | | |
| Characeae | 1 | |
| Filamentous algae | 3 | 3 |

[a] Number of tows of a total of 126 in which each listed taxon was taken.

Table 3. Mean biomass (in grams AFDW · 1000 m$^{-3}$ ± SE; $n = 6$) of live and dead macrophytes at transects II and III combined and volume-weighted drift (in tonnes, AFDW) that entered Lake Erie, May to October 1985[a].

| Month | Biomass | | Drift | | |
|---|---|---|---|---|---|
| | Live | Dead | Live | Dead | Total |
| May | 1.3 ± 0.6 | 1.8 ± 1.2 | 15.9 | 20.8 | 36.7 |
| June | 4.6 ± 3.2 | 1.1 ± 0.3 | 24.9 | 11.2 | 36.1 |
| July | 2.3 ± 1.5 | 0.3 ± 0.2 | 16.9 | 2.5 | 19.4 |
| August | 2.2 ± 1.8 | 0.5 ± 0.2 | 12.4 | 5.1 | 17.5 |
| September | 0.7 ± 0.2 | 0.2 ± 0.1 | 7.6 | 2.8 | 10.4 |
| October | 5.6 ± 3.5 | 0.3 ± 0.1 | 26.8 | 3.7 | 30.5 |
| All months combined | | | 104.5 | 46.1 | 150.6 |

[a] Ash-free dry weight (AFDW) was assumed to equal 0.05 × wet weight, a conversion derived for Detroit River submersed macrophytes from the data of Hudson *et al.* (1986).

drifted early in the growing season in May and June. The taxonomic composition of live and dead drifting macrophytes did not vary with depth (Schloesser *et al.*, 1989); all live and dead taxa occurred more frequently at the water surface than at depths of 3–12 m. Emergent macrophytic and terrestrial plant materials were insignificant in our drift samples. Such materials are large energy sources in small streams and rivers (Dawson, 1980; Hill & Webster, 1983), particularly in autumn (Rodgers *et al.*, 1983), but decrease proportionally in the drift as the size of the river increases (Paul *et al.*, 1983).

## 3.2. Biomass of the drift

The mean monthly biomass of live and dead macrophytes was greatest in October and May, respectively (Table 3). The estimated drift of live and dead macrophytes into Lake Erie during our 6-month study was 151 tonnes AFDW. Because plant drift during the other half of the year (November–April) is probably equal to or greater than that during the period of our study (Hill & Webster, 1983), we estimated that the total drift per year into Lake Erie was at least twice that estimated during our 6-month study, or 302 tonnes. This annual amount of plant drift for the 51 km length of the Detroit River is equal to 21 percent of the biomass produced annually by aquatic macrophytes in a 135-km reach of the New River in Virginia (1,435 tonnes AFDW · yr$^{-1}$; Hill & Webster, 1983) and represents only 2 percent of the estimated 12380 tonnes of submersed macrophytes produced each year in the Detroit River (Edwards *et al.*, 1989).

Most of the macrophytic biomass produced in the Detroit River was either not adequately represented in our samples or did not drift as large plant fragments during the period of our study, for three possible reasons: plant drift on the surface of Lake St. Clair and the Detroit River is highly episodic, sometimes greatly exceeding that recorded in our study (Hunt, 1957; Hunt, 1962; Hiltunen, 1971); 14 (67%) of the submersed macrophyte species and 10 (91%) of the emergent macrophyte species that grow in the Detroit River were not collected in our drift samples; and coarse particulate organic matter (>1 mm)

338

*Table 4.* Concentrations of selected metals in digested blanks, National Bureau of Standards (NBS) spinach leaves, and macrophyte samples.

| Sample | Metal | | | | | | | |
|---|---|---|---|---|---|---|---|---|
| | Cr | Co | Ni | Cu | Zn | Cd | Pb | Hg |
| Blank | ($\mu$g l$^{-1}$ of digestate) | | | | | | | |
| Mean (N = 6) | 3.16 | 2.31 | 2.07 | 2.27 | 7.41 | 0.98 | 4.85 | |
| Std. dev. | 0.17 | 0.01 | 0.14 | 0.09 | 4.01 | 0.74 | 5.86 | |
| Est. MDL[a] | 3.67 | 2.34 | 2.49 | 2.54 | 19.44 | 3.20 | 22.43 | |
| NBS1573 | | | | | | | | |
| Mean (N = 7) | 14.15 | 4.38 | 8.19 | 49.28 | 271.04 | 12.77 | 28.21 | |
| Std. dev. | 0.59 | 0.09 | 0.43 | 1.71 | 4.60 | 0.14 | 1.44 | |
| CV[b] | 0.04 | 0.02 | 0.05 | 0.03 | 0.02 | 0.01 | 0.05 | |
| Macrophytes | | | | | | | | |
| Mean (N = 6) | 45.17 | 17.28 | 187.34 | 137.61 | 546.73 | 23.81 | 72.41 | |
| Std. dev. | 32.73 | 5.83 | 65.56 | 82.74 | 195.71 | 9.65 | 19.22 | |
| NBS1573 | ($\mu$g g$^{-1}$ dry weight) | | | | | | | [d] |
| Mean (N = 7) | 2.84 | 0.88 | 1.64 | 9.88 | 54.23 | 2.56 | 5.68 | |
| 95% CI[c] | ± 0.26 | ± 0.04 | ± 0.20 | ± 0.78 | ± 2.07 | ± 0.08 | ± 0.70 | |
| Certified value | 4.50 | [e] | [e] | 11.00 | 62.00 | 3.00 | 6.30 | |
| % Difference | − 36.89 | | | − 10.18 | − 12.53 | − 14.67 | − 9.84 | |
| Macrophytes | | | | | | | | |
| Mean (N = 6) | 10.34 | 3.77 | 41.40 | 31.14 | 117.82 | 5.19 | 16.10 | 0.09 |
| 95% CI | ± 21.02 | ± 2.81 | ± 37.49 | ± 54.32 | ± 83.81 | ± 5.48 | ± 13.11 | ± 0.07 |

[a] Estimated method detection limit.
[b] Coefficient of variation.
[c] 95% confidence interval.
[d] Mercury content of NBS standard was not determined.
[e] NBS standard not certified for that element.

seldom exceeds 10 percent of the total particulate organic matter drifting in streams because it is rapidly broken down by microbes and immature insects (McIntire, 1983; Paul *et al.*, 1983). These factors should be addressed in future investigations of macrophytic contributions to the organic matter budget of large rivers.

### 3.3. Contaminants in the plant drift

Concentrations in method blanks of all metals reported, except mercury, significantly exceeded zero (Table 4). Calculation of a method detection limit equal to the mean concentration of each metal in the blanks plus three standard deviations (MacDougall & Crummett, 1980; Keith *et al.*, 1983) showed that concentrations of all metals in our macrophyte samples were significantly higher than the mean blank value and that background interference was negligible relative to the samples. This test also showed that precision of the metal analyses (expressed as the coefficient of variation of replicate analyses of National Bureau of Standards spinach leaves; Table 4) was acceptable.

Results for chromium, copper, zinc, and cadmium in the standard spinach leaves were biased significantly low, compared with the certified value for each metal (Table 4). The extent of these biases ranged from − 10.18 percent for copper to

Table 5. Mean metals concentrations in summer in shoots of submersed macrophytic plants from polluted and unpolluted freshwater sources, in $\mu g\,g^{-1}$ dry weight.

| Sources | Co | Cr | Ni | Cu | Zn | Cd | Pb | Hg | Reference |
|---|---|---|---|---|---|---|---|---|---|
| Unpolluted | | | | | | | | | |
| Numerous | 3 | | | | 140 | | | | Hutchinson (1975) |
| Lake Erie marsh | | 8 | 3 | 40 | 16 | 1.5 | 11 | | Mudroch (1980) |
| Saginaw Bay shoreline | 0.7 | 4 | 10 | 6 | 66 | 1.1 | 21 | | Estabrook et al. (1985) |
| Saginaw Bay marsh | 0.8 | 5 | 6 | | 70 | 4.4 | | <0.1 | Wells et al. (1980) |
| Michigan wilderness lakes | 0.2 | 1 | 11 | | 1 | 3.8 | | <0.1 | Wells et al. (1980) |
| South Carolina pond | | | 3 | 41 | 73 | | | | Boyd (1970) |
| Hindon River, India[b] | 4 | 14 | 11 | 9 | 60 | <0.8 | | | Ajmal et al. (1987) |
| Great Lakes marshes | 6 | 9 | 8 | 6 | 20 | 1.5 | 6 | | Mudroch (1981) |
| New Jersey waters | | | | | 181 | | 25 | | Riemer & Toth (1968) |
| English streams | | | | 33 | 64 | | | | Cumbus et al. (1980) |
| Polluted | | | | | | | | | |
| River Leine, west Germany | | | 13 | 17 | 190 | 0.7 | 13 | 0.5 | Abo-Rady (1980) |
| Ganges River, India[b] | 6 | 237 | 11 | 16 | 84 | <0.1 | 8 | | Ajmal et al. (1983) |
| Yamuna River, India[b] | 8 | 24 | 26 | 42 | 154 | <0.1 | 14 | | Ajmal et al. (1985) |
| Malaysian stream | | 12 | 57 | 37 | 498 | | | | Lee et al. (1984) |
| English lakes | | | | 25 | 33 | 10 | 176 | | Welsh & Denny (1980) |
| Mine effluent | 230 | 4 | 93 | 16 | | 1.4 | 22 | | Mudroch & Capobianco (1979) |
| | | | | | | | | | |
| Mean of unpolluted | 2.4 | 7 | 7 | 22 | 69 | 2.2 | 16 | <0.1 | |
| Mean of polluted | 81 | 69 | 40 | 26 | 192 | 2.5 | 47 | 0.5 | |
| | | | | | | | | | |
| Detroit River | 4 | 10 | 41 | 31 | 118 | 5 | 16 | 0.1 | Present study |

[a] Blanks indicate no data.
[b] Data are for the free-floating hydrophyte Eichhornia crassipes, with submersed roots.

– 36.89 percent for chromium, in comparison with the known metals content of the spinach standard (Table 4). The accuracy of cobalt, mercury, and nickel values could not be estimated because the content of these metals in our spinach standard was not known. The metals content of another NBS spinach standard that was analyzed by neutron activation (Wells *et al.*, 1980: 189) was nearly equal to cadmium and zinc values but was two to three times higher than values for cobalt, chromium, and nickel measured in our spinach standard (Table 4). Therefore, our data probably underestimated the true metals content of Detroit River macrophytes.

The concentrations of all measured metals, except lead, were higher in macrophytes drifting in the Detroit River than in those growing in unpolluted waters elsewhere (Table 5). The concentrations of nickel, copper, and cadmium were also higher in Detroit River macrophytes than in those growing in polluted waters elsewhere (Table 5). These comparisons suggest that Detroit River macrophytes are contaminated with heavy metals; however, many factors affect the heavy metal content of aquatic macrophytes (Murdoch & Capobianco, 1979; Welsh & Denny, 1980; Lee *et al.*, 1984; Estabrook *et al.*, 1985). For example, lead apparently absorbs to the shoots of some submersed macrophytes, whereas copper is taken up by the roots of these plants and translocated to the shoots (Welsh & Denny, 1980).

Our study was not designed to determine if rooted submersed macrophytes take up heavy metals from Detroit River sediments. However, because the metals content of macrophytes is proportional to the metals content of surrounding water and sediment (Mudroch & Capobianco, 1979; Welsh & Denny, 1980; Lee *et al.*, 1984) and metals concentrations in water and sediments are lower above the City of Detroit than below it (EC & EPA, 1988; Nichols *et al.*, 1991) we evaluated whether or not plants collected above the city contained lower concentrations of metals than plants collected below the city (Table 6). Similar data for macrophytes from a west German river above and below the medium-size city of Göttingen (population 106 000), which discharges sewage containing heavy metals into the River Leine (Abo-Rady, 1980), were also included in Table 6 for comparison with our data. The mean concentrations of all metals, except mercury, in macrophytes collected in the two rivers was indeed lower above than below the city; however, none of the upstream-downstream differences in plant metal concentrations in either river were statistically significant.

The movement of mercury in the St. Clair-Detroit River waterway has been a subject of

*Table 6.* Comparison of mean summer metals concentrations in submersed macrophytes[a] collected in the Detroit River (present study) and the River Leine in west Germany (from Abo-Rady, 1980) above and below a city. Units are $\mu g\ g^{-1}$ dry weight ($\pm 95\%$ confidence interval).

| River and Location | Metal | | | | | |
|---|---|---|---|---|---|---|
| | Ni | Cu | Zn | Cd | Pb | Hg |
| Detroit | | | | | | |
| Above | $40 \pm 31$ | $20 \pm 10$ | $97 \pm 27$ | $3.5 \pm 0.8$ | $16 \pm 24$ | $0.11 \pm 0.07$ |
| Below | $42 \pm 70$ | $41 \pm 94$ | $138 \pm 124$ | $6.9 \pm 6.3$ | $16 \pm 12$ | $0.10 \pm 0.11$ |
| Difference (%) | $+5$ | $+105$ | $+42$ | $+97$ | $0$ | $0$ |
| Leine | | | | | | |
| Above | $9 \pm 8$ | $6 \pm 25$ | $130 \pm 204$ | $0.4 \pm 0.4$ | $5 \pm 9$ | $0.4 \pm 0.3$ |
| Below | $17 \pm 21$ | $20 \pm 10$ | $249 \pm 268$ | $1.0 \pm 0.4$ | $21 \pm 77$ | $0.5 \pm 0.4$ |
| Difference (%) | $+89$ | $+233$ | $+92$ | $+150$ | $+320$ | $+25$ |

[a] *Potamogeton pectinatus*, *Zannichellia palustris*, and the filamentous green alga, *Cladophora glomerata* only in the River Leine.

special interest for some time (Thomas & Jaquet, 1976). Mercury levels in Detroit River sediments declined substantially from 1970 to 1980 (Hamdy & Post, 1985); however, the average mercury concentration in Detroit River sediments in 1985 of $1.6 \mu g\, g^{-1}$ (Nichols et al., 1991) still exceeded the mercury guideline for disposal of heavily polluted sediments of $> 1.0 \mu g\, g^{-1}$ (International Joint Commission, 1982). The mercury content of the plants drifting in the Detroit River of $0.09\, \mu g\, g^{-1}$ (Table 4) was not only much lower than the mean mercury concentration of Detroit River sediments but also nearly the same as that of unpolluted sediments in North America of $0.08\, \mu g\ g^{-1}$ (Hutchinson, 1975: 328). Thus drifting macrophytes move little mercury into western Lake Erie, compared to sediments moved by Detroit River water currents (Thomas & Jaquet, 1976).

### 3.4. Contaminants transported by plant drift

Our estimate of metals transported by macrophytes drifting in the Detroit River per year, based on samples we collected from May to October 1985, was much smaller than that we calculated on the basis of total estimated net annual production of submersed macrophytes in the river (Table 7). The true mass of each metal transported probably falls into the range of annual values shown in Table 7 because our measured metals concentrations in Detroit River macrophytes were biased low and the lower estimate does not include the metals content of detritus and emergent macrophytes that we collected. As

calculated, live drifting macrophytes transport out of the Detroit River much more zinc, nickel, copper, lead, and chromium than cobalt, cadmium, and mercury. In this respect, Detroit River macrophytes resemble macrophytes in wastewater treatment ponds and tidal marshes that also accumulate and sometimes transport heavy metals, especially zinc and copper (Bulthuis et al., 1974; Gallagher & Kibby, 1980).

The amount of each metal annually transported out of the Detroit River by macrophytes (Table 7) is small relative to that transported daily by Detroit River water (Table 8). Thus the mass of heavy metals transported out of the Detroit River by drifting plants represents only a small fraction of the total mass of each metal that enters Lake Erie each year.

Drifting macrophytes were significantly enriched in all heavy metals we measured relative to the concentration of those metals in Detroit River water (Table 9). The ratio of each metal in macrophytes to that metal in filtered river water ranged from 3 800 to 161 000. Remarkably high in the macrophytes was the enrichment of cadmium and lead (relative to water). With respect to lead, macrophytes in the Detroit River resembled plants in the River Leine in west Germany, which were enriched with lead by a factor of 2 400, relative to filtered river water (Abo-Rady, 1980). Relative to water, Detroit River macrophytes were more enriched with copper and less enriched with zinc, than aquatic plants elsewhere (Hutchinson, 1975: 404). Relative to sediments, Detroit River macrophytes were only slightly enriched in cadmium and nickel.

Ecologically, our results are significant. Ducks,

Table 7. Estimated annual mass of heavy metals (in kg $yr^{-1}$), based on (A) drift of live submersed macrophytes from May to October 1985 and (B) estimated annual net production of submersed macrophytes in the Detroit River[a].

| Basis | Metal | | | | | | | | Total |
|-------|-------|------|-----|-----|------|-----|-----|------|-------|
|       | Cr    | Co   | Ni  | Cu  | Zn   | Cd  | Pb  | Hg   |       |
| A     | 2     | 1    | 9   | 6   | 24   | 1   | 3   | 0.02 | 46    |
| B     | 128   | 48   | 512 | 386 | 1458 | 64  | 199 | 1.1  | 2,796 |

[a] Basis in A equals 104 tonnes 6 mo$^{-1}$ × 12 mo yr$^{-1}$ = 208 tonnes yr$^{-1}$ and in B equals 12,380 tonnes yr$^{-1}$ (Edwards et al., 1989).

Table 8. Mass of heavy metals (range in kg d$^{-1}$) transported in water of the Detroit River each day (EC & EPA, 1988).

| Metal | Mass |
|-------|------|
| Cd | 14–22 |
| Cu | 663–920 |
| Hg | 5–9 |
| Pb | 93–94 |
| Ni | 644–747 |
| Zn | 1016–1840 |

*Table 9.* Mean metal concentrations in drifting macrophytes ($\mu g\, g^{-1}$; present study), filtered Detroit River water ($\mu g\, ml^{-1} \times 10^{-6}$; EPA, 1988a), and sediments ($\mu g\, g^{-1}$; Nichols *et al.*, 1990) and enrichment ratios of each metal in macrophytes relative to water and sediment.

| Metal | Metal Concentration | | | Macrophyte enrichment ratio | |
|---|---|---|---|---|---|
| | Macrophytes | Water | Sediment | vs. water | vs. sediment |
| Cd | 5.2 | 60 | 2.0 | 86,700 | 2.6 |
| Co | 3.8 | 1000[a] | 6.2[b] | 3,800 | c |
| Cr | 10.3 | 2000[a] | 37.0 | 5,200 | c |
| Cu | 31.1 | 1786 | 38.0 | 17,400 | c |
| Hg | 0.1 | 8 | 1.6 | 12,500 | c |
| Pb | 16.1 | 100[d] | 65.6 | 161,000 | c |
| Ni | 41.4 | 1924 | 27.2 | 21,500 | 1.5 |
| Zn | 117.8 | 7078 | 272.7 | 16,600 | c |

[a] From EPA Storet for station numbers 820414 and 820018 in the Detroit River, 1980-88.

[b] From Bertram *et al.* (1989).

[c] No enrichment.

[d] From EC & EPA (1988).

benthic macroinvertebrates, and fish in the Detroit River and Lake Erie rely heavily on macrophytes for food (Manny *et al.*, 1988; Schloesser & Manny, 1990). Submersed macrophytes typically decompose rapidly into energy-rich detritus that is rapidly used by benthic macroinvertebrates in secondary production (Hill & Webster, 1983; McIntire, 1983). Macrophytes produced in the Detroit River are thus an important food source for benthic production in Lake Erie (Manny *et al.*, 1988; Edwards *et al.*, 1989). Because macrophytes in the Detroit River are contaminated with heavy metals, they are a contaminated food source for animals in the river and in Lake Erie. Each level of the food chain in Lake Erie already contains a greater concentration of a potent, organic contaminant (polychlorinated biphenyls) than the next lower trophic level (International Joint Commission, 1989). The cumulative effects of organic and heavy-metal contaminants in the food chain may reduce ecosystem stability by preventing once-abundant, pollution-intolerant, benthic animals from recolonizing the Detroit River and western Lake Erie (Rapport, 1984).

## 4. Acknowledgements

This work was performed as a cooperative project between the U.S. Fish and Wildlife Service, National Fisheries Research Center-Great Lakes, and the U.S. Environmental Protection Agency, Great Lakes National Program Office (EPA), under the terms of Interagency Agreement No. DW 14931214–01–0. Macrophyte chemistry services were provided by EPA under the general supervision of EPA Project Officer Paul Bertram. We thank Michael D. Mullin of the EPA Large Lakes Research Station in Grosse Ile, Michigan, for providing the heavy-metal analyses of our samples; Jan Derecki of the Great Lakes Environmental Research Laboratory (National Oceanic and Atmospheric Administration) in Ann Arbor for river discharge data, and Anthony Frank and Marilyn Murphy of the National Fisheries Research Center-Great Lakes for statistical review and typing of the manuscript, respectively.

## References

Abo-Rady, M. D. K., 1980. Makrophytische Wasserpflanzen als Bioindikatoren für die Schwermetallbelastung der oberen Leine. [Aquatic macrophytes as indicator for heavy metal pollution in the River Leine (West Germany).] Arch. Hydrobiol. 89: 387–404.

Admiraal, W. & R. Van Zanten, 1988. Impact of biological activity on detritus transported in the lower river Rhine: an exercise in ecosystem analysis. Freshwat. Biol. 20: 215–225.

Ajmal, M., A. A. Nomani & M. A. Khan, 1983. Pollution of the Ganges River, India. Wat. Sci. Technol. 16: 347–358.

Ajmal, M., M. A. Khan & A. A. Nomani, 1985. Distribution of heavy metals in plants and fish of the Yamuna River (India). Envir. Monit. Assess. 5: 361–367.

Ajmal, M., R. Kahn & A. U. Khan, 1987. Heavy metals in water, sediments, fish and plants of river Hindon, U.P., India. Hydrobiologia 148: 151–157.

Beeton, A. M., 1961. Environmental changes in Lake Erie. Trans. Amer. Fish. Soc. 90: 153–159.

Bertram, P., T. A. Edsall, B. A. Manny, S. J. Nichols & D. W. Schloesser, 1989. Physical and chemical characteristics of sediments in the upper Great Lakes connecting channels. U.S. Fish Wildl. Serv., Natl. Fish. Res. Center-Great Lakes, Ann Arbor, Mich. Unpub. MS.

Boyd, C. E., 1970. Chemical analyses of some vascular aquatic plants. Arch. Hydrobiol. 67: 78–85.

Brown, C. L. & B. A. Manny, 1985. Comparison of methods for measuring surface area of submersed aquatic macrophytes. J. Freshwat. Ecol. 3: 61–68.

Bryant, W. C. & A. J. Nuhfer, 1984. Monthly report of the netting and creel census study of the St. Clair River, Lake St. Clair, and the Detroit River. November report to U.S. Army Corps of Engineers by the Michigan Department of Natural Resources, Mt. Clemens, Michigan.

Bulthuis, D. A., J. R. Craig & C. D. McNabb, 1974. Metal dynamics in municipal stabilization ponds. In; D. D. Hemphill (ed.), Trace substances in environmental health – VII. Univ. Missouri, Columbia: 117–125.

Cumbus, I. P., L. W. Robinson & R. G. Clare, 1980. Mineral nutrient availability in watercress bed substrates. Aquat. Bot. 9: 343–349.

Dawson, F. H., 1980. The origin, composition, and downstream transport of plant material in a small chalk stream. Freshwat. Biol. 10: 419–435.

Denny, P., 1987. Mineral cycling by wetland plants – a review. Arch. Hydrobiol. Beih. 27: 1–25.

EC & EPA (Environment Canada & U.S. Environmental Protection Agency), 1988. Upper Great Lakes connecting channels study. Final report. Vol. II. Great Lakes National Program Office, Chicago, IL 626 pp.

Edwards, C. J., P. L. Hudson, W. G. Duffy, S. J. Nepszy, C. D. McNabb, R. C. Haas, C. R. Liston, B. A. Manny & W.-D. N. Busch, 1989. Hydrobiological, morphometrical, and biological characteristics of the connecting rivers of the international Great Lakes: A review. In; D. P. Dodge (ed.) Proc. Internat. Large Rivers Symp., Can. J. Fish. aquat. Sci. 106: 240–264.

EPA (U.S. Environmental Protection Agency), 1988a. Input-output mass loading studies of toxic and conventional pollutants in the Trenton Channel, Detroit River. Large

Lakes Res. Stat. Grosse Ile, Mich. EPA/600/3–88/004. 236 pp.

Estabrook, G. F., D. W. Burk, D. R. Inman, P. B. Kaufman, J. R. Wells, J. D. Jones & N. Ghosheh, 1985. Comparison of heavy metals in aquatic plants on Charity Island, Saginaw Bay, Lake Huron, USA, with plants along the shoreline of Saginaw Bay. Am. J. Bot. 72: 209–216.

Gallagher, J. L. & H. V. Kibby, 1980. Marsh plants as vectors in trace metal transport in Oregon tidal marshes. Am. J. Bot. 67: 1069–1074.

Galtsoff, P. S., 1924. Limnological observations in the upper Mississippi, 1921. U.S. Dep. Commerce, Bull. Bur. Fish. 39: 347–438.

Hamdy, Y. & L. Post, 1985. Distribution of mercury, trace organics, and other heavy metals in Detroit River sediments. J. Great Lakes Res. 11: 353–365.

Harding, J. P. C. & B. A. Whitton, 1981. Accumulation of zinc, cadmium, and lead by field populations of *Lemanea*. Wat. Res. 15: 301–319.

Haslam, S. M., 1978. River plants. The Macrophytic Vegetation of Watercourses. Cambridge Univ. Press, New York. 396 pp. ISBN 0-521-29172-0.

Hill, B. H. & J. R. Webster, 1983. Aquatic macrophyte contribution to the New River organic matter budget. In; T. D. Fontaine III & S. M. Bartell (eds), Dynamics of Lotic Ecosystems. Ann Arbor Science, Ann Arbor, Mich. 273–282.

Hiltunen, J. K., 1971. Limnological data from Lake St. Clair, 1963 and 1965. NOAA, Natl. Mar. Fish. Serv., Data Rep. No. 54, 45 pp.

Hudson, P. L., B. M. Davis, S. J. Nichols & C. M. Tomcko, 1986. Environmental studies of macrozoobenthos, aquatic macrophytes, and juvenile fishes in the St. Clair-Detroit River system, 1983–1984. U.S. Fish Wildl. Serv., Natl. Fish Res. Cent.-Great Lakes, Ann Arbor, Mich. Admin. Rep. No. 86–7. 303 pp.

Hunt, G. S., 1957. Causes of mortality among ducks wintering on the lower Detroit River. Ph.D. Dissert., Univ. Mich., Ann Arbor. 296 pp.

Hunt, G. S., 1962. Seasonal aspects of Berchtold's pondweed. Mich. Bot. 1: 35.

Hutchinson, G. E., 1975. A Treatise on Limnology. Vol. 3. Limnological Botany. Wiley, New York: 328–329.

International Joint Commission, 1982. Guidelines and register for evaluation of Great Lakes Dredging projects. Windsor, Ontario, Canada, 365 pp.

International Joint Commission, 1989. 1987 Report on Great Lakes water quality. Appendix B, Great Lakes Surveillance. Vol. I. Windsor, Ontario, Canada.: 2.4–11 to 2.4–15.

Jude, D. J., M. Winnell, M. S. Evans, F. J. Tesar & R. Futyma, 1986. Drift of zooplankton, benthos, and larval fish and distribution of macrophytes during winter and summer, 1985. U.S. Army Corps Engin., Detroit, Mich. Rep. DACW–35–85–c–0005, 174 pp. + appendices.

Keith, L. H., W. B. Crummett, J. Deegan, Jr., R. A. Libby, J. K. Taylor & G. E. Wentler, 1983. Principles of environmental analysis. Analyt. Chem. 55: 2210–2218.

344

Lau, Y. L., B. G. Oliver & B. G. Krishnappan, 1989. Transport of some chlorinated contaminants by the water, suspended sediments, and bed sediments in the St. Clair and Detroit Rivers. Env. Toxicol. Chem. 8: 293–301.

Lee, C. K., K. S. Low & S. H. Tan, 1984. The accumulation of heavy metals in *Hydrostemma Motleyi* Hook. f. Mabberlay and *Hydrilla verticillata* Casp. Pertanika 7: 119–123.

MacDougall, D. & W. B. Crummett, 1980. Guidelines for data acquisition and data quality evaluation in environmental chemistry. Analyt. Chem. 52: 2242–2249.

Manny, B. A. & D. Kenaga, 1991. The Detroit River: effects of contaminants and human activities on aquatic plants and animals and their habitats. Hydrobiologia 219: 269–279.

Manny, B. A., T. A. Edsall & E. Jaworski, 1988. The Detroit River, Michigan: An ecological profile. U.S. Fish Wildl. Ser., Biol. Rep. 85(7.17). 86 pp.

McIntire, C. D., 1983. A conceptual framework for process studies in lotic ecosystems. In: T. D. Fontaine III and S. M. Bartell (eds), Dynamics of Lotic Ecosystems. Ann Arbor Science, Ann Arbor, Mich.: 43–68.

Michigan Water Resources Commission, 1967. Water resource uses: present and prospective for St. Clair River, Detroit River, Lake Erie, Maumee River Basin, water quality standards, and plan of implementation. Michigan Dep. Conserv. Lansing. 153 pp.

Mudroch, A., 1980. Biogeochemical investigation of Big Creek Marsh, Lake Erie, Ontario, J. Great Lakes Res. 6: 338–347.

Mudroch, A., 1981. A study of selected Great Lakes coastal marshes. Can. Dept. Env., Can. Centre Inl. Wat., Burlington, Ontario. Sci. Ser. No. 122. 44 pp.

Mudroch, A., 1985. Geochemistry of the Detroit River sediments. J. Great Lakes Res. 11: 193–200.

Mudroch, A. & J. A. Capobianco, 1979. Effects of mine effluent on uptake of Co, Ni, Cu, As, Zn, Cd, Cr, and Pb by aquatic macrophytes. Hydrobiologia 64: 223–231.

National Oceanographic and Atmospheric Administration, 1979. National Ocean Survey Chart Number 14853, April 14.

Nichols, S. J., B. A. Manny, D. W. Schloesser & T. A. Edsall, 1991. Heavy metal contamination of sediments in the upper connecting channels of the Great Lakes. Hydrobiologia 219: 307–315.

Paul, R. W., Jr., E. F. Benfield & J. Cairns, Jr., 1983. Dynamics of leaf processing in a medium-sized river. In: T. D. Fontaine III & S. M. Bartell (eds), Dynamics of Lotic Ecosystems. Ann Arbor Science, Ann Arbor, Mich. 403–423.

Pocklington, R. & F. C. Tan, 1987. Seasonal and annual variations in the organic matter contributed by the St. Lawrence River to the Gulf of St. Lawrence. Geochim. Cosmochim. Acta 51: 2579–2586.

Poe, T. P. & T. A. Edsall, 1982. Effects of vessel-induced waves on the composition and amount of drift in an ice environment in the St. Marys River. U.S. Fish Wildl. Serv. Fish. Res. Center-Great Lakes, Ann Abor, Mich. Admin. Rep. 82–6. 45 pp.

Quinn, F. H., & J. C. Hagman, 1977. Detroit and St. Clair River transient models. NOAA Technical Memorandum ERL GLERL-14. Great Lakes Environ. Res. Lab., Ann Arbor, Mich. 45 pp.

Rapport, D. J., 1984. State of ecosystem medicine. In; V. W. Cairns, P. V. Hodson & J. O. Nriagu (eds.), Contaminant Effects on Fisheries. John Wiley & Sons, New York: 315–324.

Riemer, D. N. & S. J. Toth, 1968. A survey of the chemical composition of aquatic plants in New Jersey. N.J. Agric. Exp. Stn., Coll. Agric. Environ. Sci., Rutgers Univ., Bull. 820, 14 pp.

Rodgers, J. H., Jr., M. E. McKevitt, D. O. Hammerlund, K. L. Dickson & J. Cairns, Jr. 1983. Primary production and decomposition of submergent and emergent aquatic plants of two Appalachian rivers. In; T. D. Fontaine III & S. M. Bartels (eds), Dynamics of Lotic Ecosystems. Ann Arbor Science, Ann Arbor, Mich.: 283–301.

Say, P. J. & B. A. Whitton, 1983. Accumulation of heavy metals by aquatic mosses. 1. *Fontinalis antipyretica* Hedw. Hydrobiologia 100: 245–260.

Schloesser, D. W. & B. A. Manny, 1986. Distribution of submersed aquatic macrophytes in the St. Clair-Detroit River system. J. Freshwat. Ecol. 3: 537–544.

Schloesser, D. W. & B. A. Manny, 1990. Decline of wild-celery buds in the lower Detroit River, 1950–1985. J. Wildl. Manage. 54: 72–76.

Schloesser, D. W., T. A. Edsall & B. A. Manny, 1985. Growth of submersed macrophyte communities in the St. Clair-Detroit River system between Lake Huron and Lake Erie. Can. J. Bot. 63: 1061–1065.

Schloesser, D. W., S. J. Nichols & B. A. Manny, 1989. Use of a horizontal, unobstructed tow-net to collect macrophyte drift. U.S. Fish Wildl. Serv., Natl. Fish. Res. Center-Great Lakes, Ann Arbor, Mich., Unpubl. MS.

Sculthorpe, C. D., 1967. The Biology of Aquatic Vascular Plants. Edward Arnold, London, 610 pp. ISBN 0-713-12135-1.

Thomas, R. L. & J. M. Jaquet, 1976. Mercury in the surficial sediments of Lake Erie. J. Fish. Res. Bd. Can. 33: 404–412.

Wefring, D. R. & A. J. Hopwood, 1981. Method for collecting invertebrate drift from the surface and bottom of large rivers. Prog. Fish-Cult. 43: 108–110.

Wells, J. R., P. B. Kaufman & J. D. Jones, 1980. Heavy metal contents in some macrophytes from Saginaw Bay (Lake Huron, U.S.A.). Aquat. Bot. 9: 185–193.

Welsh, R. P. H. & P. Denny, 1980. The uptake of lead and copper by submerged aquatic macrophytes in two English lakes. J. Ecol. 68: 443–455.

*Hydrobiologia* **219**: 345–352, 1991.
*M. Munawar & T. Edsall (eds),*
*Environmental Assessment and Habitat Evaluation of the Upper Great Lakes Connecting Channels.*
© *1991 Kluwer Academic Publishers.*

# Distribution of *Hexagenia* nymphs and visible oil in sediments of the Upper Great Lakes Connecting Channels[1]

D. W. Schloesser, T. A. Edsall, B. A. Manny & S. J. Nichols
*National Fisheries Research Center-Great Lakes, U.S. Fish and Wildlife Service, Ann Arbor, MI 48105,*
*USA*

*Key words:* Hexagenia, oil, Great Lakes, channels, pollution

## Abstract

As part of the study of the Upper Great Lakes Connecting Channels sponsored by the U.S. Environmental Protection Agency, the U.S. Fish and Wildlife Service examined the occurrence of *Hexagenia* nymphs and visible oil in sediments at 250 stations throughout the St. Marys River and the St. Clair-Detroit River system from May 14 to June 11, 1985. The mean density of *Hexagenia* nymphs per square meter averaged 194 for the total study area, 224 in the St. Marys River, 117 in the St. Clair River, 279 in Lake St. Clair, and 94 in the Detroit River. The maximum density of nymphs ranged from 1,081 to 1,164 $m^{-2}$ in the three rivers and was 3,099 $m^{-2}$ in Lake St. Clair. A comparison of nymph density at 46 stations where oil was observed in sediments physically suitable for nymphs showed that densities were lower in oiled sediments (61 $m^{-2}$) than in sediments without oil (224 $m^{-2}$). Densities of nymphs were relatively high at only four stations where oil was observed in sediments. In general, oiled sediments and low densities of nymphs occurred together downstream from industrial and municipal discharges.

## 1. Introduction

Over the past century, industrial and urban discharges have substantially degraded sediment and water quality in connecting channels of the upper Great Lakes such that each channel is now designated as an Area of Concern (GLWQB, 1983). Degraded benthic invertebrate communities are a common feature of these areas of concern (Limno-Tech, Inc., 1985; Everitt *et al.*, 1985; EC & USEPA, 1988), and in these communities the burrowing mayfly *Hexagenia*, which is sensitive to environmental degradation (Fremling,

1964; National Academy of Sciences, 1973), is usually absent or low in abundance.

*Hexagenia* has been eliminated from most Great Lakes waters directly affected by industrial and urban discharges, including portions of Green Bay (Howmiller & Beeton, 1971; Mozley & Howmiller, 1977), the St. Marys River (Veal, 1968; Hamdy *et al.*, 1978; Hiltunen & Schloesser, 1983), Saginaw Bay (Schneider *et al.*, 1969), the Detroit River (Fallon & Horvath, 1985; Thornley & Hamdy, 1984), and portions of Lakes Erie, Superior, Huron, and Michigan (Britt, 1955; Veal & Osmond, 1967; Mozley & LaDronka, 1988). Hiltunen & Schloesser (1983) linked the absence or reduced abundance of *Hexagenia* nymphs in portions of the St. Marys River with the presence of visible oil in the sediments. The present study,

---
[1]Contribution number 736 of the National Fisheries Research Center-Great Lakes, U.S. Fish and Wildlife Service, 1451 Green Road, Ann Arbor, MI 48105.

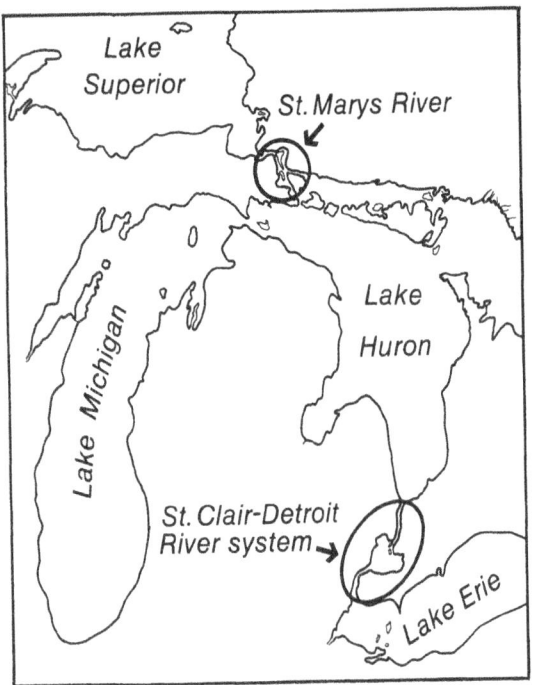

*Fig. 1.* The Upper Great Lakes Connecting Channels: the St. Marys River and the St. Clair-Detroit River system.

which was conducted as part of a multi-year, joint undertaking by the United States and Canada to determine the existing environmental condition of these waters (Everitt *et al.*, 1985; EC & USEPA, 1988), was designed to determine if a linkage between oiled sediments and the abundance of *Hexagenia* nymphs could be demonstrated throughout the Upper Great Lakes Connecting Channels (Fig. 1).

## 2. Methods

Samples of *Hexagenia* nymphs and sediment were systematically collected from May 14 to June 11, 1985 at one station in each cell of a 250-cell grid covering the Upper Great Lakes Connecting Channels (Fig. 2). Cell dimensions in each of the water bodies were (east to west and north to south); $2.2 \times 2.2$ km in the St. Marys River, $2.2 \times 2.2$ km in the St. Clair River, $6.9 \times 4.8$ km in Lake St. Clair, and $2.8 \times 1.8$ km in the Detroit River. All samples were collected before nymph

emergence in July when densities of nymphs in Great Lakes waters are relatively high (Schloesser & Hiltunen, 1984).

Stations were located where *Hexagenia* nymphs had been collected previously or where the substrate was texturally suitable for them, i.e., consisted of a mixture of silt, clay, and fine sand with a sticky consistency that would support nymph burrows (Hunt, 1953; Eriksen, 1963; Wright & Mattice, 1981).

Three samples were collected at each station with a standard ($0.048$ m$^2$ jaw opening) Ponar grab. Each sample was examined for the presence of visible oil as it was transferred from the grab to a metal tub and washed through a standard U.S. No. 30 sieve ($0.65$ mm mesh opening). When present, oil appeared as a thin film or sheen that formed on the surface of the water surrounding the sample during washing. Oil observed during the washing process was recorded and such sediments were classified as 'oiled'. The screen residue from each sample was placed in a jar and preserved with 10 percent formalin-phloxine B, a preservative-dye (Mason & Yevich, 1967).

In the laboratory, the screen residue from each sample was divided into subsamples of a size convenient for processing (5 to 10 cm$^3$) and placed in white enamel pans with a concentrated sugar solution (Anderson, 1959). Subsamples were examined at $7 \times$ magnification and nymphs were removed manually from each sample. A total of 7,045 *Hexagenia* nymphs were identified. Oil observed while examining sample residue in the laboratory was also recorded.

Data were summarized and analyzed using 18 connecting channel sections of the study area (Fig. 2). Designation of channel sections was based on knowledge of pollution sources, natural geographical features, previous studies (e.g., Hiltunen & Schloesser, 1983), and international boundaries. Differences ($P \leq 0.05$) between the mean number of nymphs m$^{-2}$ ($\log[x + 1]$) at stations with oil and oil-free sediments were tested for significance by student's t-distribution (Snedecor & Cochran, 1973).

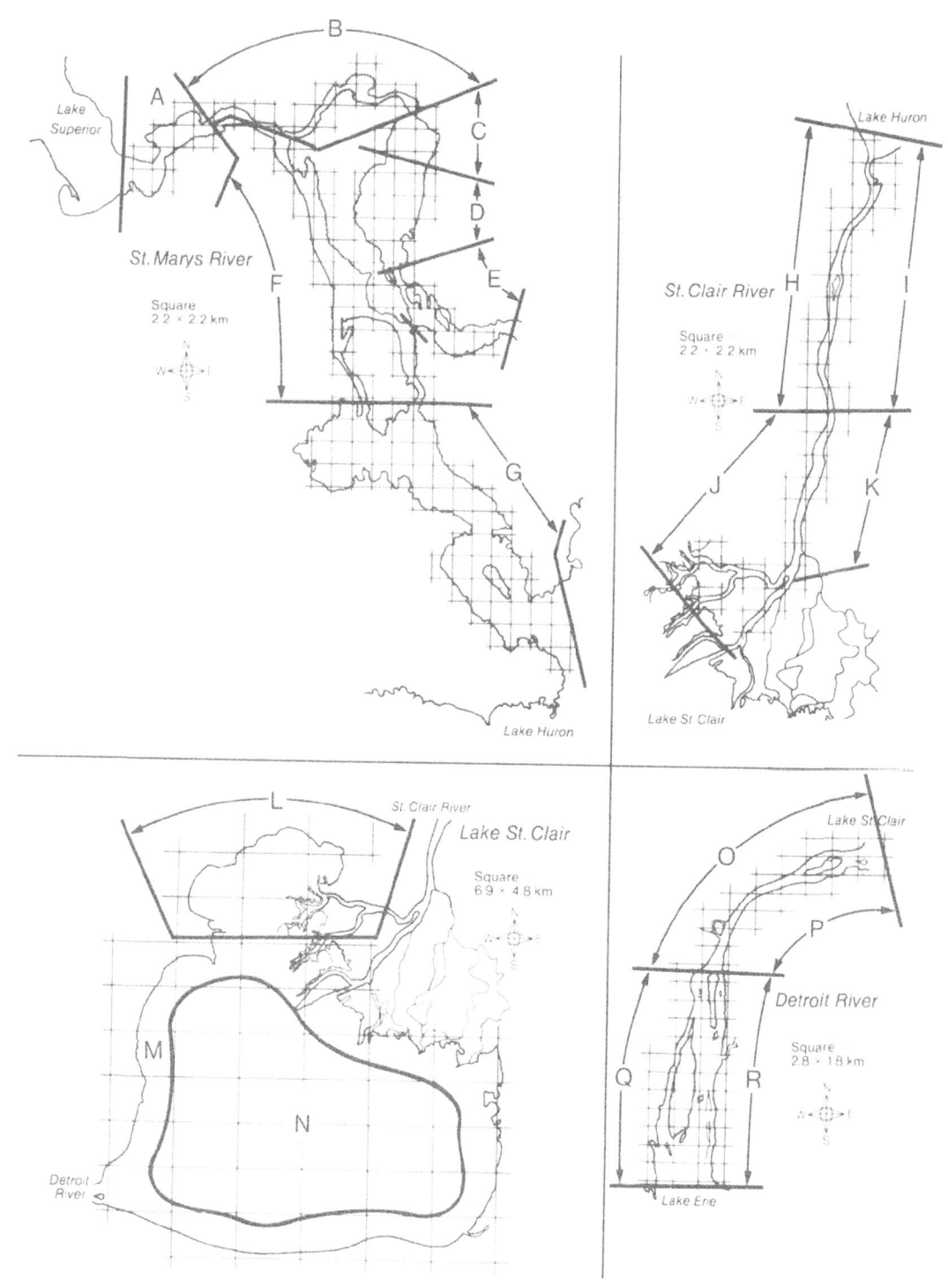

*Fig. 2.* Locations of 250 grid squares (in each of which one station was sampled) and 18 geographical sections (A–R) used to summarize the occurrence of *Hexagenia* nymphs and oiled sediments in the Upper Great Lakes Connecting Channels May 14–June 11, 1985.

Table 1. Mean number, standard error (SE), and range of *Hexagenia* nymphs (per square meter) at all stations and at stations with visibly oiled and oil-free sediments in 18 geographical sections of the Upper Great Lakes Connecting Channels, May 14-June 11, 1985.

| Connecting channel section | | All stations | | | | Oiled sediments | | | | Oil-free sediments | | | |
|---|---|---|---|---|---|---|---|---|---|---|---|---|---|
| Description | Desig- nation[a] | Number stations | Mean | SE | Range | Number stations | Mean | SE | Range | Number stations | Mean | SE | Range |
| **St. Marys River** | | | | | | | | | | | | | |
| Upriver of Sault Ste. Marie | A | 8 | 193 | 84.0 | 0–668 | 0 | | | | 8 | 193 | 84.0 | 0–668 |
| Channel above Sugar Island | B | 14 | 11 | 5.6 | 0–69 | 9 | 12 | 7.5 | 0–69 | 5 | 11 | 9.3 | 0–48 |
| Northern Lake George | C | 6 | 75 | 34.7 | 0–186 | 2 | 4 | 3.5 | 0–7 | 4 | 110 | 41.9 | 34–186 |
| Southern Lake George | D | 15 | 129 | 31.7 | 0–461 | 1 | 0 | | | 14 | 138 | 32.5 | 0–461 |
| Below Lake George | E | 8 | 232 | 53.0 | 34–530 | 0 | | | | 8 | 232 | 53.0 | 34–530 |
| Lake Nicolet | F | 24 | 331 | 63.7 | 0–1164 | 1 | 34 | | | 23 | 344 | 65.1 | 0–1164 |
| Lake Munuscong | G | 50 | 282 | 32.8 | 0–964 | 0 | | | | 50 | 282 | 32.8 | 0–964 |
| Total | A–G | 125 | 224 | 29.3 | 0–1164 | 13 | 11 | 5.6 | 0–69*b | 112 | 249 | 22.6 | 0–1164 |
| **St. Clair River** | | | | | | | | | | | | | |
| Upper U.S. waters | H | 10 | 23 | 7.6 | 0–76 | 3 | 39 | 22.0 | 0–76 | 7 | 16 | 5.8 | 0–41 |
| Upper Canadian waters | I | 9 | 47 | 30.7 | 0–282 | 7 | 10 | 6.0 | 0–44 | 2 | 179 | 103.0 | 76–282 |
| Lower U.S. waters | J | 9 | 359 | 121.0 | 7–1081 | 0 | | | | 9 | 359 | 121.0 | 7–1081 |
| Lower Canadian waters | K | 7 | 31 | 7.9 | 0–62 | 2 | 28 | 13.5 | 14–41 | 5 | 34 | 10.7 | 0–62 |
| Total | H–K | 35 | 117 | 39.3 | 0–1081 | 12 | 20 | 7.0 | 0–76*b | 23 | 168 | 57.3 | 0–1081 |
| **Lake St. Clair** | | | | | | | | | | | | | |
| Anchor Bay | L | 6 | 343 | 84.1 | 62–530 | 0 | | | | 6 | 343 | 84.1 | 62–530 |
| Perimeter of lake | M | 19 | 22 | 16.1 | 0–310 | 0 | | | | 19 | 22 | 16.1 | 0–310 |
| Open waters of lake | N | 18 | 528 | 170.9 | 41–3099 | 0 | | | | 18 | 528 | 170.9 | 41–3099 |
| Total | L–N | 43 | 279 | 80.2 | 0–3099 | 0 | | | | 43 | 279 | 80.2 | 0–3099 |
| **Detroit River** | | | | | | | | | | | | | |
| Upper U.S. waters | O | 7 | 20 | 7.9 | 0–62 | 3 | 5 | 4.7 | 0–14 | 4 | 31 | 11.4 | 14–62 |
| Upper Canadian waters | P | 9 | 253 | 115.3 | 41–1129 | 4 | 338 | 265.3 | 28–1129 | 5 | 184 | 55.0 | 41–310 |
| Lower U.S. waters | Q | 21 | 16 | 6.5 | 0–130 | 12 | 17 | 11.0 | 0–130 | 9 | 15 | 6.0 | 0–48 |
| Lower Canadian waters | R | 10 | 167 | 64.5 | 0–606 | 2 | 431 | 175.5 | 255–606 | 8 | 101 | 50.0 | 0–420 |
| Total | O–R | 47 | 94 | 28.8 | 0–1129 | 21 | 116 | 59.2 | 0–1129 | 26 | 76 | 21.8 | 0–420 |
| All sections | A–R | 250 | 194 | 19.4 | 0–3099 | 46 | 61 | 27.8 | 0–1129*b | 204 | 224 | 22.4 | 0–3099 |

a See Figure 2.

b Means of oiled and oil-free sediments significantly different at P ≤ 0.05.

## 3. Results

Mean densities of nymphs (number m$^{-2}$) varied substantially in the Upper Great Lakes Connecting Channels (Table 1, Fig. 2): means of 224 were found in the St. Marys River, 117 in the St. Clair River, 279 in Lake St. Clair, and 94 in the Detroit River. Maximum densities of nymphs in the St. Marys (1,164 m$^2$), St. Clair (1,081 m$^2$), and the Detroit (1,129 m$^{-2}$) rivers were about one-third of that in Lake St. Clair (3,099 m$^{-2}$). In the St. Marys River, mean densities of nymphs were relatively high in connecting channel sections upriver from Sault Ste. Marie (193 m$^{-2}$), below Lake George (232 m$^{-2}$), and in Lakes Nicolet (331 m$^{-2}$) and Munuscong (282 m$^{-2}$). Densities were low in the channel above Sugar Island (11 m$^{-2}$), northern Lake George (75 m$^{-2}$), and southern Lake George (129 m$^{-2}$). In the St. Clair River, densities of nymphs were low in upper U.S. waters (23 m$^{-2}$) and upper and lower Canadian waters (47 and 31 m$^{-2}$), and high in lower U.S. waters (359 m$^{-2}$). Densities were high in open waters of Anchor Bay and Lake St. Clair (343 and 528 m$^{-2}$), but not along the perimeter of the lake (22 nymphs m$^{-2}$). In the Detroit River, the lowest mean densities (20 and 16 m$^{-2}$) occurred in U.S. waters and the highest mean densities (253 and 167 m$^{-2}$) were in Canadian waters.

In general, the densities of nymphs in sediments were lower at stations where oil was observed than at stations where no oil occurred in sediments (Table 1). The mean and maximum density of nymphs was significantly lower at stations where oil was present (61 and 1,129 m$^{-2}$). Of the total stations sampled, oiled sediments were found at 13 (10%) in the St. Marys, 12 (34%) in the St. Clair River, and 21 (45%) in the Detroit River. No oiled sediments were found in Lake St. Clair. In the St. Marys and St. Clair rivers, mean densities of nymphs were significantly lower at stations where sediments were oiled (11 and 20 m$^{-2}$, respectively) than at stations without oil (249 and 168 m$^{-2}$, respectively). In the Detroit River, mean density of nymphs was nominally higher in oiled sediments (116 m$^{-2}$) than in oil-free sediments (76 m$^{-2}$). However, this difference was not significant. Mean densities of nymphs were relatively low (< 40 m$^{-2}$) in oiled sediments in 9 of 11 channel sections and relatively high (> 192 m$^{-2}$) in 6 of 7 channel sections where no oil was observed. Densities of nymphs were high in only 2 of 11 channel sections (P and R; Fig. 2) where sediments were oiled. High densities of nymphs in these two sections are attributable to the relatively high densities of nymphs (152, 255, 606, and 1,129 m$^{-2}$) at four stations where oil was found in both hard clay and soft mud sediments.

## 4. Discussion

Densities of nymphs at stations in connecting channel sections indicated an inverse relation between visibly oiled sediment and the abundance of *Hexagenia* (Table 1). Densities of nymphs were low in oiled sediments in the St. Marys and St. Clair rivers and in the U.S. waters of the Detroit River, but not in Canadian waters of the Detroit River. However, high mean densities (338 and 431 m$^{-2}$) in two channel sections and maximum density at a single station (1 129 m$^{-2}$) where oiled sediments were found in the Detroit River indicated that oiled sediments did not prevent the occurrence of nymphs as they apparently did in the St. Marys River (Hiltunen & Schloesser, 1983). In addition, nymph densities were similar in oiled and oil-free sediments in 5 of 11 channel sections (B, H, K, O, and Q) where sediments were oiled. For example, in the channel above Sugar Island, the mean density of nymphs was 12 m$^{-2}$ at stations with oiled sediments and 11 m$^{-2}$ at stations with oil-free sediments.

In general, the densities of *Hexagenia* nymphs and the occurrence of oiled substrates documented in the present study were similar to those previously reported in these areas (Thornley & Hamdy, 1984; Hamdy *et al.*, 1978; Veal, 1968). Densities of nymphs and the occurrence of oiled sediments were high near urban and industrial discharges (Table 1). In the St. Marys River, industrial discharges in Sault Ste. Marie, Ontario have been identified as probable causes for degraded benthic habitat in the north channel above Sugar Island (section B, Table 1 and Fig. 2) and into Lake George (sections C and D); (Veal, 1968; Hamdy *et al.*, 1978; Hiltunen & Schloesser, 1983; Burt *et al.*, 1988). Since at least 1968, water and sediment quality have been impaired in the St. Clair River from Sarnia, Ontario, along the upper and lower Canadian shore (sections I and K) to Lake St. Clair (OME, 1979, 1986). Low densities of nymphs at ten stations in U.S. waters of the upper St. Clair River (section H) were not documented in previous studies. In the Detroit River, low densities of *Hexagenia* in U.S. waters (sections O and Q) were reported by Hiltunen and Manny (1982) and relatively high densities in Canadian waters (sections P and R) by Thornley & Hamdy (1984). The perimeter of Lake St. Clair is the only area in the present study where a low density of nymphs was not associated with the presence of industrial pollution. We observed no oil in the sediments of Lake St. Clair.

Hiltunen & Manny (1982) reported that mean and maximum nymph densities were substantially lower in Lake St. Clair in spring 1977 (183 and 654 m$^{-2}$, respectively) than in the present study (279 m$^{-2}$ and 3,099 m$^{-2}$, respectively). The maximum density of 3,099 *Hexagenia* nymphs m$^{-2}$ found near the middle of Lake St. Clair is among the highest *Hexagenia* densities reported in Great Lakes waters. In 1977, the benthic environment of Lake St. Clair was not severely impaired by pollution (Hiltunen & Manny, 1982). The low density of nymphs in the perimeter of Lake St. Clair has been attributed to unsuitable, natural substrates (Hiltunen & Manny, 1982; Thornley, 1985). In the present study, the substrate at 17 of 19 stations in this area consisted of loose sand, often on hard pan substrate. Empirically established criteria indicate that suitable substrates for *Hexagenia* nymphs consist of mixtures of clay, silt, and mud, and that unsuitable substrates are composed of coarse sand and larger-grained substrates (Hunt, 1953; Wright & Mattice, 1981). Laboratory studies of substrate size preference have shown that the optimal substrate for *Hexagenia limbata* nymphs consists of fine materials such as silt and clay (Eriksen, 1963).

Survival of a relatively large number of nymphs in the presence of oiled sediments at four stations in Canadian waters of the Detroit River may be attributed to relatively low oil concentrations in the upper 3 cm of sediments (Bertram *et al.*, in press). Substrates at these four stations were soft and the mean oil concentration in the top 3 cm of sediments was 366 (range 38 to 716) mg kg$^{-1}$ of dry weight sediments. The mean oil concentration in the upper 3 cm of sediment at the other 17 stations in the Detroit River where sediments with visible oil were found was 3,695 (range 14 to 24,100) mg kg$^{-1}$. In addition, the four stations where densities of nymphs were high in oiled sediments were located in sediment deposition zones where water velocities were relatively low (Fallon & Horvath, 1985). We believe that high densities of *Hexagenia* nymphs can occur in the upper 3 cm layer of sediment in these deposition zones, if that layer is oxidized and oil-free, even when underlying sediments are heavily oiled. However, we are unable to demonstrate vertical stratification of oil or nymphs in sediments in the present study because each sample was a composite of sediments to a depth of about 15 cm.

The occurrence and relatively high density of *Hexagenia* in oiled sediments in the Detroit River but not in oiled sediments of the St. Marys River may reflect the greater diversity of oils spilled and discharged into the Detroit River. Most oils discharged into the St. Marys River for the past 25 years originated from three sources: 1) rolling mills in a steel plant (Veal, 1968; Anonymous, 1979); 2) machines in a paper mill (Anonymous, 1980); and 3) oil separators in a chemical plant (Hamdy *et al.*, 1978). In addition, heavy fuel or bilge oils are sometimes spilled into the St. Marys River by vessels (Anonymous, 1985a, 1985b). In contrast, hundreds of industries discharge oil of various kinds into the St. Clair and Detroit rivers (GLWQB, 1983) and large amounts of more than then ten kinds of light and heavy oils entered the St. Clair-Detroit River system from 1974 to 1985 (Edsall *et al.*, 1988; Manny *et al.*, 1988). The high density of nymphs that we observed in some visibly oiled sediments in the Detroit River suggest some of these oils may be relatively non-toxic.

Crude petroleum and most refined petroleum products are complex mixtures of organic compounds that have been used as gross indicators of hydrocarbon pollution. The visible oil reported in the present study is probably the lighter fraction of oil in sediments, because it floats easily in water and is often aromatic. In general, lighter oil fractions are more toxic than heavier fractions to aquatic organisms because the lighter fractions contain water-soluble compounds, such as benzene, toluene, and naphthalene, that in low concentration (about $1 \text{ mg l}^{-1}$) reduce the growth, reproduction, and survival of many aquatic plants and animals (McCauley, 1966; Emery, 1972; Anderson, 1977; Burk, 1977). However, in view of our findings that some visible oils appear to be toxic to *Hexagenia* nymphs, whereas others do not, we recommend that laboratory and field bioassays be performed to demonstrate the relative toxicity of oiled sediments from representative segments of the Great Lakes Connecting Channels.

## 5. Conclusion

Our study indicates that the density of *Hexagenia* nymphs in texturally suitable substrates varied substantially throughout the Upper Great Lakes Connecting Channels. In general, oiled sediments and low densities of nymphs were found in areas downstream from industrial and municipal discharges. These results indicate the objectives of 'no visible oil' and 'no impairment of benthic communities' stated in the Water Quality Agreement of 1978 (IJC, 1988) have not been met in many areas of the Upper Great Lakes Connecting Channels.

## 6. Acknowledgements

This work was performed as a cooperative project between the U.S. Fish and Wildlife Service, National Fisheries Research Center-Great Lakes, and the U.S. Environmental Protection Agency, Great Lakes National Program Office (EPA), under the terms of Interagency Agreement No. DW 14931213–01–0.

## References

Anderson, R. O., 1959. A modified floatation technique for sorting bottom fauna samples. Limnol. Oceanogr. 4: 223–225.

Anderson, J. W., 1977. Responses to sublethal levels of petroleum carbons: Are they sensitive indicators and do they correlate with tissue contamination? In; D. A. Wolfe (ed). Proc. Symp. on Fate and Effects of Petroleum Hydrocarbons in Marine Ecosystems and Organisms, Nov. 10–12, 1976. pp. 95–114 Seattle, Washington Pergamon Press.

Anonymous, 1979. River oil pollution complaints climb. The Sault Star. Sault Ste. Marie, Ontario. August 7.

Anonymous, 1980. Abitibi faces $1.5 million order. The Sault Star. Sault Ste. Marie, Ontario. August 29.

Anonymous, 1985a. St. Marys River oil slick concentrated near Birch Point. The Evening News. Sault Ste. Marie, Mich. August 2.

Anonymous, 1985b. Coast Guard finishes off $19 000 oil slick clean-up. The Evening News. Sault Ste. Marie, Mich. August 13.

Bertram, P., T. A. Edsall, B. A. Manny, S. J. Nichols & D. W. Schloesser. Physical and chemical characteristics of sediments in the upper Great Lakes connecting channels 1985. U.S. Env. Protect. Agency Tech. Rep., Chicago, Illinois. (In press).

Britt, N. W., 1955. Stratification of western Lake Erie in summer 1953: effects on the *Hexagenia* (Ephemeroptera) population. Ecology 36: 239–244.

Burk, C. J., 1977. A four year analysis of vegetation following an oil spill in a freshwater marsh. J. Appl. Ecol. 14: 515–522.

Burt, A. J., D. R. Hart & P. M. McKee, 1988. Benthic invertebrate survey of the St. Marys River, 1985. Rep. Prep. for Ont. Min. Env. by Beak Consultants Ltd. Mississanga, Ontario.

EC & USEPA (Environment Canada & U.S. Environmental Protection Agency), 1988. Upper Great Lakes Connecting Channels Study. Draft Final Report. Vol. II. Chicago, Ill. & Toronto, Ontario. 583 pp.

Edsall, T. A., B. A. Manny & C. N. Raphael, 1988. The St. Clair River and Lake St. Clair, Michigan: an ecological profile. U.S. Fish Wildl. Serv. Biol. Rep. 85(7.3). 130 pp.

Emery, A. R., 1972. A review of the literature of oil pollution with particular reference to the Canadian Great Lakes. Ont. Min. Ntrl. Resour., Res. Inform. Pap. (Fish.) No. 40. 52 pp.

Eriksen, C. H., 1963. The relation of oxygen consumption to substrate particle size in two burrowing mayflies. J. Exp. Biol. 40: 447–453.

Everitt, R. R., G. Cunningham, M. L. Jones & D. R. Marmorek, 1985. Upper Great Lakes Connecting Channel Study Planning Workshop, Final Rep., Envir. Social Sys. Anal. Ltd., Toronto, Ontario.

Fallon, M. E. & F. J. Horvath, 1985. Preliminary assessment of contaminants in soft sediments of the Detroit River. J. Great Lakes Res. 11: 373–378.

Fremling, C. R., 1964. Mayfly distribution indicates water quality on the Upper Mississippi River. Science 146: 1164–1166.

GLWQB (Great Lakes Water Quality Board), 1983. Report on Great Lakes Water Quality; Appendix A: Areas of concern in the Great Lakes Basin; 1983 update of class A areas, 1984. Rep. to International Joint Commission, Windsor, Ontario. 113 pp.

Hamdy, Y., J. D. Kinkead & M. Griffiths, 1978. St. Marys River water quality investigations 1973–74. Ont. Min. Env., Toronto, Ontario. 53 pp.

Hiltunen, J. K. & B. A. Manny, 1982. Distribution and abundance of macrozoobenthos in the Detroit River and Lake St. Clair, 1977. Admin. Rpt. 82–2. Great Lakes Fish. Lab., Ann Arbor, Michigan. 87 pp.

Hiltunen, J. K. & D. W. Schloesser, 1983. The occurrence of oil and the distribution of Hexagenia (Ephemeroptera: Ephemeridae) nymphs in the St. Marys River, Michigan and Ontario. Freshwat. Invert. Biol. 2: 199–203.

Howmiller, R. P. & A. M. Beeton, 1971. Biological evaluation of environmental quality, Green Bay, Lake Michigan. J. Wat. Pollut. Cont. Fed. 43: 123–133.

Hunt, B. P., 1953. The life history and economic importance of a burrowing mayfly Hexagenia limbata in southern Michigan lakes. Mich. Dep. Conserv. Bull. Inst. Fish. Res. 4: 1–151.

IJC (International Joint Commission), 1988. Revised Great Lakes Water Quality Agreement of 1978 as amended by Protocol November 18, 1987. Annex 1, P. 40. Windsor, Ontario. 130 pp.

Limno-Tech, Inc., 1985. Summary of the existing status of the Upper Great Lakes Connecting Channels Data. Report to Upper Great Lakes Connecting Channels Study. Ann Arbor, Michigan. 156 pp.

Manny, B. A., T. A. Edsall & E. Jaworski, 1988. The Detroit River, Michigan: an ecological profile. U.S. Fish Wildl. Serv. Biol. Rep. 85(7.17). 86 pp.

Mason, W. & P. Yevich, 1967. The use of phloxine B and rose bengal stains to facilitate sorting benthic samples. Trans. Amer. Microscop. Soc. 86: 221–223.

McCauley, R. N., 1966. The biological effects of oil pollution in a river. Limnol. Oceanogr. 11: 475–486.

Mozley, S. C. & R. P. Howmiller, 1977. Environmental status of the Lake Michigan region: zoobenthos of Lake Michigan. Argonne Ntnl. Lab. Rep. No. ANL/ES-40. Vol. 6. U.S. Energy Res. Develop. Admin. Argonne, Illinois. 148 pp.

Mozley, S. C. & R. M. LaDronka, 1988. Ephemera and Hexagenia (Ephemeridae, Ephemeroptera) in the Straits of Mackinac, 1955–56. J. Great Lakes Res. 14: 171–177.

National Academy of Sciences, 1973. Water quality criteria 1972. U.S. Env. Protect. Agency, Washington, D.C. EPA Ecol. Res. Ser. EPS-R3-73-033.

OME (Ontario Ministry of Environment), 1979. St. Clair River organics study, biological surveys, 1968 and 1977. Toronto, Ontario, 99 pp.

OME (Ontario Ministry of Environment), 1986. St. Clair River pollution investigation (Sarnia area). Toronto, Ontario. 135 pp. + appendices.

Schloesser, D. W. & J. K. Hiltunen, 1984. Life cycle of a mayfly Hexagenia limbata in the St. Marys River between Lakes Superior and Huron. J. Great Lakes Res. 10: 435–439.

Schneider, J. C., F. F. Hooper & A. M. Beeton, 1969. The distribution and abundance of benthic fauna in Saginaw Bay, Lake Huron. Proc. 12th Conf. Great Lakes Res. 12: 80–90.

Snedecor, G. W. & W. G. Cochran, 1973. Statistical methods. 6th Edition. Ames, Iowa State Univ. Press. 593 pp. ISBN 0-8138-1560-6.

Thornley, S., 1985. Macrozoobenthos of the Detroit and St. Clair Rivers with comparisons to neighboring waters. J. Great Lakes Res. 11: 290–296.

Thornley, S. & Y. Hamdy, 1984. An assessment of the bottom fauna and sediments of the Detroit River. Ont. Min. Env. Toronto, Ontario. 48 pp.

Veal, D. M., 1968. Biological survey of the St. Marys River. Ont. Wat. Resour. Commiss., Toronto, Ontario. 53 pp.

Veal, D. M. & D. S. Osmond, 1967. Bottom fauna of the western basin and nearshore Canadian waters of Lake Erie, Ontario. Ont. Wat. Resour. Commiss., Toronto, Ontario.

Wright, L. L. & J. S. Mattice, 1981. Substrate selection as a factor in Hexagenia distribution. Aquat. Insects 3: 13–24.

*Hydrobiologia* **219**: 353–361, 1991.
*M. Munawar & T. Edsall (eds),*
*Environmental Assessment and Habitat Evaluation of the Upper Great Lakes Connecting Channels.*
© *1991 Kluwer Academic Publishers.*

# Production of *Hexagenia limbata* nymphs in contaminated sediments in the Upper Great Lakes Connecting Channels [1]

Thomas A. Edsall, Bruce A. Manny, Donald W. Schloesser, Susan J. Nichols & Anthony M. Frank
*National Fisheries Research Center-Great Lakes, U.S. Fish and Wildlife Service, Ann Arbor MI 48105,*
*USA*

*Key words:* Great Lakes, burrowing mayfly, production, contaminated sediments, *Hexagenia*

## Abstract

In April through October 1986, we sampled sediments and populations of nymphs of the burrowing mayfly, *Hexagenia limbata* (Serville), at 11 locations throughout the connecting channels of the upper Great Lakes, to determine if sediment contaminants adversely affected nymph production. Production over this period was high (980 to 9231 mg dry wt m$^{-2}$) at the five locations where measured sediment levels of oil, cyanide, and six metals were below the threshold criteria of the U.S. Environmental Protection Agency and the Ontario Ministry of Environment for contaminated or polluted sediments, and also where the criterion for visible oil given in the Water Quality Agreement between the U.S.A. and Canada for connecting waters of the Great Lakes was not exceeded. At the other six locations where sediments were polluted, production was markedly lower (359 to 872 mg dry wt m$^{-2}$). This finding is significant because it indicates that existing sediment quality criteria can be applied to protect *H. limbata* from oil, cyanide, and metals in the Great Lakes and connecting channels where the species fulfills a major role in secondary production and trophic transfer of energy.

## 1. Introduction

We conducted a study to determine if contaminated sediments were adversely affecting the production of nymphs of the burrowing mayfly, *Hexagenia limbata* (Serville), in the St. Marys River, the St. Clair River, Lake St. Clair, and the Detroit River – the waterbodies that collectively compose the Upper Great Lakes Connecting Channels. Our study was an integral component of a broader investigation (EC & EPA, 1988) initiated in 1984 by U.S. and Canadian agencies to identify and quantify the effects of conventional

pollutants and toxic substances on the biota of these channels. The Upper Great Lakes Connecting Channels were singled out for study because they were identified by the International Joint Commission (IJC, 1983) as Areas of Concern, where beneficial uses of water, sediments, or both were significantly impaired by human activities.

We selected *H. limbata* for study because, during its 2-yr nymphal life stage, it lives in burrows in the sediment, indiscriminately ingests sediment while feeding (Smock, 1983), and is sensitive to environmental pollutants, including oil and metals concentrated in the sediments (Fremling, 1964, 1970; NAS, 1973; Hiltunen & Schloesser, 1983; Malueg *et al.*, 1984a, 1984b; Burt *et al.*, 1988). *H. limbata* typically is abundant in the shallow, soft-bottomed portions of the Great Lakes and

---

[1] Contribution 733, of the National Fisheries Research Center-Great Lakes, U.S. Fish and Wildlife Service, 1451 Green Road, Ann Arbor, MI 48105.

their connecting channels where water and sediment quality have not been substantially degraded by human activities (Veal, 1968; Hiltunen, 1971, 1980; Hamdy *et al.*, 1978; Fallon & Horvath, 1983; Hiltunen & Schloesser, 1983; Thornley & Hamdy, 1984; Schloesser & Hiltunen, 1986; Burt *et al.*, 1988; Mozley & La Dronka, 1988); in these areas it can have a central role in trophic relations, converting plant detritus into food for fish (Duffy *et al.*, 1987).

## 2. Materials and methods

We collected sediment and *H. limbata* nymphs at 11 stations in the Upper Great Lakes Connecting Channels in 1986 (Fig. 1). Some stations were established in areas remote from major pollution sources where nymphal densities were high, and others were in areas closer to pollution sources where nymphal densities were lower. Sediments at all 11 stations were primarily a mixture of clay and silt, or clay and sandy silt. We did not collect particle-size data or measure the organic content of the sediment, but such sediment was described by Hunt (1953) and Wright & Mattice (1981) as a preferred habitat for *H. limbata* because it was soft enough to permit the nymphs to burrow easily, and also was sufficiently cohesive to prevent the burrows from collapsing.

One or two samples of sediment for contaminant analysis were collected with a Ponar grab (0.05 m² jaw opening) at each station in June. Each sample was placed carefully in a large metal tub in a manner that preserved the *in situ* profile of the sediment and prevented the upper layers representing the sediment-water interface portion of the sample from contacting the tub or any other potential source of contamination. About 500 g of sediment in the upper 3 cm of the sample was scooped off with a clean, stainless steel spoon and placed in an acid-washed, acetone-rinsed glass jar. The jar was then capped with acetone-rinsed aluminum foil and placed on ice, in darkness, in an insulated container and transported to the laboratory. The samples were refrigerated and held in darkness in the laboratory until they were

analyzed. Heavy metals were measured by Inductively Coupled Argon Plasma analysis with standard methodology (EPA, 1979a; Jarrell-Ash, 1979); cyanide was measured by dry weight basis following EPA (1979b) and Technicon Corporation (1980); and oil was measured by dry weight basis according to APHA (1980) and EPA (1974). We collected 15 samples monthly April through October (except July) at each station with the Ponar grab to provide nymphs for production estimates. The samples were washed over a U.S. Standard No. 30 sieve (0.65-mm mesh) and the nymphs and other sample residue on the screen were preserved in formalin-phloxine B solution. In the laboratory we placed each preserved sample in a shallow pan and extracted the nymphs manually. We then identified the nymphs according to Edmunds *et al.* (1976) and measured them to the nearest 0.5 mm at $7 \times$ magnification.

One additional sample was collected at each station with the Ponar grab and No. 30 sieve, to provide data for a length-weight relation and a wet weight-dry weight relation for *H. limbata* in the study area. Live nymphs and the other sample residue on the screen were placed on ice, in a small amount of water and transported to the laboratory. The nymphs were extracted from the residue, blotted with a paper towel for 10 to 15 seconds, and weighed immediately. Nymphs used to provide data on live (wet) weight versus dry weight were placed in a drying oven at 100 °C for at least six hours, and then reweighed; the remaining live nymphs were measured, placed in formalin-phloxine B solution for at least 30 days and then re-measured to provide data on live versus preserved length. Weights were determined to the nearest milligram and length measurements were made to the nearest 0.5 mm under $7 \times$ magnification.

Preserved lengths in millimeters ($L$) were converted to wet weights in grams ($W$) by applying the length-weight equation,

$$\ln(W) = 2.82 \ln(L) - 11.09$$
$$(r^2 = 0.9655, \ n = 186),$$

and wet weights were converted to dry weight ($D$)

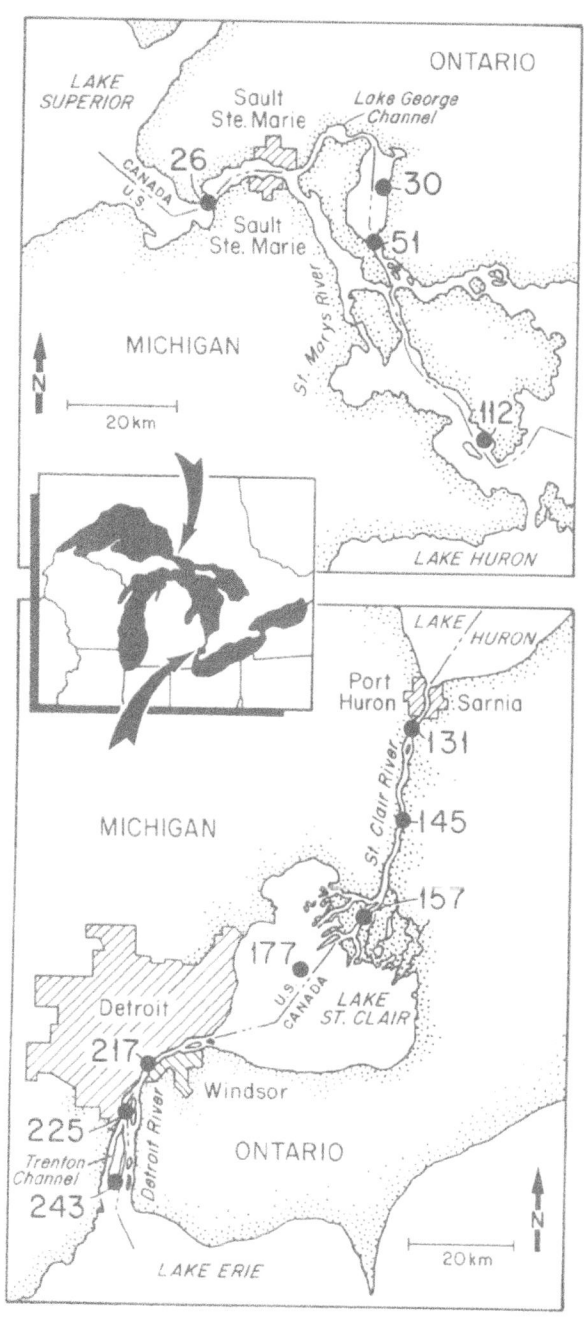

*Fig. 1.* Locations of the 11 stations at which *Hexagenia* nymphs and sediments were collected for study.

with the equation

$$\ln(D) = -1.5167 + 1.1189 \ln(W)$$
$$(r^2 = 0.954, n = 100).$$

Production $(Pw)$ and its variance $v(Pw)$ were estimated by using the size frequency method of Hynes as modified by Hamilton (1969) and equations from Krueger & Martin (1980).

Because Schloesser & Hiltunen (1984), and Schloesser *et al.* (1988) indicated that the cohort production interval (CPI) is about 730 days in the Upper Great Lakes Connecting Channels, we followed Benke (1979) in multiplying *Pw* by 365/CPI to obtain the estimate of production and Krueger & Martin (1980) in multiplying $v(Pw)$ by $(365/CPI)^2$ to obtain variance for this production estimate.

To reduce the variation associated with small sample sizes in some of the size groups, we grouped the nymphs collected throughout the study into 4-mm size classes before we computed production.

## 3. Results and discussion

### 3.1. Production

Production for the April to October period varied widely throughout the study area (Table 1) and was markedly higher at stations 26, 51, and 112 in the St. Marys River and at station 177 in Lake St. Clair (3333 to 9231 mg m$^{-2}$) than at station 30 in the St. Marys River and the six stations in the St. Clair and Detroit rivers (359 to 980 mg m$^{-2}$). The values in Table 1 are valid estimates of the portion of the annual production that occurred in April to October 1986 and can be used to examine the effects of contaminated sediments on the performance of the *H. limbata* populations sampled throughout the study area.

### 3.2. Sediment contaminants

In the St. Marys River none of the measured concentrations of contaminants in sediments at stations 26, 51, and 112 (Table 2) exceeded the Ontario Ministry of Environment (OME) or U.S. Environmental Protection Agency (EPA) guidelines for the disposal of dredged materials in the Great Lakes and their connecting channels (IJC, 1982). These guidelines (Table 3) are expressed as the concentrations of various contaminants in sediments that if exceeded could cause the sedi-

*Table 1.* Production of *H. limbata* nymphs in the Upper Great Lakes Connecting Channels, April through October, 1986.

| Channel and station | Production | |
|---|---|---|
| | Dry wt (mg m$^{-2}$) | 95% confidence interval |
| **St. Marys River** | | |
| 26 | 3,481 | 3,082–3,890 |
| 30 | 403 | 287–518 |
| 51 | 3,333 | 3,021–3,644 |
| 112 | 3,375 | 2,935–3,815 |
| Average | 2,648 | |
| **St. Clair River** | | |
| 131 | 741 | 605–877 |
| 145 | 359 | 288–430 |
| 157 | 980 | 770–1,189 |
| Average | 693 | |
| **Lake St. Clair** | | |
| 177 | 9,231 | 8,757–9,705 |
| **Detroit River** | | |
| 217 | 376 | 264–489 |
| 225 | 708 | 558–857 |
| 243 | 872 | 708–1,035 |
| Average | 652 | |

ments to be classified as contaminated (OME) or polluted (EPA). In this paper, we considered sediments to be polluted if one or more of the contaminants we measured equalled or exceeded the numerical values given in Table 3.

In the St. Marys River, the measured concentrations of oil, cyanide, Cd, Cr, Cu, Ni, Pb, and Zn in one sample at station 30 exceeded the OME and EPA guidelines, and the concentrations of oil, cyanide, Cr, Cu, and Zn were the highest we measured during this study. Oil was visible in the other sample collected at station 30, indicating that the sediment was polluted according to the criterion for oil and petrochemicals in the revised Great Lakes Water Quality Agreement of 1978 (IJC, 1988).

In the St. Clair River at station 131, oil was visible in one sample and the measured concentration of oil exceeded EPA and OME guidelines in the other. Furthermore, Cu exceeded the OME and EPA guidelines in both samples and Zn

*Table 2.* Sediment contaminant levels (mg kg$^{-1}$ dry wt) in the Upper Great Lakes Connecting Channels, 1986.

| Channel and station | Contaminant[1] | | | | | | | |
| --- | --- | --- | --- | --- | --- | --- | --- | --- |
| | Oil | Cyanide | Cd | Cr | Cu | Ni | Pb | Zn |
| St. Marys River | | | | | | | | |
| 26 | – | – | 0.3 | 7.9 | 9.8 | 7.7 | 12.0 | 33.0 |
| | | – | 0.3 | 7.9 | 8.2 | 9.4 | 8.6 | 26.0 |
| 30 | + | – | – | 9.0 | 6.8 | 6.4 | 12.0 | 43.0 |
| | 3,170* | 2.1* | 1.1* | 49.0* | 44.0* | 27.0* | 59.0* | 210.0* |
| 51 | – | – | 0.3 | 11.0 | 8.6 | 7.2 | 10.0 | 48.0 |
| | 869 | – | 0.3 | 14.0 | 11.0 | 9.2 | 11.0 | 51.0 |
| 112 | – | – | 0.3 | 18.0 | 12.0 | 10.0 | 11.0 | 30.0 |
| St. Clair River | | | | | | | | |
| 131 | + | – | 0.6 | 13.0 | 46.0* | 15.0 | 34.0 | 93.0* |
| | 1,670* | – | 0.6 | 13.0 | 52.0* | 15.0 | 33.0 | 100.0* |
| 145 | + | – | 0.3 | 6.7 | 14.0 | 8.9 | 7.5 | 46.0 |
| | | – | 0.7 | 7.7 | 18.0 | 14.0 | 9.4 | 63.0 |
| 157 | 892 | – | 0.4 | 11.0 | 18.0 | 15.0 | 18.0 | 57.0 |
| | 793 | – | 0.4 | 11.0 | 18.0 | 14.0 | 18.0 | 55.0 |
| Lake St. Clair | | | | | | | | |
| 177 | – | – | 0.6 | 14.0 | 24.0 | 18.0 | 28.0 | 63.0 |
| Detroit River | | | | | | | | |
| 217 | + | – | 1.0* | 37.0* | 43.0* | 42.0* | 71.0* | 170.0* |
| 225 | – | – | 0.4 | 32.0* | 32.0* | 39.0 | 19.0 | 83.0 |
| | – | – | 0.4 | 30.0* | 33.0* | 36.0 | 26.0 | 83.0 |
| 243 | +, 881 | – | 3.2* | 39.0* | 48.0* | 31.0 | 55.0* | 160.0* |
| | 1,020* | – | 1.6* | 30.0* | 38.0* | 29.0 | 46.0* | 140.0* |

[1] Dash ( – ) indicates no data; cross ( + ) indicates oil was detected by smell or as a visible sheen in sample and exceeded the pollution criterion (IJC, 1987); asterisk (*) indicates measured value exceeds U.S. Environmental Protection Agency or Ontario Ministry of Environment guideline for polluted dredged sediments (IJC, 1982).

exceeded the EPA guideline in one sample and the OME and EPA guidelines in the other. At station 145, oil was visible in one sample. At station 157 in the St. Clair River and at station 177 in Lake St. Clair, none of the contaminants exceeded the OME or EPA guidelines.

In the Detroit River, oil was visible in the single sample taken at station 217 and the concentrations of Cu, Cr, and Zn exceeded the OME and EPA guidelines; the concentrations of Ni and Pb also exceeded the guidelines and were the highest measured for those contaminants in this study. At station 225, concentrations of Cr and Cu exceeded the OME and EPA guidelines. At station 243, oil was visible in one sample and measured oil exceeded the EPA guideline in the other; concentrations of Cd, Cr, Cu, Pb, and Zn exceeded OME and EPA guidelines in both samples.

*Table 3.* Guidelines of the Ontario Ministry of the Environment (OME) and U.S. Environmental Protection Agency (EPA) for the disposal of dredged materials (IJC, 1982). Sediments with contaminant values equal to or larger than the tabular values are considered to be contaminated (OME) or polluted (EPA).

| Contaminant ($mg\ kg^{-1}$) | OME | EPA |
|---|---|---|
| Cadmium | 1 | 6 |
| Chromium | 25 | 25 |
| Copper | 25 | 25 |
| Cyanide | 0.1 | 0.1 |
| Lead | 50 | 40 |
| Nickel | 25 | 20 |
| Zinc | 100 | 90 |
| Oil | 1,500 | 1,000 |

### 3.3. Effect of contaminated sediments on production

Comparison of production data (Table 1) with sediment contamination data at our 11 stations (Table 2) showed that at stations 26, 51, 112, 157, and 177, where *H. limbata* production was 980 to 9231 $mg\ m^{-2}$, sediment contaminant levels did not exceed the OME and EPA guidelines, and sediments had no oily odor or visible sheen of oil. In contrast, at the other six stations, where the sediments were polluted, the production of *H. limbata* was 359 to 872 $mg\ m^{-2}$. Although production was higher at station 157 than at the six stations where sediments were polluted, it did not differ significantly from production at two of them (stations 225 and 243) – suggesting that the April to October production of about 700 to 1000 $mg\ m^{-2}$ may be typical of the boundary area between polluted and unpolluted sediments.

We believe our results indicate that contaminated sediments are adversely affecting the production of *H. limbata* in portions of these connecting channels; support for this interpretation is provided by the results of a 21-day laboratory bioassay (Henry, 1987) conducted in 1986 with *H. limbata* nymphs and sediments from the connecting channels. The bioassay showed that sediments from a portion of the Lake George Channel in the upper St. Marys River and from a portion of the Trenton Channel in the lower Detroit River (Fig. 1) were acutely toxic to the nymphs. No *H. limbata* nymphs were collected at these two

*Table 4.* Sediment contaminant levels ($mg\ kg^{-1}$ dry wt) in the Upper Great Lakes Connecting Channels and in the Keweenaw Waterway.

| Location | Contaminant[1] | | | | | | | |
|---|---|---|---|---|---|---|---|---|
| | Oil | Cyanide | Cd | Cr | Cu | Ni | Pb | Zn |
| St. Marys River Upper Lake George Channel[2] | 4,720* | 2.5* | 1.0* | 40* | 40* | 19 | 42* | 160* |
| Lower Detroit River Trenton Channel[2] | 16,200* | 8.9* | 9.3* | 230* | 130* | 120* | 330* | 1,400* |
| Keweenaw Waterway[3] Torch Lake | – | – | 2.5* | 180* | 1,800* | 150* | 110* | 310* |
| North waterway | – | – | 1.0* | 73* | 140* | 63* | 17 | 100* |
| | – | – | 2.0* | 130* | 930* | 120* | 40* | 240* |
| South waterway | – | – | 0.2 | 18 | 13 | 24* | 1.9 | 53 |
| | – | – | 0.7 | 51* | 37* | 34* | 27 | 91* |

[1] Asterisk (*) indicates measured value exceeds guidelines of the U.S. Environmental Protection Agency or Ontario Ministry of Environment for polluted dredged sediments (IJC, 1982); dash indicates no data.

[2] Henry (1987).

[3] Malueg *et al.* (1984a, 1984b); waterway is located in Michigan in the Lake Superior watershed.

locations, even though the sediment was of the 'preferred' type described by Hunt (1953) and Wright & Mattice (1981).

Sediment concentrations of oil, cyanide, Cd, Cr, Cu, Ni, Pb, and Zn were far higher in the Trenton Channel than in the upper Lake George Channel or at any of the 11 stations in the present study; they also exceeded the OME and EPA guidelines for polluted sediment by a wide margin. Sediment concentrations of oil, cyanide, Cr, Cu, and Zn in the upper Lake George Channel (Table 4) substantially exceeded OME and EPA guidelines for polluted sediments; concentrations of Cd and Pb barely exceeded the guidelines; and Ni was slightly below the EPA guideline value. Sediment concentrations of oil, cyanide, and metals other than Ni in the upper Lake George Channel substantially exceeded the values measured in the present study at stations 26, 51, 112, 157, and 177, where the production of *H. limbata* was highest (Tables 2 & 4). In the Lake George Channel, sediment concentrations of most contaminants were within the range of values we measured at our six stations where production was lower; the exception was oil, which was substantially higher in the Lake George Channel (Tables 2 & 4). The bioassay (Henry, 1987) also revealed that sediments from stations 26, 30, 51, 131, 145, 157, 225, and 243, where we measured production, were not acutely toxic to *H. limbata* nymphs. Sediment toxicity was not determined by Henry (1987) at stations 112 and 177, but the high production we measured and the low concentrations of contaminants in the sediments at those stations indicate the sediments there probably were not toxic to *H. limbata*.

Additional evidence for an impact of metals on the production of *H. limbata* in the connecting channels is provided by the results of studies in the Keweenaw Waterway (Malueg et al., 1984a, 1984b), where mining activities created extensive deposits of tailings that contaminated the waterway. Concentrations of metals substantially exceeded OME and EPA guidelines, except for Cd in Torch Lake, and Cd and Pb in the north waterway. In the south waterway, on the other hand, only the concentrations of Cr, Cu, Ni, and Zn exceeded OME and EPA guidelines (Table 4). Ten-day bioassays (Malueg et al., 1984a; 1984b) showed that the sediments from Torch Lake were acutely toxic to nymphs of *Hexagenia* (probably *H. limbata*), but that sediments from the north and south waterways were not. Field studies by these investigators showed nymphs present in the south waterway but none in Torch Lake and the north waterway. Although the absence of nymphs in the north waterway is not explained by the results of the bioassay by Malueg et al. (1984b), the absence of nymphs is perhaps not surprising, because sediment metal concentrations in the north waterway were generally within the range shown in the longer, 21-day bioassay by Henry (1987) to be acutely toxic in the Upper Great Lakes Connecting Channels (Table 4). In the south waterway, where *Hexagenia* nymphs were present, sediment concentrations of all metals were generally within the range found in the present study at locations where *H. limbata* nymphs were present (Tables 1, 2 & 4). In the south waterway, concentrations of Cd and Pb were similar to those at our productive stations 26, 51, 112, 157, and 177, whereas the concentrations of Cr, Cu, Ni, and Zn more closely resembled values we measured at our other six stations, where production was lower.

## 4. Conclusions

Our results strongly suggest that the considerably lower nymphal production measured in portions of the Upper Great Lakes Connecting Channels, where oil, cyanide, and metals in sediments exceeded pollution criteria, can be attributed to those contaminants operating in a manner that adversely affected the health of individual nymphs and the performance of the population. At five stations where sediments were not polluted, the production of nymphs was up to nine times higher than the highest production measured at six other stations, where sediments were polluted. Sediment bioassays and related field studies by Henry (1987) and Malueg et al. (1984a, 1984b) demonstrate the lethality of sediment contaminants to *Hexagenia* nymphs and provide independent support for the interpretation that the lower production we observed at 6 of 11 stations in the study area can be attributed to sediment contamination.

No numerical criteria have been developed specifically to permit an evaluation of the effect of sediment contaminants on the performance of populations of *H. limbata* nymphs, and none were developed in the course of this study. However, our results indicate that the OME and EPA guidelines for dredged sediments, together with the Water Quality Agreement criterion for visible oil (IJC, 1988), can probably be applied to protect *H. limbata* in Great Lakes habitats. This finding is noteworthy because *H. limbata* tends to assume a major role in secondary production and trophic transfer of energy in soft-bottomed, mesotrophic habitats in shallow portions of the Great Lakes system that have not been polluted by oils, metals, and other toxic substances. Furthermore, because *H. limbata* is among the most pollution-sensitive members of the macrozoobenthos community in the Great Lakes system, the OME and EPA guidelines for disposal of dredged materials (Table 3) can probably be used as sediment pollution criteria to protect not only *H. limbata*, but also the other macrozoobenthos in these waters.

## 5. Acknowledgements

This work was performed as a cooperative project between the U.S. Fish and Wildlife Service, National Fisheries Research Center-Great Lakes, and the U.S. Environmental Protection Agency, Great Lakes National Program Office (EPA), under the terms of Interagency Agreement No. DW 14931214-01-0. Sediment chemistry services were provided by EPA under the general supervision of EPA Project Officer Paul Bertram.

## References

APHA (American Public Health Association), 1980. Standard methods for the examination of water and wastes. 15th edition. pp. 460–465. APHA-AWWA-WPCF. ISBN 0-8553-091-5.

Benke, A. C., 1979. A modification of the Hynes method for estimating secondary production with particular significance for multivoltine populations. Limnol. Oceanogr. 24: 168–171.

Burt, A. S., D. R. Hart & P. M. McKee, 1988. Benthic invertebrate survey of the St. Marys River. Vol. 1, Main Report. Prep. for Ont. Min. Env. by Beak Consultants, Ltd., Brampton, Ontario, 88 pp.

Duffy, W. G., T. R. Batterson & C. D. McNabb, 1987. The St. Marys River, Michigan: an ecological profile. U.S. Fish Wildl. Serv. Biol. Rep. 85(7.10), 138 pp.

Edmunds, G. F., Jr., S. L. Jensen & L. Burner, 1976. The mayflies of North and Central America. Univ. Minnesota Press, Minneapolis, 330 pp. ISBN 0-8166-0759-1.

EC & EPA (Environment Canada & U.S. Environmental Protection Agency), 1988. Upper Great Lake Connecting Channels Study. Final Report, Vol. 2. Chicago, Illinois & Toronto, Ontario, 626 pp.

EPA (U.S. Environmental Protection Agency), 1974. Methods for chemical analysis of water and wastes. pp. 226–228. Office of Technology Transfer, Washington, D.C.

EPA, 1979a. Simultaneous analysis of liquid samples for metals by inductively coupled argon plasma atomic emission spectroscopy (ICAP-AES). U.S. Env. Protect. Agency, Region V, Chicago, Illinois. Unpublished.

EPA, 1979b. Methods for chemical analysis of water and wastes. U.S. Env. Protect. Agency, Cincinnati, Ohio. Pub. No. 600/4-79-02.

Fallon, M. E. & F. J. Horvath, 1983. Preliminary assessment of contaminants in soft sediments of the Detroit River. J. Great Lakes Res. 11: 373–378.

Fremling, C. R., 1964. Mayfly distribution indicates water quality on the Upper Mississippi River. Science 146: 1164–1166.

Fremling, C. R., 1970. Mayfly distributions as a water quality index. U.S. Env. Protect. Agency, Wat. Qual. Off., Wat. Pollut. Control Res. Serv. 16030 DQH 11/70, 39 pp.

Hamdy, Y., J. D. Kinkead & M. Griffiths, 1978. St. Marys River quality investigations 1973–74. Ont. Min. Env., Toronto, Ontario, 53 pp.

Hamilton, A. L., 1969. On estimating annual production. Limnol. Oceanogr. 14: 771–782.

Henry, M. G., 1987. Toxicity of upper connecting channels sediments to *Hexagenia* nymphs exposed under laboratory conditions. U.S. Fish Wildl. Serv., Ntnl. Fish. Res. Center-Great Lakes, Ann Arbor, Michigan. Unpublished manuscript.

Hiltunen, J. K., 1971. Limnological data from Lake St. Clair, 1963 and 1965. Ntnl. Mar. Fish. Serv. Washington, D.C. COM-71-00644. 45 pp.

Hiltunen, J. K., 1980. Composition, distribution, and density of benthos in the lower St. Clair River, 1976–1977. Administrative Report 80-4. Great Lakes Fish. Lab., Ann Arbor, Michigan, 28 pp.

Hiltunen, J. K. & D. W. Schloesser, 1983. The occurrence of oil and distribution of *Hexagenia* nymphs in the St. Marys River, Michigan and Ontario. Freshwat. Invert. Biol. 2: 199–203.

Hunt, B. P., 1953. The life history and economic importance of a burrowing mayfly *Hexagenia limbata* in southern Michigan lakes. Michigan Dep. Conserv., Bull. Inst. Fish. Res. No. 4., Lansing, Michigan, 151 pp.

IJC (International Joint Commission), 1982. Guidelines and register for evaluation of Great Lakes dredging projects. Windsor, Ontario, 365 pp.

IJC, 1983. Report on Great Lakes Water Quality; Appendix A: Areas of concern in the Great Lakes Basin; 1983 update of class A areas, 1984. Rep. Great Lakes Wat. Qual. Bd. 113 pp.

IJC, 1988. Revised Great Lakes Water Quality Agreement of 1978. Windsor, Ontario, 130 pp.

Jarrell-Ash, 1979. Mark III Atomcomp interim operator's manual. Jarrell-Ash Division, Fisher Scientific Co., Waltham, Massachusetts. March 1979, M 79-1.

Krueger, C. C. & F. B. Martin, 1980. Computation of confidence intervals for the size frequency (Hynes) method of estimating secondary production. Limnol. Oceanogr. 25: 773–777.

Malueg, K. W., G. S. Schuytema, J. H. Gakstatter & D. F. Krawczyk, 1984a. Toxicity of sediments from three metal-contaminated areas. Envir. Toxicol. Chem. 3: 279–291.

Malueg, K. W., G. S. Schuytema, D. F. Krawczyk & J. H. Gakstatter, 1984b. Laboratory sediment toxicity tests, sediment chemistry and distribution of benthic macro-invertebrates in sediments from the Keweenaw Waterway, Michigan. Envir. Toxicol. Chem. 3: 233–242.

Mozley, S. C. & R. M. La Dronka, 1988. *Ephemera* and *Hexagenia* (Ephemeridae, Ephemeroptera) in the Straits of Mackinac, 1955–56. J. Great Lakes Res. 14: 171–177.

NAS (National Academy of Science), 1973. Water quality criteria, 1972. U.S. Government Printing Office. Washington, D.C. U.S. Env. Protect. Agency Ecol. Res. Ser. EPS-R3-73-033, 594 pp.

Schloesser, D. W. & J. K. Hiltunen, 1984. Life cycle of a mayfly, *Hexagenia limbata* in the St. Marys River between Lakes Superior and Huron. J. Great Lakes Res. 10: 435–439.

Schloesser, D. W. & J. K. Hiltunen, 1986. Distribution and abundance of mayfly nymphs and caddisfly larvae in the St. Marys River. Administrative Report 86-3. Great Lakes Fish. Lab., Ann Arbor, Michigan, 18 pp.

Schloesser, D. W., M. A. Ford & T. A. Edsall, 1988. Age structure of the mayfly, *Hexagenia*, in the St. Clair-Detroit River system. Pap. Pres. 31st Conf. Internat. Assoc. Great Lakes Res. Hamilton, Ontario, May 17–20, 1988.

Smock, L. A., 1983. The influence of feeding habits on whole-body metal concentrations in aquatic insects. Freshwat. Biol. 13: 301–311.

Technicon Corporation, 1980. Operations manual for Technicon Autoanalyzer II. C system. Technicon Industrial Systems, Tarrytown, N.Y. Tech. Pub. TA9-0460-00.

Thornley, S. & Y. Hamdy, 1984. An assessment of the bottom fauna and sediments of the Detroit River. Ont. Min. Env., Toronto, 48 pp.

Veal, D. M., 1968. Biological survey of the St. Marys River. Ont. Wat. Resourc. Commiss. Toronto, Ontario, 23 pp.

Wright, L. L. & J. S. Mattice, 1981. Substrate selection as a factor in *Hexagenia* distribution. Aquat. Insects 3: 13–24.

The manufacturer's authorised representative in the EU is Springer
Nature Customer Service Centre GmbH, Europaplatz 3, 69115 Heidelberg,
Germany. If you have any concerns regarding our products, please
contact ProductSafety@springernature.com

Printed and bound by CPI Group (UK) Ltd, Croydon, CR0 4YY

23/04/2026

02095657-0003